W9-CEF-633

Evolutionary Quantitative Genetics

Evolutionary Quantitative Genetics

Derek A. Roff

Department of Biology, McGill University, Montreal, Quebec, Canada

CHAPMAN & HALL

I(T)P® International Thomson Publishing

New York • Albany • Bonn • Boston • Cincinnati • Detroit • London • Madrid • Melbourne
Mexico City • Pacific Grove • Paris • San Francisco • Singapore • Tokyo • Toronto • Washington

Cover design: Trudi Gershenov
Cover photos: © 1997 PhotoDisc, Inc.

Printed in the United States of America

For more information, contact:

Chapman & Hall
115 Fifth Avenue
New York, NY 10003

Chapman & Hall
2-6 Boundary Row
London SE1 8HN
England

Thomas Nelson Australia
102 Dodds Street
South Melbourne, 3205
Victoria, Australia

Chapman & Hall GmbH
Postfach 100 263
D-69442 Weinheim
Germany

International Thomson Editores
Campos Eliseos 385, Piso 7
Col. Polanco
11560 Mexico D. F.
Mexico

International Thomson Publishing - Japan
Hirakawacho-cho Kyowa Building, 3F
1-2-1 Hirakawacho-cho
Chiyoda-ku, 102 Tokyo
Japan

International Thomson Publishing Asia
221 Henderson Road #05-10
Henderson Building
Singapore 0315

1 2 3 4 5 6 7 8 9 10 XXX 01 00 99 98 97

Library of Congress Cataloging-in-Publication Data

Roff, Derek A.
    Evolutionary quantitative genetics / Derek A. Roff.
        p.   cm.
    Includes bibliographical references and index.
    ISBN 0-412-12971-X (alk. paper)
    1. Quantitative genetics   2. Evolutionary genetics.   I. Title
QH452.7. R64   1997
576.5--dc21   97-47079
                                                          CIP

British Library Cataloguing in Pubication Data available

To order this or any other Chapman & Hall book, please contact **International Thomson Publishing, 7625 Empire Drive, Florence, KY 41042.** Phone: (606) 525-6600 or 1-800-842-3636. Fax: (606) 525-7778. e-mail: order@chaphall.com.

For a complete listing of Chapman & Hall's titles, send your request to **Chapman & Hall, Dept. BC, 115 Fifth Avenue, New York, NY 10003.**

Dedicated to Mum, Dad,

Daphne, Robin and Graham,

who have had much to put up with.

Also Boris and Yerdle—victims of unnatural selection.

'The thing can be done,' said the Butcher, 'I think.
The thing must be done, I am sure.
    The thing shall be done! Bring me paper and ink,
The best there is time to procure.'

    The Beaver brought paper, portfolio, pens,
And ink in unfailing supplies:
    While strange creepy creatures came out of their dens,
And watched them with wondering eyes.

    So engrossed was the Butcher, he heeded them not,
As he wrote with a pen in each hand,
    And explained all the while in a popular style
Which the Beaver could well understand.

    'Taking Three as the subject to reason about—
A convenient number to state—
    We add Seven and Ten, and then multiply out
By One Thousand diminished by Eight.

    'The result we proceed to divide, as you see,
By Nine Hundred and Ninety and Two:
    Then subtract Seventeen, and the answer must be
Exactly and perfectly true.

    'The method employed I would gladly explain,
While I have it so clear in my head,
    If I had but the time and you had but the brain—
But much yet remains to be said.

    'In one moment I've seen what has hitherto been
Enveloped in absolute mystery,
    And without extra charge I will give you at large
A lesson in Natural History.'

The Hunting of the Snark
Lewis Carrol

Contents

Preface

The impetus for this book arose out of my previous book, *The Evolution of Life Histories* (Roff, 1992). In that book I presented a single chapter on quantitative genetic theory. However, as the book was concerned with the evolution of life histories and traits connected to this, the presence of quantitative genetic variation was an underlying theme throughout. Much of the focus was placed on optimality theory, for it is this approach that has proven to be extremely successful in the analysis of life history variation. But quantitative genetics cannot be ignored, because there are some questions for which optimality approaches are inappropriate; for example, although optimality modeling can address the question of the maintenance of phenotypic variation, it cannot say anything about genetic variation, on which further evolution clearly depends. The present book is, thus, a natural extension of the first. I have approached the problem not from the point of view of an animal or plant breeder but from that of one interested in understanding the evolution of quantitative traits in wild populations. The subject is large with a considerable body of theory: I generally present the assumptions underlying the analysis and the results, giving the relevant references for those interested in the intervening mathematics. My interest is in what quantitative genetics tells me about evolutionary processes; therefore, I have concentrated on areas of research most relevant to field studies. Although I have attempted to standardize the use of symbols, I have sometimes erred on the side of clarity rather than consistency. A list of the major symbols used is given in a glossary at the end of the book.

Without the suggestion of Greg Payne or the continued encouragement of my wife and colleague Dr. Daphne Fairbairn, this book would not have come to fruition. I cannot thank them enough for their support.

Evolutionary Quantitative Genetics

1

Introduction

1.1 Introducing the Problem

The history of domestication of animals and plants, and the selective breeding of the enormous variety of animals and plants for aesthetic purposes, attests very clearly to the presence of genetic variation within species. This has been demonstrated further by numerous common garden experiments in which individuals from different populations are grown under the same set of conditions [Table 1.1; note that nongenetic, maternal effects require that the populations pass at least two generations under the common garden conditions; see Nelson et al. (1970), Baskin and Baskin (1973), Quinn and Colosi (1977)]; differences between populations are almost invariably found. These differences occur in traits that vary continuously (Table 1.1) or show discrete variation [e.g., diapause propensity, wing dimorphism in insects, dimorphism in "weaponry," behavioral dimorphisms; see Roff (1996a) for a review] that cannot be reconciled with simple Mendelian models (e.g., single locus with two alleles).

For Darwin, the question of the genetic basis of quantitative traits presented a problem that he was unable to solve. During the latter half of the nineteenth century, two schools of opinion arose, the Biometricians and the Mendelians, the history of which is well documented by Provine (1971) in his book *The Origins of Theoretical Population Genetics*. The Biometricians saw continuous variation as the "stuff of evolution," whereas the Mendelians favored evolution by discrete jumps. The problem was that Mendel had provided a mechanism for the transmission of discrete traits such as color or seed morphology (e.g., wrinkled versus round), but it was not clear that this mechanism could be used to account for continuous variation. On the other hand, the Biometricians were confronted with the problem of the regression to the mean. Suppose one plots the mean offspring value on that of the mean parental value; there is for many traits, human height being the classic example (Fig. 1.1), a clear regression between the mean offspring value and the mean parental value. The slope of such a relationship is almost always less than unity and the intercept is greater than zero, say $X_O = c + bX_P$.

Table 1.1   Some Examples of Common Garden Experiments Demonstrating the Presence of Genetic Variation Between Populations

Species	Common Name	Trait	References
Mesocyclops edax	Copepod	Development time, body size, clutch size, dormancy	Wyngaard (1986, 1988)
Drosophila robusta	Fruit fly	Morphology	Stalker and Carson (1947)
Drosophila subobscura	Fruit fly	Body size, development time, survival	Prevosti (1955), Misra and Reeve (1964)
Drosophila pseudoobscura	Fruit fly	Morphology	Sokoloff (1965)
Drosophila melanogaster	Fruit fly	Morphology	David and Bocquet (1975)
Heliothis virescens	Tobacco budworm	Ovipostion preference, larval performance	Schneider and Rouse (1986), Waldvogel and Gould (1990)
Pararge aegeria	Speckled wood butterfly	Body size, development time, survival	Gotthard et al. (1994)
Limnoporus notabilis	Water strider	Body size, development time	Fairbairn (1984)
Aquarius remigis	Water strider	Morphology, reproductive traits	Blanckenhorn and Fairbairn (1995)
Allonemobius sp.	Striped ground cricket	Morphology, diapause	Mousseau and Roff (1989a, 1989b, 1995)
Melandrium sp.	Plant	Flowering time, plant weight, sex ratio	Lawrence (1964)
Lolium perenne	Ryegrass	Morphology	Thomas (1967)
Danthonia	Grass	Biomass accumulation	Scheiner et al. (1984)
Arabis serrata	Perennial	Morphology	Oyama (1994a)

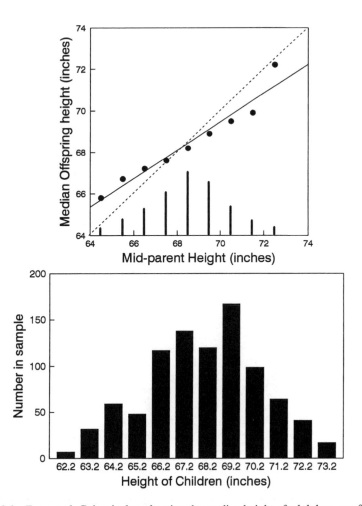

Figure 1.1   Top panel: Galton's data showing the median height of adult human offspring as a function of the mean height of their parents. Each point is the median value of all children from parents with the particular mid-parent value (today the values per family would be plotted separately). The vertical lines on the *x* axis show the relative number of mid-parent values (from left to right, 7, 32, 59, 48, 117, 138, 120, 167, 99, 64, 41, 17). *Bottom panel:* Distribution of height of children. Note that the distribution is approximately normal. To correct for differences in height between men and women Galton multiplied the height of women by 1.08. [Data from Table 11 of Galton (1889).]

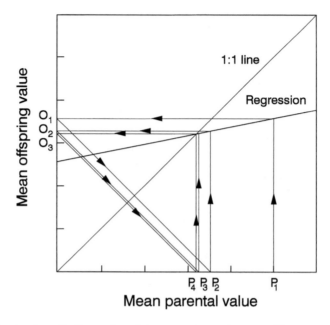

Figure 1.2 A schematic illustration of the regression to the mean. The parents in the first generation (P_1) are larger than average but produce offspring (shown by the regression line) that are smaller than themselves. This process is continued until the parents in the fourth generation produce offspring that are, on average, the same height as themselves. The same process, in reverse, occurs if the initial size of parents is below the point of intersection of the regression line and the line of equality.

This regression line crosses the line of unity at $c/(1 - b)$. Now assume that the mean value of parents that produce the next generation is $X_P^* > c/(1 - b)$, as shown in Fig. 1.2. Using the regression equation, we have that the predicted value of their offspring is $X_O^* = c + bX_P^*$, and, as can be seen from Fig. 1.2, $X_O^* < X_P^*$; that is, the mean offspring value is less than the mean parental value. Assuming no further selection and random mating, the mean value of the offspring will decrease (regress) until the phenotypic value $c/(1 - b)$ is reached, at which point $X_O^* = X_P^*$ (this takes four generations in Fig. 1.2). Taking this approach, it does not appear possible that the evolution of quantitative traits is stable, continuous selection being required to prevent the regression to the mean. The solution to this dilemma was suggested by Yule (1902, 1906) and formally worked out by Fisher (1918).

The solution did not require any exotic genetic mechanism, simply a greater appreciation of the Mendelian mechanism of inheritance. We suppose that (1) the quantitative trait is controlled by many loci segregating according to the Mendelian rules and (2) each allele contributes some small amount to the trait value, the overall value being a function of all the allelic contributions. These two conditions are sufficient to generate the observed offspring–parent regressions and to prevent the regression to the mean. The reason that selection of parents leads to a permanent shift in the phenotypic value is that the frequency of the alleles is changed by the selection and the new frequency will remain as such in the absence of selection, thereby maintaining the new phenotypic value (this is formally shown in Chapter 2). The error of the Biometricians was in assuming that the regression line itself remains fixed; under the Mendelian model, the slope of the line is a function of allelic frequencies and, hence, remains constant only if the allelic frequencies remain constant (and the environment does not change, as this may affect genotypes differentially). The working out of this model and its ramifications for the understanding of the process of evolution is the subject of this book.

1.2 Overview

Chapter 2 deals with the basic mathematical framework of the biometrical model of quantitative genetics, most particularly the genetic interpretation of the slope of the regression line. For obvious reasons, the initial discussion centers on single traits, but it is readily apparent that traits are neither inherited nor acted on by selection as separate units. Chapter 3 extends the model to include interactions between traits, introducing the concept of the genetic correlation for this purpose. Much of the theory of quantitative genetics has been generated by breeders who wished to make use of the method to improve the rate at which they can bring about change in economically valuable characters. Consequently, much attention has been paid to directional selection (i.e., the continued selection of a trait or traits in a fixed direction). Directional selection is also an important component of the evolution of traits in natural populations. Furthermore, an empirical test of the predictions from quantitative genetics for the case of directional selection provides a solid foundation on which to judge the robustness of the mathematical assumptions underlying quantitative genetics. Chapter 4 first outlines the theory and then assesses its worth using empirical studies. As in the case of Chapter 2, Chapter 4 focuses on a single trait; extension to the case of multiple traits is made in Chapter 5.

Environments are continually changing, leading to selection for a different phenotype. There are two possible responses to such variation: First, there may be a genetic change producing a new phenotype; second, there may be an interaction between the genotype and its environment such that the phenotype is a function of the environment. The second process is termed phenotypic plasticity and is evident in virtually all quantitative traits (although this need not necessarily mean that all response are adaptive). The quantitative genetics of this phenomenon are described in Chapter 6. Phenotypic plasticity demonstrates that the partitioning of genetic and environmental effects is not always a simple matter. This is perhaps even more readily seen in the case of maternal and other sex-related

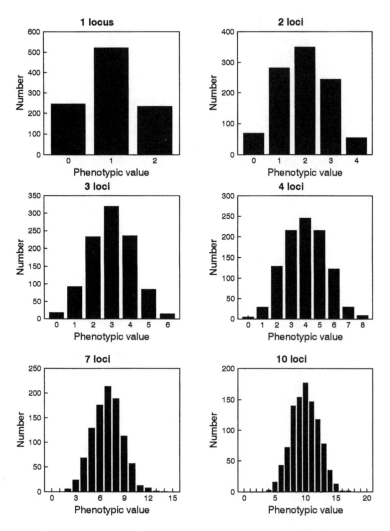

Figure 1.3 Distributions of phenotypic values for a genetic model in which there are *n* loci with two alleles per locus, one allele contributing 0 to the phenotypic value and the other contributing 1, the overall genotypic (= phenotypic, as no environmental effects are assumed) value being the sum of the allelic values. Each distribution is generated by drawing at random 1000 individuals with *n* loci with allelic frequency per locus of 0.5. The theoretical distribution can be generated by using the coefficients of the binomial expansion $(p + q)^n$, where p is the frequency of one allele and $q = 1 - p$.

Table 1.2 Phenotypic Means and Variances of a Quantitative Trait for the Castle–Wright Method of Estimating the Number of Loci

Population	Mean	Variance
P_1, P_2	m_1, m_2	$V_{G,1} + V_E, V_{G,2} + V_E$
$F_1 = P_1 \times P_2$	$\frac{1}{2}(m_1 + m_2)$	$\frac{1}{2}(V_{G,1} + V_{G,2}) + V_E$
$F_2 = F_1 \times F_1$	$\frac{1}{2}(m_1 + m_2)$	$\frac{1}{2}(V_{G,1} + V_{G,2}) + V_E + V_{Extra}$

influences. These effects, discussed in Chapter 7, can have profound influences on evolutionary trajectories and pose difficult experimental problems, because their separation from genetic effects requires both large samples and frequently complex breeding designs. Most of the theoretical analyses presented in Chapters 2–7 assume implicitly an effectively infinite population size. Natural populations are frequently very small, either permanently or because they pass through bottle-necks. The consequences of finite population size on the predictions of quantitative genetics and tests of these predictions are the subject of Chapter 8. The observation that there is considerable genetic variation in natural populations raises the obvious issue of what factors are maintaining it. Chapter 9 reviews the theory and data pertinent to this question. Finally, Chapter 10 gives an overview of the previous chapters and makes suggestions for future work.

1.3 Two General Approaches to Quantitative Genetic Modeling

There are two general models used in quantitative genetics: the single locus model and the infinitesimal model. The former model focuses on a single locus and typically assumes that there are two alleles segregating at this locus with values of a, d, $-a$ for genotypes A_1A_1, A_1A_2, and A_2A_2, respectively. Without loss of generality we can divide throughout by a to give the values 1, d, and -1, (where, again for simplicity and without loss of generality, d has been redefined to be equal to d/a). The value of d establishes the dominance relationship between the alleles. The analysis proceeds by considering changes at this single locus and then expands to multiple loci by assuming that the contributions of loci to the genotype are independent (no epistasis, as defined below) and that the allele frequencies are the same at all loci (or, alternatively, that we are considering the value averaged over all relevant loci). The second approach is based on normal distribution theory. As with the single locus approach, epistatic effects are assumed not to exist, allowing the genotypic value, G, to be represented as the sum of the contributions at each of n loci,

$$G = m + G_1 + G_2 + \cdots + G_n \tag{1.1}$$

where m is the population mean. Now, instead of assuming a specific number of

Table 1.3   *Some Estimates of the Number of Loci Controlling Quantitative Traits*

Organism	Trait	Number of Loci	Reference
Mouse	60-Day weight	11	Chai (1956)
	6-Week weight	~20	Roberts (1966a)
	6-Week weight	32	Falconer (1989)
	Weight gain (18–40 days)	66–237	Comstock and Enfield (1981)
	Litter size	2	Falconer (1989)
	Litter size	164	Eklund and Bradford (1977)
Rat	Coat color	5–9	Lande (1981a)
Rabbit	Weight	14	Wright (1968)
	Ear length	19	Wright (1968)
	Skull length	5	Wright (1968)
	Femur length	6	Wright (1968)
Human	Skin color	5	Lande (1981a)
Fish	Eye diameter	5–7	Lande (1981a)
Drosophila	Female head shape	6–9	Lande (1981a)
	Abdominal bristles	98	Falconer (1989)
	Sternopleural bristles	5[a]	Spickett and Thoday (1966)
	Sternopleural bristles	18[b]	Shrimpton and Robertson (1988)
	Sternopleural bristles	"Few"	Gallego and López-Fanjul (1983)
	Longevity	1, >3	Luckinbill et al. (1987, 1988)
Tribolium	Pupal weight	157–485	Comstock and Enfield (1981)
Maize	Oil content in kernels	17–22	Lande (1981a)
Goldenrod	Date of anthesis	6–7	Lande (1981a)
Nicotiana rustica	Flowering time	>16[c]	Jinks and Towey (1976)
	Height at flowering	>9	Jinks and Towey (1976)
	Final height	>10	Jinks and Towey (1976)

[a] Accounts for 87.5% of difference between two inbred lines.

[b] Based on a chromosomal mapping approach.

[c] Genotype assay. The estimate will generally be biased downwards (Hill and Avery, 1978).

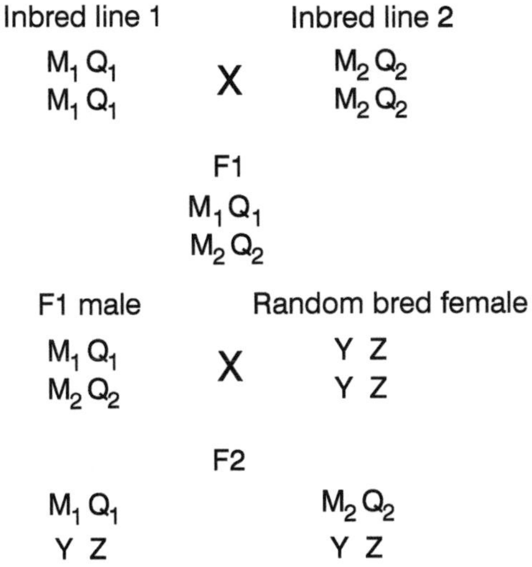

Inbred line 1   Inbred line 2

$$M_1 Q_1$$
$$M_1 Q_1$$ X
$$M_2 Q_2$$
$$M_2 Q_2$$

F1

$$M_1 Q_1$$
$$M_2 Q_2$$

F1 male   Random bred female

$$M_1 Q_1$$
$$M_2 Q_2$$ X
Y Z
Y Z

F2

$$M_1 Q_1$$ $$M_2 Q_2$$
Y Z   Y Z

Analysis by ANOVA

Source of variation	DF
Total	N-1
Between Groups	1
Error	N-2

Figure 1.4 A schematic illustration of QTL estimation. The marker locus is designated *M* and the QTL as *Q*. The particular design shown is given for purposes of illustration only. A more usual design is to backcross the F_1 to produce the F_2. The backcross breeding design is possible only if heterozygotes can be distinguished, which is possible with molecular markers.

alleles per locus, we assume that there are sufficient number of loci that the genotypic value can, by the central limit theorem, be represented by a normal distribution. Further, by assuming a normally distributed environmental contribution that is independent of the genotypic value, we have a normally distributed phenotypic distribution (Fig. 1.1). The genotypic value is itself broken down into separate additive (*A*) and dominance (*D*) contributions:

$$G = m + (A_1 + D_1) + (A_2 + D_2) + \cdots$$
$$+ (A_n + D_n) = m + A + D \quad (1.2)$$

which will be bivariate normal for a sufficiently large *n* (Bulmer, 1985, p. 123).

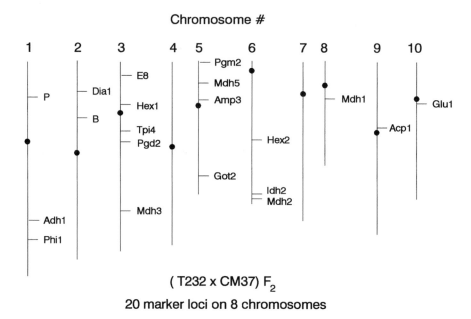

Figure 1.5 Distributions of 20 electrophoretically detectable marker loci in an F_2 maize hybrid. [Redrawn from Edwards et al. (1987).]

Analysis then proceeds by considering the changes in means and variances using the mathematical machinery of statistical theory.

If the number of alleles per locus is small, then the predictions of quantitative genetics are dependent on the number of loci being reasonably large; how large is dependent on the type of prediction. If traits were generally controlled by just a few loci with a few alleles per locus, then quantitative genetic theory might be seriously compromised. Before continuing the development of quantitative genetic ideas, we shall first address the question of whether there is any reason to suppose that we need such a theory. At first glance, this may appear to be a trivial question, for does not the continuous distribution of trait values such as height (Fig. 1.1) argue for the action of many loci? The answer is that it does not (Thoday and Thompson, 1976). The reason for this is illustrated in Fig. 1.3; the number of possible genotypic values increases very rapidly with the number of loci, thereby producing an apparently continuous distribution even without the additional smoothing factor of environmental variation. With only four loci, one would have great difficulty in statistically distinguishing the true four-loci model from one of an infinite number of alleles.

However, if only a few loci were involved, strong directional selection should lead to a rapid erosion of genetic variation; for example, simulations show that

Table 1.4   Power Analysis for the Detection of QTL in the Model Analyzed by Carbonell et al. (1993); Data Ranked in Decreasing Order with Respect to the Percentage of Phenotypic Variance Accounted for by the QTL

Linkage Group[a]	Genotypic Component[b]		$h^2 = 0.5$[c]		$h^2 = 0.2$	
	a	d	%V_P[d]	$P(\%)$[e]	%V_P	$P(\%)$
2	1.5	1.5	22.25	100	8.86	99
1	1.5	0.0	14.76	91	5.91	44
4	1.0	1.0	9.84	99	3.94	65
3	1.0	0.0	6.56	43	2.62	12
5	0.75	0.75	5.51	94	2.20	51
6	0.75	0.75	5.51	87	2.20	47
7	0.0	0.0	0	3	0	3
8	0.0	0.0	0	1	0	4

[a]Each marker is separated by 20 cM (i.e., equidistant from the adjacent markers).

[b]Homozygotes for QTL take values $+a$ and $-a$, whereas heterozygotes are d.

[c]h^2 is the proportion of the phenotypic variance that can be explained by the additive effect of the QTLs.

[d]Percentage of the phenotypic variance that is determined by the particular QTL.

[e]Proportion of simulations (100 replicates) in which a QTL was identified within the designated linkage group. For linkage groups 1–6, this represents statistical power (the probability of detecting a QTL given that one exists), whereas for linkage groups 7 and 8, it is the probability of a false positive (detection of a nonexistent QTL).

with 12 diallelic loci, fixation at all loci can be expected within 20 generations (Bulmer, 1976). This is not observed in selection experiments, although a decline in genetic variance is typically apparent after 20 generations (see Chapter 4). This does not definitively rule out a low number of loci, but it does suggest that more loci are likely. Various statistical methods have been proposed by which the number of loci might be determined, the Wright method and quantitative trait locus (QTL) mapping being the most often used.

1.4.   The Wright Method for Estimating the Number of Loci

This method was developed primarily by Wright (1968) but appeared first in a paper by Castle (1921); hence, it is sometimes referred to as the Castle–Wright method. It makes use of two inbred lines, assuming that one line has been fixed for alleles that increase the trait value and the other line has been fixed for alleles that decrease the trait value. Further, it is assumed that all loci are independent and act additively, and each makes an equal contribution to the phenotype. For

Table 1.5 Estimated Phenotypic Variability Explained by QTL for Plant Height Using a Sample Size of 400 Individuals and 4 Subsets of 100 Individuals; Empty Cells Indicate No QTL Detected

Chromosome	Flanking Markers	Complete	Subset 1	Subset 2	Subset 3	Subset 4
1	php1122/bnl7.21	3				
3	bnl8.35/umc10	4		10	8	
6	umc62/php20599					17
8	bnl12.30/bnl10.24	7		15	13	13
9	wx1/css1	8	17		16	23

Source: Modified from Beavis (1994).

example, letting the number of loci be *n* and the per locus genetic contribution to the phenotypic value for the two lines be 0 and 1, the genotypic values are equal to 0 ($n \times 0$) and *n* ($n \times 1$). These two lines are crossed to form an F_1 population which is then crossed to produce an F_2 population. The statistical relationships of these three generations can be derived from Mendelian inheritance and statistical theory. By definition, the means and variances in the parental generation are

$$m_1 = 2 \sum_{i=1}^{n} a_{i,1} = 2na_1, \qquad m_2 = 2 \sum_{i=1}^{n} a_{i,2} = 2na_2$$

$$V_{G,1} = 2 \sum_{i=1}^{n} V_{i,1}, \qquad V_{G,2} = 2 \sum_{i=1}^{n} V_{i,2} \tag{1.3}$$

where m_j and V_{Gj} are respectively the phenotypic mean and total genetic variance in the *j*th parental line, $a_{i,j}$ and $V_{i,j}$ are the contributions of the *i*th locus of the *j*th parental line (note the factor 2 is present because we are considering a diploid organism). The phenotypic variance is the sum of the genetic variance and the environmental variance, $V_P = V_G + V_E$. The statistics for the other two generations are given in Table 1.2. The difference between the means of the parental populations is

$$m_2 - m_1 = 2n(a_2 - a_1) = 2n\delta \tag{1.4}$$

and the extra genetic variance, V_{Extra}, in the F_2 populations is

$$V_{Extra} = \frac{1}{2} \sum_{i=1}^{n} \delta_i^2 = \frac{n}{2} (V_\delta + \delta^2) \tag{1.5}$$

From the above two equations we can obtain

$$n = \frac{(m_2 - m_1)^2}{8V_{\text{Extra}}} \left(1 + \left[\frac{V_\delta}{\delta}\right]^2\right) \tag{1.6}$$

The term V_δ/δ will not generally be known, but because its squared value must be positive, a minimum estimate of the number of loci is

$$n > \frac{(m_2 - m_1)^2}{8V_{\text{Extra}}} \tag{1.7}$$

Wright (1968) provides an approximate correction factor if there is dominance and other types of crosses are used. Lande (1981a) showed that the formulas can be applied to genetically heterogeneous or wild populations. Violation of the assumptions of additivity, no linkage between loci, and equality of allelic effects per locus leads to an underestimation of the number of loci. The estimated number of loci cannot exceed the haploid number of chromosomes plus the mean number of recombination events per gamete (Lande, 1981a). Because in higher plants and animals there is usually only one to a few recombinations per chromosome, the estimated number of loci cannot be much greater than the diploid number of chromosomes. This is illustrated in Table 1.3 in which the estimates taken from Wright (1968) and Lande (1981a) are typically small. Falconer (1989) used the same formula in the context of long-term response to divergent selection

$$n = \frac{R^2}{8V_A} \tag{1.8}$$

where R is the divergence between the two selected lines and V_A is the additive genetic variance (= total genetic variance if dominance and epistasis are absent). With the exception of mouse litter size, Falconer's estimates are considerably larger than those obtained by Wright or Lande. Using the same approach as Falconer, Eklund and Bradford (1977) estimated that 164 loci were segregating for litter size in the mouse: The contrast between the two estimates is reason to be skeptical of this method of estimation. Comstock and Enfield (1981) developed an adaptation of Wright's method assuming multiplicative effects; their estimates of the number of loci for weight gain in the mouse (66-237) and pupal weight in *Tribolium* (157-485) are extremely large. Following a detailed analytical and numerical evaluation of Wright's method, Zeng et al. (1990, p. 236) concluded that the method "is more apt to be misleading than illuminating." Given that virtually all of the estimates displayed in Table 1.3 are based on the general conceptual framework of Wright's method, it is probably reasonable to conclude that these estimates have little value.

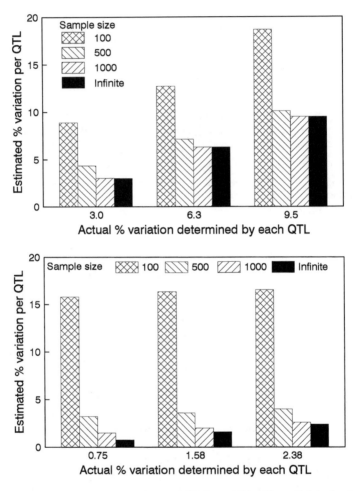

Figure 1.6 Average estimated phenotypic variability explained by a QTL in the simulation model of Beavis (1994). Data in the three blocks show results for a model that assumes the QTLs explain 30%, 63%, or 95% of the phenotypic variance. The total number of QTLs are 10 and 40 in the upper and lower panel, respectively. [Data from Table 5 of Beavis (1994).]

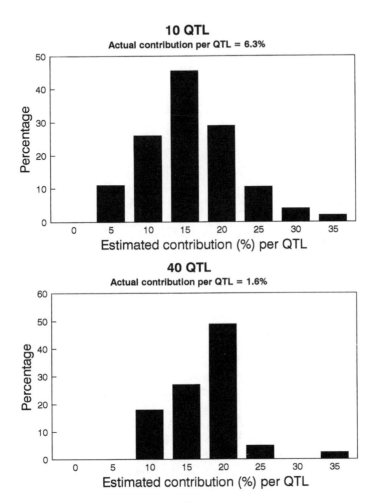

Figure 1.7 Frequency distribution of the estimated amount of phenotypic variability explained by a correctly identified simulated QTL. In one model, there are 10 QTLs, each contributing 6.3% of the phenotypic variance; in the other, there are 40 QTLs, each contributing 1.6% of the phenotypic variance. [Redrawn from Beavis (1994).]

Table 1.6 Estimates of Power for the Simulation Model of Beavis (1994)

No. of QTLs	Sample Size	Power (%)		
		30	63	95
40	100	2	4	6
40	500	11	29	46
40	1000	25	59	77
10	100	12	33	39
10	500	57	86	93
10	1000	84	99	100

1.5 QTL Mapping

The concept behind this method goes back to Sax (1923), who suggested that differences in seed weight of different color morphs of bean seeds was due to loci linked to the visual marker locus. Thoday (1961) suggested that this phenomenon could be used to map loci coding for quantitative traits and quantify their effects. The principle is illustrated in Fig 1.4: A locus coding for a quantitative trait is linked to a marker that produces a visible effect. By inbreeding two lines, it is possible to produce a line that has allele M_1 at the marker locus and allele A_1 at the quantitative trait locus (QTL) and a line that comprises M_2 and A_2. These two lines are crossed forming a heterozygote. The F_1 are crossed to a line that is outbred, in which it can be assumed that all effects are randomly distributed. Assuming complete linkage between the marker locus and the QTL, the relative contribution of the QTL to the phenotypic value can be ascertained using analysis of variance (Fig. 1.4). The basic method can be extended to cover incomplete linkage and alternate breeding designs [for a review, see Tanksley (1993)]. Obviously, there is not a wealth of visible phenotypic markers and, hence, Thoday's suggestion could not be implemented to a significant degree. The situation changed, however, with the advent of molecular markers, first using electrophoretically detectable markers (Fig. 1.5) and more recently with restricted fragment length polymorphisms (RFLPs) and those based on the polymerase chain reaction (RAPDs and microsatellites), which enabled relative saturation of chromosomes with markers.

A QTL is not a locus per se but "a region of a chromosome (usually defined by linkage to a marker gene) that has a significant effect on a quantitative trait" (Tanksley, 1993, p. 211). Detecting QTLs is extremely expensive and labor-intensive. Further, there are significant statistical assumptions associated with the analyses, many of which have not been fully explored [for a brief review, see Weller and Ron (1994)]. There are two important statistical questions with respect to sampling intensity: (1) How large a sample is required to detect a QTL?

Table 1.7   Estimates of the Effect of Detected Quantitative Trait Loci

Organism	Trait(s)	N	nQTL[a]	% Total[a]	Per QTL	Ref[e]
Loblolly pine	Wood specific gravity	48	5	23	5	1
Potato	Tuber shape	50	1	60	60	2
Populus	Height	55	2	26	13	3
	Spring bud flush		5	85	17	
Soybean	Seed hardness	60	5	57	11	4
Common bean	Resistance (2)	70	4, 4	52, 77 (64)	16	5
Maize	Morphology (3)	90	3, 3, 3	46, 47, 45	15	6
Mouse	Epilepsy	112	2	50	25	7
Rat	Blood pressure (4)	115	2–2 (2)	18–30 (24)	12	8
Eucalyptus	Seedling traits (3)	122	4, 6, 10	33, 28, 52	7	9
Maize	Height	112	4, 6 (5)	53, 73 (63)	13	10
		144	3, 3 (3)	34, 45 (39.5)	13	
Mosquito	Plasmodium resistance	138	2	67	34	11
Maize	Yield traits (8)	150	1–6 (4.1)	35–71 (55.5)	14	12
Barley	Various (8)	150	6–10 (7.7)	57–72 (64.9)	8	13
Rice	Kernel elongation	170	1	14[b]	14	14
Maize	Plant height (5)	187	3–11 (7.0)	24–114 (74.8)[c]	11	15
	Yield (8)		3–13 (6.4)	10–87 (47.6)	7	
Maize	Disease resistance (2)	200	1, 1	47, 48	47	16
Potato	Tuber dormancy	220	6	58	10	17
Maize	Various (3)	232	3, 4, 6	26, 35, 39	8	18
Tomato	Fruit traits	237	6, 4, 5	58, 44, 48 (50)	10	19
Barley	Various (12)	250	1–5 (2.7)	5–52 (26.7)[b]	10	20
Maize	Kernel weight	260	4	79	20	21
		290	6	97	16	
Maize	Yield	264	6	60	10	22
Maize	Various (9)	290	4–6 (5.6)	29–62[d] (50)	9	23
Tomato	Fruit traits	350	7, 4, 5	72, 44, 34 (50)	10	24
Cattle	Weight (3)	~400	4, 3, 4	6, 2, 2	3	25
	Milk production		5	5	5	
Mouse	12-wk body weight, epididymal fat pad	424	5, 4	2, 3	3	26
Tomato	Morphology (11)	432	3–10 (7)	7–60 (37)	7	27
Mouse	Weight at age (10)	535	7–17 (14)	29–76 (55)	4	28
Maize	Various (25)	1930	3–17 (11[b])	8–37 (18)	2	29

[a]Mean value shown in parentheses.

[b]Percentage of genetic variance.

[c]The percentage of the phenotypic variance accounted for was estimated for each QTL separately. Values greater than 100% indicate that QTL on the same chromosome (there are 10) must overlap in the effect.

[d]Estimated using multiple regression; interval mapping gave 42–87%, with a mean of 63.5%.

[e]References: (1) Groover et al. (1994); (2) Eck et al. (1994); (3) Bradshaw and Settler (1995); (4) Keim et al. (1990); (5) Nodari et al. (1993); (6) Reiter et al. (1991); (7) Rise et al. (1991); (8) Jacob et al. (1991); (9) Grattapaglia et al. (1995); (10) Beavis et al. (1991); (11) Severson et al. (1995); (12) Velboom and Lee (1994); (13) Hayes et al. (1993); (14) Ahn et al. (1993); (15) Edwards et al. (1992); (16) Jung et al. (1994); (17) Freyre et al. (1994); (18) Ajmone-Marsan et al. (1995); (19) Paterson et al., (1988); (20) Backes et al. (1995); (21) Doebley et al. (1994); (22) Stuber et al. (1992); (23) Doebley and Stec (1993); (24) Paterson et al. (1991); (25) Moody et al. (1994); (26) Pomp et al. (1994); (27) DeVincente and Tanksley (1993); (28) Cheverud et al. (1996); (29) Edwards et al. (1987).

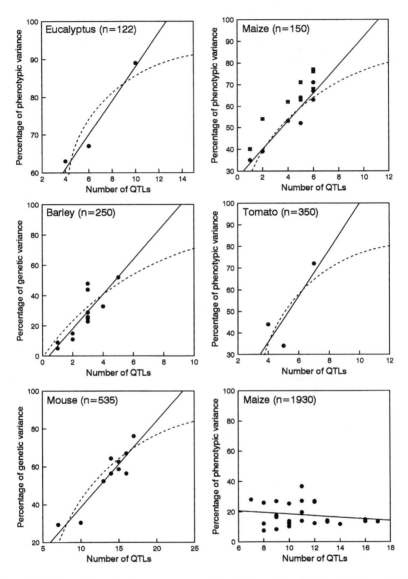

Figure 1.8 Percentage of phenotypic or genetic variance in a variety of traits explained by the number of QTLs detected with varying sample sizes. As shown in the maize plot (upper right), there is little difference between the use of the phenotypic (dots) and the genetic (squares) variances. Solid lines show fitted regressions; dotted lines illustrate an alternative model in which the magnitude of QTLs decreases as the number increases (curves fitted "by eye"). [Data from Grattapaglia et al. (1995)—*Eucalyptus*; Velboom and Lee (1994)—maize n = 150; Backes et al. (1995)—barley; Paterson et al. (1991)—tomato; Cheverud et al. (1996)—mouse; Edwards et al. (1987)—maize).]

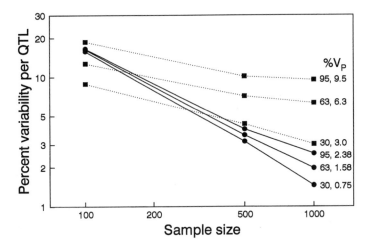

Figure 1.9. Percentage of the phenotypic variability explained by the average QTL in the simulation model of Beavis (1994). Solid lines show results for a model in which there are 40 QTLs and dotted lines the results for a model in which there are 10 QTLs. The numbers on the right show the total amount of phenotypic variation actually determined by the QTLs (total, per QTL).

(2) How large a sample is required to quantify the effect of the detected QTL? These are related questions, but only the first has received much attention (Ooijen, 1992; Carbonell et al., 1993; Beavis, 1994; Jansen, 1994).

Carbonell et al. (1993) analyzed a model in which there were eight linkage groups, six of which contained QTLs (one per linkage group) and an associated marker. The genotypic contribution to the phenotypic value at each QTL was a for one homozygote (say Q_1Q_1), $-a$ for the other homozygote (say Q_2Q_2), and d for the heterozygote (Q_1Q_2). If d does not equal 0, the heterozygotic value lies closer to one homozygote than the other and, thus, one allele shows partial dominance. The simulated breeding procedure differed from that shown in Fig. 1.4 in that a backcross between the F_1 and the parental lines was assumed (this is actually one of the most common designs). For each of 100 replicates, 250 "individuals" were analyzed according to the procedure given in Carbonell et al. (1992). Two primary cases were analyzed: one in which the additive genetic component (heritability) comprised 50% of the total phenotypic variance and one in which it comprised 20% of the variance. (A detailed discussion of the concept of heritability is given in Chapter 2.) Some typical results are presented in Table 1.4; roughly speaking, the more a particular QTL contributed to the phenotypic value, the more likely it was to be detected. Note that not all QTLs will be detected in any single experiment. The probability of making a false detection was within the 5% level, usually taken as desirable in statistical analysis. Thus, a sample size of

250 individuals seems suitable for the detection of QTLs which have a large or modest effect (>5%) on the phenotypic value. Similar results were obtained by Ooijen (1992) and Jansen (1994).

Beavis (1994) expressed doubt about the efficacy of sample sizes of a few hundred individuals. This doubt stems from two sources: a comparison of sub-samples from a larger sample and a simulation study. Subsampling of 100 individuals from a data set comprising 400 individuals produced a very different picture of the contribution of the identified QTL (Table 1.5). In every case, the percentage of the phenotypic variance explained by the identified QTL in a subset exceeded by approximately 100% the value obtained in the complete set. Because all QTLs occurred on different chromosomes, this discrepancy cannot be explained by the possible overlap of several QTLs (e.g., the larger sample detects two QTLs on the same chromosome, whereas the smaller sample detects only one and attributes the variation ascribable to both QTLs to the single detected QTL). To further investigate this problem, Beavis constructed a simulation model. In one population, genotypic variation was due to the additive action of 10 loci of equal effect, and in a second population, 40 loci contributed additively. In both cases, each QTL was on a separate linkage group, there being 75 linkage groups in total. The sum of the additive effects accounted for 30%, 63%, or 95% of the phenotypic variability (*i.e.*, h^2 = 0.30, 0.63, or 0.95). The phenotypic value was thus determined by summing the QTL values and then adding an appropriate random normal variate. Sample size per replicate was 100, 500, or 1000 individuals, with 200 replicates per QTL number/sample size combination. The results were striking; with a sample intensity of 100 individuals, the average effect of a QTL was overestimated by more than twofold when 10 QTLs determined the trait and 20-fold when there were 40 QTLs (Fig. 1.6). Even with a sample size of 1000, the effect was twofold for the 40 QTL case. This considerable bias in the estimated contribution per QTL is not due to a few estimates with extraordinary effects but is characteristic of the majority of estimates (Fig. 1.7). Furthermore, the bias was due to overestimates of both additive and dominance effects, the latter being simulated as zero! As might be expected from the previous power analyses, the probability of correctly identifying a QTL was small unless the sample size was large (>500) and the contribution per QTL greater than 3% (the 10 QTL case, Table 1.6).

Typically, QTL studies use sample size of about 200 individuals and have found a per-QTL contribution of approximately 10% (Table 1.7). Further, plotting the percentage phenotypic variance accounted for versus number of QTLs using either independent traits from a single study or the same trait from different studies suggests that approximately 10–20 QTLs account for the vast majority of the phenotypic variance (Fig. 1.8). Thus, taken at face value, these analyses indicate that relatively few loci are segregating for quantitative traits. This conclusion is seriously undermined by the previous simulation study. Interpolation of the data of Beavis shows that when the per-QTL estimate is about 12%, a sample size of

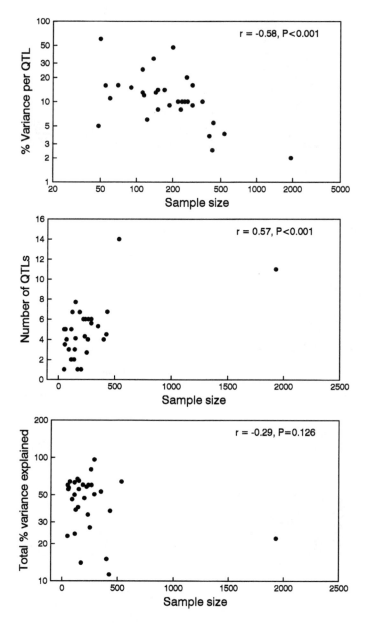

Figure 1.10. Observed relationship between sample size and the statistics estimated from QTL analysis. Data taken from Table 1.7.

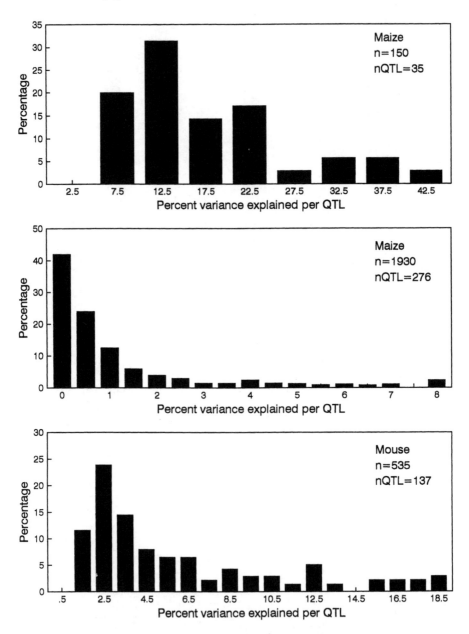

Figure 1.11. Distribution of the magnitude of individual QTL effects. The number of traits examined are 8 [maize, upper plot; Velboom and Lee (1994)], 25 [middle plot; Edwards et al. (1987)], and 10 [bottom plot; Cheverud et al. (1996)]. Because of overlap of QTL effects, the summed value per trait can exceed 100% (the data shown in Table 1.7 correct for this).

200 is not capable of distinguishing between a trait determined by 40 QTLs from one determined by 10 QTLs with a heritability of 0.63 (Fig. 1.9). The simulation study predicts that as the sample size is increased, the per-QTL effect should decrease, which indeed it does (Fig. 1.10). However, we might expect an increase due to the increased statistical power of the tests picking out additional QTLs of smaller magnitude. In this case, we would expect the number of detected QTLs and the total variance explained to increase with sample size. Although the number of detected QTLs does increase with sample size, the total variance explained does not increase (Fig. 1.10). From this, I conclude that the per-QTL contribution is probably greatly overestimated in most QTL studies and, consequently, the number of loci determining a trait may be much greater than indicated by these studies. The potentially misleading picture that can be obtained using an insufficient sample size is illustrated in Fig. 1.11. With a sample size of 150, the majority of individual QTLs appear to explain greater that 12% of the variance, whereas with larger samples, the values drops to less than 3%.

Whether the number of loci determining quantitative traits is in the tens or hundreds (or even thousands) remains to be shown. Nevertheless, we can conclude that the data support the hypothesis of numerous loci determining quantitative traits and, hence, to suppose that the most fundamental assumption of quantitative genetic theory is correct.

1.6 Summary

Evolution requires genetic variation. However, Darwin was unable to provide an adequate genetic model to explain continuous variation. The Mendelian model was rejected because it was initially thought to apply only to simple discrete characters. The biometrical model of offspring on parent regression produced the paradox of the regression to the mean, suggesting that continuous selection was necessary to maintain a new selected trait value. The two schools were shown to be compatible with the introduction of the idea of many loci, inherited in a Mendelian manner, that acted additively. The consequences of this model are the subject of this book. Two different mathematical approaches are considered: first, analyses based on a single locus; second, an approach that ignores individual loci and uses the statistical distribution of effects (the infinitesimal model). The fundamental assumption that there are a large number of loci has not been adequately tested because there are significant statistical problems with estimation. QTL mapping is a promising approach, but present sample sizes are typically insufficient to provide useful estimates of QTL effects. Overall, the data are consistent with the hypothesis of a large number of loci of small effect.

2

Heritability

A principle focus of quantitative genetics is the relationship between offspring and parents; without such a relationship, evolution may still proceed but not in a predictable fashion. Offspring may resemble their parents for three reasons:

1. Both parents and offspring may experience the same environmental conditions. Such a situation could arise if there are strong preferences of the offspring for their natal environment; for example, in birds, the offspring may return to the same general area for nesting in which they were raised.

2. The phenotype of the offspring may be determined by the phenotype but not the genotype of the parents. An example is when the ability of parents to provide food to their offspring depends on their condition, which is determined by environmental factors. As a consequence, "high-quality" parents produce "high-quality" offspring and "low-quality" parents produce "low-quality" offspring, but this has little to do with the genetic value of the parents.

3. The correspondence between the phenotype of the offspring and the parents is a consequence, in part, of the genes shared by the individuals.

A fairly obvious approach to examining the relationship between offspring and parents is to plot the mean phenotypic value of the offspring on the mean value of the parents (hereafter referred to as the mid-parent value). Provided that reasons 1 and 2 can be discounted, the offspring on mid-parent regression reflects the influence of the genes transmitted from parents to offspring plus environmental effects that can be viewed statistically as noise. Genetic effects can themselves be divided into two statistical components: additive effects that contribute to the resemblance between offspring and parents, and nonadditive effects, which contribute to the variance about the regression line but not to its slope.

The fact that there exists a significant regression between offspring and parents might encourage one to utilize the regression equation as a means of predicting

changes across generations. Although this is appropriate across the two generations from which data are taken, it is not valid beyond this point because, as will
be shown (and noted in Chapter 1), the value of the intercept of the regression is
a function of the mean value of the population and the slope of the regression. If
the parents are not random samples from the population, then the mean value in
the subsequent generation will, if the trait is heritable (in the sense of a significant
slope), change and, thus, so will the intercept. (This point is discussed in detail
later in the chapter.) The general relationship between mean offspring and mid-
parent is expressed by the equation

$$\text{Mean offspring value} = (1 - h^2)m + h^2(\text{mid-parent value}) \qquad (2.1)$$

where m is the mean value of the parental population and h^2 is the heritability of
the trait in question, to be defined more precisely in Section 2.1. Heritability is a
measure of the genetic determination of the trait and, because it is a function of
allelic frequencies, will itself change when a nonrandom set of parents contributes
to the next generation; however, the change is often sufficiently small and slow
that h^2 can be assumed constant during at least short-term evolutionary changes.
It is, thus, a useful index of the degree to which traits are determined by genetic
effects of evolutionary significance.

How Eq. (2.1) arises from considerations of Mendelian genetics is the subject
of the first section of this chapter. In the second section, I consider the most
commonly used methods of estimating heritability. Section 2.3 examines how
heritability varies among different types of traits. Section 2.4 examines the relative
contribution of dominance to the genetic variance of these different traits. Finally,
Section 2.5 addresses the problem of variation of heritability in natural populations.

2.1   The Meaning of Heritability

2.1.1   Single-Locus Model

We begin with the simplest possible model: a single locus with two alleles, A and
a. Without loss of generality, we can assign values $+1$ to genotype AA and -1
to genotype aa. (Typically, the values assigned are $+a$ and $-a$, but without loss
of generality and to simplify the formulas, I have divided throughout by a to
obtain the values used here.) The heterozygote can, in principle, take any value;
to maintain our model as simple as possible, I shall assume initially that there is
complete additivity of the allelic values, in which case the heterozygote has the
value 0 (i.e., allele A contributes $+0.5$ to the trait value and allele a contributes
-0.5). There is assumed to be no environmental influence, so the phenotypic
value is equal to the genotypic value. As described above, we are interested in
the relationship between parent and offspring—in this case, the relationship be-

tween the mid-parent value and the mean offspring value. There are five possible mid-parent phenotypes (Table 2.1, Fig. 2.1). We require the slope of the relationship between offspring and parents, which is also called **heritability in the narrow sense**, hereafter simply referred to as heritability, and designated h^2 (heritability in the broad sense will be discussed later). From standard regression theory we have

$$\text{Slope} = h^2 = \frac{\text{Cov}_{OP}}{V_{PAR}} \tag{2.2}$$

where Cov_{OP} is the covariance between the mid-parent and mean offspring values and V_{PAR} is the phenotypic variance of the mid-parent values. As is readily apparent from Table 2.1 and Fig. 2.1, the heritability is equal to 1. The actual value of the numerator and the denominator can be obtained very simply by noting that, provided males and females do not differ, the covariance between parents and offspring is equal to one-half of the variance of the genotypic values, as is the variance in the mid-parent values. Because from standard statistical theory, the variance is equal to $\mu_2 - \mu^2$, where μ_2 is the expected value of the squared values and μ is the expected mean value, we have

$$\mu = (1)p^2 + (0)2pq + (-1)q^2 = p - q$$
$$\mu_2 = (1^2)p^2 + (0^2)2pq + (-1^2)q^2 = p^2 + q^2 \tag{2.3}$$
$$\text{Variance} = p^2 + q^2 - (p - q)^2 = 2pq$$

Thus, the covariance between mean offspring and mid-parent values, Cov_{OP}, and the variance in mid-parent values, V_{PAR}, is simply pq. Because, by the Hardy–Weinberg law, the allele frequency remains constant, the offspring–parent regression will also remain constant (changes in the regression will occur whenever the allele frequency is changed).

Now, suppose there are environmental influences such that the phenotypic value is equal to the genotypic value $(-1, 0, 1)$ plus an amount E that is normally distributed with mean zero and some variance V_E. The phenotypic variance will be increased by an amount V_E, but as E has a mean of zero, the mean offspring value will not be changed. Therefore, the slope of the offspring–mid-parent regression (the heritability) will be equal to $pq/(pq + V_E)$. Three points should be noted: (1) Heritability is less than 1 and decreases as the environmental variance (V_E) increases (Fig. 2.1). (2) Heritability changes as the allele frequency changes, being maximal when $p = q = 0.5$. This means that the similarity between offspring and parents declines as heritability declines. As a consequence, the force of natural selection declines; that is, as heritability declines, the response to selection, all other things being equal, will likewise decline. (3) The offspring–parent regression can be used to predict the next-generation value even if a se-

Table 2.1   Distribution of Offspring Values for a Simple Single-Locus, Two-Allele Model in Which the Phenotypic Value Is Equal to the Sum of the Allelic Values

Genotype of Parents	Frequency of matings f_i	Mid-parent Value X_i	Proportion of Each Type of Offspring			Mean Offspring Value Y_i
			AA	Aa	aa	
AA, AA	p^4	1	1	0	0	1
AA, Aa	$4p^3q$	½	½	½	0	½
AA, aa	$2p^2q^2$	0	0	1	0	0
Aa, Aa	$4p^2q^2$	0	¼	½	¼	0
Aa, aa	$4pq^3$	−½	0	½	½	−½
aa, aa	q^4	−1	0	0	1	−1

lected group of parents is used, but it applies only to that generation because the selection of parents will change the allele frequency.

The alleles at a given locus may not interact in a strictly additive manner. This effect is the phenomenon termed **dominance** and can be incorporated by designating the value of *Aa* by *d*, a value which specifies the strength of the dominance. When $d = 1$, allele *A* is dominant to allele *a*, whereas $d = -1$ specifies that allele *a* is dominant to allele *A*, and if $d > 1$ or $d < -1$, there is **overdominance**. Assuming no environmental influence, there are six different mid-parent values; in four cases, the mean offspring value is equal to the mid-parent value, whereas in two cases, the mean offspring values differ by an amount *d* or *d*/2 (Fig. 2.2). The relationship between parents and offspring can thus been seen to be composed of two components: an **additive component** and a **dominance deviation**.

The value of an individual as measured by the mean value of its progeny is called its **breeding value**. For the single-locus case, the genotypic value of an individual can be decomposed into

$$\text{Genotypic value} = \text{Breeding value} + \text{Dominance deviation} = A + D \quad (2.4)$$

For mathematical simplicity and without loss of generality it is convenient to express values as deviations from the population mean (i.e., set the population mean equal to zero). To obtain the required covariance between parents and offspring, we proceed as follows: First we have

$$\text{Cov}_{OP} = E\{Y(X_\delta + X_\varphi)/2\} \quad (2.5)$$
$$= \frac{1}{2}(E\{\text{Cov}_{YX\delta}\} + E\{\text{Cov}_{YX\varphi}\})$$

where E { } denotes the expected value, *Y* is the mean value of offspring, X_δ and

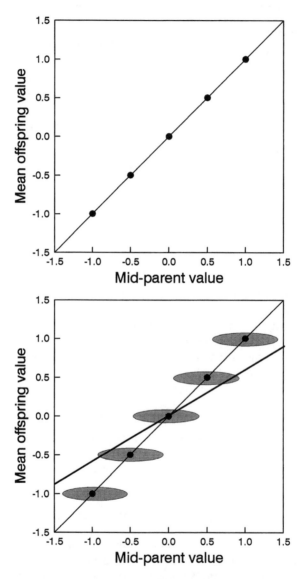

Figure 2.1   Top: The relationship between the mean offspring value and the mid-parent value when the trait value is determined by a single locus with two alleles *A* and *a* which contribute 0.5 and −0.5 to the trait value and act additively. *Bottom:* The effect of introducing an environmental factor, normally distributed with mean zero and variance V_E (schematically shown by the ellipses). Note that the line still passes through 0,0 but that the slope is now less than 1.

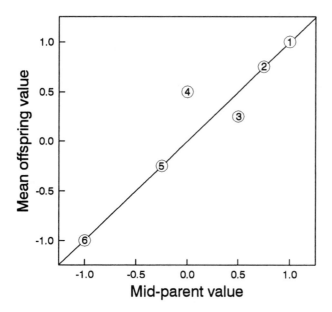

Symbol #	Parent Cross	Mid-parent value	Offspring frequency			Mean Offspring value
			AA	Aa	aa	
1	AA x AA	1	1.00	0.00	0.00	1
2	AA x Aa	0.5(1+d)	0.50	0.50	0.00	0.5(1+d)
3	Aa x Aa	d	0.25	0.50	0.25	0.5d
4	AA x aa	0	0.00	1.00	0.00	d
5	Aa x aa	0.5(-1+d)	0.00	0.50	0.50	0.5(-1+d)
6	aa x aa	-1	0.00	0.00	1.00	-1

Figure 2.2 Mean offspring value on mid-parent value when there is dominance. The values of the three genotypes are $AA = 1$, $Aa = d$, $aa = -1$. Below the figure is a table giving the required calculations. For the purpose of illustration, the value of d used in the plot is 0.5.

$X_♀$ are the phenotypic values of male and female, respectively, and $\text{Cov}_{YX_♂}$ and $\text{Cov}_{YX_♀}$ are the covariances between the mean value of the offspring and each parent separately. Assuming that the sexes have equal variances, then $E\{\text{Cov}_{YX_♂}\} = E\{\text{Cov}_{YX_♀}\} = E\{\text{Cov}_{YX}\}$, where X is the value for a single parent. Because mating is at random, the expected value of the offspring is, by definition, one-half the breeding value of one parent (the half enters because an offspring receives genes from each parent). The value of the parent is equal to its breeding value plus its dominance deviation $(A + D)$. Hence,

$$E\{\text{Cov}_{YX}\} = E\{A(A + D)/2\} = \tfrac{1}{2}(E\{A^2\} + E\{AD\}) \qquad (2.6)$$

The term $E\{A^2\}$ is the variance in breeding values, generally termed the **additive genetic variance**, V_A. The term $E\{AD\}$ is the covariance between the breeding value and the dominance deviation and is equal to zero (Falconer, 1989, p. 129). Thus, $E\{\text{Cov}_{YX}\} = \tfrac{1}{2}V_A$ and $E\{\text{Cov}_{OP}\} = \tfrac{1}{2}(E\{\text{Cov}_{YX}\} + E\{\text{Cov}_{YX}\}) = \tfrac{1}{2}V_A$. The covariance of mean offspring and mid-parent is equal to half the additive genetic variance. The variance of the mid-parent values, V_{PAR} is equal to one-half the phenotypic variance, V_P (i.e., the variance among individuals); thus, the heritability is given by

$$h^2 = \frac{\tfrac{1}{2}V_A}{\tfrac{1}{2}V_P} = \frac{V_A}{V_P} \qquad (2.7)$$

Equation (2.7) is the general definition of heritability: **Heritability is the ratio of the additive genetic variance to the total phenotypic variance**. The proportion of the total phenotypic variance attributable to both additive and nonadditive genetic variance is termed **heritability in the broad sense**; it is a useful measure in setting an upper limit to heritability in the narrow sense. For the simple dominance model presently being considered, the additive genetic variance is (Falconer, 1989, p. 129)

$$V_A = 2pq[1 + d(q - p)]^2 \qquad (2.8)$$

The total phenotypic variance is the sum of the additive genetic variance and the variance of the dominance deviations, V_D,

$$V_D = (2pqd)^2 \qquad (2.9)$$

Hence, the heritability is

$$h^2 = \frac{2pq[1 + d(q - p)]^2}{2pq[1 + d(q - p)]^2 + (2pqd)^2} \qquad (2.10)$$

From the above equations, it can be seen that V_A and h^2 are functions of allele frequency and d. This is illustrated in Fig. 2.3 for four different values of d and p ranging from 0 to 1. When $d = 0$, there is complete additivity, and although the additive genetic variance changes, heritability remains constant at 1. With a moderate value of d (0.5), heritability remains high, but with complete dominance ($d = 1$), heritability declines with p. When there is overdominance ($d = 5$), the additive genetic variance is bimodal and heritability is a U-shaped function with extremes of 1 and a minimum of 0. As in the simple additive model, environmental variance can be added, the result being no change in the numerator and an increase in the denominator (i.e., a reduction in h^2).

The presence of dominance may result from the particular choice of scale. Consider for example, the following two cases: (1) $AA = 10$, $Aa = 3.16$, $aa = 1$; (2) $AA = 1$, $Aa = 0.5$, $aa = 0$. In the first case, AA appears to be dominant over aa, but in the second case, there is apparent additivity. However, the second is merely the logarithm (base 10) of the first.

There is one further level of interaction, that among loci, which is termed **epistasis**. When there is interaction between two loci, the total phenotypic variance can be divided into three components, V_A, V_D, and V_I, the last being the variance due to interactions among loci, which can be divided further into additive $\times$ additive interactions (V_{AA}), additive $\times$ dominance interactions (V_{AD}), and dominance $\times$ dominance interactions (V_{DD}). Two examples of offspring–parent relationships with epistatic interactions are shown in Fig. 2.4: In the top panel, the dominant alleles are complementary, giving rise to the "classical" Mendelian ratio of 9:7, whereas in the bottom panel, the dominant alleles are duplicate, giving a Mendelian ratio of 15:1 [for the method of estimating variance components in a two-locus epistatic model, see Crow and Kimura (1970, pp. 124–128)]. Despite the epistatic effects, there is a clear overall correspondence between the mid-parent values and the mean offspring values. The relationship between heritability and allele frequencies for the preceding two models is shown in Fig. 2.5. The important point to note is that much of the epistasis appears not in the epistatic variance component but in the additive term. Thus, as illustrated in Figs. 2.3 and 2.5, although the relationship between offspring and parent depends on the additive, dominance, and epistatic effects, the existence of dominance and epistatic effects in the Mendelian sense does not necessarily lead to an insignificant parent–offspring correlation. Because the epistatic effects are effectively distributed among the additive and dominance components, detection of epistatic interactions is in practise very difficult (Barker, 1979).

2.1.2 The Infinitesimal Model

Epistatic effects greatly complicate quantitative genetic theory and, hence, they are generally assumed to be absent. As shown above, even if present, their contribution may be primarily through the additive component. Ignoring epistasis,

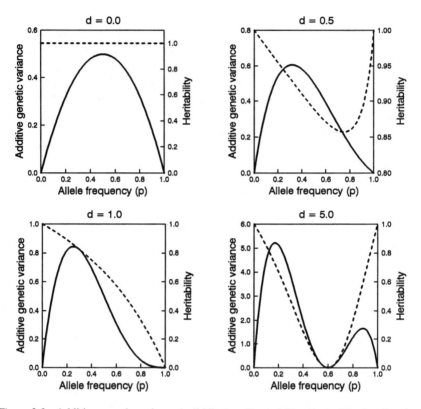

Figure 2.3 Additive genetic variance (solid line) and heritability (dashed line) as functions of allele frequency p when there are both additive and dominance effects.

the effect of different loci are independent and the genotypic value, G, of a trait determined by n loci can be decomposed into

$$G = m + \sum_{i=1}^{n} \sum_{j=1}^{2} (A_{i,j} + D_{i,j})$$
$$= m + A + D$$

$$(2.11)$$

where m is the population mean and $A_{i,j}$ and $D_{i,j}$ are respectively the additive and dominance deviations due to locus i. The second subscript, j, is necessary because there are two alleles per locus in a diploid organism. From the central limit theorem, A and D will become bivariate normal when the number of loci, n, is large (Bulmer, 1985, p. 123); as a consequence, the relationship between offspring and parents can be described using standard statistical methods. The variance of the genotypic value is simply

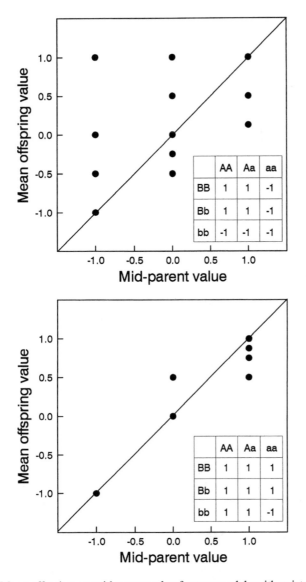

Figure 2.4 Mean offspring on mid-parent value for two models with epistatic and dominance effects.

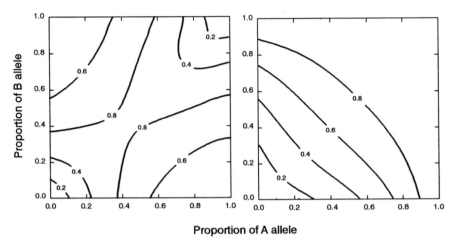

Figure 2.5 Isoclines of heritability for the two genetic models shown in Fig. 2.4.

$$V_G = V_A + V_D \tag{2.12}$$

where the subscripts refer to the relevant components of Eq. (2.11). In general, there will be environmental influences; hence, the general relationship for the variance of the phenotypic value is

$$V_P = V_A + V_D + V_E \tag{2.13}$$

It follows from this that the regression of the mean offspring value, X, on the mid-parent value, Y, is

$$Y = (1 - h^2)m + h^2X \tag{2.14}$$

As is expected from the single-locus case, the slope of the regression of mean offspring on mid-parent is equal to the heritability of the trait.

2.2   Estimation of Heritability for Continuous Traits

In this section, I examine some of the more common methods of estimating heritability for traits in which variation at the phenotypic level is continuous; in the next section, I examine the problem of estimation for threshold traits. Because, as will be shown, the estimation of standard errors may not be possible with nonstandard designs or may be very large, it is preferable to use a simple, statistically well-behaved method. From the previous derivation, an obvious method of estimating heritability is the mean offspring on mid-parent regression, or mean offspring on one parent. However, because heritability is equal to the ratio of additive to phenotypic variance, it is also possible to use other breeding designs, the most commonly used being full-sib and half-sib families. To use such designs to estimate heritability we must know the covariance (Cov_{XY}) between individuals (X and Y) of different relatedness. Crow and Kimura (1970, pp. 132–140) present the algebraic derivation, but here I give only the final result, which is for two loci with epistasis,

$$\text{Cov}_{XY} = (k_1 + k_2)V_A + k_2 V_D + (k_1 + k_2)^2 V_{AA}$$
$$+ (k_1 + k_2)k_2 V_{AD} + k_2^2 V_{DD} \quad \textbf{(2.15)}$$

where the coefficients k_i are the probabilities of two genes being identical by descent. Two genes are said to be **identical by descent** if they are descended from the same mutational event. Suppose we label the allele in a particular individual as A_i: Two offspring from this individual may carry an identical copy of this allele in the sense that both A_i's are derived by duplication from the A_i in the parent. Note that identity by descent does not mean that two individuals have alleles that have the same function (e.g., an allele A_i may occur in two individuals but not be identical by descent). The coefficient k_i is one-half the probability that one gene in individual X is identical by descent to one gene in Y, but not both, and k_2 is the probability that both genes in X are identical by descent to those in Y. Assuming that in the parental population no genes are identical by descent, the values of k_i can be deduced from probability theory for any given relationship: Values for full-sibs, half-sibs, and offspring on parent relationships are shown in Table 2.2. Substituting the values from Table 2.2 into Eq. (2.15), we obtain the covariances

$$\text{Full-sib covariance} = \frac{V_A}{2} + \frac{V_D}{4} + \frac{V_{AA}}{4} + \frac{V_{AD}}{8} + \frac{V_{DD}}{16}$$
$$\text{Half-sib covariance} = \frac{V_A}{4} + \frac{V_{AA}}{16} \quad \textbf{(2.16)}$$
$$\text{Parent-offspring covariance} = \frac{V_A}{2} + \frac{V_{AA}}{4}$$

Note that epistatic effects enter into all three relationships. However, as stated above, epistatic effects are generally ignored and, thus, the latter two sets of relationships provide direct estimates of the additive genetic variance. The heritability can then be calculated by dividing V_A by the phenotypic variance, which is easily obtained. The full-sib covariance contains, in addition to epistatic effects, effects due to dominance deviations. Thus, estimates of h^2 obtained from a full-sib breeding design are potentially biased upward. Assortative mating may also

Table 2.2   Probabilities of Identity by Descent for Different Relationships

Genotype					Parent–
X	Y	k_i	Full-sibs	Half-sibs	Offspring
$A_1 A_2$	$A_1 A_2$	k_2	¼	0	0
$A_1 A_2$	A_1 not A_2	k_1	¼	¼	½
$A_1 A_2$	Not $A_1 A_2$	k_1	¼	¼	½

bias heritability estimates. The estimation of this effect, first analyzed by Fisher (1918), is complex and I provide in the relevant sections simply the results [for more tractable derivations than those of Fisher, see Crow and Kimura (1970) or Bulmer (1985)].

In all analyses, it is assumed that the data are normally distributed (as is required by linear regression theory and analysis of variance). If the data are markedly non-normal, the data should be transformed if possible. (In the case of threshold traits, this transformation actually takes place after the analysis of variance—see Section 2.3.)

2.2.1   Mean Offspring on Mid-parent

In this design, the mean offspring value, Y, is regressed on the mid-parent value, X; two examples are shown in Fig. 2.6. Provided the number of offspring per family is the same, the data can be analyzed using simple linear regression methods. There is no bias introduced by assortative mating; hence, it is worthwhile to mate parents assortatively to increase the range of values. The heritability is equal to the slope of the regression, b, and the associated standard error equal to the standard error of the slope,

$$\text{SE}(h^2) = \sqrt{\frac{V_{PY}/V_{PX} - b^2}{N - 2}} \tag{2.17}$$

where the V's refer to the respective phenotypic variances and N is the number of families (X-Y pairs). The above equation is generally not necessary, as standard statistical packages provide the standard error of the regression slope as part of the output. If the number of offspring per family is not constant, then the mean value for each value is estimated with differing precision. This is not a problem if the number is reasonably large (>10) but may cause problems if sample sizes are small and very variable (Bohren and McKean, 1961). In this case, we proceed as follows (Kempthorne and Tandon, 1953; Bulmer, 1985, p. 79): The model is

$$Y_i = m + h^2 X_i + e_i \tag{2.18}$$

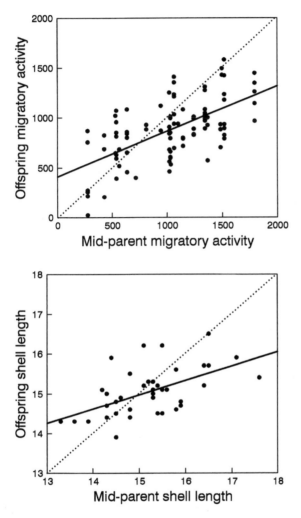

Figure 2.6 Two examples of mean offspring on mid-parent regressions. Upper graph shows data on migratory activity (measured as half hours of night activity) in a bird, the blackcap (*Sylvia atricapilla*). Heritability = 0.45 (SE = 0.08, $n = 94$). Lower graph shows data on shell length (mm) in the snail, *Partula taeniata*. Heritability = 0.36 (SE = 0.17, $n = 40$). [Data from Berthold and Pulido (1994) and Murray and Clarke (1967)].

where m is a constant (the mean), Y_i is the mean value from family i, X_i is the mid-parent value for family i, and e_i is the error, assumed to be normally distributed with zero mean with variance

$$\text{Var}(e_i) = V_{PY}\left\{\left(\rho - \frac{h^2}{2}\right) + \frac{1 - \rho}{n_i}\right\} \tag{2.19}$$

where ρ is the correlation between full-sibs and n_i is the number of individuals in family i. The individual data points are weighted by the reciprocal of $\text{Var}(e_i)$; thus, defining $w_i = 1/\text{Var}(e_i)$, we have

$$h^2 = \frac{\sum w_i(X_i - \bar{X})(Y_i - \bar{Y})}{\sum w_i(X_i - \bar{X})^2} \tag{2.20}$$

with the associated standard error

$$\text{SE}(h^2) = \frac{1}{\sum w_i(X_i - \bar{X})^2} \tag{2.21}$$

To obtain the weights, we need ρ and h^2. The former can be estimated as $h^2/2$ (see Section 2.2.3), leaving only h^2 to be estimated. This parameter is the one of central interest and occurs on both the left- and right-hand sides of Eq. (2.20); therefore, an iterative procedure must be adopted to find the value of h^2 which satisfies both sides of the equation.

If the variance in males differs from that of females, then one cannot, in principle, use the mean offspring on mid-parent regression. Separate heritabilities can be estimated for each sex using sons on sires and daughters on dams. Heritabilities estimated from the regressions of sons on dams and daughters on sires must be corrected to take into account the different variances: For the former, $h^2 = b(V_\female/V_\male)^{1/2}$, where b is the slope of the regression and the Vs are the respective phenotypic variances. As described in the following section, the standard error is considerably inflated when only a single parent is used; it is, therefore, worthwhile to search for a transformation that will remove the differences in variance. A typical reason for differences in variance is that the sexes vary in size and there is a relationship between the mean and the variance. In such a case, a logarithmic transformation may be sufficient to remove the difference.

An important question is, "What sample size is required to detect a particular heritability?" Assuming a constant family size, the standard error of heritability is equal to (Latter and Robertson, 1960)

$$\text{SE}(h^2) = \left(\frac{h^2(1 - h^2) + (2 - h^2)/n}{nN/(n + 2) - 3}\right)^{1/2} \tag{2.22}$$

The standard error is very sensitive to the number of families (N) and family size (n) when n is less than about 6 (Fig. 2.7). If the total number of individuals to be measured is fixed [i.e., $N(n + 2)$ = constant], then the optimum number of offspring per family is approximately

$$n = \left(\frac{2(1 - h^2/2)}{h^2(1 - h^2)}\right)^{1/2} \tag{2.23}$$

For h^2 between 0.25 and 0.75, a total family size of four (two parents + two offspring) is optimal when the total number of individuals is fixed. However, the potential for a great increase in the standard error if by chance family size is reduced suggests a more conservative family size between 6 and 10 offspring (see Fig. 2.7).

It has been implicitly assumed above that parents and offspring were raised under the same environmental conditions. This generally means raising both generations in the laboratory, although in the case of ectotherms such as birds, it may be possible to measure both parents and offspring in the wild. A lower bound of heritability in nature may be obtained by taking parents from the wild and raising their offspring under controlled conditions. Under these conditions, we have (Riska et al., 1989)

$$h_N^2 = \frac{b^2 \, V_{PN}}{r_A^2 \, V_{AL}} \tag{2.24}$$

where h_N^2 is the heritability under natural conditions, b is the slope of the mean offspring on mid-parent regression, r_A is the genetic correlation between the two environments, V_{PN} is the phenotypic variance of the parents under the natural conditions, and V_{AL} is the additive genetic variance of the offspring under the laboratory conditions. The concept of genetic correlation is discussed in detail in Chapter 3; at present, it is sufficient simply to note that like all correlation coefficients it lies between 0 and 1. Because it is generally not known, a minimum estimate of h_N^2 is made by assuming that $r_A = 1$. The additive genetic variance is calculated from the offspring data using the full-sib design described in Section 2.2.3. There is no fully satisfactory method of placing confidence limits on h_N^2: Riska et al. (1989) suggested the use of the bootstrap method, whereas Simons and Roff (1994) estimated the upper and lower 95% confidence bounds by substituting the upper and lower bounds of the regression slope.

2.2.2 Mean Offspring on One Parent

There are two reasons for using a regression of mean offspring on one parent: first, if the phenotypic variance in the trait under consideration differs between males and females and, second, if there are likely to be effects attributable to one

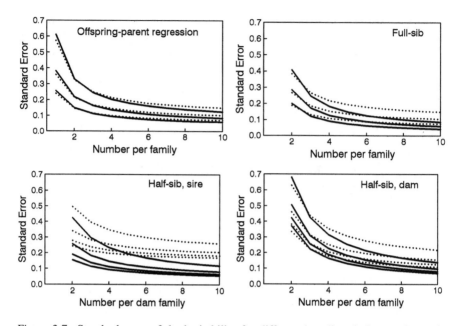

Figure 2.7 Standard error of the heritability for different breeding designs and sample size. $h^2 = 0.1$, dotted line; $h^2 = 0.5$, solid line. For the mean offspring on mid-parent regression and full-sib methods, the number of families, from top curve to bottom, is 25, 50, and 100. For the half-sib design, the number of sires is constant at 25, and the number of dams per sire from top curve to bottom is 2, 3, 4, and 5.

parent. The latter is particularly likely in mammals where maternal effects are common (see Chapter 7 for a detailed discussion of sex-related effects). It follows from Eq. (2.16) that the slope of the regression of the mean offspring value on one parent is equal to one-half of the heritability. Thus, to obtain the heritability, the slope must be multiplied by 2. Similarly, the standard error of the heritability is equal to twice the standard error of the regression slope. Therefore, it is clear that use of a single parent considerably increases the confidence region of the heritability estimate. Differences between the heritabilities estimated using fathers or mothers may be attributable to non-Mendelian factors such as maternal effects, but the relative imprecision of the estimate makes it difficult to detect significant differences.

Assortative mating biases the estimate of heritability, the appropriate correction being (Falconer, 1989, p. 178)

$$h^2 = \frac{2b}{1 + r} \qquad (2.25)$$

where r is the phenotypic correlation between parents.

If parents come from the wild and offspring are raised in the laboratory a minimum estimate of h_N^2 can be obtained using equation 2.24 multiplied by 4 to take into account the use of only a single parent ($h_N^2 < 4b^2 V_{PN}/V_{AL}$).

To obtain a preliminary estimate of the standard error given a fixed family size one can use (Latter and Robertson, 1960)

$$SE(h^2) = \sqrt{2} \left(\frac{h^2 (1 - h^2/2) - (2 - h^2)/n}{nN/(n + 1) - 3} \right)^{1/2} \tag{2.26}$$

which is essentially Eq. (2.22) multiplied by $\sqrt{2}$. Thus, the optimum family size remains the same, but the associated standard error is increased.

2.2.3 Full-sib Design

The offspring families in the previous two designs are full-sib families and can be used themselves, without reference to the parents, to estimate heritability. However, there are two potentially important sources of error: dominance variance and common environment. As shown previously, the covariance between full-sibs contains one-quarter of the variance due to dominance deviations, V_D. Thus, h^2 estimates from full-sib analyses may be somewhat inflated. If members of a family are raised together (e.g., in the same cage), then they all experience the same environment and thus resemblance may be due to common environment rather than genetic effects. Therefore, it is imperative that a split family design be used so that environmental effects can be estimated and the estimated heritability corrected for such effects. For simplicity, I shall first consider the analysis where there are no environmental effects and V_D is assumed to be negligible. The covariance between full-sibs is equal to $\frac{1}{2}V_A$ and can be obtained by a simple one-way analysis of variance as shown in Table 2.3. Heritability is then given by

$$
\begin{aligned}
h^2 &= \frac{2V_{AF}}{V_{AF} + V_{AP}} \\
&= \frac{2(MS_{AF} - MS_{AP})}{MS_{AF} + (k - 1)MS_{AP}}
\end{aligned} \tag{2.27}
$$

where the components are as defined in Table 2.3. The proportion of the total variance attributable to the among families (groups) component is also known as the intraclass correlation, designated by the symbol t (hence, $h^2 = 2t$). For equal family size, the standard error is

$$SE(h^2) = 2\left(1 - \frac{h^2}{2}\right)\left(1 + (k - 1)\frac{h^2}{2}\right)\left(\frac{2}{k(k - 1)(N - 1)}\right)^{1/2} \tag{2.28}$$

and for unequal family size, an approximate formula is (Swiger et al., 1964)

Table 2.3 Analysis of Variance Table for a Full-sib Design in Which There Are No Effects Due to Common Environment

Source of Variation	Degrees of Freedom	Mean Squares	Expected Mean Squares
Among families (AF)	$N-1$	MS_{AF}	$V_{AP} + kV_{AF}$
Among progeny (AP), within families	$T-N$	MS_{AP}	V_{AP}

Note: N is the number of families, T is the total number of individuals, and k is equal to family size if these are equal, otherwise it is the weighted estimate

$$k = \frac{\sum_{i=1}^{N} n_i - \left(\sum_{i=1}^{N} n_i^2\right)\left(\sum_{i=1}^{N} n_i\right)^{-1}}{N-1}$$

where n_i is the size of the ith family.

Variance components are estimated from $V_{AP} = MS_{AP}$ and $V_{AF} = (MS_{AF} - MS_{AP})/k$.

$$SE(h^2) = 2\left(1 - \frac{h^2}{2}\right)\left(1 + (k-1)\frac{h^2}{2}\right)\left(\frac{2(T-1)}{k^2(T-N)(N-1)}\right)^{1/2} \quad (2.29)$$

Arveson and Schmitz (1970) suggested the jackknife procedure as a method of estimating variance components in the estimation of heritabilities. Such robust procedures should not, however, be used unless it can be demonstrated that, under the presumed conditions pertaining to the particular problem at hand, these procedures are appropriate (Miller, 1974; Potvin and Roff, 1993). Simons and Roff (1994) tested the utility of the jackknife as a method of estimating the heritability and its associated standard error for the full-sib design: It was found to be superior to the approximate formula given above, but the difference is not great and for most purposes, the approximate formula is probably satisfactory.

The analysis of variance (ANOVA) layout for the nested design is illustrated in Table 2.4. For a general description of nested ANOVA, see Chapter 10 in Sokal and Rohlf (1995). From the variance component estimates, heritability is calculated as

$$h^2 = \frac{2V_{AF}}{V_{AF} + V_{AC} + V_{WC}} \quad (2.30)$$

with the standard error calculated according to Eq. (2.29) (note that k is computed as the adjusted number per cage not family). The standard error can also be calculated using the jackknife where a whole family is deleted per pseudovalue (Simons and Roff, 1994). Although the estimate is not sensitive to variable family size, it may be sensitive to lack of balance at the level of cage, and it is, therefore,

Table 2.4   Analysis of Variance Table for the Full-sib Design in Which Families Are Divided Among c Cages

Source of Variation	Degrees of Freedom	Mean Squares	Expected Mean Squares
Among families (AF)	$N-1$	MS_{AF}	$V_{WC} + kV_{AC} + kcV_{AF}$
Among cages (AC), within families	$N(c-1)$	MS_{AC}	$V_{WC} + kV_{AC}$
Within cages (WC)	$Nc(k-1)$	MS_{WC}	V_{WC}

Note: N is the number of families and k is equal to number per cage if these are equal, otherwise it is the weighted estimate

$$k = \frac{\sum\limits_{i=1}^{Nc} n_i - \left(\sum\limits_{i=1}^{Nc} n_i^2\right)\left(\sum\limits_{i=1}^{Nc} n_i\right)^{-1}}{Nc - 1}$$

where n_i is the size of the *i*th cage.

Variance components are estimated from $V_{WC} = MS_{WC}$, $V_{AC} = (MS_{AC} - MS_{WC})/k$, and $V_{AF} = (MS_{AF} - MS_{AC})/(kc)$.

advisable to ensure full balance at this level, even dropping cages or families to achieve this.

If parents are mated assortatively the intraclass correlation, t, is not $\frac{1}{2}h^2$ but $\frac{1}{2}(1 + rh^2)h^2$, where r is the correlation between parents (Bulmer, 1985, p. 129). From this quadratic relationship ($rh^4 + h^2 + t = 0$), the heritability estimate is estimated using the formula (Falconer, 1989, p. 179)

$$h^2 = \frac{-1 + \sqrt{1 + 4rH^2}}{2r} \tag{2.31}$$

where H^2 is the uncorrected heritability (i.e., $2t$).

The relationship among sample size, heritability, and the standard error can be estimated using Eq. (2.28) (Fig. 2.7). As with the mean offspring on mid-parent regression, a family size of between six and 10 is advisable, unless one can be assured of maintaining family size at four or above. If the total sample size (nN) is fixed, the optimal family size is approximately equal to $\frac{1}{2}h^2$ (Robertson, 1959a). There is little difference between the optimal design for the full-sib and mean offspring on mid-parent regression (Hill and Nicholas, 1974).

2.2.4   Half-sib Design

In this design, each male is mated to several dams, the advantage being that the covariance between half-sibs does not contain any contribution from V_D (Table

2.5). It is sometimes called the North Carolina Design 1 (Comstock and Robinson, 1952). The statistical design is basically the same as for the nested one-way ANOVA described above. There are three estimates of heritability: first, that obtained from the sire intraclass correlation (t_S),

$$h_S^2 = 4t_S = \frac{4V_{AS}}{V_{AS} + V_{AD} + V_{AP}} \tag{2.32}$$

This is the best estimate in that it is free from possible maternal and common environment effects. That obtained from the dam intraclass correlation (t_S)

$$h_D^2 = 4t_D = \frac{4V_{AD}}{V_{AS} + V_{AD} + V_{AP}} \tag{2.33}$$

contains in its numerator four times the maternal effects and all the dominance variance. Common environment effects appear as maternal effects. Finally, both sire and dam components can be used jointly:

$$h_{S+D}^2 = 2(t_S + t_D) = \frac{2(V_{AS} + V_{AD})}{V_{AS} + V_{AD} + V_{AP}} \tag{2.34}$$

Becker (1985) refers to this estimate as the genotypic heritability; it contains twice the maternal effects and half the dominance variance. A comparison of the sire and dam estimates may reveal the potential influence of maternal or common environment, but these can also be confounded by the dominance variance. Use of a split family design, as discussed for the full-sib design, is recommended to at least eliminate common environment effects.

If parents are mated assortatively, then the intraclass correlation is (Bulmer, 1985, p. 129), $t = 0.25h^2 (1 + 2rh^2 + r^2h^2)$, from which the appropriate adjustment is

$$h^2 = \frac{-1 + \sqrt{1 + 4(2r + r^2)t}}{2(2r + r^2)} \tag{2.35}$$

Approximate estimates of the standard errors are (Robertson, 1959a)

$$\mathrm{Var}(t_S) = \frac{2(1 - t_S)^2 \, [1 - 2t_S + k_1 \, (D + 1)t_S]^2}{(S - 1)D^2k_1^2}$$
$$+ \frac{2[1 + (D - 1)t_S]^2 \, [1 + (k_1 - 2)t_S]^2}{SD^2k_1^2 \, (D - 1)} + \frac{2t_S(k_1 - 1)(1 - 2t_S)^2}{SDk_1^2 \, (k_1 - 1)}$$

Table 2.5 Analysis of Variance Table for a Half-sib Design

Source of Variation	Degrees of Freedom	Mean Squares	Expected Mean Squares
Among sires (AS)	$S-1$	MS_{AS}	$V_{AP} + k_2 V_{AD} + k_3 V_{AS}$
Among dams (AD), within sires	$D_T - S$	MS_{AD}	$V_{AP} + k_1 V_{AD}$
Among progeny (AP), within dams	$T - D_T$	MS_{AP}	V_{AP}

Note: S is the number of sires, D_T is the total number of dams, and T is the total number of individuals. If the design is balanced, or at least not too unbalanced, k_1 and k_2 can be set equal to the mean number of offspring per dam family. Otherwise,

$$k_1 = \frac{T - \sum_{i=1}^{S} n_i \left(\sum_{j=1}^{D} n_{i,j}^2 \right) n_{i.}}{D_T - S}$$

$$k_2 = \frac{\sum_{i=1}^{S} \sum_{j=1}^{D} n_{i,j}^2 \, (1/n_{i.} - 1/T)}{S - 1}$$

where $n_{i,j}$ is the number of progeny in the family of the jth female mated to the ith sire and $n_{i.}$ is the number of progeny of the ith sire. The number of dams per sire are assumed equal in the above (D); if this is not the case, then D should be replaced by a subscripted term.

k_3 is the number of progeny per sire. For unequal numbers,

$$k_3 = \frac{T - \left(\sum_{i=1}^{S} n_{i.}^2 \right)/T}{S - 1}$$

Variance components are estimated from $V_{AP} = MS_{AP}$, $V_{AD} = (MS_{AD} - MS_{AP})/k_1$, and $V_{AS} = [MS_{AS} - (MS_{AP} + k_2 V_{AD})]/k_3$.

$$SE(h_S^2) = 4\sqrt{Var(t_S)}$$

$$Var(t_D) = \frac{2t_D^2 \, [1 - 2t_D + k_1 \, (D + 1)t_D]^2}{(S - 1)D^2 \, k_1^2}$$
$$+ \frac{2[D - (D - 1)t_D]^2 \, [1 + (k_1 - 2)t_D]^2}{SD^2 \, k_1^2 \, (D - 1)} + \frac{2[1 + (k_1 - 1)t_D]^2 \, (1 - 2t_D)^2}{SDk_1^2 \, (k_1 - 1)}$$

$$SE(h_D^2) = 4\sqrt{Var(t_D)} \tag{2.36}$$

where D is the number of dams per sire (assumed constant). For an unbalanced design, the mean number of dams per sire could be used, although its reliability has not been investigated. Results from my own analyses suggest that when the above formulas give confidence limits than encompass zero but the ANOVA indicates a significant sire effect, a significant heritability can be assumed. Standard errors estimated by Eq. (2.35) are shown in Fig. 2.7 for various sample sizes, keeping the number of sires constant at 25. It can be seen that, as with the other designs discussed thus far, a family size greater than four is advisable. A. Robertson (1959a, 1960a) investigated the optimal design when the total number of individuals to be measured is kept constant. The recommendations from this study are as follows:

If the magnitude of h^2 is known then,

- (1a) If only the sire component is to be used, then the number of offspring per dam (k_1) should be set at $k_1 = 4/h^2$,
- (1b) If both the sire and dam components are to be used, then $k_1 = 2/h^2$ with three or four dams per sire.

If no a priori estimate of h^2 is available, then

- (2a) If only the sire component is to be used, then $20 < k_1 < 30$.
- (2b) If both components are desired, then $k_1 = 10$ with three or four dams per sire.

Note that the optimum family size is very similar to that for the mean offspring on mid-parent and full-sib designs. Likewise, family sizes as low as two to three should be avoided (Robertson, 1959a). If there is no dominance variance or maternal effects, the full-sib design is preferable to the half-sib because the standard error of the former is approximately $4(h//T)^{1/2}$, whereas that of the latter is $4(2h^2/T)^{1/2}$, where T is the total number of individuals measured. This is illustrated in Table 2.6. The standard errors from the mean offspring on mid-parent regression are as small as those from the full-sib design, although, of course, the offspring–parent design requires measurements on 200 additional individuals (i.e., the parents). However, the advantage is that the estimate is not potentially confounded by dominance variance, and one still has the full-sib estimate obtainable from the offspring. For relatively large heritabilities ($h^2 = 0.5$ in Table 2.6), the standard errors from the half-sib design are considerably larger than those from the other two methods; to obtain comparable estimates, the number of sires must be quadrupled (Table 2.6). As the heritability decreases, the difference between the standard errors predicted for the full-sib and half-sib methods decreases ($h^2 = 0.10$ in Table 2.6). If h^2 is less than 0.25, then for a given number of individuals measured, the half-sib method is more accurate than the mean offspring on midparent regression, the reverse being true for $h^2 > 0.25$ (Robertson, 1959a).

Table 2.6   Comparison of Expected Standard Errors for Three Breeding Designs

Breeding Design	No. of Sires	No. of Individuals Measured	Heritability	
			0.10	0.5
Offspring–parent	100	700	0.07	0.09
Full-sib	100	500	0.07	0.10
Half-sib (sire, dam)	25	500	0.10, 0.15	0.19, 0.15
	50	1000	0.07, 0.11	0.14, 0.13
	100	2000	0.05, 0.07	0.10, 0.09

Note: The number of offspring per family (= dam family for the half-sib) is kept constant at five; for the half-sib design, there are four dams per sire.

2.2.5   Restricted Maximum Likelihood Methods

If the above designs (offspring on parent regression, full- or half-sib) are followed and statistical imbalance minimized (equal number of cages per family in split family designs and equal numbers of dams per sire), then the above formulas should be adequate [ANOVA is quite robust to imbalance at the lowest level, the number of individuals per family; see, for example, Swallow and Monahan (1984)]. Even for these simple designs, very large sample sizes are typically required to obtain reasonable estimates (e.g., Fig. 2.7). Thus, more complex designs should not be embarked upon without considerable reason. If, however, one does follow such a course, the usual analysis of variance techniques are likely to prove inadequate. An alternative approach is restricted maximum likelihood (Hill and Nicholas, 1974; Thompson, 1977a, 1977b; Thompson and Shaw, 1990; Yu et al, 1993; Knott et al., 1995). The conceptual basis of this technique is simple, but its implementation is not. To illustrate the approach, consider the problem of estimating the heritability with data from a single full-sib family. (With a single family, there is actually insufficient information to estimate the parameters, and the present example is given simply to illustrate the principle). Assuming no dominance or epistasis, the phenotypic value of an individual can be decomposed into the linear function (Ronningen, 1974)

$$X_i = m \sqrt{\frac{h^2}{2}} + E_i \sqrt{1 - \frac{h^2}{2}} \qquad (2.37)$$

where X_i is the value of trait X in the ith individual, m is the genotypic mean, and E_i is the environmental deviation, which is distributed as a random standard normal, $N(0, V_E)$. From standard probability theory, the term $E_{ij}(1 - h^2/2)^{1/2}$ is normally distributed with mean 0 and variance $V = V_E(1 - h^2/2)$.

The problem is to estimate h^2 given n values of X_i (i.e., the n phenotypic values of a set of individuals from a full-sib family). Consider some particular value of h^2 and m; given a value of X_i and m, then E_i is known, as

$$E_i = \frac{X_i - m\sqrt{h^2/2}}{\sqrt{1 - h^2/2}} \tag{2.38}$$

The probability of observing some value X_i, $P(X_i)$, is

$$P(X_i) = \frac{1}{\sqrt{2\pi V}} \exp\left(-\frac{E_i^2}{2V}\right) \tag{2.39}$$

The probability of obtaining the observed set of values is called the likelihood, L, and is equal to

$$L = P(X_1)\, P(X_2) \cdots = \prod_{i=1}^{n} P(X_i) \tag{2.40}$$

It is generally convenient to work with the logarithm (base e) of the likelihood

$$\ln(L) = -n\ln(\sqrt{2\pi V}) - \frac{1}{2V}\sum_{i=1}^{n} E_i^2 \tag{2.41}$$

There are three unknown parameters, m, V_E, and h^2. It is intuitively reasonable to select that set of parameter values which gives the highest probability of giving rise to the observed data set; this is called the maximum likelihood solution. It is clear from the log-likelihood equation that, in the case of the full-sib problem, we are dealing with a one-way analysis of variance problem. It is obviously much easier to use the ANOVA approach in this instance than the likelihood approach. The advantage of the maximum likelihood approach is that any breeding design can be accommodated [for examples, see Thompson and Shaw (1990), Cheverud et al. (1994), and Cheverud (1995)]. Considering a single trait and assuming multivariate normality, the log-likelihood of the observed data is (Hopper and Mathews, 1982)

$$\ln(L) = -C - \tfrac{1}{2}(\mathbf{X} - \mathbf{M})^{\mathrm{T}}\, \mathbf{V}^{-1}\, (\mathbf{X} - \mathbf{M}) - \tfrac{1}{2}\ln|\mathbf{V}| \tag{2.42}$$

where C is a constant that need not be evaluated [cf. the first term in Eq. (2.41)], $\mathbf{X}$ is a vector of the trait values, $\mathbf{M}$ is a vector of the trait's mean value, T designates the transpose of the vector, and $\mathbf{V}$ is the phenotypic variance–covariance matrix. Assuming no dominance or epistasis, $\mathbf{V}$ can be decomposed into

$$\mathbf{V} = 2V_A \mathbf{\Theta} + \mathbf{I}V_E \tag{2.43}$$

where V_A is the additive genetic variance, V_E is the environmental variance, $\mathbf{I}$ is the identity matrix, and $\mathbf{\Theta}$ is the matrix of expected additive genetic covariances between individuals (e.g., 0.5 for parents and offspring, 0.25 for half-sibs, assuming no inbreeding). Using numerical methods, values of M, V_A, and V_E are found that maximize the log-likelihood; heritability is then estimated as $V_A/(V_A + V_E)$. The model can readily be extended to consider several traits at once, dominance, or other complicating factors (Hopper and Mathews, 1982; Shaw, 1987).

Because maximum likelihood estimators are biased, Patterson and Thompson (1971) introduced the restricted maximum likelihood (REML) method, so-called because parameter values are restricted in the values they can take (in genetical analysis, this means requiring all variance estimates to be greater than zero). However, although eliminating bias, REML methods are more difficult to implement. A general review of REML is given by Kennedy (1981), Shaw (1987), and Thompson and Shaw (1990). Misztal (1994) and Spilke and Groeneveld (1994) review several public-domain computer packages which implement REML techniques. Statistical packages such as SAS are capable of maximum likelihood estimation. It is worth repeating that the present difficulties of using such techniques strongly favor the adoption of simple experimental designs.

2.2.6 Other Methods

The above methods are those that are most commonly used for non-domestic species. There are, however, a number of other techniques that have applicability under particular circumstances. Here I present a brief overview of these techniques giving the reference from which the relevant equations can be obtained.

2.2.6.1 Offspring–Parent Regression with Half-sib Offspring

In this design, parental values are measured and several dams are mated to each sire. Because of the half-sib relationship between some offspring, a standard mean offspring on mid-parent regression cannot be used. The effect of sire is removed by regression of the offspring values on dam within the sire (Becker, 1985, p. 93).

2.2.6.2 Diallel Cross

"A diallel cross is the set of all possible matings between several genotypes. The genotypes may be defined as individuals, clones, homozygous lines, etc., and, if there are n of them, there are n^2 mating combinations, counting reciprocals separately" (Hayman, 1954a). This design is used most frequently by agronomists working with inbred lines of plants where such crosses are generally feasible. It has, however, also been used to assess genetic variation in animal traits. Mather and Jinks (1982, Chap. 9), Becker (1985), and Bulmer (1985) provide an account of its theory and implementation. The Some papers providing examples of its use are Lawrence (1964: morphology and flowering time in *Melandrium album*), Thomas (1967, 1969a, 1969b: morphology and flowering time in perennial ryegrass), Ecker and Barzily (1993: growth rate and flowering time in the ornamental plant *Lisianthus*), Robertson et al. (1994: morphology and reproductive traits in *Mimulus guttatus*, a weedy perennial), Jinks and Broadhurst (1963: litter size and weight in the rat), Parsons (1964: mating speed in *D. melanogaster*), Underhill (1968: morphology in the frog *Rana pipiens*), Caligari and Mather (1980: sternopleural bristles in the fruit fly), Henderson (1981: locomotor activity in the house mouse), Crusio et al. (1984: detailed treatment with respect to behavioral genetics), Gerlai et al. (1990: behavior in the paradise fish), Mangan (1991: behavior in the screwworm), Levin et al. (1991: morphology and reproductive traits in a polychaete), and Antolin (1992b: sex ratio in a parasitic wasp). Despite considerable publication on the theory of the diallel cross (e.g., Yates, 1947; Hayman, 1954a, 1954b, 1957, 1960; Jinks, 1954; Kempthorne, 1956; Griffing, 1956a, 1956b; Gilbert, 1958; Eisen et al., 1966; Wearden, 1969), there is still controversy over its utility and statistical interpretation (Wright, 1985). Hayman (1954a, p. 808) claimed that "experiments with diallel crosses provide a powerful method of investigating polygenic systems," a view supported by Mather and Jinks (1982, p. 251): "the Hayman (1954a) analysis of variance . . . of a complete diallel crosses including selfs is probably the most sensitive means available of detecting non-additive variation, and maternal sources of reciprocal differences and dominance, if the model is adequate, in a randomly mating population." On the other hand, Pooni et al. (1984, p. 252) concluded that if "the primary purpose of an investigation is to measure the genetical components of variation and to test the assumptions on which estimates are based, the diallel should not be the preferred design. The triple test cross (Kearsey and Jinks, 1968) in one or more of its many forms (Jinks et al., 1969; Pooni et al., 1978, 1980) will always be preferred." For a discussion of the use of the diallel cross applied to natural populations, see Gebhardt (1991).

2.2.6.3 The Triple Test Cross

Devised by Kearsey and Jinks (1968) the triple test cross is a very rarely used design for nondomestic species. The design consists of crossing a sample of males from a population under study with the same three testers. These testers are two inbred lines and the F_1 produced from them. The method provides a test for the presence of epistatic, dominance, and additive variances. However, quantitative estimates of heritability can be made only if the two inbred lines are extreme selection lines (Kearsey and Jinks, 1968). Hewitt and Fulker (1981) discuss the use of this method in behavioral genetics and used the technique to dissect the genetic architecture of behavior in the rat *Rattus norvegicus*. Henderson (1981) used the triple test cross to examine locomotor behavior in house mice, and Goodwill and Walker (1978a, 1978b) estimated epistatic components in morphological and life history traits in *Tribolium casteneum*.

2.2.6.4 North Carolina Design 2

This is similar to the diallel cross in that all possible combinations are used, but in this case, the crosses are between individuals: Specifically, *m* males are each mated to *f* females to produce *mf* families. There are very limited cases in which this design could be utilized for animals, but it is readily achieved for hermaphroditic plants. See Edwards and Emara (1970), Cockerham and Weir (1977), or Mather and Jinks (1982, pp. 243–251) for a discussion of its implementation.

2.2.6.5 North Carolina Design 3

This design is the most efficient of the three North Carolina designs for the detection of dominance and is described in Bulmer (1985, pp. 69–71). The method is very restrictive and consists of first producing an F_2 generation from a cross between 2 inbred lines, then crossing these back to the original inbred lines. Another group of F_2 individuals are each crossed to each of the two parental lines. Kearsey (1980) discusses the merits of this design and the sample sizes required.

Comparing the North Carolina designs 2 and 3 and the diallel cross for the detection of dominance variation among populations, Kearsey (1970) concluded that although NC 3 is best, the sample sizes are exceedingly large (see Table 4 in his paper), and hence (p. 542), "Over most of the situations considered, the minimum experimental size of all designs is large, usually too large to make them practical for comparing different populations."

2.2.6.6 Incomplete Factorial

Instead of a complete diallelic cross, only a particular set of combinations are employed (Kempthorne and Curnow, 1961; Cockerham, 1963); for example, in an analysis of genetic variation in *Rana lessonae*, Semlitsch (1993) used

		Male Parent									
		1	2	3	4	5	6	7	8	9	10
F	1	U	U	U	U	.	.	.	.	.	.
e	2	.	.	U	U	U	U	.	.	.	.
m	3	.	.	.	.	U	U	U	U	.	.
a	4	.	.	.	.	.	.	U	U	U	U
l e	5	U	U	.	.	.	.	.	.	U	U

where *U* designates a cross utilized. This design permits the separation of additive, dominance, maternal, and environmental effects. Because it maximizes the number of sires represented among the progeny, this design maximizes the ability to detect genetic effects when only a limited number of families can be raised (Travis et al., 1987). It is possible for many plant species but only a relatively limited number of animals, most particularly anurans and fishes in which sperm and ova can be stripped from the parents and fertilization done artificially. Examples illustrating its use are Travis et al. (1987: *Hyla crucifer*, an anuran), Semlitsch (1993: *Rana lessonae*, an anuran), and Kelly (1993: *Chamaecrista fasciculata*, a legume).

2.3 Estimation of Heritability for Threshold Traits

A great many traits in natural populations show discontinuous variation (Roff, 1996a) (Fig. 2.8). Most commonly there are two morphs, which are distinguishable on the basis of morphology, life cycle, and/or behavior. In some cases, the different morphs are determined by simple Mendelian mechanisms such as a single diallelic locus. Color polymorphisms are particularly likely to be so determined (Roff, 1996a). An alternate model when simple Mendelian models do not fit is the threshold model (Falconer, 1989). Consider a trait that shows dimorphic variation. The threshold model posits that the determination of this variation is a consequence of some underlying character, called the **liability** by Falconer, that is itself continuously distributed: Individuals in which this character exceeds a particular value, the threshold, develop into one morph, whereas individuals below the threshold develop into the alternate (Fig. 2.9). Because the underlying character is continuously distributed, it can be treated using the usual quantitative genetic approach (see below). Although the threshold and single-locus models may appear disparate, they can be subsumed under a single causal model (Roff, 1986a). In the threshold model, the value of the underlying trait is assumed to be a consequence of the additive action of alleles at several loci, whereas in the single-locus case the value of the underlying trait can be assumed to be due to the additive action of the two alleles at the single locus (Fig. 2.9).

Throughout this section, I shall consider only the simplest case of dimorphic variation [for an example of a three-class, two-threshold model, see Falconer (1989, p. 305)]. Obviously, if the value of the underlying trait could itself be measured, then we could proceed as described above for continuously distributed traits. However, this trait may not be measurable either because it is technically not feasible or because it may not actually exist. By the latter statement, I mean the following: The production of a particular morph may be the outcome of the interaction of a number of physiological processes, the combined action of which is summarized in a "hypothetical" underlying continuously distributed trait. A simple example in the context of a continuously distributed trait is that of shape: Shape can be described mathematically by a principal component score that integrates a large number of morphological measurements. There is no problem

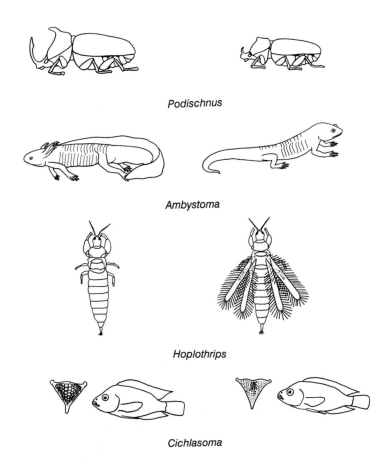

Figure 2.8 Some examples of dimorphic variation. From bottom to top and left to right: dental dimorphism shown by the molariform and papilliform dentition in *Cichlasoma* (Meyer, 1990); life-cycle and male mating dimorphism shown by wing and foreleg dimorphism in *Hoplothrips* (Crespi, 1986); life-cycle dimorphism illustrated by the paedomorphic and terrestrial forms of *Ambystoma* (Duméril, 1867); male mating dimorphism displayed by the major and minor males in *Podischnus* (Eberhard, 1980).

with considering shape as a trait, but it is hypothetical in the sense that it cannot be directly measured.

The problem is to estimate the heritability of the underlying trait from the relative proportion of the two phenotypic classes among relatives. We proceed in two steps: First we compute the heritability measured on the 0,1 scale of the manifested phenotypes, and second, convert this to the heritability of the trait on the underlying scale by the formula (Dempster and Lerner, 1950)

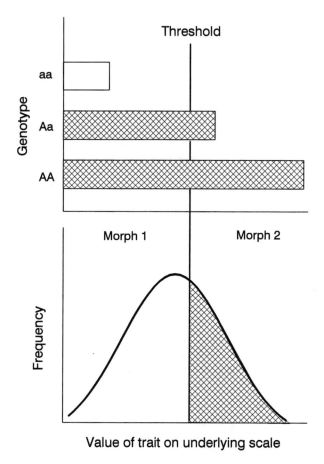

Figure 2.9 Schematic illustration of the threshold model. The upper graph shows the single-locus case in which the value of the underlying trait is determined by the additive action of the alleles at a particular locus. The bars indicate the value of the underlying trait, not its frequency in the population. Genotypes *AA* and *Aa* have values that exceed the threshold, and therefore develop into morph 2. Because it does not exceed the threshold, the alternate homozygote, *aa*, develops into morph 1. Note that allele *A* appears dominant to *a*. The lower graph shows the polygenic model: Many loci act additively to determine the value of the underlying trait, which is thus normally distributed in the population. The hatched and unhatched regions indicate the frequency in the population. Individuals lying to the left of the threshold develop into morph 1, wheras individuals to the right (cross-hatched region) develop into morph 2.

$$h^2 = h_{0,1}^2 \frac{p(1-p)}{z^2} \tag{2.44}$$

where h^2 is the heritability measured on the underlying scale, $h_{0,1}^2$ is the heritability measured on the 0,1 scale, p is the mean proportion in the population, and z is the ordinate on the standardized normal curve which corresponds to a probability p. The last can be obtained from a table of the normal distribution [e.g., in Zar (1984)] or from the very exact approximation derived by Hamaker (1978):

$$z = \frac{\exp^{(-1/2\ x^2)}}{\sqrt{2\pi}}$$

$$\text{where } x = [\text{Sign}(0.5 - p)]\ [1.238c(1 + 0.0262c)] \tag{2.45}$$

$$c = \sqrt{-\ln[4p(1-p)]}$$

The approximate standard error is obtained in the same manner, viz.

$$\text{SE}(h^2) = \text{SE}(h_{0,1}^2) \frac{p(1-p)}{z^2} \tag{2.46}$$

As with the former methods, the jackknife can be used, the jackknife being applied directly to the heritability estimate on the underlying scale. For convenience, and by convention, I shall henceforth refer to the heritability on the underlying scale simply as the heritability of the dimorphic trait.

2.3.1 Full-sib Design

The task is to calculate the intraclass correlation on the 0,1 scale. Three methods have been proposed: maximum likelihood, χ^2, and ANOVA (Robertson, 1951; Elston, 1977). In practice, these generally give more or less the same result [e.g., Roff (1986b): note that in this paper there are several typographical errors in the formulas; see Mousseau and Roff (1989b) for the correct formulas]. The ANOVA method is simpler to implement because it is available in statistical packages: It is simply an analysis of variance using the data coded as 0 or 1. Because of coding by 0 and 1, the usual formulas can be somewhat simplified to

$$t = \frac{\text{MS}_{AF} - \text{MS}_{AP}}{\text{MS}_{AF} + (k-1)\text{MS}_{AP}}$$

$$\text{MS}_{AF} = \frac{\sum (m_i^2/n_i) - (\sum m_i)^2/T}{N-1} \tag{2.47}$$

$$\text{MS}_{AP} = \frac{\sum m_i - \sum (m_i^2/n_i)}{T-N}$$

$$k = \frac{T - \sum (n_i^2/T)}{N-1}$$

where N is the number of families, T is the total number of individuals, m_i is the number of individuals of a given morph in family i, and n_i is the number of individuals in family i. The proportion p is estimated (Roff, 1986b) as

$$p = \frac{\sum p_i}{N} = \frac{\sum (m_i/n_i)}{N} \tag{2.48}$$

The heritability of the trait is then estimated as

$$h^2 = 2t \frac{p(1 - p)}{z^2} \tag{2.49}$$

with approximate standard error of

$$\mathrm{SE}(h^2) = \frac{p(1 - p)}{z^2} 2(1 - t)[1 + (k - 1)t] \left(\frac{2(T - 1)}{k^2(T - N)(N - 1)} \right)^{1/2} \tag{2.50}$$

Note that the above is simply Eq. (2.28) with the correction for the transformation between scales. The above assumes unequal sample size; for equal sample size, use Eq. (2.29) with the scale transformation correction.

Because of the possibility of cage effects, at least two cages per family should be used and a nested ANOVA employed to separate cage effects from family effects. If cage effects are detected and the number per cage varies greatly, it is probably preferable to estimate p using the mean proportion per family estimated from the mean proportion per replicate cage. Because the use of 0,1 data clearly violates the assumption of normality required in the analysis of variance, a randomization test is a preferable means of establishing the statistical significance of family effects. In an analysis of wing dimorphism [= a long-winged (macropterous) morph capable of flight and a short-winged (micropterous) flightless morph] in the cricket *Gryllus firmus*, I used three methods (Roff, unpublished):

1. A nested ANOVA using all individuals categorized as 0 (= macropterous) or 1 (micropterous).

2. A one-way ANOVA using the mean proportion per cage (two estimates per family, both the raw proportions and arcsine square root transformed values were used; the results did not differ).

3. A randomization test conducted as follows. First, the heritability was computed by pooling the two cages per family. Next, cages were paired at random and the heritability computed for this sample: 999 such randomized heritabilities were computed. The probability of obtaining a value of h^2 as large or larger than that observed was estimated by the

proportion of heritabilities from the randomized set plus the observed h^2 that were as large as or larger than the observed value. A somewhat better method would have been to compute the heritability using the nested ANOVA for each set rather than pooling the cages.

Sixteen different comparisons were made (proportions differ between the sexes, two rearing environments, and eight lines, Table 2.7). The three methods of statistically testing for differences attributable to family (nested ANOVA, one-way ANOVA, randomization) give very similar results, despite the fact that in some cases the proportions are close to 0 or 1, producing a highly skewed distribution. In 13 of the 16 tests, all three methods indicate highly significant variation among families. All three tests indicate no significant effect due to family in L1 female offspring or S1 male offspring. Phenotypic variation is low (proportion macropterous, $p = 0.92$ and 0.03, respectively) in both cases; hence, the lack of significance probably reflects the low power of the tests under these conditions. In one case, S2 females, the nested ANOVA indicates no significant variation among families ($P = .113$), whereas the other two tests indicate significant variation ($P < .001$ and $P = .039$ for the ANOVA and randomization methods, respectively). As with the previous cases, phenotypic variability is very low (proportion macropterous, $p = 0.04$). These results suggest that the analysis of variance is very robust to fairly extreme skew. Nevertheless, a randomization test is a useful additional test.

2.3.2 Half-sib Design

The approach is exactly the same as above, using the 0,1 data and running the analysis of variance as indicated in Table 2.5 and Eqs. (2.32)–(2.34) to obtain $h^2_{0,1}$. Heritability on the underlying scale is then computed using Eq. (2.44), the value of p being estimated as

$$p = \frac{\sum\limits_{i=1}^{S} \left(\sum\limits_{j=1}^{D_i} p_{i,j} \right) D_i^{-1}}{S} \tag{2.51}$$

where D_i is the number of dams of the ith sire and p_{ij} is the proportion of a given morph in the family of the jth dam mated to the ith sire. If there are maternal effects, these will be present in p and, hence, will contaminate the heritability estimate.

2.3.3 Mean Offspring on Mid-parent Regression

Because the value of the parent is known only on the 0,1 scale, it is not possible to make use of the usual mean offspring on mid-parent regression. The value of

Table 2.7   *Probability Values from Three Methods of Statistical Analysis for an Effect of Family on the Distribution of Macroptery in G. firmus*

Env.[a]	Line[b]	Sex	p[c]	A	R	NA
15/25	L1	F	0.92	0.350	0.578	0.820
15/25	L1	M	0.77	<0.001	0.004	<0.001
15/25	C1	F	0.74	<0.001	0.006	<0.001
15/25	C1	M	0.53	0.002	0.002	0.001
15/25	L2	F	0.91	0.013	0.016	0.002
15/25	L2	M	0.79	0.030	0.017	0.006
15/25	C2	F	0.60	<0.001	0.001	<0.001
15/25	C2	M	0.30	<0.001	0.001	<0.001
17/30	S1	F	0.08	0.002	0.003	0.002
17/30	S1	M	0.03	0.622	0.430	0.618
17/30	C1	F	0.50	0.001	0.001	<0.001
17/30	C1	M	0.27	<0.001	0.001	<0.001
17/30	S2	F	0.04	<0.001	0.039	0.113
17/30	S2	M	0.02	<0.001	0.009	0.008
17/30	C2	F	0.53	<0.001	0.001	<0.001
17/30	C2	M	0.38	<0.001	0.001	<0.001

Note: A = one-way ANOVA, R = randomization, and NA = nested ANOVA.

[a]A/B, A = photoperiod (number of hours per day of light), B = temperature (°C).

[b]L = line selected for increased proportion macropterous, S = line selected for decreased proportion macroptery, C = control line. The number refers to replicate (two per line).

[c]Proportion macropterous.

the parent on the underlying scale can be approximated by the transformed value of the parent's family (Roff, 1986b). Thus, if the proportions in the families of the male and female parents are p_δ and $p_\circ$, respectively, then the mid-parent value is approximated by $\frac{1}{2} (z_\delta + z_\circ)$. The value for the mean offspring is found in the same manner. Analysis then proceeds as for a normal continuous trait. However, because there is error in the measurement of the independent variable (the mid-parent value), the heritability will be underestimated.

2.3.4   Falconer's Proband Method

Falconer (1965a) developed this method for cases in which both the incidence in the population and the incidence among relatives of some specified sort are known. The method is most easily understood using the mean offspring on mid-parent regression. From the offspring–parent regression we can obtain the predicted value in offspring (Y) from a particular set of parents (X) by

$$Y = (1 - h^2)m + h^2 X \qquad (2.52)$$

where m is the population mean. Rearranging the above so that h^2 is the dependent variable gives

$$h^2 = \frac{Y - m}{X - m} \tag{2.53}$$

If the offspring are from a group of related individuals, such as full- or half-sibs, the above equation must be corrected for the assortative mating:

$$h^2 = \frac{Y - m}{(X - m)r} \tag{2.54}$$

where r is the coefficient of relatedness ($= 0.5$ for full-sibs, 0.25 for half-sibs). Therefore, from the above equation, the heritability on the $0,1$ scale can be estimated, given the proportional representation of the two morphs in the population in general and in a group of related individuals (Fig. 2.10). Without loss of generality, let the variance of the underlying trait (liability) be 1 and the threshold value be 0: The values of Y and m are then simply the abscissas on the standardized normal curve which correspond to the observed probabilities of affected individuals in the group of related individuals and the general population, respectively, and can be calculated using a table of the normal curve [e.g., appendix A in Falconer (1989) or Eq. (2.43)]. The denominator, designated i by Falconer (1989), is the difference between the mean of the population and the mean of the affected individuals in the general population, the latter being simply the mean from a truncated normal distribution

$$X - m = i = m + \frac{z}{P_{pop}} - m \tag{2.55}$$

$$= \frac{z}{P_{pop}}$$

where

$$z = \frac{\exp(-\frac{1}{2} m^2)}{\sqrt{2\pi}}$$

and P_{pop} is the proportion of affected individuals in the general population. For quick calculation, values of i corresponding to a given proportion p are given in appendix A of Falconer (1989). It is evident from Fig. 2.10 that the variances among affected individuals from the general population and from the group of

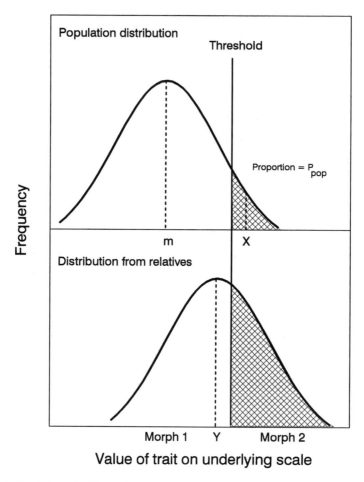

Figure 2.10 Schematic illustration of Falconer's proband method. The upper panel shows the distribution in the general population, where m is the mean liability in the population and X is the mean value of the parents which produce the offspring distribution shown in the lower panel.

related individuals is not exactly the same, as implicitly assumed above. To correct for this, Reich et al. (1972) derived the modified formula

$$h_{0,1}^2 = \frac{m - Y\left[1 - (m^2 - Y^2)(1 - Y/i)\right]^{1/2}}{(i + Y^2\left[i - m\right])r} \tag{2.56}$$

The sign of the square root is taken as that which produces a positive heritability.

Assuming that P_{pop} is estimated with negligible error, the standard error is approximately (Falconer, 1989, p. 303)

$$SE(h_{0,1}^2) = \frac{1 - P_{rel}}{i_{pop}^2 \, i_{rel}^2 \, m_{rel}} \qquad (2.57)$$

where "pop" refers to the general population, "rel" refers to the group of related individuals; P_{rel} and m_{rel} are the proportion and number of affected individuals, respectively, in the group of relatives.

Simulation analysis indicates that any of the above methods is suitable when the incidence in the population is not extreme (95% > proportion > 5%), but if this is the case, then Falconer's proband method is preferable (Olausson and Ronningen, 1975; Mercer and Hill, 1984).

2.4 Heritability Values Among Different Types of Trait

In a brief survey (19 estimates) of heritabilities in domestic animals (humans and *Drosophila* included), Falconer (1989, p. 164) observed that, "On the whole, the characters with the lowest heritabilities are those most closely connected with reproductive fitness, while the characters with the highest heritabilities are those that might be judged on biological grounds to be the least important as determinants of natural fitness." Does such a pattern hold for nondomestic organisms?

Gustafsson (1986) measured the heritabilities of a large number of traits in the collared flycatcher (*Ficedula albicollis*) and over a five-year period, estimated individual lifetime reproductive success. Using analysis of variance, Gustafsson estimated the proportion of variance in lifetime reproductive success accounted for by each trait: There is a highly significant negative correlation between the heritability of the trait and its proportional contribution to the variance in lifetime reproductive success (Fig. 2.11); those traits such as life span and number of fledged young have low heritabilities (for females, -0.0160 and -0.0161, respectively, and for males -0.0001 and -0.0052, respectively, all estimates being not significantly different from zero) but make relatively large contributions to variance in lifetime reproductive success (18.9%, 7.4%, 26.9%, and 15.2%, respectively). The heritability of fitness itself (lifetime reproductive success) is very low (-0.0142 for females, 0.0083 for males) and not significantly different from zero.

The traits measured by Gustafsson can be conveniently divided into two classes: morphological and life history traits, the latter being traits such as fecundity, viability, and development rate that are invariably connected to fitness. Morphological traits may themselves be closely connected to fitness; for example, in the collared flycatcher, the height of the white patch on males accounts for 8.4% of the variance in lifetime reproductive success, which is more than that contributed

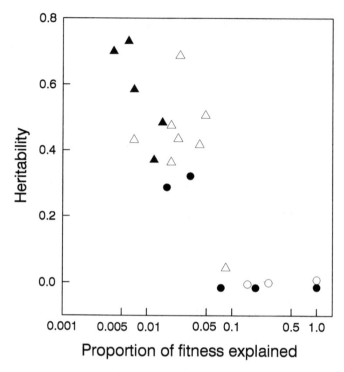

Figure 2.11 Heritability estimates of life history (circles) and morphological (triangles) traits in the collared flycatcher (males = closed symbols, females = open symbols). Excluding the two points at the far right (because they are the composite trait lifetime reproductive success), the correlation between heritability and log (proportion of fitness explained) is highly significant ($r = 0.85$, $n = 19$, $P < 0.0001$). [Data from Gustaffson (1986).]

by the two life history traits clutch size (3.2%) or time of laying (1.7%). Nevertheless we might expect, in general, morphological traits to be further removed from fitness than life history traits. To test the hypothesis that traits most closely connected to fitness will have the lowest heritabilities, Roff and Mousseau (1987) surveyed estimates for a single genus, *Drosophila*, whereas Mousseau and Roff (1987) examined estimates for animals in general (75 species). In addition to the categories of morphology and life history, we defined two further classes: behavioral traits, which comprise such traits as activity level, alarm reaction, and courtship behavior, and physiological traits such as oxygen consumption, resistance to heat stress, and body temperature. It is difficult to rank these two categories with respect to the others, but it seems intuitively reasonable that behavioral traits should lie between life history and morphological traits. Because of potential statistical biases due to nonindependence of estimates, we analyzed both the entire data set and that consisting of the medians of the estimates reported for each character of each species. Life history traits consistently have significantly lower heritabilities than morphological traits, with behavioral and physiological traits lying between (Table 2.8). A third method of comparing life history and morphological traits is to do a paired comparison using data from individual species (Fig. 2.12): Of the 15 different species for which such data are available, there is only one case in which the median heritability of life history traits exceeds that of the morphological traits, a distribution which is significantly different from the null hypothesis of $50 : 50$ ($\chi^2 = 11$, df $= 1$, $P < 0.001$).

What might explain the above pattern of heritabilities of life history traits being lower than morphological traits? We (Roff and Mousseau, 1987; Mousseau and Roff, 1987) suggested that this pattern is consistent with the expected action of selection. Consider some trait such as fecundity, which is closely connected to fitness; suppose that this trait is determined by the additive action of n loci at which there are two alleles 0 and 1, fecundity increasing as the sum increases. Obviously, over time, selection will favor loss of the 0-type alleles with a consequent decrease in additive genetic variance and, thus, heritability. The rate of fixation of alleles will depend on the intensity of selection with traits such as morphological traits approaching fixation much more slowly in general than life history traits because they will be under relatively weaker selection. Therefore, if we begin with the same heritability, in general, life history traits will have lower heritabilities over their evolutionary trajectories than morphological traits. The situation at equilibrium is less clear and depends on the mechanism maintaining genetic variance (discussed in Chapter 9). However, it still seems likely that, overall, selection will reduce the heritabilities of that constellation of traits which contribute primarily to fitness more than those which are further removed.

Price and Schluter (1991) suggested an alternative hypothesis for the difference in heritabilities between life history and morphological traits. Their hypothesis is summarized in the upper panel of Fig. 2.13. They note that many, if not all, life history traits are directly connected to morphology—examples being a positive relationship between fecundity and body size, or longevity and body size. Now let the additive genetic variance in the morphological trait be V_{AM} and the environmental variance be V_{EM}. Further, assume that the life history trait is determined solely by the morphological trait plus an additional environmental component E with variance V_E. For simplicity, assume that the value of the life history variable is a simple linear function of the morphological trait and the second environmental factor (i.e., life history value = morphological value + E). The heritability of the morphological trait is thus $V_{AM}/(V_{AM} + V_{EM})$, whereas that of the life history trait is $V_{AM}/(V_{AM} + V_{EM} + V_L)$. This scenario leads to the observed pattern. However, an alternate and possibly more reasonable scenario is that illustrated in the lower panel of Fig. 2.13, in which the life history trait is determined in part by the morphological trait and in part by other genes, which contribute an additive genetic variance V_{AO}. The two heritabilities are then $V_{AM}/(V_{AM} + V_{EM})$ and

Table 2.8   Summary of Heritability Estimates for Nondomestic Animals

Comparison	Life History	Behavior	Physiology	Morphology
Drosophila	0.12	0.18	—	0.32
All animals	0.26	0.30	0.33	0.46
Medians	0.26	0.37	0.31	0.51

Notes: The heritability estimates for *Drosophila* were estimated using the median value of each character from each study. "All animals" does not include the *Drosophila* data (which some would assert is not a true animal in any case) and consists of 1120 estimates. "Medians" is based on the median heritability for a given species and character and consists of 283 separate values. Kolmogorov–Smirnov tests of the "medians" indicate that the heritability of morphology is significantly different from the other three categories but that the remaining three do not differ from each other [for details see Table 3 of Mousseau and Roff (1987)].

Source: Data from Roff and Mousseau (1987) and Mousseau and Roff (1987).

$(V_{AM} + V_{AO})/(V_{AM} + V_{AO} + V_{EM} + V_E)$. If V_{AO} is much greater than V_{AM}, then the heritability of the life history trait will be more dependent on the former source of additive genetic variance than the latter. Selection on the life history trait will proceed by reducing V_{AO} at a faster rate than V_{AM}, again producing the observed pattern, but because of the effect of selection, not the "downstream" nature of the relationship between the life history and morphological traits.

There is no theoretical way to resolve these two hypotheses and, indeed, both may be playing significant roles. What is required is a greater understanding of the genetical and "nongenetical" (e.g., physiological, mechanical, ecological) architecture underlying life history and morphological variation. The importance of such a more holistic approach than simply examining the genetic architecture is illustrated by the traits body size and development time in insects. In a wide variety of insect species, metamorphosis into the adult form is triggered by the attainment of a critical size [*Rhodinus prolixus* (Wigglesworth, 1934); *Manduca sexta* (Nijhout and Williams, 1974a, 1974b); Nijhout, 1975); *Oncopeltus* spp (Blakley and Goodner, 1978; Nijhout, 1979); *Acheta domestica* (Woodring, 1983)]. As a consequence, variation in development time may be primarily a result of genetic variation in the critical size for metamorphosis and environmental variation in the rate of growth. A similar situation also appears to occur in the plant *Cynoglossum officinale* in which there is an inherited threshold size for flowering (Wesselingh et al., 1993; Wesselingh and de Jong, 1995). In both cases, the situation is precisely that suggested by Price and Schluter (1991), the heritability of the life history trait (development time) being deflated by environmental factors and thus smaller than the morphological trait (body size). The situation is, in fact, likely to be much more complicated than this, with genetic variation in rate of growth and possibly an interaction between development time and body size. It is only with a detailed analysis of the genetic basis of the underlying traits in

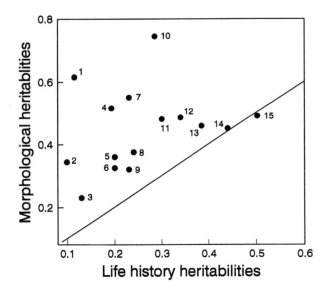

Figure 2.12 A comparison of morphological and life history heritabilities for a variety of species. Where multiple values were available, I used the median (means gave qualitatively the same result). List of species: (1) *Eurytemora herdmani* [copepod: McLaren (1976)]; (2) *Drosophila melanogaster* (Tantawy and El-Helw, 1970); (3) *Drosophila simulans* (Tantawy and Rakha, 1964); (4) *Dysdercus bimaculatus* [cotton stainer bug; Derr (1980)]; (5) *Homarus americanus* [lobster: Fairfull et al. (1981)]; (6) *Crassostrea gigas* [oyster: Lannan (1972)]; (7) *Callosobruchus maculatus* [beetle: Moller et al. (1989)]: (8) *Oncopeltus fasciatus* [milkweed bug: Hegmann and Dingle (1982)]: (9) *Salmo gairdneri* [rainbow trout: Gjerde and Gjedrem (1984)]; (10) *Gambusia affinis* [mosquitofish: Busack and Gall (1983)]; (11) *Gryllus pennsylvanicus* [field cricket: Simons and Roff (1994)]; (12) *Gryllus firmus* [sand cricket: Webb and Roff (1992) and Roff (1995a)]; (13) *Pseudocalanus* [copepod: McLaren and Corkett (1978)]; (14) *Salmo salar* [Atlantic salmon: Gjerde and Gjedrem (1984)]; (15) *Mimulus guttatus* [plant: Carr and Fenster (1994)].

conjunction with a study of the mechanisms underlying the determination of metamorphic events that the various influences can be disentangled.

2.5 Dominance Variance in the Different Types of Traits

If selection is eroding the additive genetic variance of life history traits at a faster rate than morphological traits, then the relative proportion of dominance variance should be greater in the former than the latter. The model of Price and Schluter (1991) does not make such a prediction. Crnokrak and Roff (1995) compared relative amounts of dominance and additive variance using data from 17 non-domestic species and 21 domestic species. The relative contribution of dominance

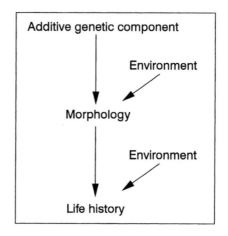

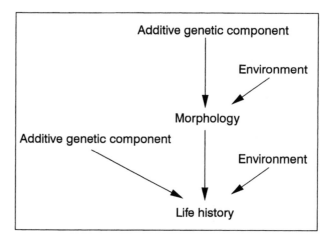

Figure 2.13 *Top:* Schematic representation of the model postulated by Price and Schluter (1991) for the observation that heritabilities of life history traits are typically less than those of morphological traits. *Bottom:* An alternative path diagram in which the genetic variability in the life history trait is a consequence of genetic variation in a morphological trait and independent genetic effects.

variance was assessed in two ways: (1) as a proportion of the additive and dominance variance, $D_G = V_D/(V_D + V_A)$, and (2) as a proportion of the phenotypic variance, $D_P = V_D/V_P$. To correct for possible bias due to overrepresentation by some species or studies, the analysis was conducted using both all the data and just mean values. The results were qualitatively the same. The statistical analyses revealed that, as predicted by the "erosion of variance" hypothesis, in nondomestic species, dominance variance comprises a greater fraction of the genetic vari-

Table 2.9 Comparison of Values of D_G and D_P Among Different Types of Traits for Non-domestic Species

	Life History	Behavior	Physiology	Morphology
Life history	0.54	0.03	0.04	0.0001
	0.31			
Behavior	ns	0.24	ns	ns
		0.04[a]		
Physiology	0.02	ns	0.27	ns
			0.21	
Morphology	ns	ns	ns	0.17
				0.13

Note: Values along the diagonals show the mean values of D_G (upper) and D_P (lower), whereas the values in the upper triangle show significant probabilities from the Kolmogorov–Smirnov test for D_G, and in the lower those for D_P. For complete details of the analysis, see Crnokrak and Roff (1995).

[a]This estimate based only on two estimates, all others are based on at least 12.

ance in life history traits than in morphological traits, with behavioral and physiological traits being intermediate (Table 2.9).

Domestic animals have typically been under strong selection for morphological traits such as body size; therefore, we would expect that in domestic animals the difference between life history traits and morphological traits should be substantially reduced. This is exactly what is observed, D_G being approximately the same for morphological and life history traits in domestic animals (Table 2.10). Further, as expected, there is a highly significant difference between nondomestic and domestic species in the value of D_G for morphological traits (Table 2.10).

The above analysis is also important with respect to the estimation of heritability. Because of the relatively large component of dominance variance in life history traits, it would be unwise to use full-sib estimates for this category. However, for morphological traits, there would appear to be little error in general in using full-sib estimates. This is supported by a comparison by Mousseau and Roff (1987) between estimates from mean offspring on mid-parent regression and full-sib designs. Most of the estimates in this comparison were for morphological traits: We found no significant difference between the two types of estimates. Estimates of heritabilities of behavioral or physiological traits from full-sib designs can be expected to be moderately contaminated by dominance variance (see Table 2.9).

2.6 Heritability Values in Nature

With few exceptions, the heritability values reported in Mousseau and Roff (1987) were made up of laboratory estimates from animals recently brought in from the

Table 2.10 Comparison of Values of D_G in Nondomestic and Domestic Animals

Trait Category	Nondomestic Mean (SE, n)	Domestic Mean (SE, n)	P (t-test)
Morphology	0.17 (0.035, 26)	0.42 (0.028, 101)	<.0001
Life history	0.54 (0.073, 22)	0.39 (0.058, 24)	.089

Note: There were insufficient data to test D_P.

Source: After Crnokrak and Roff (1995).

field. For these estimates to be of any use, we must be able to extrapolate them to wild populations. Organisms raised under relatively constant conditions as are to be found in the laboratory or greenhouse are generally expected to exhibit lower levels of phenotypic variation than under the more variable field conditions (Mitchell-Olds and Rutledge, 1986; Riska et al., 1989; Bull et al., 1982; Janzen, 1992). The increased phenotypic variance in field populations is postulated to be a result of increased environmental variance, leading to reduced heritabilities (Bull et al., 1982; Coyne and Beecham, 1987; Falconer, 1989; Prout and Barker, 1989; Riska et al., 1989; Schoen et al., 1994). Assuming this, then

$$\frac{h^2_{\text{Field}}}{h^2_{\text{Lab}}} = \frac{V_{\text{P,Field}}}{V_{\text{P,Lab}}} \tag{2.58}$$

That is, the field heritability will be increased by the ratio of the phenotypic variances. The ratio of the phenotypic variances are typically fairly modest, lying between 1 and 2, although extreme values do occur (Table 2.11). The median ratio for a variety of species is only 1.40 ($n = 24$, range 0.4–9.2, data from Table 2.11). At 2.51, the mean is considerably higher and results from extremely high ratios observed in several *Drosophila* species; this is illustrated in Fig. 2.14 for *D. obscura*, along with what appears to be a more typical distribution. When the genus *Drosophila* is excluded, the mean ratio drops to 1.22 and the median ratio is 1.30 ($n = 13$, range 0.4–1.7), with four estimates actually being less than 1. This suggests that field heritabilities will not generally be greatly reduced by a relative inflation of the environmental variance in the field.

A comparison of estimates from the laboratory and field confirm this prediction, with the mean estimates of life history and morphological traits actually being larger from field studies than laboratory studies (Table 2.12). Pairwise comparisons using a Mann–Whitney test showed no differences between lab and field estimates (Weigensberg and Roff, 1997). Eight studies have measured heritabilities simultaneously in the laboratory and the field, producing a total of 22 esti-

Table 2.11 Comparison of Phenotypic Variances Measured in the Laboratory and in Natural Populations

| Species | Trait[a] | Field/Lab | | | Reference |
		Mean	Range	n	
Drosophila subobscura	Thorax	8.3	—	1	McFarqhar & Robertson (1963)
Drosophila pseudoobscura	Wing	4.8	—	1	Sokoloff (1966)
Drosophila melanogaster	Wing	1.4	1.1–1.7	2	Tantawy (1964)
	Wing	3.8	—	1	Coyne & Beecham (1987)
	Bristle #	1.4	—	1	Coyne & Beecham (1987)
Drosophila simulans	Wing	1.1	1.1–1.2	2	Tantawy (1964)
Drosophila mimica	Ovariole #	3.3	1	1	Kambysellis & Heed (1971)
Drosophila buzzatii	2 lengths	9.2	4.9–16.7	6	Thomas (1993)
	Thorax	8.6	7.9–9.3	2	Ruiz et al. (1991)
Drosophila montana	Song	1.8	1.1–2.7	6	Aspi & Hoikkala (1993)
Drosophila littoralis	Song	0.7	0.3–1.6	4	Aspi & Hoikkala (1993)
Limnoporus notabilis	Body	1.1	0.8–1.4	6	Fairbairn (1984)
Aquarius remigis	Body	1.4	—	1	Preziosi (unpubl.)
Choristoneura rosaceana	Wing	1.3	—	1	Carrière & Roitberg (1994)
Allonemobius socius	Song	1.7	1.2–4.2	15	Mousseau & Howard (unpubl.)
Allonemobius fasciatus	Song	1.3	0.6–3.2	24	Mousseau & Howard (unpubl.)
Melanoplus sanguiinipes	3 Lengths	1.3	0.1–5.2	57	Mousseau & Dingle (unpubl.)
Gryllus pennsylvanicus	5 Lengths	1.1	1–1.6	9	Simons & Roff (1994)
	Gonad	0.8	0.8–0.8	2	Simons & Roff (1994)
	Dtime	1.6	1.6–1.6	2	Simons & Roff (1994)
Heterandria formosa	Body	0.4	0.1–0.6	2	Leips (unpubl.)
	Brood #	1.5	1.0–1.9	2	Leips (unpubl.)
	Gonad	1.6	0.7–2.4	2	Leips (unpubl.)
	Offspring size	0.8	0.5–1.0	2	Leips (unpubl.)

[a]Trait definitions: Thorax, Wing, and Body are all lengths; Song = song parameters such as pulse length; Dtime = development time. To avoid scale effects, if several laboratory measurements were available, we chose that for which the mean value most closely approximated the field means.

Source: From Weigensberg and Roff (1997).

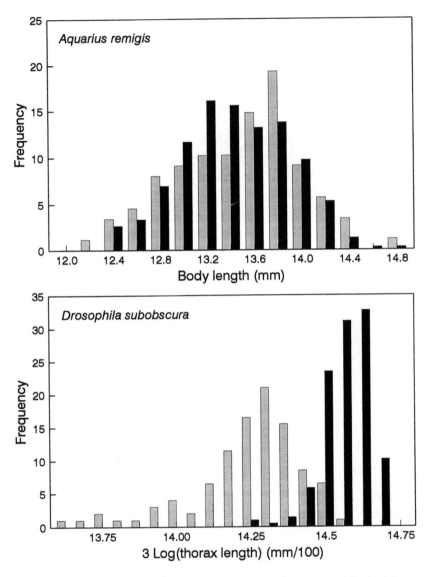

Figure 2.14 Phenotypic distributions of two insect species measured in the laboratory (solid bars) and field (hatched bars). That for *D. subobscura* illustrates the high ratios frequently found in *Drosophila* species, whereas that for the water strider, *Aquarius remigis*, shows a more typical pattern.

Table 2.12 Means of Heritabilities (Sample Size) Estimated Under Lab or Field Conditions

	Life History	Behavior	Morphology
Laboratory estimates	0.27 (75)	0.36 (24)	0.50 (90)
Field estimates	0.32 (12)	0.22 (3)	0.56 (150)

Source: Modified from Weigensberg and Roff (1997).

mates (Fig. 2.15). A paired t-test of these data indicates no significant difference (mean difference = 0.11, t = 1.90, df = 21, P = .07; note that the difference is significant if a one-tailed test is used). The confidence limits of the slope of the reduced major axis regression also includes a slope of 1 (0.61–1.16). It is apparent from Fig. 2.15 that more estimates fall below the 1:1 line than above it (although the number is not significant); however, this is somewhat misleading because multiple estimates from the same species may not be independent. For example, morphological traits are generally highly correlated and, thus, the seven heritability estimates of morphological traits in *G. pennsylvanicus* should perhaps be averaged. The important point is that there is a strong correlation between laboratory and field estimates (r = 0.63, n = 22, P < .002) and that the field estimates are by no means insignificant. Laboratory estimates of heritability are useful indicators of heritability in field populations.

2.7 Summary

Because it measures the genetic basis of the resemblance between offspring and parents, heritability is a central parameter in quantitative genetics. Heritability in the narrow sense is defined as the additive genetic variance divided by the total phenotypic variance. Nonadditive genetic variance contributes to heritability in the broad sense but not to the linear portion of the mean offspring on mid-parent regression. However, it is important to distinguish nonadditive gene action from nonadditivity in the statistical sense. Thus, for example, a trait which is controlled by a single locus with two alleles with one being dominant over the other may be decomposed into additive and nonadditive components (Fig. 2.2). Dominance may appear because of the particular scale of measurement; for example, if additivity is present on a logarithmic scale, then there will appear to be dominance on the linear scale. Changes in allele frequency will alter the heritability of a trait, the change depending on both the additive and nonadditive components (Figs. 2.3, 2.4, and 2.5). Heritability can be estimated by a variety of experimental designs, the simplest and statistically the most satisfactory being mean offspring on mid-parent regression and the half-sib designs. Other methods may not exclude nonadditive effects or may be experimentally difficult to implement in nondomes-

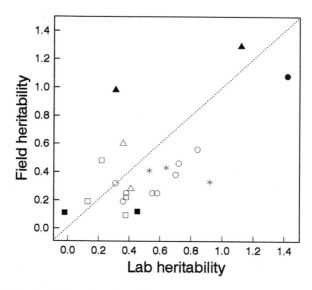

Figure 2.15   The distribution of heritabilities of traits measured in both the laboratory and the field. Open circles: *Gryllus pennsylvanicus* [cricket: Simons and Roff (1994)]; closed circle: *Poecilia reticulata* [guppy: Houde (1992)]; open squares: *Drosophila melanogaster* and *D. simulans* [fruit fly: Coyne and Beecham (1987), Prout and Barker (1989), and Jenkins and Hoffman (1994)]; closed squares: *Drosophila montana* [fruit fly: Aspi and Hoikkala (1993)]; asterisk: *Musca domestica* [housefly: Bryant (1977)]; open triangles: *Partula taeniata* [snail: Murray and Clark (1967)]; closed triangles, *P. suturalis* [Murray and Clark (1967)].

tic species. There is a clear pattern of difference in heritability values among trait categories, the ranking being Morphology > Physiology ≥ Behavior > Life history. This pattern is consistent with the hypothesis that selection will erode the additive genetic variance in traits associated directly with fitness (e.g., fecundity) more quickly than in traits more distantly related (e.g., morphology). An alternative hypothesis is that life history traits have lower heritabilities because such traits are functions of morphological traits. Analysis of dominance variance supports the former hypothesis because this predicts, as observed, that life history traits will have higher amounts of dominance variance than morphological traits. Finally, a comparison of heritabilities of traits measured in both laboratory and field indicate that, contrary to conventional wisdom, the former are good indicators of the latter.

3

The Genetic Correlation

The last chapter examined the genetic basis of the resemblance between parents and offspring for a single trait. However, it is common experience that traits are not inherited as independent units, but that several traits tend to be associated with each other. This phenomenon can arise in two ways: first, a subset of the genes that influence one trait may also influence another trait, a phenomenon known as **pleiotropy**, and second, the genes may act independently on the two traits, but because of nonrandom mating, selection, or drift they may be associated, a phenomenon called **linkage disequilibrium**. The latter causes only a transitory association between traits and it is the former which is of greatest interest to evolutionary biologists for pleiotropic effects may greatly alter the rate and direction of evolution. This is discussed in greater detail in Chapter 5. The present chapter is concerned with the theory of the genetic correlation, its estimation, and the comparison of genetic correlations among populations.

3.1 Theory

3.1.1 Genetic Correlations Arising from Linkage Disequilibrium

Although it is generally not likely to be an important evolutionary factor, the generation of a genetic correlation by linkage disequilibrium does need to be considered. This is particularly true from an experimental perspective, as the estimation of genetic correlations are frequently made in relatively small populations in which drift could generate transitory correlations. A further reason for its consideration is that it is the theoretical basis for the evolution of mate preference.

According to the Hardy–Weinberg law, in a randomly mating population under no selection, genotypic frequencies will come into equilibrium (i.e., p^2, $2pq$, q^2) after one generation. However, although this is true for the individual loci, it is not true for the multilocus genotypes. Suppose a population consisting of individuals homozygous at two loci, say *AABB*, and another population consisting entirely of the genotype *aabb* are randomly mixed. In the first generation, only three parental crosses are possible, giving rise to only three offspring genotypes, although, at equilibrium, nine genotypes will be found (Table 3.1). The distribution of the individual loci will be at Hardy–Weinberg equilibrium (e.g., frequencies of 0.25, 0.5, and 0.25 for *AA*, *Aa*, and *aa*, respectively). In the second generation of random mating, all nine genotypes will appear but they will not occur at the frequency that is expected at equilibrium (Table 3.1). The equilibrium frequencies will be approached asymptotically. The phenomenon of nonrandom association of loci is called **linkage disequilibrium** or **gametic phase disequilibrium**. The term is somewhat misleading because the disequilibrium does not imply that the loci are linked in the sense of being on the same chromosome. It is intuitively obvious that in the absence of retarding forces, loci, whether linked or not, will eventually attain linkage equilibrium, the time taken decreasing with the crossover rate. It can readily be shown that (Crow and Kimura, 1970, p. 48; Falconer, 1989, p. 19)

$$P_t\,(AB) \,-\, p_A q_B \,=\, (P_0(AB) \,-\, p_A q_B)(1 \,-\, c)^t \qquad (3.1)$$

where $P_t(AB)$ is the frequency of genotypes containing at least one A and one B at time t, p_A is the frequency of the A allele, p_B is the frequency of the B allele (thus, $p_A p_B$ is the equilibrium probability of an individual containing at least one A and one B), and c is the crossover rate. For unlinked loci, $c = 0.5$; thus, for unlinked loci, the disequilibrium value is halved each generation: as linkage (c) is increased the time taken to attain linkage equilibrium increases (Fig. 3.1).

If a small sample of individuals is drawn from a population and the genetic correlation estimated, it is not certain to what extent this correlation results from pleiotropy (discussed in the next section) or from linkage disequilibrium. It is only if the correlation is maintained for several generations or is found in several populations can one be reasonably sure that linkage is not the predominant factor.

One phenomenon in which a genetic correlation is predicted to arise specifically by linkage disequilibrium is the evolution of mating preference. Two theories for how mating preference (hereafter, for simplicity, referred to as female mate choice because it is in this sex that the phenomenon principally occurs) evolves have been advanced: Fisher's runaway selection process and the good-genes (or handicap) model. In the first case, it is postulated that females have a genetic propensity to choose males with exaggerated sexual traits, the result being that the two traits will coevolve in a runaway process (Fisher, 1930; Lande, 1981b). The good-genes model assumes that females are selected to choose males with particular adornments because these indicate that the males are genetically superior. In either case, a covariance between the female preference and the male trait will build up due to linkage disequilibrium (Lande, 1981b; Kirkpatrick, 1982; Heisler, 1984; Pomiankowski, 1988; Pomiankowski and Iwasa, 1993). At equilibrium, the ge-

Table 3.1 Genotype Frequencies Following the Mixture of Two Populations, One Homozygous as AABB and the Other Homozygous as aabb

Offspring Genotypes	Parental Genotypes						Freq. ×256	Equil. Freq. ×256
	AaBb AaBb	AaBb AABB	AaBb aabb	AABB AABB	AABB aabb	aabb aabb		
Generation 1								
P^a =				4	8	4		
AABB				16	0	0	64	
AaBb				0	16	0	128	
aabb				0	0	16	64	
Generation 2								
P^a =	4	4	4	1	2	1		
AABB	1	4	0	16	0	0	36	16
AABb	2	4	0	0	0	0	24	32
AAbb	1	0	0	0	0	0	4	16
AaBB	2	4	0	0	0	0	24	32
AaBb	4	4	4	0	16	0	80	64
Aabb	2	0	4	0	0	0	24	32
aaBB	1	0	0	0	0	0	4	16
aaBb	2	0	4	0	0	0	24	32
aabb	1	0	4	0	0	16	36	16

Note: For simplicity, frequencies are multiplied by 16 or 256. In the far column is shown the frequency of genotypes at equilibrium.

[a]Probability (×16) of the indicated parental cross.

netic covariance, $Cov_A (X_f Y_m)$, between female preference X_f and the male trait Y_m is approximately (Barton and Turrelli, 1991; Pomiankowski and Iwasa, 1993),

$$Cov_A(X_f Y_m) \approx c V_A(X_f) V_A(Y_m) \tag{3.2}$$

where c is the effectiveness of female preference and male signaling in creating nonrandom mating. There appears to be good evidence for additive genetic variance in female preference and male sexual traits (Bakker and Pomiankowski, 1995), but the evidence for a genetic correlation between these two traits is equivocal (Breden and Hornaday, 1994; Bakker and Pomiankowski, 1995).

3.1.2 Genetic Correlations Arising from Pleiotropy

From standard probability theory, the phenotypic correlation, r_P, between two traits, X and Y, is

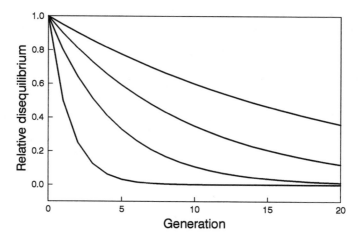

Figure 3.1   Relative disequilibrium, $[P_t(AB) - p_A q_B]/[(P_0(AB) - p_A q_B)]$, as a function of crossover rate, c, and generation. Values of c, from top to bottom, are 0.05, 0.1, 0.2, and 0.5, the last corresponding to unlinked loci.

$$r_P = \frac{\text{Cov}_{PXY}}{\sqrt{V_{PX}V_{PY}}} \tag{3.3}$$

where Cov_{PXY} is the covariance between X and Y, and V_{PX} and V_{PY} are the phenotypic variances of X and Y, respectively. The phenotypic correlation is made up of two components: a component that is ascribable to the additive action of the two sets of overlapping genes, and another which comprises the correlation of environmental effects plus the nonadditive genetic effects. The first is equivalent to the correlation between breeding values and is called the **genetic correlation**, whereas the second is termed, somewhat misleadingly, the **environmental correlation**. In the formulas that follow, I shall use the subscript P to designate phenotypic values, A to designate additive genetic values, and E to designate environmental values. Thus,

$$r_P = \frac{\text{Cov}_{PXY}}{\sqrt{V_{PX}V_{PY}}} \tag{3.4a}$$

$$r_A = \frac{\text{Cov}_{AXY}}{\sqrt{V_{AX}V_{AY}}} \tag{3.4b}$$

$$r_E = \frac{\text{Cov}_{EXY}}{\sqrt{V_{EX}V_{EY}}} \tag{3.4c}$$

Because environmental and genetic effects are assumed to be uncorrelated, the phenotypic covariance between X and Y is the sum of the genetic and environmental covariances

$$\text{Cov}_{PXY} = \text{Cov}_{AXY} + \text{Cov}_{EXY} \tag{3.5}$$

thus, from algebraic manipulation

$$r_P = r_A \left(\frac{V_{AX}}{V_{PX}} \frac{V_{AY}}{V_{PY}}\right)^{1/2} + r_E \left(\frac{V_{EX}}{V_{PX}} \frac{V_{EY}}{V_{PY}}\right)^{1/2} \tag{3.6}$$

Noting that $h^2 = V_A/V_P$ and that $V_E = V_P - V_A = [1 - (V_A/V_P)]V_P$, the above can be rewritten as

$$r_p = r_A \sqrt{h_X^2 h_Y^2} + r_E \sqrt{(1 - h_X^2)(1 - h_Y^2)} \tag{3.7}$$

If the two heritabilities are very small, the phenotypic correlation is determined primarily by the environmental correlation, whereas if they are both high, it is the genetic correlation that is most important. Note that the phenotypic correlation does not by itself give any idea of the importance of the genetic relationship between the two traits.

Because it depends by definition on the genetic architecture, the genetic correlation will, like the heritability, vary with allele frequencies. This can be illustrated with the following simple model. Consider two traits, X and Y, that initially are genetically uncorrelated. Now assume that a mutation occurs at a locus that has opposite pleiotropic effects on the two characters, decreasing X and increasing Y. The wild-type allele is designated M_1 and the mutant allele as M_2. Two cases are considered: (1) the additive model in which the change in the value of the traits is a simple additive function of the number of mutant alleles present in the individual and (2) the dominance model in which the value of the traits is the same for both the mutant homozygote, M_2M_2, and the heterozygote, M_1M_2. The genotype frequencies and the phenotypic values added to X and Y by the mutant allele are shown in Table 3.2. It is sufficient to derive the additive and dominance variances, and heritability for trait X, the results for trait Y being obtained simply by substituting $-ka$ for a in the formulas presented below.

Let the heritability of X before the mutation be $h_x^2 = V_{AX}/V_{PX}$, where, as above, A and P refer to the additive genetic and phenotypic variances, respectively. For algebraic convenience, let $V_{AX} = c_x a^2$, where c_x is a constant. The additive genetic variances resulting from locus M are (see Falconer, 1989, p. 129)

$$\text{Additive model:} \quad V_{AM} = 2pqa^2 \tag{3.8a}$$
$$\text{Dominance model:} \quad V_{AM} = 8p^3qa^2 \tag{3.8b}$$

Table 3.2 Genotypic Frequencies and Phenotypic Values After the Mutation of an Allele That Affects Two Traits X *and* Y

Genotype	Frequency	Change in the Phenotypic Value of	
		Trait X	Trait Y
Additive model			
$M_1 M_1$	p^2	0	0
$M_1 M_2$	$2pq$	$-a$	$+ka$
$M_2 M_2$	q^2	$-2a$	$+2ka$
Dominance model			
$M_1 M_1$	p^2	0	0
$M_1 M_2$	$2pq$	$-a$	$+ka$
$M_2 M_2$	q^2	$-a$	$+ka$

The variance of dominance deviations are likewise 0 and $(2pqa)^2$. After the mutation, the total additive genetic variance is

$$V_{AX} + V_{AM} = c_x\, a^2 + V_{AM} \tag{3.9}$$

and the total phenotypic variance of X is equal to the original phenotypic variance, $V_{PX} = a^2 c_X/h_X^2$, plus the additive and dominance contributions from the M locus. Therefore, after the appearance of M_2, the heritabilities are

Additive model:

$$h_I^2 = \frac{2pq + c_I}{2pq + c_I/h_I^2} \tag{3.10a}$$

Dominance model:

$$h_I^2 = \frac{8p^3q + c_I}{8p^3q + 4p^2q^2 + c_I/h_I^2} \tag{3.10b}$$

where $I = X$ or Y. Note that the terms a and ka in trait X and Y, respectively, cancel out (their effects are actually absorbed in the terms c_X and c_Y).

To obtain the genetic correlation between X and Y after the mutation, we need the covariance in breeding values. Because Y differs from X solely due to the M locus by the multiplier $-k$, the covariance is equal to $-k$ times the additive genetic variance due to M (effects due to the other loci are independent and, therefore, do not contribute to the covariance). As described above, the genetic correlation is equal to the covariance divided by the square root of the additive genetic variances. Noting that the additive genetic variance of trait Y due to the mutation is equal to $k^2 V_A$, the genetic correlation for both models can be written as

$$r_A = \frac{-kV_{AM}}{[(V_{AM} + V_{AX})(V_{AM} k^2 + V_{AY})]^{1/2}} \qquad (3.11)$$

where V_{AY} is the additive genetic variance of trait Y before the mutation ($= h_Y^2 V_{PY} = c_Y a^2$, the last for convenience only). From the above equation, it can be seen that the effect of the mutation is to generate a negative genetic correlation between the two traits. The effect of the mutation on the heritabilities and genetic correlations is illustrated in Fig. 3.2 in which it is assumed that before the mutation, the heritability of each trait is 0.4, and $c_X = c_Y = k = 1$. Note that if the mutant allele is recessive, the curve is simply a mirror image of that obtained for the dominance model.

The above model, although simple, may in fact be applicable to the evolution of pesticide resistance in animals and plants (Carrière and Roff, 1995). It is a general observation that resistance to pesticides is typically a consequence of a single major locus (McKenzie and Batterham, 1994; Jasieniuk et al., 1996). Further, such loci may have pleiotropic effects on other traits, as hypothesized in the above model [reviewed in Carrière and Roff (1995)]. In the lepidopteran, the oblique-banded leaf roller, *Choristoneura rosaceana*, the incidence of diapause and larval weight at day 16 after hatching are changed by the presence of insecticide resistance, the former increasing and the latter decreasing (Carrière et al., 1994). According to the above model, the heritabilities of the two traits and the magnitude of the genetic correlation should be increased in partially resistant populations, as has been found (Carrière and Roff, 1995). Because the actual number of loci contributing to resistance has not yet been determined for the oblique-banded leaf roller, these results should be considered encouraging but not definitive.

3.2 Estimation of the Genetic Correlation Between Traits Within an Individual

The theory developed above applies specifically to traits within a particular individual; for example, head width and leg length, or head width at age t and head width at age $t + i$. In all these cases, there is no difficulty in pairing up the traits. But consider the same trait, such as head width, measured on, say, males and females or on individuals taken from two different environments. As pointed out by Falconer (1952), it is possible to view these traits as being genetically correlated; that is, we can speak of the genetic correlation between sexes or between environments. However, although conceptually simple, we face the problem that we cannot, in these cases, unambiguously assign pairings. Before considering this

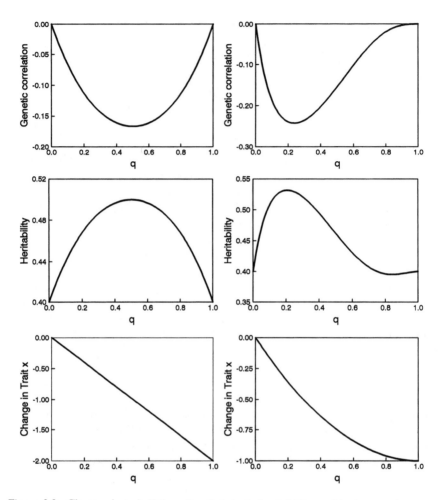

Figure 3.2 Changes in trait X (in units of a), in its heritability, and in the genetic correlation between traits X and Y as a function of the frequency, q, of the mutant allele M_2. Left panels show the results for the additive model; the right panels show the results for the dominance model. The change in trait X is given by the formulas: additive model, $-2aq$; dominance model, $-aq(2 - q)$.

problem, I shall deal with the simpler situation in which the two traits are present in the same individual.

3.2.1 Mean Offspring on Mid-parent Regression

If two traits, X and Y, are genetically correlated, then we should find that the value of either in the offspring is correlated with the other in the parents. It follows from the theory presented in Chapter 2 and above that the genetic correlation can be estimated by the offspring on mid-parent correlation. There are two such estimates, one obtained using X in the offspring and Y in the parents and the other using Y in the offspring and X in the parents. The genetic correlation is given by

$$r_A = \frac{\text{Cov}_{XY}}{\sqrt{\text{Cov}_{XX}\text{Cov}_{YY}}} \tag{3.12}$$

where Cov_{XY} refers to the covariance across traits and the denominator is the product of the covariances within traits. An overall estimate is obtained using the arithmetic average of the two separate estimates (Falconer, 1989, p. 817; Bulmer, 1985, p. 93). Becker (1985) also suggests computing the geometric average, although he does not provide a rationale for such a choice.

The estimation of confidence limits for the genetic correlation is difficult and even in the restricted cases where estimation methods have been worked out, the statistical behavior is not well understood (Robertson, 1959b, 1960; Van Vleck and Henderson, 1961; Van Vleck, 1968; Hammond and Nicholas, 1972; Grossman and Norton, 1974; Becker, 1985). As an estimate of the standard error, Falconer (1989) suggested the approximate formula derived by Reeve (1955) and Robertson (1959b):

$$\text{Est SE } (r_A) = \frac{1 - r_A^2}{\sqrt{2}} \left(\frac{\text{SE}(h_X^2) \, \text{SE}(h_Y^2)}{h_X^2 h_Y^2} \right)^{1/2} \tag{3.13}$$

where $\text{SE}(h_X^2)$ and $\text{SE}(h_Y^2)$ are the estimated standard errors of the heritabilities, h_X^2 and h_Y^2, respectively. However, this approximation produces confidence intervals that are smaller than the required 95% [approximately 80–90% instead of 95%; Roff and Preziosi (1994)]. A recent empirical evaluation by Koots and Gibson (1996) suggests that it may be much worse in some cases. For the ANOVA methods described below, the jackknife method gives exact confidence intervals (Roff and Preziosi, 1994) and will probably also do so in the present case (although this should be checked with simulation). Aastveit (1990) suggested using the bootstrap to estimate the confidence interval of the genetic correlation but did not demonstrate that such a method produces the appropriate intervals.

3.2.2 Full-sib Design

As with the use of the full-sib design to estimate heritability, there is a potential bias due to dominance variance and common environment effects. Assuming these to be negligible, the covariance terms can be estimated as shown in Table 3.3; the nested design to remove common environment effects is given in Table 3.4. Also required are the phenotypic variances for the two traits (V_{AFX}, V_{AFY}), which can

Table 3.3 Analysis of Covariance Table for a Full-sib Design in Which There Are No Effects Due to Common Environment

Source of Variation	Degrees of Freedom	Mean Cross-products	Expected Mean Cross-products
Among families (AF)	$N-1$	MC_{AF}	$Cov_{AP} + kCov_{AF}$
Among progeny (AP), within families	$T-N$	MC_{AP}	Cov_{AP}

Note: N is the number of families, T is the total number of individuals, k is equal to family size if these are equal, otherwise it is the weighted estimate

$$k = \frac{\sum_{i=1}^{N} n_i - \left(\sum_{i=1}^{N} n_i^2\right)\bigg/\left(\sum_{i=1}^{N} n_i\right)^{-1}}{N-1}$$

where n_i is the size of the ith family. The sums of cross-products are computed as

$$SC_{AF} = \sum_{i=1}^{N} \left(\frac{\sum_{j=1}^{n_i} x_{ij} \sum_{j=1}^{n_i} y_{ij}}{n_i} - \frac{\sum_{i=1}^{N}\sum_{j=1}^{n_i} x_{ij} \sum_{i=1}^{N}\sum_{j=1}^{n_i} y_{ij}}{T} \right)$$

$$SC_{AP} = \sum_{i=1}^{N}\sum_{j=1}^{n_i} x_{ij} y_{ij} - \sum_{i=1}^{N} \left(\frac{\sum_{j=1}^{n_i} x_{ij} \sum_{j=1}^{n_i} y_{ij}}{n_i} \right)$$

and thus

$MC_{AF} = SC_{AF}/(N-1);$ $Cov_{AP} = MC_{AP}$

$MC_{AP} = SC_{AP}/(T-N);$ $Cov_{AF} = (MC_{AF} - MC_{AP})/k$

be obtained from the one-way ANOVA given in Table 2.3. The three types of correlation are estimated from the following:

Genetic correlation:

$$r_A = \frac{Cov_{AF}}{\sqrt{V_{AFX} V_{AFY}}} \tag{3.14a}$$

Environmental correlation:

$$r_E = \frac{Cov_{AP} - Cov_{AF}}{\sqrt{(V_{APX} - V_{AFX})(V_{APY} - V_{AFY})}} \tag{3.14b}$$

Table 3.4 Analysis of Covariance Table for the Full-sib Design in Which Each Family is Divided Among c Cages

Source of Variation	Degrees of Freedom	Mean Cross-products	Expected Mean Cross-products
Among families (AF)	$N-1$	MC_{AF}	$Cov_{WC} + kCov_{AC} + kcCov_{AF}$
Among cages (AC), within families	$N(c-1)$	MC_{AC}	$Cov_{WC} + kCov_{AC}$
Within cages (WC)	$Nc(k-1)$	MC_{WC}	Cov_{WC}

Note: N is the number of families, k is equal to number per cage if these are equal, otherwise it is the weighted estimate

$$k = \frac{\sum_{i=1}^{Nc} n_i - \left(\sum_{i=1}^{Nc} n_i^2\right)\Big/\left(\sum_{i=1}^{Nc} n_i\right)^{-1}}{Nc - 1}$$

where n_i is the size of the ith cage. Sum of cross-products computed from (i = family, j = cage within family)

$$SC_{AF} = \sum_{i=1}^{N} \left(\frac{\sum_{j=1}^{c}\sum_{k=1}^{n_{ij}} x_{ijk} \sum_{j=1}^{c}\sum_{k=1}^{n_{ij}} y_{ijk}}{n_{i.}} - \frac{\sum_{i=1}^{N}\sum_{j=1}^{c}\sum_{k=1}^{n_{ij}} x_{ijk} \sum_{i=1}^{N}\sum_{j=1}^{c}\sum_{k=1}^{n_{ij}} y_{ijk}}{T} \right)$$

$$SC_{AC} = \sum_{i=1}^{N}\sum_{j=1}^{c} \left(\frac{\sum_{k=1}^{n_{ij}} x_{ijk} \sum_{k=1}^{n_{ij}} y_{ijk}}{n_{ij}} \right) - \sum_{i=1}^{N} \left(\frac{\sum_{j=1}^{c}\sum_{k=1}^{n_{ij}} x_{ijk} \sum_{j=1}^{c}\sum_{k=1}^{n_{ij}} y_{ijk}}{n_{i.}} \right)$$

$$SC_{WC} = \sum_{i=1}^{N}\sum_{j=1}^{c}\sum_{k=1}^{n_{ij}} x_{ijk}y_{ijk} - \sum_{i=1}^{N}\sum_{j=1}^{c} \left(\frac{\sum_{k=1}^{n_{ij}} x_{ijk} \sum_{k=1}^{n_{ij}} y_{ijk}}{n_{ij}} \right)$$

and thus

$MC_{AF} = SC_{AF}/(N-1);$ $Cov_{AF} = (MC_{AF} - MC_{AC})/kc$

$MC_{AC} = SC_{AC}/N(c-1);$ $Cov_{AC} = (MC_{AC} - MC_{WC})/k$

$MC_{WC} = SC_{WC}/Nc(k-1);$ $Cov_{WC} = MC_{WC}$

Phenotypic correlation:

$$r_P = \frac{Cov_{AP} + Cov_{AF}}{\sqrt{(V_{APX} + V_{AFX})(V_{APY} + V_{AFY})}} \tag{3.14c}$$

As described above, the best method of estimating the standard error is to use the jackknife. Bias due to common environmental effects can be eliminated by use of a nested design, the covariance table for which is given in Table 3.4. Correlations are then estimated as above. The genetic correlation between a threshold trait and another trait (threshold or continuous) can be estimated using Eq. (3.14a), where the threshold character is coded as 0 or 1 (Mercer and Hill, 1984; Roff and Bradford, 1996).

An alternative approach suggested by Via (1984) for the estimation of the genetic correlation is to use the Pearson product–moment correlation between family means, for which the usual methods of estimating confidence intervals on correlations can be applied. The correlation of family means is given by

$$r_m = \frac{\text{Cov}_m}{\sqrt{V_{mX} V_{mY}}} \tag{3.15}$$

where Cov_m and V_m are calculated from family means of phenotypic values. This is an approximation because the variance and covariance terms contain a fraction of the within-family error term:

$$\text{Cov}_m = \text{Cov}_{\text{among}} + \frac{\text{Cov}_{\text{within}}}{n} \tag{3.16}$$

where n is the family size. The potential advantage of the above method is that because they are simply standard product–moment correlations the usual significance tests can be applied and confidence intervals computed. Specifically, if the number of families is N, the confidence interval is computed by first transforming the correlation to the z scale:

$$z = \frac{1}{2} \ln \left(\frac{1 + r}{1 - r} \right) \tag{3.17}$$

The standard error is then approximately $(N - 3)^{-1/2}$, and confidence limits on r_m can be estimated by computing the confidence limits on z and back-transforming [for more, details see Sokal and Rohlf (1995)]. Note that the family size, n, does not appear in the formula; thus, any bias due to an insufficiently large family size will not be reflected in the confidence interval.

It can be shown (Roff and Preziosi, 1994) that the family mean correlation is equal to

$$r_m = \frac{\left[r_A + \frac{1}{n} \left(\frac{2 r_P}{h_X h_Y} - r_A \right) \right]}{\left\{ \left[1 + \frac{1}{n} \left(\frac{2}{h_X^2} - 1 \right) \right] \left[1 + \frac{1}{n} \left(\frac{2}{h_Y^2} - 1 \right) \right] \right\}^{1/2}} \tag{3.18}$$

Unlike the estimated standard error which contains only N, the family mean correlation contains only n, the family size. For family sizes less than 20, there is a considerable bias in the family mean correlation as an approximation to the genetic correlation. Further, the estimated 95% confidence intervals are far from correct when family sizes are small (Roff and Preziosi, 1994). The bias and error in the confidence interval will decline as the genetic correlation approaches the phenotypic correlation. However, given that these are the parameters we are attempting to estimate, it would certainly be unwise to make any a priori assumptions. The formula given above can be used as a guide for the bias resulting from a particular family size.

3.2.3 Half-sib Design

Because this design is not biased by dominance variance it is superior to the full-sib design. From the components estimated from Tables 2.5 and 3.5 the correlations are given by the following:

Genetic correlation (sire component):

$$r_{A(\text{Sire})} = \frac{\text{Cov}_{AS}}{\sqrt{V_{ASX}V_{ASY}}} \qquad (3.19\text{a})$$

Genetic correlation (dam component):

$$r_{A(\text{Dam})} = \frac{\text{Cov}_{AD}}{\sqrt{V_{ADX}V_{ADY}}} \qquad (3.19\text{b})$$

The dam component, being derived from full-sibs, is potentially contaminated by dominance effects. If these effects are negligible, the two components can be combined to give

Genetic correlation (sire + dam):

$$r_{A(\text{Sire + Dam})} = \frac{\text{Cov}_{AS} + \text{Cov}_{AD}}{[(V_{ASX} + V_{ADX})(V_{ASY} + V_{ADY})]^{1/2}} \qquad (3.19\text{c})$$

All estimates of environmental correlation are potentially biased by dominance variances, the dam component less so than the sire:

Environmental correlation (sire component):

$$r_{E(\text{Sire})} = \frac{\text{Cov}_{AP} - 2\text{Cov}_{AS}}{[(V_{APX} - 2V_{ASX})(V_{APY} - 2V_{ASY})]^{1/2}} \qquad (3.20\text{a})$$

Table 3.5 Analysis of Covariance Table for a Half-sib Design

Source of Variation	Degrees of Freedom	Mean Cross-products	Expected Mean Cross-products
Among sires (AS)	$S-1$	MC_{AS}	$Cov_{AP} + k_2 Cov_{AD} + k_3 Cov_{AS}$
Among dams (AD), within sires	$D_T - S$	MC_{AD}	$Cov_{AP} + k_1 Cov_{AD}$
Among progeny (AP), within dams	$T - D$	MC_{AP}	Cov_{AP}

Note: S is the number of sires, D_T is the total number of dams, and T is the total number of individuals. If the design is balanced, or at least not too unbalanced, k_1 and k_2 can be set equal to the mean number of offspring per dam family. Otherwise,

$$k_1 = \frac{T - \sum_{i=1}^{S}\left(\sum_{j=1}^{D} n_{ij}^2\right)\Big/ n_{i.}}{D_T - S}$$

$$k_2 = \frac{\sum_{i=1}^{S}\sum_{j=1}^{D} n_{ij}^2 \, (1/n_{i.} - 1/T)}{S - 1}$$

where n_{ij} is the number of progeny in the family of the jth female mated to the ith sire, and $n_{i.}$ is the number of progeny of the ith sire. The number of dams per sire are assumed equal in the above (D); If this is not the case, then D should be replaced by a subscripted term. k_3 is the number of progeny per sire. For unequal numbers,

$$k_3 = \frac{T - \left(\sum_{i=1}^{S} n_{i.}^2\right)\Big/ T}{S - 1}$$

Sums of cross-products are computed from

$$SC_{AS} = \sum_{i=1}^{S}\left(\frac{\sum_{j=1}^{D}\sum_{k=1}^{n_{ij}} x_{ijk}\sum_{j=1}^{D}\sum_{k=1}^{n_{ij}} y_{ijk}}{n_{i.}} - \frac{\sum_{i=1}^{S}\sum_{j=1}^{D}\sum_{k=1}^{n_{ij}} x_{ijk}\sum_{i=1}^{S}\sum_{j=1}^{D}\sum_{k=1}^{n_{ij}} y_{ijk}}{T}\right)$$

$$SC_{AD} = \sum_{i=1}^{S}\sum_{j=1}^{D}\left(\frac{\sum_{k=1}^{n_{ij}} x_{ijk}\sum_{k=1}^{n_{ij}} y_{ijk}}{n_{ij}}\right) - \sum_{i=1}^{S}\left(\frac{\sum_{j=1}^{D}\sum_{k=1}^{n_{ij}} x_{ijk}\sum_{j=1}^{D}\sum_{k=1}^{n_{ij}} y_{ijk}}{n_{i.}}\right)$$

$$SC_{AP} = \sum_{i=1}^{S}\sum_{j=1}^{D}\sum_{k=1}^{n_{ij}} x_{ijk}y_{ijk} - \sum_{i=1}^{S}\sum_{j=1}^{D}\left(\frac{\sum_{k=1}^{n_{ij}} x_{ijk}\sum_{k=1}^{n_{ij}} y_{ijk}}{n_{ij}}\right)$$

and thus

$$MC_{AS} = SC_{AS}/(S-1); \qquad MC_{AD} = SC_{AD}/(D_T-S); \qquad MC_{AP} = SC_{AP}/(T-D_T)$$

Covariance components are estimated as

$Cov_{AP} = MC_{AP}$
$Cov_{AD} = (MC_{AD} - MC_{AP})/k_1$
$Cov_{AS} = [MC_{AS} - (MC_{AP} + k_2 Cov_{AD})]/k_3$

Environmental correlation (dam component):

$$r_{E(\text{Dam})} = \frac{\text{Cov}_{AP} - 2\text{Cov}_{AD}}{[(V_{APX} - 2V_{ADX})(V_{APY} - 2V_{ADY})]^{1/2}} \tag{3.20b}$$

Environmental correlation (sire + dam):

$$r_{E(\text{Sire} + \text{Dam})} = \frac{\text{Cov}_{AP} - \text{Cov}_{AS} - \text{Cov}_{AD}}{[(V_{APX} - V_{ASX} - V_{ADX})(V_{APY} - V_{ASY} - V_{ADY})]^{1/2}} \tag{3.20c}$$

Phenotypic correlation:

$$r_P = \frac{\text{Cov}_{AP} + \text{Cov}_{AS} + \text{Cov}_{AD}}{[(V_{APX} + V_{ASX} + V_{ADX})(V_{APY} + V_{ASY} + V_{ADY})]^{1/2}} \tag{3.21}$$

The formula for the standard error is very complex (Table 3.6): As described above for the estimation of heritability, standard errors might be obtained using the jackknife procedure (the validity of this needs to be checked with simulation).

3.2.4 Precision of Estimates

The standard error of the genetic correlation is at least approximately proportional to the geometric mean of the standard errors of the heritabilities [Eq. (3.13)]. Consequently, the optimal design for the estimation of the heritabilities is also the optimal design for the estimation of the genetic correlation. Whether the standard error of the genetic correlation exceeds that of the heritabilities depends on the value of the standard error. For simplicity, consider the case in which the two heritabilities are equal: Using the approximate formula [Eq. (3.13)], we have

$$\text{Est SE}(r_A) = \left(\frac{1 - r_A^2}{h^2 \sqrt{2}}\right) \text{SE}(h^2) \tag{3.22}$$

As r_A approaches 1 (or -1), the standard error decreases to zero (Fig. 3.3). When r_A is very small, its standard error is approximately $0.71/h^2$ times the standard error of the heritability, which can lead to a considerable increase; for example, if h^2 equals 0.25, the standard error of the genetic correlation is 2.8 times that of the heritability (Fig. 3.3). Figure 3.3 shows the standard errors for the three major breeding designs, assuming that the total number of individuals that can be measured is 2000 and the two heritabilities are either both 0.5 or 0.1. For values of r_A greater than about 0.5, the standard error of the genetic correlation is actually less than that of the heritability. In the absence of prior information on the genetic correlation, very large sample sizes are required to ensure reasonably small confidence limits.

Table 3.6   Estimation of the Standard Errors for the Half-sib Design

Component (Z)	Variance of Mean Squares		
	$X\ [=\mathrm{Var}(MS_{ZX})]$	$Y\ [=\mathrm{Var}(MS_{ZY})]$	Cross-product $[=\mathrm{Var}(MCP_Z)]$
Sire (S)	$2MS^2_{ASX}/(S+1)$	$2MS^2_{ASY}/(S+1)$	$(MS_{ASX}MS_{ASY} + MC^2_{AS})/(S+1)$
Dam (D)	$2MS^2_{ADX}/(D_T-S+2)$	$2MS^2_{ADY}/(D_T-S+2)$	$(MS_{ADX}MS_{ADY} + MC^2_{AD})/(D_T-S+2)$
Error (E)	$2MS^2_{APX}/(T-D_T+2)$	$2MS^2_{APY}/(T-D_T+2)$	$(MS_{APX}MS_{APY} + MC^2_{AP})/(T-D_T+2)$

Component (Z)	Covariance of Mean Squares		
	$XY\ [=\mathrm{Cov}(XY_Z)]$	X*Cross-product $[=\mathrm{Cov}(XC_Z)]$	Y*Cross-product $[=\mathrm{Cov}(YC_Z)]$
Sire (S)	$2MC^2_{AS}/(S+1)$	$2MS_{ASX}MC_{AS}/(S+1)$	$2MS_{ASY}MC_{AS}/(S+1)$
Dam (D)	$2MC^2_{AD}/(D_T-S+2)$	$2MS_{ADX}MC_{AD}/(D_T-S+2)$	$2MS_{ADY}MC_{AD}/(D_T-S+2)$
Error (E)	$2MC^2_{AP}/(T-D_T+2)$	$2MS_{APY}MC_{AP}/(T-D_T+2)$	$2MS_{APY}MC_{AP}/(T-D_T+2)$

Component (Z)	Variances $(=V_{ZI})$		Covariance $(=\mathrm{Cov}_Z)$
Sire (S)	V_{ASX}	V_{ASY}	Cov_{AS}
Dam (D)	V_{DX}	V_{ADY}	Cov_{AD}
Sire+dam (SD)	$V_{ASX} + V_{ADX}$	$V_{ASY} + V_{ADY}$	$\mathrm{Cov}_{AS} + \mathrm{Cov}_{AD}$
Phenotypic (P)	$V_{ASX} + V_{ADX} + V_{APX}$	$V_{ASY} + V_{ADY} + V_{APY}$	$\mathrm{Cov}_{AS} + \mathrm{Cov}_{AD} + \mathrm{Cov}_{AP}$

Component (Z)	K	c_0	c_1	c_2
Sire (S)	k_3^2	1	$(-k_2/k_1)^2$	$[(k_2-k_1)]/k_1]^2$
Dam (D)	k_1^2	0	1	1
Sire+dam (SD)	$(k_1k_3)^2$	k_1^2	$(k_3-k_2)^2$	$(k_2-k_1-k_3)^2$
Phenotypic (P)	$(k_1k_3)^2$	k_1^2	$(k_3-k_2)^2$	$[k_2-k_1+k_3(k_1-1)]^2$

$$\mathrm{Var}(r_z) = (r_z^2/K)\,[(c_0\,\mathrm{Var}(MCP_S) + c_1\,\mathrm{Var}(MCP_D) + c_2\,\mathrm{Var}(MCP_E)/\mathrm{Cov}_Z^2$$
$$+\ c_0\,\mathrm{Var}(MS_{SX}) + c_1\,\mathrm{Var}(MS_{DX}) + c_2\,\mathrm{Var}(MS_{EX})/4V_{ZX}^2$$
$$+\ c_0\,\mathrm{Var}(MS_{SY}) + c_1\,\mathrm{Var}(MS_{DY}) + c_2\,\mathrm{Var}(MS_{EY})/(4V_{ZY}^2)$$
$$-\ c_0\,\mathrm{Cov}(XC_S) + c_1\,\mathrm{Cov}(XC_D) + c_2\,\mathrm{Cov}(XC_E)/(V_{ZX}\mathrm{Cov}_Z)$$
$$-\ c_0\,\mathrm{Cov}(YC_S) + c_1\,\mathrm{Cov}(YC_D) + c_2\,\mathrm{Cov}(YC_E)/(V_{ZY}\mathrm{Cov}_Z)$$
$$+\ c_0\,\mathrm{Cov}(XY_S) + c_1\,\mathrm{Cov}(XY_D) + c_2\,\mathrm{Cov}(XY_E)/(2V_{ZX}V_{ZY})]$$

$$\mathrm{SE}(r_z) = \sqrt{\mathrm{Var}(r_z)}$$

Note: Symbols are as defined in Tables 2.5 and 3.5, subscripted with X or Y where necessary. For further explanation, see Becker (1985).

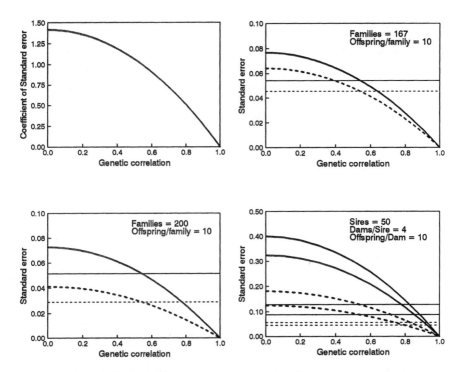

Figure 3.3 The standard error of r_A as function of r_A, h^2 ($=0.5$, solid line; 0.1, dotted line), and breeding design (upper right, mean offspring on mid-parent; lower left, full-sib; lower right, half-sib). The upper left panel shows the coefficient for $h^2 = 0.5$ [i.e., $(1 - r_A^2)/h^2\sqrt{2}$]. The horizontal lines show the standard error for the heritability. In the half-sib plot the upper curves give the estimates using the sire component, the lower curves the estimates using the dam component. The total number of individuals measured is approximately 2000.

3.3 Estimation of the Genetic Correlation Between Different Environments

Falconer (1952) noted that a character measured in two different environments, which might be different habitats or different morphs such as sexes, can be viewed as different characters that are genetically correlated. Conceptually elegant and making the theoretical analysis of evolution among different environments fall within the same framework as characters within an individual, this approach does raise the problem of how to actually make the estimate. A common approach is to use the family means, which as discussed above may be erroneous unless family sizes are large. Further, unless a full-sib design is used, information must necessarily be lost when a more complex design such as a half-sib is collapsed to a full-sib. Recently, Fry (1992), building on an analysis by Yamada (1962), provided an approach making use of the mixed-model analysis of variance. In this model, "families" are random effects because we wish to make inferences about the population as a whole, but the "environments" (habitats, hosts, morphs, etc.) are fixed because we restrict attention only to the specific environments used. Here, I present only the recipe; for technical details, refer to Fry (1992).

A test for the existence of a nonzero genetic correlation is given by the F-test constructed as

$$F = \frac{\text{Mean square among families}}{\text{Interaction mean square}} = \frac{\text{MS}_{AF}}{\text{MS}_{F \times E}} \tag{3.23}$$

This test is only exact for a balanced design and assumes the following:

1. Errors are independent and normally distributed with constant variance.

2. The family means, M_{ij} (where i is the environment and j the family), are normally distributed with the same variance in each environment. These two assumptions imply that heritabilities do not differ between environments. If the variances do differ between the environments, they should both be standardized to a common variance [e.g., 1; for an alternative approach based on family means, see Dutilleul and Potvin (1995)]. This standardization is necessary for the statistical testing of the interaction but not for the estimation of the genetic correlation if, as described below, the denominator in the appropriate formula comes from separate analyses of variance.

3. If there are more than two environments, then $\text{Cov}(M_{ij}, M_{i'j})$ is the same for every pair of environments i and i'. This is automatically satisfied for two environments. If the alternate hypothesis is one-tailed (e.g., $r_A < 0$), a two-tailed F-test is employed (Sokal and Rohlf, 1995). For unbalanced data, the above test is only approximate, and other methods such as maximum likelihood may be better. However, care should be taken that the method is not restrictive in the sense of only permitting positive values. The genetic correlation between two environments, 1 and 2, for character j is estimated from

$$r_A = \frac{\text{Cov}(M_{1j}, M_{2j})}{\sqrt{V(M_{1j})V(M_{2j})}} \tag{3.24}$$

where Cov and V signify covariances and variances, respectively. For a full-sib breeding design, the covariance is estimated as

$$\text{Cov} = \frac{\text{MS}_{AF} - \text{MS}_{F \times E}}{2n} \tag{3.25}$$

where n is the family size per environment. For a half-sib design, a nested ANOVA is used and the two estimates are

$$\text{Cov}_{\text{Dam}} = \frac{\text{MS}_{\text{AF}(D)} - \text{MS}_{F \times E(D)}}{2n} \qquad (3.26a)$$

$$\text{Cov}_{\text{Sire}} = \frac{\text{MS}_{\text{AF}(S)} - \text{MS}_{F \times E(S)} - 2n_D \, \text{Cov}_{\text{Dam}}}{2nn_D} \qquad (3.26b)$$

where the S and D refer to sire and dam components, respectively, and n_D is the number of dams per sire. In all cases, the variances are best estimated from separate one-way ANOVAs in each environment. Standard errors may be approximated using the jackknife, although the validity of this procedure requires testing. If the design is highly unbalanced, the numerator should be estimated using the method of regression with dummy variables [see appendix to Fry's (1992) paper].

3.4 The Distribution of Genetic Correlations

Unlike the case of heritability estimates, it is difficult to make a priori predictions about how genetic correlations should be distributed. Previously (Chapter 2), it was argued that life history traits, being subject in general to stronger selection than morphological traits, would have lower heritabilities. This hypothesis was supported both by the data for *D. melanogaster* (Roff and Mousseau, 1987) and from the general survey of animals (Mousseau and Roff, 1987), but see Price and Schluter (1991) for an alternate explanation for the pattern. A similar argument can be applied to the distribution of genetic correlations: Correlations between traits that have been subject to strong selection in the same direction are predicted to be typically negative. This prediction arises from the following argument: Alleles which produce positive correlations, for example, between early and late fecundity, will be selected, resulting in a fixation of such alleles; hence, at equilibrium, the only loci still segregating will be those which produce an increase in one fitness component but a decrease in another [e.g., increased early fecundity but decreased late fecundity; see Hazel (1943), Bell and Burris (1973), and Falconer (1989)]. Theoretical analysis of the hypothesis that such alleles will be maintained at equilibrium in the population by antagonistic pleiotropy has indicated that the conditions necessary for this to occur are quite restrictive (Rose, 1982, 1985; Curtsinger et al., 1994; for further details see Chapter 9). However, even if the necessary conditions are not fulfilled, there are two other reasons why we might still expect to find a preponderance of antagonistically pleiotropic alleles: First, the rate of fixation of such alleles will be less than for alleles showing positive pleiotropy, and second, new mutations are likely to show antagonistic rather than positive pleiotropy, because, as argued by Prout (1980, p. 57), "most new mutations are deleterious in *all* manifestations; so it would seem to follow from this, that the next most abundant class would be those where one of the pleiotropic effects results in a net advantage to the genotype, but the remaining pleiotropic effects are deleterious" [see also Simmons et al. (1980)]. Thus, we can replace the equilibrium antagonistic pleiotropy hypothesis by a "weaker" version which predicts that the rate of loss of antagonistically pleiotropic alleles is less and their influx via mutation greater than alleles showing positive pleiotropy.

It follows from the above arguments that genetic correlations between life history traits should be more frequently negative than those between morphological traits. A number of authors have pointed out that positive genetic correlations may frequently be found because the testing situation represents a novel environment (Gould, 1988; Rausher, 1988; Jaenike, 1990). Service and Rose (1985) demonstrated that a novel environment led to a reduction in the negative genetic correlation between fecundity and starvation time in *Drosophila melanogaster* (from −0.91 to −0.45), but I know of no work that has actually demonstrated a change in sign. Nevertheless, this phenomenon could lead to a reduction in the number of negative genetic correlations; however, in testing the hypothesis presented at the beginning of this paragraph, we are concerned with relative rather than absolute numbers. Unfortunately, we lack a theoretical basis to go beyond the general prediction made by this hypothesis. Because of this, in the present survey I take an empirical approach, using the above hypothesis as a point of departure.

I scanned the literature for estimates from nondomestic species, finding data for 51 species [summary in Table 3.7; for a list of the species see Roff (1996b)]. I restricted the analysis to traits that appear in the same individual, omitting, for example, correlations between sexes, morphs, or environments. Of the 1798 estimates obtained, the majority are between morphological traits (M $\times$ M, $n = 1210$, 67%), the remainder falling primarily into three categories: morphological $\times$ life history (M $\times$ L, $n = 175$, 9.7%), behavior $\times$ behavior (B $\times$ B, $n = 166$, 9.2%), and life history $\times$ life history (L $\times$ L, $n = 152$, 8.5%). These estimates are widely distributed among the different species, except for the category B $\times$ B in which 132 estimates come from a single study—chemoreceptive response in the garter snake, *Thamnophis elegans* (Arnold, 1981). For the purpose of analysis, I first analyzed the entire data set divided among the four principal categories. Because of disparate representation among studies and species, I next analyzed the data set blocked by species as well as trait category (I shall refer to this as the "species data set"). For the species data set, I used the mean per species as the basic datum for the analysis of mean values and the median for the analysis of standard errors (although this is statistically most appropriate, both means and medians gave the same qualitative result).

Table 3.7 Summary of the Data Set Used in the Analysis of Genetic Correlations

	Trait Combinations				
	L×L	M×M	L×M	B×B	Other
Number of species	22	38	20	6	4
Minimum number of estimates/ species	1	1	2	1	2
Maximum number of estimates/ species	133	210	36	132	70
Mean number of estimates/ species	14	29	9	28	24
Median number of estimates/ species	3	10	6	8	12

The proportion of cases in which the genetic correlation is less than zero is more or less the same whether the whole data set is used or the species data set (Table 3.8). As predicted, genetic correlations between life history traits are significantly more likely to be negative than those between morphological traits (Fig. 3.4, Table 3.8; statistical difference between combinations for the entire data set was tested for using the χ^2 goodness of fit, giving $\chi^2 = 17.9$, df $= 1$, $P < .0001$. For the species data set, the data were first arcsine square root transformed and then tested using ANOVA, giving $F_{1.32} = 3.62$, $P = .033$, and also the Mann–Whitney test, giving $\chi^2 = 4.33$, df $= 1$, $P = .018$, all tests one-tailed). With respect to the proportion of negative values, the overall ranking for the four combinations is L × L > L × M > B × B > M × M (Table 3.8).

Of importance for the evolution of traits is both the sign of the genetic correlation and its magnitude. These latter show remarkable uniformity among trait combinations (Table 3.9). There is no significant difference between either the means or the medians of the two combinations L × L and M × M whether the entire data set or the species set is considered ($P > .05$, t-test, and Mann-Whitney test). Comparison of all four combinations does indicate significant heterogeneity (ANOVA, $F_{3,1699} = 6.92$, $P < .0001$; Kruskal–Wallis, $\chi^2 = 11.7$, df $= 3$, $P = .009$), but this is entirely due to the combination B × B (the Tukey HSD test gives two significant comparisons; B × B versus L × L, $P < .0001$, and B × B versus M × L, $P = .02$). These differences are not observed in the species data set. Given the small number of species contributing to the combination B × B, the significantly larger absolute genetic correlation for this combination when the entire data set is used must be viewed with caution.

Although the absolute genetic correlation is about 0.5, there is considerable spread in values and the distribution is relatively flat across the entire range of values (Figure 3.4; the Kolmogorov–Smirnov test does indicate significant dif-

Table 3.8   Proportion of Genetic Correlations That Are Negative in Different Trait Combinations

Trait Combination	Proportion of Cases in Which $r_A < 0$ (Sample Size in Parentheses)		
	By Trait Combination Only	By Species and Trait Combination[a]	
		Mean	Median
All data	0.26 (1798)	—	—
L×L	0.39 (152)	0.41 (6)	0.39 (6)
M×M	0.23 (1210)	0.24 (28)	0.16 (28)
M×L	0.34 (175)	0.37 (12)	0.40 (12)
B×B	0.27 (166)	0.23 (5)	0.26 (5)

[a]Only those cases in which sample size per species per trait combination was at least five were used. Statistics not calculated for "all data."

ference from a uniform distribution for the combinations L × L, M × M, and M × L, but not for B × B). It is, unfortunately, not clear to what extent the breadth of the distribution reflects true variation versus experimental error (but see Section 3.5).

Estimated standard errors are, on average, very substantial (Table 3.10). Although the means do not differ significantly among combinations, there is a significant difference among the medians (Kruskal–Wallis test: for the whole data set, $\chi^2 = 77.6$, df $= 3$, $P < .0001$; for the species set, $\chi^2 = 8.4$, df $= 3$, $P = .04$), the principal difference being the relatively low value for the combination M × M (Table 3.10). A further measure of variability is the coefficient of variation, CV $= SE/r_A$. When CV < 0.5, the 95% confidence limits do not overlap zero. There are no objective criteria for the optimal size of CV, but a value of less than 0.25 (i.e., the standard error is less than one-quarter of the estimate) is certainly desirable. In fact, the data do not come close to such a value, the median value ranging from 0.38 (M × M) to 0.68 (M × L, Table 3.10; medians are used because the data are log-normally distributed). There is significant variation among combinations for the whole data set (Kruskal–Wallis, $\chi^2 = 38$, df $= 3$, $P < .0001$; using log-transformed values, $F_{3,843} = 11.9$, $P < .0001$) but not for the species set ($\chi^2 = 5.46$, df $= 3$, $P = .141$). As with the standard errors, the principal factor causing the variation is the very low CV for the M × M combination.

This analysis produces two messages: First, there is significant variation associated with trait combination, with the combination M × M having significantly smaller standard errors and CVs, and second, the experimental error, as indicated by the standard errors and CVs, is so large that much of the dispersion in the genetic correlation estimates may be due to inadequate sampling.

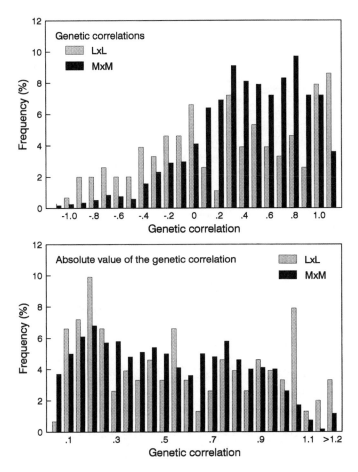

Figure 3.4 Distributions of the genetic correlation for the combinations M × M and L × L. Upper panel shows distribution for signed values; the lower panel shows the distributions of the absolute values.

3.5 Is the Phenotypic Correlation a Reasonable Estimate of the Genetic Correlation?

The phenotypic correlation is far more easily measured than either the genetic or environmental correlations and, as shown above, is a simple function of the two:

$$r_P = r_A \sqrt{h_X^2 h_Y^2} + r_E \sqrt{(1 - h_X^2)(1 - h_Y^2)} \tag{3.27}$$

The standard error of the phenotypic correlation is approximately $(T - 3)^{-1/2}$

Table 3.9   Means and Medians of the Absolute Values of the Genetic Correlations for Different Trait Combinations

Trait Combination	By Trait Combination Only			By Species and Trait Combination		
	Mean (SE)	Median	n	Mean (SE)	Median	n
All	0.48 (0.008)	0.45	1798	0.53 (0.03)	0.51	94
L×L	0.53 (0.03)	0.49	152	0.50 (0.10)	0.45	16
M×M	0.47 (0.009)	0.44	1210	0.57 (0.04)	0.59	43
M×L	0.48 (0.03)	0.48	175	0.50 (0.06)	0.50	21
B×B	0.59 (0.03)	0.56	166	0.65 (0.14)	0.56	6

Table 3.10   Means and Medians of the Estimated Standard Errors and Coefficient of Variation (CV) for Different Trait Combinations

Trait Combination	By Trait Combination Only				By Species and Trait Combination		
	Estimated Standard Errors		CV		Estimated Standard Errors	CV	
	Mean (SE)	Median	Median	n	Mean (SE)	Median	n
All	0.28 (0.01)	0.20	0.49	929	—	—	—
L×L	0.26 (0.01)	0.26	0.61	109	0.23 (0.03)	0.69	10
M×M	0.25 (0.02)	0.17	0.38	493	0.22 (0.05)	0.32	20
M×L	0.32 (0.03)	0.26	0.68	96	0.27 (0.03)	0.66	12
B×B	0.34 (0.02)	0.31	0.60	155	0.30 (0.05)	0.53	4

(Sokal and Rohlf, 1995; in practice, correlations are transformed to the z scale, but this makes little difference in the present case). For the same sample size, the standard error of the phenotypic correlation is substantially smaller than that of the genetic correlation; an example using morphological data from the cricket *Gryllus firmus* is shown in Fig. 3.5. Clearly, if we could substitute the phenotypic for the genetic correlation, we could considerably reduce experimental effort. Further, the phenotypic correlation can be obtained far more easily than the genetic correlation, as restrictions on mating design are much less.

Based on an analysis of 41 pairs of phenotypic/genetic correlation matrices, Cheverud (1988, p. 958) concluded that "phenotypic correlations are likely to be fair estimates of their genetic counterparts in many situations." The analysis of Roff and Mousseau (1987) on the phenotypic and genetic correlations in *Drosophila* suggested that the former might be reasonable estimates of the latter when

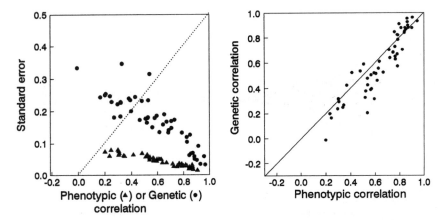

Figure 3.5   Left panel: Standard error of the estimated phenotypic and genetic correlations versus the estimate. Confidence intervals of estimates lying above the dotted line include zero. *Right panel:* Genetic versus phenotypic correlations. Data are for 11 morphological traits in male *Gryllus firmus*. [After Roff (1995b).]

only morphological traits are considered. In this regard, it is noteworthy, as pointed out by Cheverud, that almost all of the traits considered in his analysis were morphological traits (Cheverud, 1988). Koots and Gibson (1994, 1996) also found a very high correlation between the genetic and phenotypic correlations in a survey of traits related to beef production.

From Eq. (3.27) it can be seen that the maximal difference between the phenotypic and genetic correlations depends in large measure on the geometric mean of the two heritabilities [i.e., $(h_X^2 h_Y^2)^{1/2}$]. This is illustrated by the following simulation: The two heritabilities and the phenotypic and environmental correlations were selected from a uniform random distribution between 0 and 1 and the genetic correlation was then calculated from

$$r_A = \frac{r_P - r_E \sqrt{(1 - h_X^2)(1 - h_Y^2)}}{\sqrt{h_X^2 h_Y^2}} \qquad (3.28)$$

Values of r_A less than -1 or greater than $+1$ are mathematically but not genetically possible; these were discarded. Of the 5000 random combinations generated, 2257 gave acceptable values of r_A. The envelope enclosing the maximal difference between r_P and r_A decreases rapidly with the geometric mean of the heritabilities, and for means greater than 0.8, the difference is sufficiently small that the phenotypic correlation is an acceptable substitute for the genetic correlation (Fig. 3.6). However, this assumes that these parameters are estimated with-

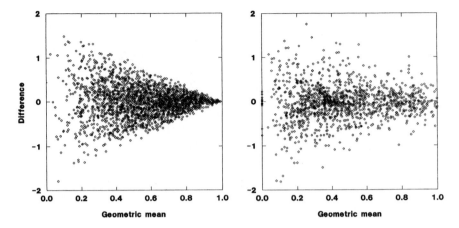

Figure 3.6 Distribution of the difference $r_P - r_A$ versus the geometric mean of the heritabilities of two traits. *Left panel:* Results from the simulation described in the text. *Right panel:* Results from the empirical data set.

out error: in practice, relatively low sample sizes will produce large sampling errors even when the estimated geometric average is greater than 0.8. In the present empirical data set, such a problem is quite evident, there still being considerable variation in the difference between r_P and r_A for geometric averages greater than 0.8 (Fig. 3.6). Thus, the geometric average can only be used as a very rough guide given the types of sampling intensities presently being utilized. A low geometric mean does not imply that the difference between the phenotypic and genetic correlations will necessarily be large, only that the difference is not constrained to be small. The median geometric average for the combination M × M is 0.52, whereas that for L × L is 0.25, values that are consistent with the estimates reported in Mousseau and Roff (1987). Thus, the maximal possible difference between r_P and r_A should, in principle, be less between morphological traits than between life history traits. However, these geometric averages are sufficiently small that in neither case is there a significant constraint.

The similarity of the phenotypic and genetic correlations can be compared in two ways. First, the genetic correlation is regressed on the phenotypic correlation (r_A is used as the dependent variable because we wish to predict its value using r_P). Because the data clearly violate the assumptions of linear regression, the probability level should be viewed with caution. Ideally, the level should be assessed using Mantel's test (Cheverud, 1988; Simons and Roff, 1994; Roff, 1995b), but its suitability has been questioned (Shaw, 1992); hence, results from this test must also be viewed with caution. In the present case, we are not so much interested in the significance or lack of in the regressions as in the values of the intercept and slope, which are still best estimated by least squares regression.

With respect to Cheverud's conjecture, we require that the estimates cluster closely around the line of equality. A trend to deviate from this line may indicate a systematic difference between the two estimates. Such a systematic bias may not be important if it is small; indeed, a small bias with the points nevertheless lying close to the $1:1$ line is preferable to one in which there is no bias but in which the points are scattered very widely about the $1:1$ line. To assess how close, on average, the two estimates were, I used the mean absolute difference, $D_{abs} = \Sigma|r_{A,i,j} - r_{P,i,j}|/n$, where i and j refer to characters i and j ($i \neq j$) and n is the total number of pairs of correlations.

The data on *Gryllus firmus* shown in Fig. 3.5 illustrate these points: There is very clearly a close fit between the phenotypic and genetic correlations, but using standard statistical tests, the slope and intercept are significantly different from 1 and 0, respectively (slope = 1.14, intercept = -0.14). The mean absolute difference is 0.09, which is reasonably small. Thus, despite a possible bias in the regression, one could substitute the phenotypic for the genetic correlations with little error. Certainly, the distribution of phenotypic correlations is a good representation of the distribution of genetic correlations.

For the larger data set, there is a highly significant correlation between the two correlations and no significant deviation from the line of equality (Table 3.11). For the trait combinations considered separately, the slope exceeds 1 in all cases, and in two cases (M × M, B × B), the slope is significantly greater than unity, although in the former case the increase is small ($<10\%$). A tendency for the slope to exceed unity is also indicated by the species set in which there are significantly more slopes that exceed 1 than are less than 1 (20 versus 9, $\chi^2 = 4.2$, df = 1, $P < .05$, only data sets containing at least 9 data points included in the analysis; Fig. 3.7).

The mean absolute difference, D_{abs}, is quite large (0.24–0.38, Table 3.11), suggesting that even though the data are relatively evenly scattered around the $1:1$ line, the scatter is large enough to introduce considerable difference between individual values of r_P and r_A. The estimated standard errors for the genetic correlations are themselves large (Table 3.10); hence, much of the scatter about the line may be a consequence of errors in estimation rather than true differences between r_P and r_A. To test this, we can proceed by considering the following hypothesis:

$$H_0: r_P = r_A + \text{Error of estimation in } r_A \qquad (3.29)$$

From this, it follows that $(\hat{r}_P - \hat{r}_A)^2 = \text{Est Var}(r_A)$, where $\hat{r}$ refers to the estimated value, and Est Var(r_A) is the estimated variance of r_A ($=$ the square of the estimated standard error). For convenience, let the former term be designated as D^2 and the latter as Var. For the entire data set, D^2 and Var are virtually identical with respect to both the mean (0.16 and 0.17, respectively) and the median (0.04 for both), and a paired t-test indicates not even a trend for a difference (Table

Table 3.11 Parameter Estimates for the Regression of the Genetic Correlation on the Phenotypic Correlation

Trait Combination	Geometric Average Heritability	Intercept (SE)	Slope (SE)	r	n	D_{abs}
All	0.46	0.004 (0.01)	1.05 (0.03)	0.64	1456	0.28
L×L	0.25	−0.01 (0.05)	1.02 (0.12)	0.58	134	0.38
M×M	0.52	−0.05[a] (0.02)	1.09[a] (0.04)	0.68	980	0.24
M×L	0.36	0.06 (0.03)	1.14 (0.07)	0.78	161	0.26
B×B	0.27	0.11 (0.06)	1.68[a] (0.24)	0.48	166	0.46

[a]Intercept significantly different from 0 and slope significantly different from 1. In all cases, the slope is significantly different from 0. Because the assumptions of linear regression are not met, these results are only approximate.

3.12). Examination of individual trait combinations shows no significant difference for the combinations M × M and L × M and significant differences in L × L and B × B (Table 3.12). In the case of L × L, although the difference is significant, it is still very small (−0.08). Because of the inordinate contribution by a single study, the results for B×B cannot be reliably generalized.

In summary, the hypothesis that the primary difference between the phenotypic and genetic correlation arises from the errors of estimation of the latter cannot be rejected for combinations involving morphological traits. Errors of estimation also appear to account for a significant part of the differences in the combination L × L but there is evidence of variation in addition to this.

The above analysis suggests that the use of the phenotypic correlation in place of the genetic correlation is justified in most cases, certainly in the case of morphological variation. Nevertheless, it would be unwise to rely entirely on the phenotypic correlations. The following protocol is therefore suggested (Roff, 1995b): suppose we wish to compare genetic correlations among a range of populations or possibly closely related species. First, we conduct a detailed genetic analysis for a single species or population. Because of the potentially confounding influences of dominance effects, a half-sib or mean offspring on mid-parent regression is preferable to the full-sib design. If the correspondence between genetic and phenotypic correlations is deemed sufficiently high, and this will be a matter of subjective judgment, then we can proceed to compare the phenotypic correlations between populations and species. Using phenotypic correlations will permit many more comparisons than are possible if genetic analyses must be carried out for each population/species. Finally, one or more disparate populations/species should be selected for detailed genetic analysis to further confirm the assumption that phenotypic correlations reflect genetic correlations, within acceptable limits. Although this research program is not a trivial undertaking, it would be, at the

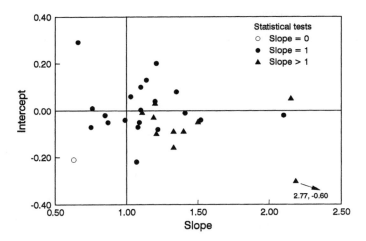

Figure 3.7 Distribution of intercepts and slopes for the regressions of the genetic on phenotypic correlations utilizing the species data set.

present time, unwise to proceed in the absence of confirmatory genetic analyses. At the same time, the suggested protocol potentially makes large-scale comparisons feasible.

3.6 Comparison of Genetic Variance–Covariance Matrices

Thus far, we have been concerned only with correlations between pairs of traits. However, evolutionary changes probably involve more than two traits changing at the same time. To deal with this, we use not the genetic correlation but the genetic variance–covariance matrix, discussed in detail in Chapter 5. In the case of three traits, the genetic variance–covariance matrix is

$$\begin{pmatrix} V_{11} & V_{12} & V_{13} \\ V_{21} & V_{22} & V_{23} \\ V_{31} & V_{32} & V_{33} \end{pmatrix} \tag{3.30}$$

where V_{ii} is the additive genetic variance of trait numbered i and $V_{ij} = V_{ji}$ is the additive genetic covariance between traits numbered i and j (note that $V_{ij} = V_{ji}$). The diagonal elements of the matrix specify the additive genetic variances and the off-diagonal elements the covariances. For convenience, the additive genetic matrix is written using bold typeface as **G**. The question that concerns us in this section is how to compare **G** matrices from different populations or species.

It is clear from theory (Section 3.1) that under strong selection allele frequen-

Table 3.12 Estimates of D^2 and Var for Different Trait Combinations, and Paired t-Tests Between These

Trait Combination	Mean				Median	
	D^2		Var			
	Mean (SE)	n	Mean (SE)	n	D^2	Var
All	0.16 (0.01)	1456	0.17 (0.02)	929	0.04	0.04
L×L	0.26 (0.04)	134	0.09 (0.01)	109	0.08	0.07
M×M	0.12 (0.01)	980	0.19 (0.04)	493	0.03	0.03
M×L	0.13 (0.02)	161	0.17 (0.04)	96	0.03	0.07
B×B	0.33 (0.04)	166	0.17 (0.03)	155	0.17	0.10

Trait Combination	Paired t-Test			
	Mean Difference	$\mid t \mid$	P	n
All	−0.012	0.52	.610	826
L×L	0.080	3.11	.002	95
M×M	0.042	1.18	.239	484
M×L	0.042	0.79	.430	84
B×B	−0.166	4.59	<.0001	155

cies will change and, thus, that **G** will eventually show significant change (this is discussed in greater detail in Chapter 4). However, Lande (1976, 1980a, 1984) has suggested that loss of variation due to weak selection will be replaced by mutation, thereby keeping **G** constant. Unfortunately, little is known about what levels of weak selection and mutation are required to satisfy the requirement of a constant **G**. If one can make the assumption that **G** remains constant, then present theory can be used to make predictions of long-term responses to selection (Lande, 1979a), inferences on past selection (Lande, 1976, 1979a), and tests of divergence of taxa by random drift (Lande, 1976, 1977, 1979a). The precise requirement is not that **G** remain constant but that only proportional changes occur (i.e., $\mathbf{G}_1 \propto \mathbf{G}_2$), which permits, for example, overall changes in body size as has occurred in many evolutionary pathways (e.g., rats and mice). Adjustment to a common mean and transformation to stabilize variances may be possible, but the consequences on the variance–covariance structure are little understood (Cowley and Atchley, 1992). For the present, I shall restrict attention to the case of constant **G**, assuming that if there are scale differences, a suitable transformation can be found that will remove them.

Turelli (1988) has argued that there are insufficient theoretical grounds for accepting or rejecting the hypothesis of a constant **G** and, therefore, the hypothesis must be addressed empirically [see also Arnold, (1992)]. I would amend this statement to say that the empirical question is not one of testing for the constancy of **G** but rather the level of divergence at different taxonomic levels and among different types of traits; that is, it is a question primarily of interval estimation rather than inference. With respect to the first comparison, Lofsvold (1986, p. 560) has observed,

> It is clear that, at some level, the pattern of genetic covariation cannot be constant. To cite an obvious example, taxa with radically different patterns of morphological organization (such as the phyla Arthropoda and Chordata) cannot possibly have similar patterns of morphological variation and covariation within populations, since the same structures are not present in both. It is equally clear that, at some lower taxonomic level, **G** must be homogeneous. An example would be two large experimental populations in a common environment which have been separated only one or a few generations without selection or inbreeding.

With respect to the comparison among different types of traits, it has already been noted that life history traits are likely to be under much stronger selection than morphological traits. Consequently, we would expect to find lack of correspondence between **G** matrices at a much lower taxonomic level for life history traits than morphological traits.

3.6.1 Methodologies for Comparing G Matrices

If, despite the foregoing arguments, one does wish to test for constancy of **G**, what procedures are appropriate? First, we might ask whether we should be examining **G** or the set of genetic correlations. The latter have the advantage of being scale-free measures of associations between traits and can be interpreted more easily, whereas the former are necessary for the prediction of change in the suite of traits (Kohn and Atchley, 1988, Chap. 5). There is obviously a mathematical relationship between the two, but, unfortunately, there is no mathematical reason to predict that exactly the same answer will be obtained. For this reason, it is best to analyze both types of data. For the purposes of discussion, I shall consider just two groups (morphs such as sexes, populations, species, etc.) referred to as G_1 and G_2, the same arguments applying for the matrix of genetic correlations.

For simplicity, the rows (or columns) of each matrix can be assembled contiguously to form a one-dimensional vector

$$\mathbf{G}_1 = (x_1, x_2, x_3, \ldots, x_n), \qquad \mathbf{G}_2 = (y_1, y_2, y_3, \ldots, y_n) \qquad (3.31)$$

For the hypothesis $\mathbf{G}_1 = \mathbf{G}_2$ (i.e., the two samples come from the same statistical population), there are the following alternate approaches.

3.6.1.1 Element-by-Element Comparison

Genetic correlations (or heritabilities) can be compared in a pairwise manner using the jackknife estimates (Knapp et al., 1989). A t-statistic is constructed as

$$t = \frac{(r_{A,1} - r_{A,2})}{\left[\left(\dfrac{(n_1 - 1)S_1^2 + (n_2 - 1)S_2^2}{n_1 + n_2 - 2} \right) \left(\dfrac{n_1 + n_2}{n_1 n_2} \right) \right]^{-1/2}} \tag{3.32}$$

where $r_{A,i}$ is the genetic correlation estimated for the ith population ($i = 1, 2$), n_i is the number of pseudovalues generated in the jackknife procedure for the ith population (e.g., if the estimate is based on full-sibs, then n is the number of families, but if it is based on half-sibs, then it is the number of sire families), and S_i is the estimated standard deviation of the pseudovalues. This is the usual formulation for the t-statistic and its significance is tested in the same manner (Sokal and Rohlf, 1995). The test assumes that the two variances are equal, which in the case of the genetic correlation may not be true; thus, results should be considered approximate.

An alternative procedure is to use randomization. First, the data sets are arranged into sampling units according to the breeding design: (a) if mean offspring on mid-parent regression, the individual unit is the offspring-mid-parent element; (b) if full-sib, the family is the individual unit; (c) if half-sib, the sire family is the sampling unit. All units are placed into a single pool and then randomly assigned (i.e., without replacement) to groups 1 and 2, the number in each group being kept the same as in the original. The data are then analyzed in the same manner as for the actual data set, and the statistic of interest, in this case the absolute difference between the two parameters (r_A, h^2, or even the covariance; absolute value because a two-tailed test is assumed), is computed and stored. The whole process is repeated a large number of times (say, 5000). Significance is then assessed by the probability of observing a value as high or higher than that obtained in the actual data ($P = (n + 1)/(N + 1)$), where n is the number of times the statistic computed from the randomized data set exceeds that actually observed and N is the total number of randomizations. The 1 is necessary to account for the observed value [see Manly (1991)]. The randomization procedure could be done using the t-value from the jackknifed estimates, which would confirm the robustness of the assumption of normality.

Paulsen (1996) used the bootstrap, a similar procedure to randomization, the principal difference being that sampling from the pooled data set is with replacement. The bootstrap method cannot be guaranteed to generate a distribution that faithfully represents the actual distribution; therefore, without simulation results or analytical justification, its use is suspect (Miller, 1974; Potvin and Roff, 1993). For this reason, the randomization procedure is the preferred approach.

The above methods are readily extended to multiple populations using one-way ANOVA [see Edgington (1987) or Manly (1991) for programs implementing the randomization technique].

Element-by-element comparison necessarily entails making multiple tests which reduces the nominal level of significance (Rice, 1989). One solution to this is to apply the sequential Bonferroni correction (Brodie, 1993). However, if there are a large number of elements, application of the Bonferroni adjustment may reduce the significance level to a value that is lower than the precision of the computer program (Paulsen, 1996). The loss of statistical power entailed by this procedure is likely to be so great that failure to find significant differences is meaningless. An alternate method is to consider whether the number of significant results at the 5% level is greater than expected by chance. For example, out of 990 trait pairs, Paulsen (1996) obtained 191 significant differences; the expected number is 198. Based on the binomial test, there is no difference between observed and expected.

In summary, element-by-element comparison is a useful first approach in that it may indicate very striking differences (e.g., 900 of the 990 comparisons above might have been significant, which is far greater than expected by chance and strongly suggests a difference) but should be accompanied by a test based on assessment of the matrices as a whole. The following four methods attempt this.

3.6.1.2   Tests Based on the Assumption of Multivariate Normality

The concept of maximum likelihood estimation has been described in Section 2.2.5; its use with respect to comparing G matrices is described in detail by Shaw (1991). The essentials of the approach can be illustrated by considering the comparison of the genetic variance of a trait estimated from two populations. First, the variances are estimated separately for each population: If the likelihoods are L_1 and L_2, respectively, then the joint probability is simply L_1L_2 (i.e., the product of the probabilities) $= L_S$ (S = separate). The second step is to estimate the variance assuming that there are no differences between the populations, the likelihood value obtained for this single estimate being L_C (C = common). Asymptotically, the log-likelihood ratio statistic, $D = 2 [\log(L_S) - \log(L_C)]$, is distributed as χ^2_{S-C}, where S is the number of parameters in the model from which L_S was derived (two in this case because two separate variances were estimated) and C is the number of parameters in the model from which L_C was estimated (one in this case, because only a single variance was estimated). For a succinct discussion of likelihood and model fitting, see Dobson (1983). Because of the additive property of log-likelihoods, the above model can be immediately extended to any number of components. Suppose, for example, that there are k traits, leading to $k(k + 1)/2$ variances and covariances. The log-likelihood ratio statistic calculated as described above has $k(k + 1)/2$ degrees of freedom. A discussion of the power of the maximum likelihood test is given in Section 3.6.1.4.

Several tests are available that begin with the two matrices of variances and covariances. Holloway et al. (1993) used the Mahalanobis D^2 test, a multivariate distance test. For a description of this test, see Manly (1986) or Flury and Riedwyl (1988). Under the null hypothesis of no difference, D^2 is distributed as χ^2_s, where s is the number of separate variances and covariances in each matrix [$= k(k + 1)/2$, where k is the number of traits]. Paulsen (1996) suggested using Bartlett's modified likelihood ratio test statistic [see Manly (1986)]. These two tests are not equivalent; in effect, the former tests for differences in means, whereas the latter examines differences in variances (cf. difference in means versus homogeneity of variances in the analysis of variance).

All of the above methods are sensitive to multivariate normality, and their behavior when this assumption fails or when sample sizes are comparatively low is largely unknown. How low must a sample size be before problems arise is difficult to decide, but the potential problem can be seen from Manly's warning about the Mahalanobis statistic, "It is difficult to say precisely what a 'small number of degrees of freedom' means in this context. Certainly there should be no problem with using Mahalanobis distances based on a covariance matrix with the order of 100 or more degrees of freedom" (Manly, 1986, p. 49). The number of degrees of freedom in the study by Holloway et al. (1993) was only 10 and, hence, some concern must be raised about the validity of the test in these circumstances (it should be noted that this comparison is not a pivotal part of the paper).

3.6.1.3 Tests Based on Robust Methods

Either the bootstrap or randomization might be used to compare matrices. Although these methods have fewer assumptions than those in the previous section, they are by no means "assumption-free," and interpretation must still be done with great caution.

Paulsen (1996) used the bootstrap method to compare the G matrices for two species of butterfly, *Precis coenia* and *P. evarete*. The variances and covariances were estimated from mean offspring on mid-parent regression; therefore, the offspring–midparent combinations formed the unit of resampling. To remove size effects, trait values were centered about zero (see below for a cautionary note on this). Families from both species were then pooled and a bootstrap replicate of the two G matrices was constructed by sampling with replacement from this pool. For each replicate, Bartlett's modified likelihood ratio test statistic, Λ, was calculated, and from this, the cumulative distribution of Λ was constructed. Finally, this distribution was used to test the significance of the observed value of Λ (i.e., the probability of observing a value of Λ as large or larger than that observed).

As a an alternative to the above, I suggest the following randomization test. First, the data sets are arranged into sampling units according to the breeding design, as described in Section 3.6.1.1. All units are placed into a single pool and then randomly assigned to group 1 and 2, the number in each group being kept the same as in the original. The statistic of interest is then computed, which may be Λ, D^2, or the correlation between the two matrices (see below for the use of the correlation), and stored. The whole process is repeated a large number of times (say 5000). Probability levels can then be assigned in the manner described above. The same comments concerning randomization versus bootstrap given in Section 3.1 also apply here.

If we wish to assess visually how similar the two G vectors are, then the obvious thing to do is to plot y on x (or x on y). If the hypothesis that $G_1 \propto G_2$ is correct, then the intercept should be zero and the correlation significant. Observed relationships appear to fall into one of three categories (Fig. 3.8): no relationship, a $1:1$ correspondence, and proportionality between the two variables. I have not found a case (the total number studied is few) in which the relationship was significant and the intercept very clearly differed from zero, as judged by regular regression analysis. There are four problems with the regression approach. First, there is no obvious dependent or independent variable. A solution in this case is to use functional regression (Ricker, 1973; Mousseau and Roff, 1987). Second, and more critically, because the elements in each vector are not independent, the standard statistical tests are not valid. Consider four traits A, B, C, and D: Because $Cov(AB) = Cov(BA)$, we consider only one-half of the matrix,

	A	B	C	D
A	1	2	3	4
B	.	1	2	3
C	.	.	1	5
D	.	.	.	1

where the unnumbered cells are the duplicate cells. The numbers in the cells designate those cells that form an independent series [e.g., $Cov(AB)$ is not independent of $Cov(AC)$ because the values of trait A occur in both]. Thus, for statistical independence, five separate experiments are required to estimate the matrix entries. In practice, one experiment is done and, therefore, statistical independence is lost. To circumvent this problem, Lofsvold (1986) advocated use of Mantel's test. This is simply a randomization test for correlation based directly on the elements of the two vectors; that is, a randomized data set is constructed by randomizing the elements in one vector (Manly, 1991, pp. 114–116). Unhappily, there are doubts as to the applicability of this technique (Shaw, 1992). Randomization as a technique for assessing the statistical significance of the correlation can be justified on one of three grounds (Manly, 1991, p. 92):

1. The (X, Y) pairs are a random sample from a population where X and Y are independent so that all possible pairings are equally likely; random sampling justifies the test.

2. The data are obtained from an experiment where the X-values are randomly assigned to n units and an experimental response Y is obtained. All possible pairings of X and Y are then equally likely if the distribution of Y-values is the same for all values of X; the random allocation of X-values justifies the test.

3. The circumstances suggest that if X and Y are unrelated, then all possible pairings of values are equally likely to occur as a result of the mechanism generating the data. This is a weaker justification than the previous two, but is nevertheless a possible justification in some cases.

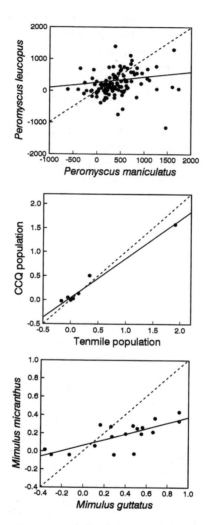

Figure 3.8 Top: A comparison of genetic variances and covariances between two species of mouse [*r* = 0.18, *n* = 120, *P* = .049; data from Lofsvold (1986)]. *Middle panel:* A comparison of genetic variances and covariances between two populations of the garter snake, *Thamnophis ordinoides* [*r* = 0.99, *n* = 10, *P* < .0001; data from Brodie (1993)]. *Bottom panel:* A comparison of the genetic correlations between two angiosperms [*r* = 0.74, *n* = 18, *P* = .0004; data from Carr and Fenster (1994)]. All statistical tests are approximate. Solid line shows fitted regression; dashed line shows the 1:1 line.

Justifications 1 and 2 do not apply in the present case, but one could argue that justification 3 does apply. However, I am uncomfortable even with this justification because randomization operates at the level of the estimates themselves, whereas the nonindependence arises one level further down from the basic data. In the absence of a clear-cut theoretical justification, a simulation study is required to empirically test the validity of randomization in the present case.

The third problem with Mantel's test in general is that the outcome is sensitive to the statistic chosen (Dietz, 1983). For example, instead of the Pearson product moment correlation, Kohn and Atchley (1988) used an association measure, K_c, based on Kendals tau-statistic proposed by Dietz (1983). For their data set comparing the rat and the mouse, the choice of the test statistic did not influence the conclusions, but in a reanalysis of the *Peromyscus* data of Lofsvold (1986), they found that the conclusions did depend on the statistic. This does not mean that one statistic was wrong and the other right, only that they were addressing the significance of different types of associations (but it is not clear what these differences are). This is a general problem in statistical analysis (for example, the *t*-test and Mann–Whitney *U*-test are frequently used interchangeably, but, in fact, one is testing for a difference in means and the other for a difference in medians) and simply points to the care with which statistical tests must be carried out. Finally, the Mantel test is suspect in the present circumstances because it does not take into account the error variance associated with each element of the matrix (Cowley and Atchley, 1992): Estimates based on say 10 families are given as much weight as those based on 1000 families. For these four reasons, results based on Mantel's test must be viewed with suspicion.

One solution to the problem of differences in sample size is to compare the observed correlation with the maximum expected correlation, using the estimated matrix repeatabilities obtained via bootstrapping (Cheverud, 1996). A comparison of the genetic correlation matrices of captive populations of the cotton-top and saddle-back tamarins gave a correlation between the two species of 0.19, which was also the maximum expected given the estimated error (Cheverud, 1996). Thus, it is not possible to conclude from the results that the differences are particularly remarkable; indeed, the low maximum expected value shows that no significant conclusion concerning the similarity of genetic correlations can be made.

3.6.1.4 Power

The *t*-test described by Eq. (3.32) can be used to examine the statistical power when only two genetic correlations are being considered. The power of the test (the probability of rejecting the null hypothesis given that it is false) is given approximately (because variances may not actually be equal) by

$$Z_{1-\beta} = Z_\alpha - \frac{|r_{A1} - r_{A2}|}{\sqrt{SE(r_{A1})^2 + SE(r_{A2})^2}} \tag{3.33}$$

where Z_α is the Z-score associated with the α level used (typically, 0.05 and two-tailed), $Z_{1-\beta}$ is the Z-score associated with power (β is the probability of a type II error) and is equal to the area under the normal curve from $Z_{1-\beta}$ to infinity [for discussion of power analysis applied to heritability, see Klein (1974)]. Three examples using a full-sib design with 10 progeny per family are presented in Fig. 3.9, where an α level of 0.05 and a two-tailed test were used. Note that power is not only a function of the difference between the two correlations but also their values; this is indicated by the isoclines not lying parallel to each other, although for values less than 0.8, the discrepancy is quite small. If the two populations are sampled at an intensity of 50 families ($= 500$ individuals) per population, the probability of detecting a difference of 0.4 is approximately 0.6 (top panel, Fig. 3.9). Doubling the sampling intensity raises the probability only to 0.8 (middle panel, Fig. 3.9), whereas a 4.5-fold increase to 250 families per population makes it virtually certain (lower panel, Fig. 3.9). Using a simulation model, Shaw (1991) investigated the statistical power of the restricted maximum likelihood method for detecting the difference in V_A between two populations. She assumed a half-sib breeding design with the number of progeny per dam family ranging from 3 to 5, the number of dams ranging from 2 to 3, and the number of sires ranging from 10 to 100; the total number of individuals measured per population ranged from 100 to 900. Dominance variance was kept constant, but both V_A and V_E were varied. Regardless of the variation in other parameters, there is a clear pattern of variation between power and the number of individuals measured for each ratio of additive genetic variances (larger/smaller, Fig. 3.10). With one variance being 2.5 times as large as the other and approximately 400 individuals per population measured, the power is disappointingly low, only 20% of tests showing statistical significance (middle line, Fig. 3.10). The results of this analysis support those obtained for the genetic correlation, namely in order to be reasonably sure of detecting even large differences between populations, very large samples sizes are required. The detection of biologically significant differences between populations is a daunting experimental task!

3.6.2 Empirical Comparisons

There are as yet comparatively few analyses of **G** (Table 3.13). The tests outlined above give two very distinct types of results. Those tests which examine the equality of the matrices (tests in Sections 3.6.1.1 and 3.6.1.2) are able to draw the conclusion that the hypothesis of equality is not supported (i.e., the two matrices are significantly different) or that the hypothesis cannot be rejected. The latter conclusion is weak in that it may simply result from low statistical power. This is evident in the two cases where maximum likelihood analyses have been applied: In both cases (*Holeus lanatus* and *Anthoxanthum odoratum*), there is evident failure of the algorithms to converge to satisfactory estimates (see footnotes 9 and 10 of Table 3.13). Pairwise comparison suggests significant difference between two populations in the genetic correlation between tail and body verte-

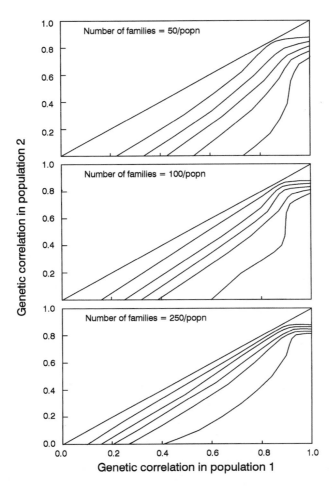

Figure 3.9 Power isoclines (solid lines) for a full-sib design (10 progeny per family) for three different sample sizes. For simplicity of display, only the lower portion is shown (i.e., r_A in population 1 > r_A in population 2). Heritability was set at 0.5 and the standard error computed using Eq. (3.13). Power values decrease from bottom to top in increments of 0.2, starting at 1.0.

brae in the snake, *Thamnophis elegans* (see footnote 4 of Table 3.13). The bootstrap test was unable to detect differences between two butterfly species (*Precis*), but element-by-element comparison did suggest significant differences in at least one element (Paulsen, 1996; Table 3.13). This study emphasizes an important statistical point: Large differences between two groups can be submerged by lack of variation among all other groups. (This is a general problem with such simple analyses as one-way analysis of variance.) It is important to attempt to restrict

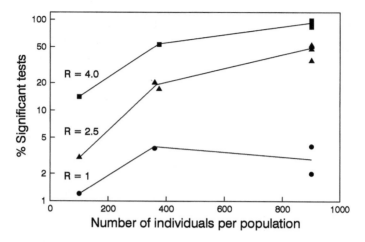

Figure 3.10 The percentage of times a significantly different additive genetic variance between two populations was detected using the restricted maximum likelihood method. The data were simulated assuming a half-sib breeding design, with 5 progeny per dam family, 2–3 dams per sire, and 10–100 sires. Three ratios of additive genetic variances (larger/smaller), R, were used. Lines have been added by eye to highlight the pattern of variation. When $R = 1$, 5% of tests should be significant: there appear to be somewhat fewer than this. Note that the y axis is a log scale. [Data from Shaw (1991).]

measurements to those traits that one might expect to vary on a priori grounds or be able to justify separate analyses of components of the matrix.

In contrast to the above methods, regression analysis (Mantel's test) is able to draw the conclusion that there is a statistically significant association between the two matrices or that no association can be detected. This is exactly opposite to the previous methods. Lack of an association is not very informative unless it can be shown that the estimates are very accurate, which generally they are not (see Section 3.4). Significant (as judged by standard regression theory or Mantel's test) correlations between matrices have been found in both interpopulation (*Gammarus minus, Drosophila melanogaster,* and *Mimulus micranthus*) and interspecific (*Peromyscus*) comparisons (Table 3.13). Significant correlations between genetic correlations have also been found at the intergenetic level (*Mus* and *Rattus*). None of the published tests considered the possibility of a significant association other than an intercept differing from zero (i.e., lack of proportionality). From visual inspection of the published graphs and plots of those for which data are provided, I cannot find a case where the intercept very clearly deviates from zero. Nevertheless, this question should be addressed in comparisons of matrices.

At the beginning of this section, I suggested that a more useful question is not whether **G** has remained constant but rather how much divergence is associated with taxonomic level and type of trait. Some of the results reported in Table 3.13 bear on this matter. That by Fong [(1986), see also Jernigan et al. (1994)] on the

Table 3.13   Survey of Tests for Similarity Between Genetic Variance–Covariance Matrices

Taxa	Method of Comparison[a]	Result[b]	Refs. and Notes[c]
Mus musculus versus *Rattus norvegicus*	Mantel's test	NSC (covariances) SC (r_A)	1
Peromyscus m. bairdii versus *P. m. nebrascensis*	Mantel's test	SC (both)	2
P. m. bairdii versus *P. leucopus*	Mantel's test	SC (both, but see note)	2
P. m. nebrascensis versus *P. leucopus*	Mantel's test	SC (both)	2
Thamnophis elegans (2 populations)	Qualitative only	SD (?)	3
T. elegans (2 populations)	None	SD (see note)	4
T. ordinoides (2 populations)	E×E	NSD	5
Gammarus minus (5 populations)	Mantel's test	SC (5 of 10, r_A)	6
Drosophila melanogaster versus *Musca domestica*	Mantel's test	SC	7
D. melanogaster (2 strains)	Mantel's test	SC (3/4 tests)	7
D. melanogaster versus *D. subobscura*	Mantel's test	NSC	7
D. melanogaster versus *D. robusta*	Mantel's test	NSC	7
Precis coenia versus *P. evarete*	Bootstrap, E×E	SD (?)	8
Holeus lanatus (2 populations)	ML	NSD	9
Anthoxanthum odoratum (2 populations)	ML	NSD	10
Mimulus micranthus (2 populations)	Standard correlation	SC	11
M. micranthus versus *M. guttatus*	Standard correlation	SC	11

[a]E×E = element-by-element comparison; ML = maximum likelihood.

Table 3.13 footnotes continue on next page

[b]NSC = Not significantly correlated; SC = significantly correlated; NSD = not significantly different; SD = significantly different.

[c]Notes and references:
1. Kohn and Atchley (1988). All morphological traits.
2. Reanalysis of morphological data of Lofsvold (1986) by Kohn and Atchley (1988). In the case of *P. m. bairdii* versus *P. leucopus*, the results for the covariance matrix depend on the statistic used (see text).
3. Arnold (1981, p. 499) writes, "Despite the striking similarity of genetic covariance structure in the two populations, there are some notable differences." All chemoreceptive responses to prey.
4. Arnold (1988, p. 634) analyzing morphological traits in this snake species writes, "Thus, one tentative conclusion emerging from the few studies done so far, including the present one, is that genetic parameters seem to be relatively constant among geographic races of the same species." However, the data on genetic correlations presented suggest that the two inland populations in California are significantly different:

	Coastal California		Inland California	
	A	B	C	D
r_A	0.48	0.30	0.37	0.22
SE	0.25	0.05	0.03	0.04

5. Brodie (1993). Morphological and behavioral measurements. See Fig 3.8 for a plot of these data.
6. Fong (1989). All measurements are of morphological traits. Fong used only 200 permutations per comparison, which is far too small (1000 should be the minimum, 5000 a good working number).
7. Cowley and Atchley (1990). Comparisons based on the phenotypic matrices of the other species and *Drosophila* strain. Comparison using the genetic or phenotypic matrix of the strain examined by Cowley and Atchley gave qualitatively the same results, except in the case of female *D. melanogaster* compared to DDT-resistant female *D. melanogaster*.
8. Paulsen (1996). Element-by-element comparison of the matrices of morphological traits in two butterfly species indicated significant differences in one element even after sequential Bonferroni adjustment of the probabilities.
9. Shaw and Billington (1991) were not able to demonstrate differences among pairwise comparison of variances of morphological and one life history variables in two populations of this plant. However, variability was very high:

Trait	Pop. 1	Pop. 2
Tiller number	0	281
Tiller dry weight	0.08	0.18
Stolon number	0	0
Stolon dry weight	0	0.01
Inflorescence number	11.7	3.2
Flowering time	0.20	−0.34

Zero values obtained using restricted maximum likelihood. (The negative value for flowering time given as the restricted maximum likelihood did not converge for this trait.) With respect to flowering time, Shaw and Billington (1991, p. 1288) observed that "the data are equivocal on the issue of whether the populations diff in V_A for flowering time."
10. In this comparison between two plant populations, Platenkamp and Shaw (1992, p. 341) concluded that "none of the matrices differed significantly between populations." Very

Table 3.13 footnotes continue on next page

few of the estimates, however, were sensible (either the estimation algorithm did not converge or the absolute value of the genetic correlation exceeded 1):

Population	Not Estimable	$\mid r_A \mid > 1$	Sensible
Xeric	7	1	3
Mesic	0	9	6

11. Carr and Fester (1994). Regression plots and standard regression statistics suggest proportionality between the matrices both between species and populations in the floral traits in these species. However, Carr and Fester (1994, p. 612) note, "We did not include covariances involving day of first flower because the genetic correlations showed highly significant differences among populations."

amphipod *Gammarus minus* is particularly interesting in this regard. Fong studied five populations, distributed between two separate drainage basins (three in one basin, two in the other). Each population could be separated into one of three ecotypes:

Ecotype I: confined to large, well-integrated cave systems in two disjunct areas of the central Appalachians. This form has the lowest number of eye ommatidia (less than six) and is the largest in body size and in the relative lengths of antenna 1, pereopod 7, and uropod 3 (these relative lengths are not simple allometric changes).

Ecotype II: occurs in small, isolated caves throughout the Appalachians. Intermediate in form to ecotypes I and III, but closer to III.

Ecotype III: found in springs and spring runs throughout the species range. This form has the highest number of ommatidia and is the smallest in body size.

Pairwise comparison of the matrices using Mantel's test indicated a significant correlation between the same ecotypes from different drainage basins (III versus III; I versus I) or ecotypes relatively similar in form (I versus II; II versus III; of the two comparisons for the latter, both between populations from different drainage basins, one was significant and the other nonsignificant). Comparisons between the two most distinct ecotypes I and III produced no significant correlations (four comparisons—one for two populations from the same drainage basin and three between populations from different drainage basins). These results suggest that in analyzing the morphological evolution of this species the assumption of a constant **G** is not warranted. Further study of this species might shed light on the rates of divergence in genetic architecture.

Comparison of the variance–covariance and the genetic correlation matrices between several rodent species and subspecies does not produce such a clean pattern as found in *Gammarus* (Table 3.14). Highly significant correlations are obtained using the genetic correlation matrix, but a comparison between **G** ma-

Table 3.14   Summary Statistics from Mantel's Test for Morphological Comparison Between Several Species and Subspecies of Rodent

		Correlation Between Taxa Using	
Taxon 1	Taxon 2	**G**	r_A
Mus	*Rattus*	0.01	0.60**
P. leucopus	*P. maniculatus bairdii*	0.16	0.59**
P. leucopus	*P. maniculatus nebrascensis*	0.58*	0.63**
P. maniculatus bairdii	*P. maniculatus nebrascensis*	0.20*	0.52**

Note: Only the results using the Pearson product–moment correlation are shown. The traits measured in the rat and mouse (pelvic components) differ from those measured in *Peromyscus* (cranial components).

*P < .05.

**P < .01.

Source: Data from Kohn and Atchley (1988).

trices produced much more disparate results. The correlation between the **G** matrices of the two different genera, *Rattus* and *Mus*, is essentially nonexistent, suggesting a relatively large divergence in comparison to that between species and subspecies of *Peromyscus* (Table 3.14; it should be remembered that different traits are involved and, hence, differences must be viewed with caution). This is consistent with the hypothesis that **G** should diverge as the taxonomic separation increases. However, the results for *Peromyscus* are inconsistent with this hypothesis, *P. leucopus* and *P. m. nebrascensis* being more alike than *P. m. bairdii* and *P. m. nebrascensis* (Table 3.14) (Lofsvold, 1986; Kohn and Atchley, 1988). Equally confusing results are found in comparisons among diptera, in which the **G** matrix of *Drosophila melanogaster* is more similar to that of *Musca domestica* than other *Drosophila* species (Table 3.15). Finally, different populations of *Mimulus guttatus* are less similar to each other than *M. guttatus* and *M. micranthus* (Carr and Fenster, 1994).

A host of reasons can be given why the above comparisons can be discounted (low sample sizes, incorrect estimates of variances and covariances, incorrect statistical comparisons, etc.), and I am not suggesting that these results should be taken very seriously. However, they do indicate the lack of data bearing on the problem of evolution of the **G** matrix. This is an area in which considerably more research can profitably be undertaken.

3.7   Summary

Genetic correlations arise as a consequence of linkage disequilibrium or pleiotropy, the latter being generally more important in the evolution of traits. One case

Table 3.15   Summary Statistics from Mantel's Test for Morphological Comparison Between Drosophila melanogaster *and Other Diptera*

Species	R_P		R_A	
	Males	Females	Males	Females
D. melanogaster	0.80**	0.81**	0.72**	0.30
D. subobscura	0.17	0.05	0.18	0.37
D. robusta	0.26	0.19	0.33	0.14
Musca domestica	0.88**	0.84**	0.72**	0.76**

Note: Test statistic is Spearman's rank-correlation coefficient. R_P are results comparing phenotypic matrices, R_A are results using the **G** matrix for *D. melanogaster* and the phenotypic matrix for the other member of the comparison pair.

*$P < .05$.

**$P < .01$.

Source: Data from Cowley and Atchley (1996).

in which linkage disequilibrium may be of considerable significance is the evolution of mate preference. The phenotypic correlation is a function of the genetic and environmental correlations and the heritabilities of the two traits. As with heritability, the genetic correlation depends on allele frequencies. Genetic correlations can be estimated from the same breeding designs as for heritability. The jackknife appears to be a valid means of estimating the standard error, at least when the data are reasonably balanced. Traits measured in different environments can be considered to be genetically correlated, although because the traits do not occur in the same individual an alternative method of analysis—the mixed model analysis of variance—is required. As predicted from consideration of selection pressure, the genetic correlation between life history traits is more often negative than that between morphological traits. The phenotypic correlation is much more easily measured than the genetic correlation; thus, it would be of considerable importance if the former could be used in place of the latter. This appears to be valid for the category morphology × morphology but not for correlations involving life history traits. The interrelationship between a suite of traits can be described by the genetic variance–covariance matrix. Because the stability of this matrix is a prerequisite for the prediction of long-term evolutionary change, a number of studies have addressed the problem of testing for similarity between the matrices of different populations or taxa. There are still serious statistical problems both because of disputes over the validity of different methods and because sample sizes required are extremely large. Empirical comparisons are still few and have produced inconsistent patterns, although this may reflect more statistical uncertainty than biological complexity.

4

Directional Selection

There are basically three types of selection: directional, stabilizing, and disruptive (Fig. 4.1). Of these, the first is the most important from the perspective of changes in the mean value of a quantitative trait (the other two types are discussed in Chapter 9). The term **directional selection** applies to selection in which the mean value of the parents which contribute to the next generation differs from that of the population from which the parents are drawn. The prediction of response to directional selection can be derived using the theoretical framework developed in Chapter 2. However, it is necessary to distinguish short-term response to selection from the long-term response, the former requiring fewer assumptions. In this chapter, I consider both types of responses with respect to a single trait. The consequences of the existence of genetic correlations between traits is explored in Chapter 5.

4.1 The Basic Equation: $R = h^2S$

The response of a trait to selection across a single generation is that given from the mean offspring on mid-parent regression (Chapter 2),

$$Y = (1 - h^2)m + h^2X \qquad (4.1)$$

where Y is the mean offspring value, m is the population mean value, and X is the mean parent value (Fig. 4.2). Rearranging Eq. (4.1), we get

$$Y - m = h^2 (X - m) \qquad (4.2)$$

which can be written as

$$R = h^2 S \qquad (4.3)$$

where R is the **response** and S is the **selection differential**. Much of the theory

118

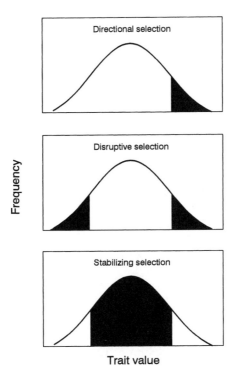

Figure 4.1 Examples of the three types of selection: the filled-in regions designate the proportion of the population that contributes to the next generation. Note that the particular form of selection is in all cases some type of truncation selection.

of directional selection is based on the idea of truncation selection in which individuals that exceed a specified value are selected (lower panels, Fig. 4.2). For comparison purposes, in such cases it is frequently convenient to write the selection differential in terms of the phenotypic standard deviation,

$$S = i \sqrt{V_P} \qquad \textbf{(4.4)}$$

where i is termed the **selection intensity**,

$$i = \frac{S}{\sqrt{V_P}} = \frac{z}{p} \qquad \textbf{(4.5)}$$

where p is the proportion selected and z is the ordinate at the point of truncation (estimated from a table of the cumulative normal or using the approximate for-

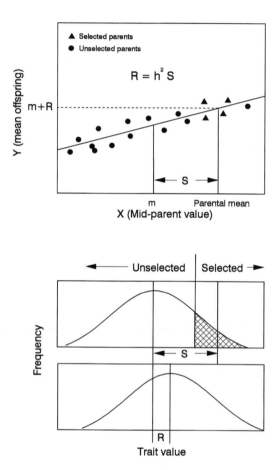

Figure 4.2 Response to directional selection illustrated using the mean offspring on mid-parent regression. The regression is shown in the top panel. The mean value of the population is m, the mean value of the parents is X, and the difference between them is the selection differential S. From the regression, the value of the offspring is $m + R$, where R is the response to selection. The lower two panels show truncation selection. All individuals (or a random selection of the specified group) greater than a certain value are selected as parents. Parents are represented by the hatched region on the right tail of the distribution. The distribution of offspring values is shown in the bottom panel, the mean value having shifted R units with respect to the mean value in the previous generation.

mula given in Eq. (2.43)). Thus, if the top 2.5% of the population is selected, the point of truncation is 1.96 and the intensity of selection is 2.3. Above 20%, the intensity of selection is roughly a linear function of the proportion selected (Fig. 4.3). The selection differential is reduced in a finite population, an approximate formula for the corrected value, i_f, being (Hill, 1985)

$$i_f \approx i - \frac{1 - p}{2iMp(1 - t_{\mathrm{fam}} + t_{\mathrm{fam}}/N)} \tag{4.6}$$

where p is the proportion selected, M is the number measured, N is the number of families, and t_{fam} is the intraclass correlation of family members. The above equation ignores the effect of restricted family size, there being no simple formula to predict this (Hill, 1977). In general, the difference between i and i_f is quite small (Fig. 4.3, Table 4.1).

Substituting $i\sqrt{V_P}$ for S in Eq. (4.3), we have $R = h^2 i \sqrt{V_P}$, and noting that $h = \sqrt{(V_A/V_P)}$, we obtain

$$R = ih \sqrt{V_A} \tag{4.7}$$

This predicted response follows from the general statistical theory of linear regression and makes no assumption about the genetical architecture of the trait. However, prediction beyond one generation does depend on genetical assumptions.

4.2 Evolvability

From Eq. (4.7) it can be seen that the response to selection depends not only on the heritability of the trait but also the intensity of selection and the square root of the additive genetic variance. Thus, for a fixed selection intensity, the relative responses of two traits to selection cannot be predicted by the heritabilities alone. Suppose there are two traits, say X and Y, in which the heritability of X is less than that of Y: Trait X may respond more for the same selection intensity than trait Y if trait X has a larger additive genetic variance. However, what do we mean by the term "more"? The problem is that R is not dimensionless. If traits X and Y were the same character, measured in different populations, then comparison is obviously possible. But suppose trait X is fecundity and trait Y is body weight; there is now no obvious common scale on which to compare them. A suggested solution is to consider not the absolute response but the proportional response (Houle, 1992).

Consider the population mean after selection, $m + R$, where m is the mean before selection: A comparison of two traits with different scales (e.g., fecundity and body weight) is not feasible using the means alone. Houle (1992) suggested

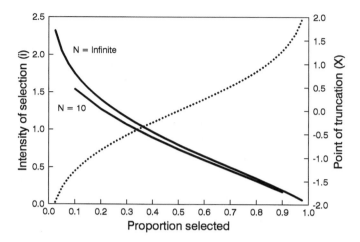

Figure 4.3 Intensity of selection (solid lines) and point of truncation (dotted line) in relation to the proportion selected. The values for the curve $N = 10$ obtained from Falconer (1989).

that a comparison can be made if the values are divided by the mean prior to selection; that is, the proportional change in the trait value, $(m + R)/m = 1 + R/m$. The ratio R/im was termed, by Houle, the **evolvability** of a trait under truncation selection. Evolvability can be written in terms of heritability and variances:

$$\text{Evolvability} = \frac{R}{im} = \frac{h^2 \sqrt{V_P}}{m} = \frac{h \sqrt{V_A}}{m} \tag{4.8}$$

From a consideration of several different types of selection, Houle (1992) suggested that a general measure of evolvability is the coefficient of variation, V_A/m. Unfortunately, the relative evolvability of two traits may still be a simple consequence of a scale transformation. (I am grateful to Dr. Ken Spitze for pointing out this problem to me.) Suppose we have two traits which are related by $Y = c + bX$ [e.g., Y might be log(weight) and X might be log(length)]. Letting the variance of X be V_X, the variance of Y is $b^2 V_X$, and the evolvabilities are

$$\text{Evolvability of } X = \frac{V_X}{m_X} \tag{4.9a}$$

$$\text{Evolvability of } Y = \frac{b^2 V_X}{c + bm_X} \tag{4.9b}$$

Table 4.1   Comparison of the Selection Differential and Response in an Infinite and Finite Population

	Selection Differential		Response	
	S	Ratio[a]	R	Ratio[a]
Infinite population	1.647	1.00	0.824	1.00
Finite population, $t_{fam} = 0$	1.525	0.93	0.762	0.92
Finite population, $t_{fam} = 0.25$	1.480	0.90	0.740	0.90
Finite population, $t_{fam} = 0.25$, with family effects included	1.480	0.90	0.687	0.83

Note: Parameter values for the finite population are $M = 16$, $n = 4$, $p = 12.5\%$ (two individuals selected), and $h^2 = 0.50$.

[a]Ratio = value for the infinite population/value for corresponding finite population.

Source: Data from Hill (1985).

where m_X is the mean value of trait X. Unless the constant c is zero, the two traits have different evolvabilities. However, the two traits may have the same heritabilities regardless of the value of the constant, as this does not influence the variance components. Further, if the parameters c and b are constant, the genetic correlation between the traits is 1!

Another example of the above problem is provided by a consideration of traits measured on a temperature scale, as, for example, development time in degree days, or the threshold temperature for development. Suppose in one instance that a Fahrenheit scale is used and in another instance the Celsius scale. Letting the trait value on the Fahrenheit scale be X, its evolvability is as given by Eq. (4.9a), whereas on the Celsius scale, it is as given by Eq. (4.9b) with $c = -32$ and $b = 4/9$. Thus the same trait has two different evolvabilities despite the fact that they are the same trait.

Because the evolvability of a trait is not scale independent it cannot be used to compare traits that are measured on different scales, although it could be used to compare, say, two weight measures.

4.3   Predicted Response in a Very Large Population

4.3.1   The Infinitesimal Model

The results for the infinitesimal model have been worked out by Bulmer [1971a; see pp. 147–154 of Bulmer (1985) for a summary of this work], and I present here a simple verbal description and the final equations. The important assumption of this model is that the number of loci and alleles and the population size is large enough that changes can be analyzed using normal distribution theory. Extension of the model to include effects due to a finite population size are discussed in the next section. Selection of a set of extreme parents changes the allele frequencies and also causes linkage disequilibrium because the process of selection will favor particular combinations of loci even if these are unlinked. Consider, for example, the simple case of two unlinked loci with two alleles at each locus, the phenotypic value being given by the sum of the effects at the two loci. For simplicity, assume that the two alleles contribute either 0 or 1 to the phenotypic value, giving contributions to the phenotypic value that range from 0 (0 at all four positions) to +4 (1 at all four positions). Selection for the largest parents will lead to the selection of genotypes in which the two loci are both homozygous for the allele contributing one unit, thereby inducing a covariance between the two loci, even though they are physically unlinked. If selection is halted, this disequilibrium will disappear as Hardy–Weinberg equilibrium is restored (recall that for a single locus, the disequilibrium disappears in a single generation, but more generations are required when there are several loci). Suppose that truncation selection is continued, a constant proportion being selected each generation. On the one hand, there is a continued decrease in the variance due to the increase in disequilibrium, but, on the other, there is a restoration due to the effects of recombination. The change from generation t to $t + 1$ is given by (Bulmer, 1985, p. 154)

$$Y(t + 1) = Y(t) + i [h^2(t)V_A(t)]^{1/2} \qquad \textbf{(4.10a)}$$

$$V_A(t + 1) = \frac{[1 - h^2(t)i(i - x)] V_A(t) + V_A(0)}{2} \qquad \textbf{(4.10b)}$$

where $h^2(t)$ is the heritability at generation t, $V_A(t)$ is the additive genetic variance at generation t, z is the ordinate at the point of truncation, and $i = z/p$, p being the proportion selected where x is the point of truncation corresponding to the proportion selected. Because only the additive genetic variance is decreasing, the heritability decreases. Eventually, a balance is reached, the equilibrium heritability being given by (Gomez-Raya and Burnside, 1990)

$$h^2(\infty) = \frac{-1 + \{1 + 4h^2(0)i(i - x)[1 - h^2(0)]\}^{1/2}}{2i(i - z)[1 - h^2(0)]} \qquad \textbf{(4.11)}$$

Two important points emerge from the above equations; (1) The approach to equilibrium is very rapid, taking only three or four generations and (2) the change in heritability is small (upper panel, Fig. 4.4).

Selection on a threshold trait is a type of truncation selection in that only individuals beyond a certain critical value, the threshold, are selected, but in this case, the proportion selected each generation does not remain constant but increases. Suppose the population originally comprises 50% of each morph; in the first generation of selection, 50% of the population will be selected. But the

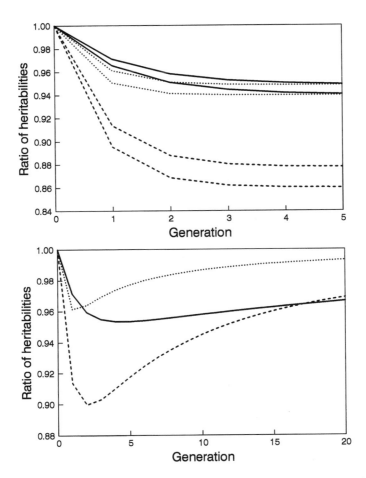

Figure 4.4 Predicted changes in heritability based on the infinitesimal model. *Top*: Change in heritability, measured as the ratio $h^2(t)/h^2(0)$, where t is the generation of selection, under truncation selection on a continuous metric trait, for three values of $h^2(0)$: 0.1 (solid line), 0.5 (dashed line), and 0.9 (dotted line), and two intensities of selection (25% and 50%; for each heritability the lower line represents the larger selection intensity). *Bottom*: Truncation selection on a threshold trait for the same three initial heritabilities as above. Because of the dichotomous phenotype, the selection intensity declines with selection and the heritability returns to its initial value. The initial population comprises 50% of each morph.

second generation consists of greater than 50% of the selected morph and, hence, the proportion of the population selected on the second generation must be greater than the first. With each subsequent generation of selection, the proportion selected increases, asymptotically approaching 100% (i.e., random mating). Allowing for the change in proportion selected, we find from Eq. (4.10) that in the first generation there is an initial decline in variance, but that as the proportion selected increases, the additive genetic variance and heritability return to their original values (lower panel, Fig. 4.4).

4.3.2 Changes in Fitness and Fisher's Fundamental Theorem

A trait in nature may be subject to a variety of selection pressures, but making use of the infinitesimal model, it can be shown that selection on the average phenotype acts to increase the mean fitness of a population; that is, selection on fitness is always directional under this model (Lande, 1976). To show this, I shall consider just a single trait, although extension to multiple traits using the phenotypic and genetic variance–covariance matrices is immediate (directional selection on multiple traits is discussed in Chapter 5). A fundamental assumption of the theoretical development is that h^2 does not change over the course of selection, which implies very weak selection. Letting the distribution of phenotypes, x, at time t be $p(x, t)$ (the symbol z has been conventionally adopted, but to be consistent with the notation of this book, I shall use x). The average phenotype before selection is

$$\mu_x = \int xp(x, t)\, dx \tag{4.12}$$

From the assumption of normality, the phenotypic distribution before selection is

$$p(x, t) = \frac{1}{\sqrt{2\pi V_P}} \exp\left(-\frac{(x - \mu_x)^2}{2V_P}\right) \tag{4.13}$$

If the fitness of an individual with phenotype x is $W(x)$, the mean fitness of individuals in the population, μ_W (for visual clarity the term t is omitted here and in the remainder of the derivation), is

$$\mu_W = \int p(x, t)W(x)\, dx \tag{4.14}$$

and the average phenotype after selection, μ_{xs}, is

$$\mu_{xs} = \frac{1}{\mu_W} \int xp(x, t)W(x)\, dx \tag{4.15}$$

Using the above equations, we can obtain the change in mean fitness with respect to the change in the mean phenotype:

$$\frac{\partial \mu_W}{\partial \mu_x} = \int \frac{\partial p(x, t)}{\partial \mu_x} W(x) \, dx$$

$$= \int \frac{(x - \mu_x)}{V_P} p(x, t) W(x) \, dx \qquad (4.16)$$

$$= \frac{\mu_W}{V_P} (\mu_{xs} - \mu_x)$$

Now the response to selection R_x is equal to $h^2(\mu_{xs} - \mu_x)$, and using Eq. (4.16), we obtain

$$R_x = \frac{h^2 V_P}{\mu_W} \frac{\partial \mu_W}{\partial \mu_x} \qquad (4.17)$$

$$= h^2 V_P \frac{\partial \ln \mu_W}{\partial \mu_x}$$

The above equation shows that, under the given assumptions (infinite population size, constant V_P, constant V_A), the evolution of the average phenotype will always be in the direction that increases the mean fitness in the population.

Recalling that $h^2 V_P = V_A$, then if the trait is itself fitness, the result of Eq. (4.17) can be restated as

$$R_{\mu_W} = \frac{V_{AW}}{\mu_W} \qquad (4.18a)$$

$$\frac{\partial \mu_W}{\partial t} = V_{AW} \qquad (4.18b)$$

where V_{AW} is the additive genetic variance in fitness. Equation (4.18b) is often call Fisher's Fundamental Theorem of Natural Selection: "the change in fitness caused by natural selection is equal to the additive variance in fitness," (Frank and Slatkin, 1992, p. 992), although Fisher's original proposition was obscure (Price, 1972; Ewens, 1989, 1992). It must be remembered that this theorem only applies to the additive portion of the genetic variance and assumes that the environment remains stable.

4.4 Predicted Response in a Finite Population

4.4.1 Continuously Distributed Traits

Alan Robertson (1960b, 1970a) developed the theory of response to selection when population size is finite. He assumed the following: (1) no epistasis or dominance, (2) no linkage [incorporated in later analyses and found to be relatively unimportant (Robertson, 1970b; Robertson and Hill, 1983)], (3) two alleles per locus, (4) an infinite number of loci, (5) a constant selection coefficient at each locus, and (6) no mutation. As a consequence of inbreeding resulting from the finite size (see Chapter 8 for a detailed discussion of inbreeding), the additive genetic variance declines according to the relationship

$$V_A(t) = V_A(0) \left(1 - \frac{1}{2N}\right)^t \tag{4.19}$$

where N is population size and t is generation. The cumulative response after t generations, R_t, is therefore [see Eq. (4.7)]

$$R(t) = \frac{iV_A(0)}{\sqrt{V_P}} \sum_{i=0}^{t-1} \left(1 - \frac{1}{2N}\right)^i \tag{4.20}$$

which can be closely approximated by (Robertson, 1970a)

$$R(t) = 2Ni \frac{V_A(0)}{\sqrt{V_P(0)}} (1 - e^{-t/2N}) \tag{4.21}$$

Hence, the total response ($t \to \infty$) is equal to

$$
\begin{aligned}
R(\infty) &= 2Ni \frac{V_A(0)}{\sqrt{V_P(0)}} \\
&= 2Nih^2(0) \sqrt{V_P(0)} \\
&= 2N(\text{Initial response})
\end{aligned}
\tag{4.22}
$$

The number of generations it takes for the population to reach halfway to the asymptotic limit, $t_{0.5}$, is given by the equally simple formula

$$t_{0.5} = 1.4N \tag{4.23}$$

To find the proportion giving the greatest total response to selection we note that $Ni = Mpz/p = Mz$, where M is the number of individuals measured per generation. For a fixed M, the only variable quantity in Eq. (4.23) is z, which has its maximum at $z = 0$, which is when $p = 0.5$. Thus, the maximal total response is predicted to be obtained when 50% of the population is selected each generation.

If there is complete dominance at each locus, the maximal response is approximately (A. Robertson, 1960b)

$$R(\infty) = \frac{2Np}{3(1-p)} \quad \text{(Initial response)} \qquad \textbf{(4.24)}$$

where p is the frequency of the dominant allele. In the above equation, effects of inbreeding depression (discussed in Chapter 8) have been ignored; they will reduce the response. As the frequency of dominant alleles approaches 1, the total response approaches infinity (remember that there is an infinite number of loci). Although these results are only approximate, they do indicate that the presence of dominance can considerably increase the total response. The equation for the half-life is more complex than that for additive alleles but obeys the relationships

$$t_{0.5} \to 2.12N \quad \text{as } p \to 1 \qquad \textbf{(4.25a)}$$

$$t_{0.5} \to 1.032N \quad \text{as } p \to 0 \qquad \textbf{(4.25b)}$$

Thus, the half-life of the response may, in general, be expected to lie between N and $2N$ (A. Robertson, 1960b).

Over the long term, some variation is going to be restored by mutation. This problem has been considered by Clayton and Robertson (1955), Hill (1982a, 1982b, 1985), Robertson and Hill (1983), and Hill and Keightley (1988). The result is intuitively simple: At the selection limit, in the absence of mutation there is no further response, and, hence, further response will be due to the additional additive genetic variance via mutation, V_M. Recall that the per-generation response in the presence of additive genetic variation is $R = iV_A/\sqrt{V_P}$; the total additive variance due to mutation is, in a diploid, sexual organism, $2NV_M$. Hence, the response expected from the input of additive mutational variance is

$$R = \frac{i2NV_M}{\sqrt{V_P}} \qquad \textbf{(4.26)}$$

Weber and Diggins (1990) reformulated the model of A. Robertson (1960b) to include mutation, arriving at

$$\frac{R_t}{R_1} = \frac{1 - C^t + \mu \sum_{i=1}^{t}(1 - C^i)}{1 - C} \qquad \textbf{(4.27)}$$

where

$$C = 1 - \frac{1}{2N}$$

and μ is the spontaneous mutation rate. In the absence of mutation, this model is identical to that of A. Robertson (1960b, 1970a).

As might be expected, for population sizes above about 50, drift is negligible, at least over 50 generations, and the response remains more or less constant (Fig. 4.5). After 50 generations, the total response for a population size of 50 is modestly lower than that of a population of size 100 or 1000. Certainly for the first 10 generations, there is a negligible difference between the relative responses of populations with sizes greater than 25. The situation is quite different if population sizes are as low as 5 or 10; there is an obvious decline in response by generation 10 and the asymptotic maximal response is attained by about generation 20 (Fig. 4.5). The relative response after 50 generations of selection increases rapidly with population size, eventually asymptoting rapidly for population sizes in excess of 100 (lower plot, Fig 4.5). When population size is small (<10), the maximum limit corresponds with that attained after 50 generations, but for larger population sizes, the relative response increases astronomically.

With a finite number of loci, the loss of additive genetic variance will be largely due to fixation of alleles; hence, the expected response will be much less than predicted by the Robertsonian model. In the absence of mutation, there must be an eventual loss of variation due to fixation at each locus. Bulmer (1976) examined the consequence of a finite population size and a finite number of loci and alleles using a simulation model. He assumed 12 loci with 2 alleles at each locus, the phenotypic value being determined by the sum of the loci plus an environmental value that was normally distributed with zero mean and variance V_E. The population consisted of 500 individuals of each sex from which the most extreme 20% were selected as parents for the next generation. In one model, Bulmer mimicked the situation found in *Drosophila*, the 12 loci being assumed to be in 3 linked groups of 4 loci with a recombination rate of 0.1, and in another case, which he termed the "Mouse" model, he assumed all loci to be unlinked. In the *Drosophila* model, the gene frequencies reached complete fixation after only 16 generations, and in the Mouse model after only 13 generations. Nevertheless, utilizing the first five generations of selection, Bulmer obtained excellent correspondence between the simulated results and model predictions. The population size used by Bulmer was probably large enough that genetic drift was not as important as the restriction on the number of loci and alleles.

The failure of the Robertsonian predictions have also been demonstrated by the simulation modeling of Hospital and Chevalet (1993). They concluded that

> even in the case of unlinked loci, the theoretical predictions have little connection with the results observed in the simulations . . . apart from the quite unrealistic situations (very small population size, a quantitative trait due to several hundred unlinked loci with equal effects), these predictions are generally unreliable except in the first few generations. We emphasize that linkage, and the joint effects of selection and drift on additive genetic variance, must not be neglected in selection theory. (pp. 778–779)

These conclusions are illustrated in Fig. 4.6, in which the predicted decay of

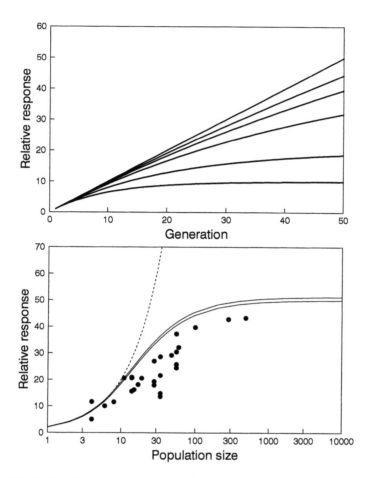

Figure 4.5 Predicted changes in the response to selection based on the Robertsonian model. *Top*: Changes per generation for population sizes of (bottom to top) 5, 10, 25, 50, 100, and 1000. *Bottom*: Predicted cumulative response relative to the initial response after 50 generations of selection in relation to population size, with (upper solid line) and without (lower solid line) mutation ($\mu = 0.001$). Dotted line shows predicted maximum response. Also plotted are the observed responses for various experiments (species used are *D. melanogaster, Zea mays, Mus domesticus,* and *T. casteneum*). [Data from Table 2 of Weber and Diggins (1990).]

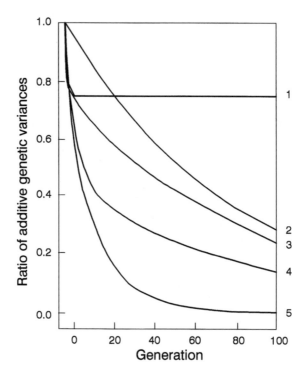

Figure 4.6 Decrease in additive genetic variance relative to the initial variance in the simulation model of Hospital and Chevalet (1993) compared to various theoretical models. The assumptions of the simulation model were (1) initial heritability = 0.5, (2) number of loci = 50, (3) two alleles per locus, (4) loci evenly distributed along a map length of 50, and (5) population size (males + females) = 200. The plotted trajectory is based on 1000 replicates. Models are (1) infinitesimal model, (2) Robertsonian model, (3) Verrier et al. (1990), (4) Chevalet (1988), and (5) simulation model of Hospital and Chevalet (1993).

additive genetic variance from various theoretical models in comparison to that observed from the simulation is shown.

4.4.2 Threshold Traits

To examine the consequences of a finite number of loci and alleles on the change in genetic variance of a threshold trait, I (Roff, 1994a) utilized a simulation model similar to that of Bulmer (1976) described above. The underlying continuously distributed character was assumed to be determined by n unlinked loci, each with two alleles contributing either 0 or 1 to the phenotypic value. I assumed no dominance or epistasis, and an initial frequency at each locus of 50%. The additive genetic variance, assuming Hardy–Weinberg equilibrium and linkage equilibrium in the unselected population, was thus equal to $0.5n$. The phenotypic value of an individual was obtained as the summed contribution of all loci plus a random normal deviate distributed with zero mean and variance V_E. The environmental variance, V_E, was determined from the heritability in the unselected population [set at 0.65 to match the value obtained for wing dimorphism in *Gryllus firmus*: Roff (1986b)] and the relationship $h^2 = V_A/(V_A + V_E)$. In the founding population, both morphs were set at equal frequency, obtained by setting the threshold value equal to n. Each generation consisted of 100 families with 5 offspring per family. All individuals with phenotypic values greater than the threshold value were selected. One hundred pairs were chosen with replacement from the selected population to form the parents of the next generation. In the initial population, each individual was heterozygous at all loci: Prior to the selection the population was passed through two generations of random mating to achieve approximate Hardy–Weinberg equilibrium. The number of loci was varied from 3 to 20, with 20 replicates per n.

The mean trajectory of the proportion of the selected morph in the population was remarkably insensitive to the number of loci determining the trait (Fig. 4.7). For all loci, heritability in the first generation of selection decreased to about 0.4, but thereafter the response depended upon the number of loci (Fig. 4.7). With 3 loci, there was a continuous decline in the genetic variance, and by generation 20, the heritability had decreased to about 0.3. However, for the first 19 generations, no loci were fixed; in the 20th generation, one locus fixed in 2 of the 20 replicates. A similar pattern was found with 4 loci, the heritability dropping to 0.4 by generation 20, and 1 locus becoming fixed in the 18th generation of one replicate. Fixation of one locus, in one replicate line also occurred with 5 loci, but in all other replicates all loci were still heterozygous at generation 20. With 5 or more loci, the heritability initially dropped but then began to increase, exceeding 0.5 by generation 20 for n equal to 10, 15, and 20.

Further analysis of the simulation model for the case of 10 loci showed that the number of loci fixed after 50 generations of selection varies with population size, N, according to the relationship (Roff, 1994a)

$$\text{Number of loci fixed} = 8.91e^{-0.005N} \qquad \textbf{(4.28a)}$$

whereas in the absence of selection

$$\text{Number of loci fixed} = 9.11e^{-0.011N} \qquad \textbf{(4.28b)}$$

Selection is not much stronger than drift alone in causing the fixation of alleles, as can be seen by considering the ratio of number fixed with selection/number fixed by drift alone, which equals $0.98(1.006)^N$. Continuous selection for 500 generations produced an approximately linear decrease in heritability, but even after 500 generations, h^2 was reduced only by about one-half (Roff, 1996a).

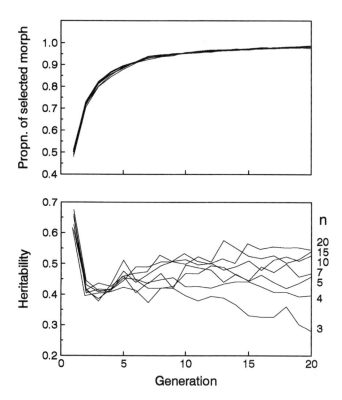

Figure 4.7 Simulation results for selection on a dimorphic trait. The underlying trait is determined by n loci, each with two alleles that take the value 0 or 1. The phenotypic value of the underlying trait is equal to the sum of the allelic values plus a normally distributed environmental deviate with mean zero and variance V_E. The value on the 0,1 (dimorphic) scale depends on the value of the underlying variable relative to the threshold—here set at n to make the initial proportion 50% for each morph. Each line is the mean of 20 replicates per number of loci (n).

The results for the continuous and threshold models indicate that at least for a few generations the assumption of a constant heritability is reasonable. This principle will be used later as a basis for estimating heritabilities and genetic correlations from response to selection.

4.5 Asymmetry of Response

According to the theory developed above, the response to selection does not depend on the direction of selection. There are a variety of reasons why this is not likely to be correct in real populations, which can be divided in to three broad categories: (1) measurement artifacts, (2) genetic asymmetry, and (3) finite population size.

4.5.1 Measurement Artifacts

(a) *Scalar effects*: Suppose the trait is normally distributed on a log scale. If measurements are taken on an arithmetic scale, the distribution will by highly asymmetrical and the basic assumption underlying Eq. (4.3) violated. The supposed section intensities, computed based on a normal distribution, will be in error and, hence, the formula will have no meaning. A simple example of this phenomenon is selection on a threshold trait. Suppose, for example, the base population consists of 25% morph A. A plot of the proportion of morph A versus generation will show an apparent asymmetry due to morph A approaching 0% faster than 100%. However, if the data are transformed to the underlying scale, the response will be symmetric, after taking into account the variation in selection intensity. Asymmetrical responses due to scalar effects will disappear once measurements are transformed to the appropriate scale.

(b) *Differences in the selection differentials*: Obviously, if the supposed selection differential is not correct, then there will be an apparent asymmetrical response. For example, suppose we select for increased and decreased adult weight, and fecundity is correlated with body size. In each group, we select two females to be parents of the next generation. Consider two females in the "large body size" group with weights w_1 and w_2 and corresponding fecundities f_1 and f_2. Ignoring effects due to the differential fecundities, the selection differential is $m - (w_1 + w_2)/2$, where m is the population mean weight. Because each female contributes differentially to the next generation, the correct selection differential is $m - (f_1 w_1 + f_2 w_2)/(f_1 + f_2)$, which, clearly, could be quite different from the uncorrected value. A second factor that can lead to an incorrect assessment of the selection differential is when the phenotypic variance changes in a different fashion in the two divergent directions of selection (recall that $S = i\sqrt{V_P}$ and thus $R = h^2 i \sqrt{V_P}$). This could occur if an incorrect scale of measurement is used. The phenotypic variance of body weight, for example, is typically correlated with the mean; therefore, selection for an increased body size may produce a greater response than selection for decreased body size.

(c) *Use of a surrogate measure*: If the method of selecting on the trait is based on a surrogate measure rather than the trait itself, and there is a nonlinear relationship between the two traits, the response of the surrogate measure may be symmetric but that of the target trait asymmetric (discussion of correlated responses to selection is presented in Chapter 5). For example, Baptist and Robertson (1976) selected for body size in *Drosophila melanogaster*, not by direct measurement of size but by having them walk through a series of slits of diminishing diameter. Thus, the actual trait being selected was not body size but a combination of body size and willingness to perform the required behavior. They found a negative correlation between size and activity which they suggested accounted for the slower response in the direction of increased body size. However, others selecting for body size (thorax length, weight, wing length, etc.) by direct measurement of the trait have also observed asymmetrical response (Robertson and Reeve, 1952; Tantawy, 1956a; Martin and Bell, 1960); thus, other factors are probably also important.

4.5.2 Genetic Asymmetry

Consider Fig. 2.3: If the allele frequency at a locus is precisely 0.5 and there is no dominance, then selection in either direction will lead to the same change in the heritability. If, on the other hand, the allele frequency is, say, 0.6, then "downward" selection will lead to a greater change in h^2 than "upward" selection. The same phenomenon occurs when there is dominance, except that the point of symmetrical change is at an allele frequency of 0.25. Because response to selection is equal to heritability times the selection differential, the response in the "up" direction will differ from that in the "down" direction. If there is dominance, there must still be eventual asymmetry in the response as the function relating h^2 and allele frequency is not itself symmetrical (Fig. 2.3). The overall effect of **genetic asymmetry** will depend on the combined action of all the loci controlling the trait. Asymmetry in response could result from asymmetry in the frequency of a few of genes with large effect or a large number of genes, each with very small effects. A single allele with large effect was shown to be responsible for the asymmetrical response to selection on abdominal bristle number in the experiment of Frankham and Nurthen (1981).

In Chapter 2, the hypothesis was advanced that traits closely related to fitness will have low heritabilities and relatively large dominance variance. From this, we can further predict that asymmetry of response will be common for traits such as fecundity, development time, and so forth that are closely related to fitness. Because selection will have tended to drive those alleles with positive effects on fitness close to fixation we predict that the asymmetry of the response will be manifested as a slow response in the direction of increased fitness [Falconer (1989); for a mathematical description of the process, see Kojima (1961)].

4.5.3 Finite Population Size

(a) *Drift*: In a finite population there will be random genetic drift from generation to generation, which could lead to asymmetry in response. If only two selected lines are used, one "up" and one "down," then it is very difficult to discount the possibility of drift being responsible for different rates of divergence from the base population. The variance in selection response, Var(R), after t generations of selection is approximately (Nicholas, 1980)

$$\text{Var}(R) \approx \left(\frac{V_P}{N}\right)(th^2 + 2p) + 2V_C \qquad (4.29)$$

where N is the number of parents, p is the proportion selected (so number actually measured $= N/p$), and V_C is the common environmental variance. Assuming that V_C is negligible, the variability in the response is proportional to the phenotypic variance and inversely proportional to the population size. Because the response to selection is proportional to $\sqrt{V_P}$, the coefficient of variation in the response to selection is independent of the phenotypic variance and inversely proportional to $\sqrt{N}$. Thus, for population sizes much less than about 100, there is likely to be considerable variation in the observed response to selection, and asymmetry in response could easily arise due solely to drift.

(b) *Inbreeding depression*: The subject of inbreeding depression is dealt with in detail in Chapter 8. It is sufficient to note here that, in a small population, there will be increased breeding between related individuals, possibly leading to a depression of the mean value of the trait. (As discussed in Chapter 8, this will occur only if there is dominance or epistasis.) As a consequence, the trait will appear to respond more quickly in the direction of decreased mean than in the direction of increased mean. (As this effect is a consequence of dominance effects, it is similar to the effect caused by genetic asymmetry.) A control population can be used to correct for this effect.

4.6 Estimating Heritability from a Directional Selection Experiment

4.6.1 Mass Selection on a Continuous Trait

Selection experiments are typically based on truncation selection, some upper or lower proportion of the population being selected as parents for the next generation (Fig. 4.1). Recall that $R = h^2 S$, from which an estimate of heritability can be obtained by rearrangement:

$$h^2 = \frac{R}{S} \qquad (4.30)$$

To obtain an estimate of the standard error of the above estimate of heritability, it is necessary to take into account variance due to drift and variance due to sampling error. Provided that differences due to environmental variation between generations is negligible, a reasonable approximation (Nicholas, 1980) is

$$\text{Var}(h^2) \approx \frac{V_P}{S^2}\left(\frac{h^2}{N} + \frac{2}{M}\right) \qquad (4.31)$$

where V_P is the phenotypic variance, N is the number of individuals selected as parents, and M is the number of individuals measured. Response in a single generation is frequently very variable, and, therefore, several generations of selection are generally required to obtain an accurate estimate of heritability.

If selection is continued over t generations, we have

$$
\begin{aligned}
R_1 &= (X_1 - X_0) = h^2[(Y_0 - X_0)] = h^2 S_1 \\
R_2 &= h^2(S_1 + S_2) \\
R_3 &= h^2(S_1 + S_2 + S_3) \\
R_t &= h^2(S_1 + S_2 + S_3 + \cdots + S_t)
\end{aligned}
\qquad (4.32)
$$

where X_i is the mean value of the trait measured in generation i, Y_i is the mean value of the selected individuals in generation i, R_i is the total or cumulative response (i.e., the difference between the mean value at the start of the experiment and that at generation i), and S_i is the selection differential applied to the parents that give rise to generation i. The sum $S_1 + S_2 + S_3 + \cdots + S_t$ is the cumulative selection differential applied. Note that in total there are $t + 1$ generations. From the above, it can be seen that the cumulative response is a straight-line function of the cumulative selection differential with a slope equal to the heritability of the trait (because the original trait value is a constant, one can replace R_i with X_i). Thus, from a selection experiment, the heritability can be estimated by a linear regression of response on cumulative selection differential (Falconer, 1989; Fig. 4.8). The heritability estimated by this method is typically termed the **realized heritability**. The realized heritability might also be estimated using the regression of individual responses on individual selection differentials or the ratio of the total response to the total selection differential. Of the three methods, the first is generally the best (Hill, 1972a, 1972b). The standard error of heritability estimated from simple linear regression theory is likely to be an underestimate because it does not take into account variance due to genetic drift. The standard error, taking into account drift, is approximately (Hill, 1972b)

$$
\begin{aligned}
\mathrm{Var}(h^2) &\approx \mathrm{Lin}\ \mathrm{Var}(h^2) \\
&+ \frac{2(3t + 4)}{5(t + 1)(t + 2)} \left(\frac{S_{\mathrm{cum}}}{t} \right)^{-2} \left(\frac{h^2(1 - h^2)}{N} + \frac{h^4}{M} \right) \mathrm{Var}(\mathrm{drift})
\end{aligned}
\qquad (4.33)
$$

where Lin $\mathrm{Var}(h^2)$ is the variance estimated from standard linear regression, S_{cum} is the cumulative selection differential (i.e., S_{cum}/t is the mean selection differential per generation), N is the effective number of parents per generation, M is the effective number measured per generation, and $\mathrm{Var}(\mathrm{drift})$ is the variance due to drift. Effective number is discussed in detail in Chapter 8; for the case of a diploid organism in which equal numbers of males and females are measured and

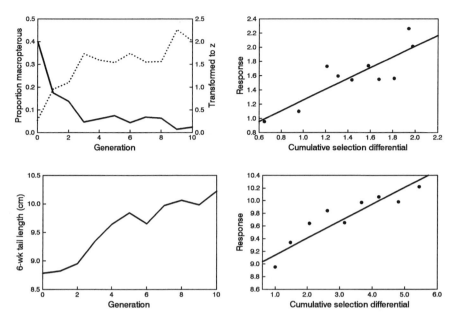

Figure 4.8 Two examples of the estimation of realized heritability. In the bottom panel is shown the result of selection for 6-week tail length in the mouse. The heritability estimated from the regression is 0.27. [Data from Cheung and Parker (1974)]. The upper panel shows the results of selection for decreased incidence of macroptery in the sand cricket (solid line). Because the trait is a threshold trait the proportions are converted to an estimate of the continuous underlying trait value by the method described in the text (dotted line: for visual clarity, the absolute value is plotted). The regression of z on the cumulative selection differential gives a heritability estimate of 0.76. [Data from Roff (1990a).]

selected, $N = \left(\dfrac{1}{4N_m} + \dfrac{1}{4N_f}\right)^{-1}$ and $M = \left(\dfrac{1}{4M_m} + \dfrac{1}{4M_f}\right)^{-1}$, where the subscripts m and f refer to male and female, respectively. The variance due to drift is estimated from

$$\text{Var(drift)} \approx h^2 V_P \left(\frac{1 - h^2}{N} + \frac{h^2}{M}\right) \tag{4.34}$$

where V_P is the phenotypic variance estimated as the within-generation or "error" variance in a one-way analysis of variance of individual measurements within generations with $(t + 1)(M - 1)$ degrees of freedom (remember that there are $t + 1$ generations in the experiment).

In practice, there should be several replicate lines plus one or more control lines to assess effects due to changing environmental conditions during the course of the experiment. First, consider the case of several replicate lines, all selected in the same direction. In this case, the best estimate of the realized heritability is the variance computed from the replicate lines [i.e., compute h^2 separately for each line, and from these, the variance in h^2: Hill (1972a)]. An alternate design is to select one line in an "up" direction and a second line in a "down" direction. Separate heritabilities can be calculated or a single response calculated using the difference between the two lines. Because of the possibility of asymmetrical response to selection, the former or both methods should be used. Details for the computation of the variance when the divergence response is used are given in Hill (1972a). A control line is always essential, as there is the possibility that changes in the selected lines may result from systematic changes in environmental conditions, or environmental fluctuations may increase the variability of the trait under selection. I shall assume that the control line is started from the same base population as the selected line [for a case in which this is not true, see Hill (1972b)]. The data from a control line can be used in two ways: First, the control line can be used simply to show that there are no systematic changes during the course of the selection, and second, the control line can be used to correct for environmental effects during the experiment. Suppose, for example, that in one or more generations, environmental conditions changed, causing an overall change in both the selected and control lines. This effect can be removed by use of the change in the control line (Fig. 4.9) or by using the control line as a covariate (Muir, 1986). Because both lines commence from the same point, the regression of cumulative response on cumulative selection differential is forced through the initial value, giving

$$h^2 = \frac{\sum_{i=1}^{t} X_i S_i}{\sum_{i=1}^{t} S_i^2} \tag{4.35}$$

where X_i is the trait value in the ith generation. The variance is then estimated from

$$\mathrm{Var}(h^2) \approx \frac{6}{t(t+1)(2t+1)} \left(\frac{S_{\mathrm{cum}}}{t}\right)^{-2} \tag{4.36}$$

$$\times \left(\frac{2t^2 + 2t + 1}{5} \mathrm{Var(drift)} + \mathrm{Var(error)}\right)$$

where

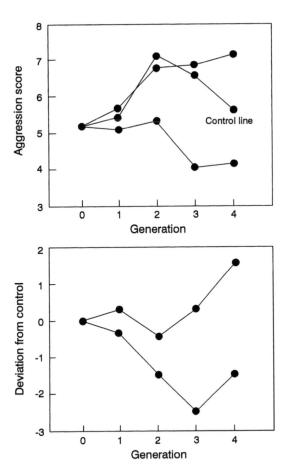

Figure 4.9 Illustration of the use of a control line to correct for environmental effects during a selection experiment. Female mice, *Mus musculus*, were selected for increased and decreased agonistic behavior. In the absence of the control line, it appears that selection during the first three generations was most effective in increasing aggression (upper panel). But after subtraction of the control line score, it is evident that the reduced aggression line actually responded more (lower panel). [Data from Ebert and Hyde (1976).]

$$\text{Var(drift)} = h^2 V_P \left(\frac{1 - h^2}{N} + \frac{1}{L} \right)$$

$$\text{Var(error)} = V_P \left(\frac{1 - h^2}{M} + \frac{h^2}{K} \right)$$

Var(error) is the variance due to measurement error, K is the effective number of individuals measured per generation in the control line (equivalent to M in the selected line), and L is the number of parents used in the control line (equivalent to N in the selected line).

Nicholas (1980) derived the following approximations for the three cases described above:

1. One selected line, no control:

$$\text{Var}(h^2) \approx S_{\text{cum}}^{-2} \left(\frac{V_P(th^2 + 2p)}{N} + 2V_C \right) \tag{4.37}$$

where p is the proportion selected each generation ($p = N/M$) and V_C is the variance due to common environmental effects among generations, which can be estimated from (Hill, 1972b)

$$V_C \approx \text{Lin Var}(h^2) \sum_{i=0}^{t} (S_i - \bar{S})^2 - \frac{(t + 3)\,\text{Var(drift)}}{15} \tag{4.38}$$

2. One selected line, one control line:

$$\text{Var}(h^2) \approx S_{\text{cum}}^{-2} \left(\frac{2V_P(th^2 + p)}{N} \right) \tag{4.39}$$

3. One line selected "up", one line selected "down" (divergent selection): The equation is exactly the same as case 2. This is, however, somewhat misleading because in this case, the selection differential is twice as large, and, therefore, the variance is one-quarter that of a selected line with a control. To examine the effect of varying different parameters, I shall use Eqs. (4.37) and (4.39), assuming that V_C is negligible. Substituting $S_{\text{cum}} = ti\sqrt{V_P}$, where i is the selection intensity, we have when no control is used

$$\text{SE}(h^2) = \left(\frac{th^2 + 2p}{t^2 N i^2} \right)^{1/2} \tag{4.40}$$

and with a control

$$\text{SE}(h^2) = \left(\frac{2(th^2 + p)}{t^2 N i^2} \right)^{1/2} \tag{4.41}$$

where p is the proportion selected each generation ($p = N/M$). As might be expected, the standard error decreases with the number of generations and the population size. Doubling the number of generations approximately halves the standard error, whereas doubling the number of individuals reduces the standard error by a factor of $1/\sqrt{2}$ ($= 0.71$). The effect of increasing the selection intensity is not immediately evident because such an increase both decreases p and $N(N = pM)$. The combined effect of these changes is to produce an approximately U-shaped relationship between the standard error and p (Fig. 4.10). Using a slightly different approximation, Soller and Genizi (1967) found that the optimum proportion is in the range $0.15-0.20$. However, if the number of loci or alleles is small, such an intensity runs the risk of fixing alleles and thus reducing the response to selection. As can be seen from Fig. 4.10, there is little difference in the range $0.1-0.5$, and, therefore, to reduce the possibility of fixation a proportion between 0.3 and 0.5 is preferable. A higher selection intensity is necessary if the number of generations is very low (e.g., two to three).

4.6.2 Mass Selection on a Threshold Trait

Mass selection on a dimorphic trait consists of selecting at each generation only one morph to be parents of the next generation. If the heritability of the trait is greater than zero, the proportion selected each generation will necessarily increase and, hence, the selection intensity decrease. Let the two morphs be designated A and B, and let B be the morph lying above the threshold and the one selected (as these designations are arbitrary there is no loss of generality). As shown earlier, the proportion, p, lying above the threshold, T, is given by

$$p = \frac{1}{\sigma\sqrt{2\pi}} \int_T^\infty \exp\left[-0.5 \left(\frac{x - \mu}{\sigma} \right)^2 \right] dx \qquad \textbf{(4.42)}$$

where μ and σ are the phenotypic mean and standard deviation of the underlying trait. Without loss of generality, we can rescale the above by setting $\sigma = 1$, and $T = 0$, in which case the mean value of the underlying trait can be estimated using a table of proportions of the normal curve or the approximation equation (Hamaker, 1978)

$$\mu = \text{Sign}(p - 0.5)\,[1.238c(1 + 0.0262c)] \qquad \textbf{(4.43)}$$

where

$$c = \sqrt{-\ln[4p(1 - p)]}.$$

Thus, for example, if $p = 0.5$, the estimated value of μ is 0.0, whereas if $p = 0.8$, the estimated value of μ is -0.84. The estimated value of the parents, X, is

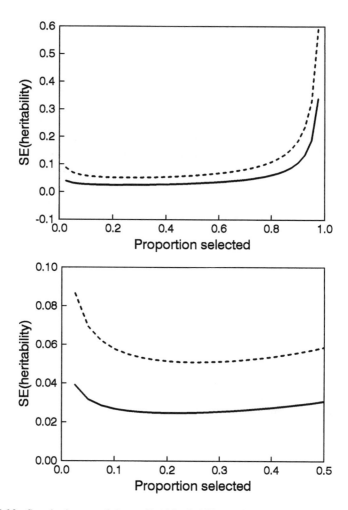

Figure 4.10 Standard error of the realized heritability estimate as a function of the proportion of the population selected after 10 generations. The number measured is kept constant at $M = 100$. Heritability values set at 0.1 (dashed lines) and 0.5 (solid line). Lower panel shows a "blowup" of the predicted standard error in the range of proportion selected from 0.0 to 0.5.

$$X = \mu + \frac{\exp(-0.5\mu^2)}{p\sqrt{2\pi}} = \mu + \frac{\phi(\mu)}{p} \qquad \textbf{(4.44)}$$

The heritability can now be estimated using the regression of cumulative response on cumulative selection differential [Eq. (4.32)]. There is no formula for the estimation of the standard error taking into account drift; however, because the number per generation should be large (>100), the effect of drift can be assumed to be small and the standard error from the simple linear regression used.

It is possible that in some generations, only a single morph will be found. In this case, it is not possible to estimate the mean of the underlying trait. The problem can be solved by using a maximum likelihood approach (Roff, 1990a). First, we note that

$$\mu_{t+1} = \mu_t(1 - h^2) + h^2 X_t \tag{4.45}$$

where t is the generation. The recursive equations are thus

$$\mu_1 = \mu_0(1 - h^2) + h^2 \left(\mu_0 + \frac{\phi(\mu_0)}{p_0}\right)$$

$$\mu_2 = \mu_1(1 - h^2) + h^2 \left(\mu_1 + \frac{\phi(\mu_1)}{p_1}\right) \tag{4.46}$$

$$\mu_{t+1} = \mu_t(1 - h^2) + h^2 \left(\mu_t + \frac{\phi(\mu_t)}{p_t}\right)$$

The likelihood ($=$ probability) of obtaining the observed number, m_i, of the selected morph from a sample of n_i individuals is given by the binomial formula

$$L = \frac{n_i!}{m_i!(n_i - m_i)!} p_i^{m_i} (1 - p_i)^{n_i - m_i} \tag{4.47}$$

The likelihood for t generations is thus

$$L_t = \prod_{i=1}^{t} \frac{n_i!}{m_i!(n_i - m_i)!} p_i^{m_i}(1 - p_i)^{n_i - m_i} \tag{4.48}$$

Taking logarithms, we get

$$LL_t = \ln\left(\sum \frac{n_i!}{m_i!(n_i - m_i)!}\right) + \sum m_i \ln(p_i) + \sum (n_i - m_i) \ln(1 - p_i) \tag{4.49}$$

The problem is thus to find the value of h^2 that, using the recursive equations given above, maximizes LL_t (note that the first term on the right is common to all estimates and, hence, can be deleted for the purposes of finding the maximum), which will be denoted as LL_{Max}. In addition to h^2, we also have to estimate the probabilities for each generation. There are thus $t + 1$ parameters to be estimated. Given that the number of individuals per generation is large, the estimates of p_i

can be taken to be the simple proportion, $m_i/n_i =$ (number of morph B)/(morph A + morph B). To obtain the 95% confidence limits, we require (Dobson, 1983)

$$LL_{Max} - LL_t = 0.5 \; \chi^2_{(0.05, \; df = t + 1)} \quad (4.50)$$

For example, with 24 generations of selection, the critical value of χ^2 is 38, and the two values of h^2 are sought that produce a log-likelihood (LL_t) that differs from the maximum likelihood value (LL_{Max}) by an amount $0.5 \times 38 = 19$. The confidence intervals may not be symmetrical, but the difference is likely to be sufficiently small that a standard error can be reported by dividing the interval into four parts. The two methods of estimating heritability were used in the estimation of the heritability of wing dimorphism in the cricket *Gryllus firmus*: Both methods gave approximately the same result, but the linear regression method could not be used in two cases because only one morph appeared in several generations (Roff, 1990a).

4.6.3 Family and Within-Family Selection

Mass selection (also sometimes called individual selection) is the most commonly used method and is preferred, in general, both because of its simplicity and its well-understood statistical behavior. Two other methods should, however, be mentioned: (1) **family selection** and (2) **within-family selection**. In the former method, entire families are selected based on the mean value of the family (usually either full-sib or half-sib), whereas in the latter, the most deviant individuals in each family are selected. For examples of family selection, see Clayton et al. (1957: a abdominal bristle number in fruit fly), Martin and Bell (1960: adult weight in fruit fly), Moav and Wohlfarth (1976: growth rate in common carp), Kincaid et al. (1977: growth rate in trout), Bondari (1983: body weight in channel catfish), and Enfield et al. (1983: postirradiation survival in the cotton boll weevil). Within-family selection has been used to select for growth rate in mice (Falconer, 1953), agonistic behavior in mice (Ebert and Hyde, 1976), body weight in guppies and mosquito fish (Ryman, 1973; Busack, 1983), and wing length in the milkweed bug (Palmer and Dingle, 1986).

Letting the response to selection be R, R_f, and R_w for mass, family, and within-family selection, respectively, it can be shown that (Falconer, 1989, pp. 233–239)

$$\frac{R}{R_f} = \frac{\sqrt{n[1 + (n - 1)t]}}{1 + (n - 1)r} \quad (4.51a)$$

$$\frac{R}{R_w} = \left(\frac{1}{1 - r}\right) \sqrt{\frac{n(1 - t)}{n - 1}} \quad (4.51b)$$

where n is family size, r is the coefficient of relationship ($= 0.5$ for full-sibs, 0.25 for half-sibs), and t is the intraclass correlation. Considering full-sibs, family selection gives a higher response for a given selection intensity than mass selection when the intraclass correlation is less than approximately 0.2, whereas within-family selection produces a greater response when t is greater than approximately 0.8 (Fig. 4.11). Family selection is thus useful when heritability is low and within-family selection is the best choice when there is a large effect of common environment. Because both methods require the rearing of separate families, family and within-family selection typically require much greater space than mass selection.

4.7 Empirical Findings on Response to Artificial Selection

4.7.1 Short-Term Response

From the theory outlined in Section 4.4 there should be little change in the additive genetic variance over the first few generations of selection. Therefore, realized heritabilities should be approximately the same as those obtained from other methods (Chapter 2). A plot of realized heritabilities on estimates from other techniques is presented in Fig. 4.12. With the exception of the heritabilities for the chicken, the heritabilities were typically estimated over $5-15$ generations of selection (the time span over which the heritabilities of the traits in the chicken were estimated was not presented in the compilation of Kinney (1969), from which the data were drawn). Although there are more estimates lying below the line, there is certainly no overwhelming trend for the realized heritability to be markedly less than that estimated from sib analysis; thus, this result supports the above prediction.

Although there is reasonable agreement, in general, between realized heritability estimates with those from sib analysis, the prediction that response should be symmetrical is frequently found not to be upheld. Some examples of asymmetrical responses are given in Table 4.2. Given the number of possible reasons why asymmetry of response may occur (see Section 4.5), it is not particularly surprising to find so many exceptions. Typically, explanations are advanced for the asymmetrical response but not verified with further experiment; thus, which of the various causes, if any, is most frequently the reason for the asymmetry cannot be ascertained. Falconer (1989, p. 215) suggested that

> If the character selected is a component of natural fitness, asymmetry should be
> expected, with selection towards increased fitness giving a slower response than
> selection towards decreased fitness. The reasons are, first that these characters usu-
> ally show inbreeding depression, which is itself a cause of asymmetry, and, second,
> that if the character has been subject to natural selection the gene frequencies are
> likely to be above the symmetrical point, i.e. nearer the upper limit, thus giving
> rise to genetic asymmetry.

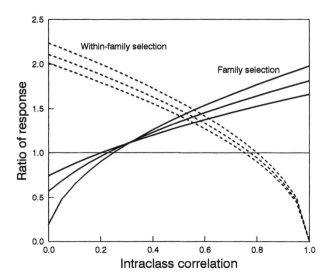

Figure 4.11 Comparison of mass selection to within-family selection and family selection. When the ratio of response exceeds 1, mass selection produces a higher response than the alternate mode of selection. Curves show response for family sizes of 5, 10, and 100 (reading top to bottom at $t = 0.0$).

As will be discussed in Chapter 8, inbreeding depression is caused by dominance, which from the theory and evidence presented in Chapter 2 is expected to be greater in traits, such as life history characters, that are closely connected to fitness. The presence of dominance causes heritability to be a skewed function of allele frequencies (see Fig. 2.3) and, hence, selection in different directions will produce an asymmetric response in heritability.

The above argument makes two predictions: first, that artificial selection on traits closely related to fitness should show asymmetry such that the response in the direction of increased fitness will be less than in the direction of decreased fitness, and second, the asymmetry should be greater in life history traits than in traits, such as morphological characters, that are less closely related to fitness. Frankham (1990) analyzed data from 30 studies of bidirectional selection on traits which are probably related to fitness; these included both life history traits (fecundity, development time, age to maturity) and behavioral traits (mating speed, mating competence, feeding rate). A significantly higher proportion of studies (80%) showed asymmetry of response toward lower fitness, verifying the first prediction. To address the second prediction, I computed the absolute proportional asymmetry in realized heritabilities $[=|h_U^2 - h_D^2|/(h_U^2 + h_D^2)$, where U and D refer to up and down lines, respectively] for the data in Table 4.2. Contrary to prediction, the mean value for the life history traits is smaller than that for mor-

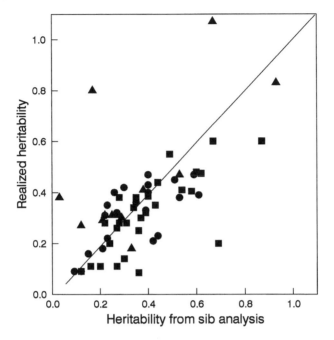

Figure 4.12 A comparison of the realized heritability with the heritability estimated from pedigree analysis. Triangles, chicken; dots, invertebrates; squares, vertebrates (other than the chicken). Invertebrate data sources: *Drosophila*—Tantawy (1956a), Clayton et al. (1957a), Latter (1964), Sen and Robertson (1964), Frahm and Kojima (1966), Frankham et al. (1968), López-Fanjul and Hill (1973), van Dijken and Scharloo (1979), Sorenson and Hill (1982, 1983), Gallego and López-Fanjul (1983), Gromko (1987), Gromko et al. (1991); *Tribolium*—Enfield et al. (1966), Martin and Bell (1960), Yamada and Bell (1969), Ruano et al. (1975); *Oncopeltus fasciatus*—Palmer and Dingle (1986); *Gryllus firmus*— Roff (1990a); *Wyeomyia smithii*—Hard et al. (1993), Bradshaw et al. (1997). Vertebrate data sources: Mouse—Eisen et al. (1970), Wilson et al. (1971), Rutledge et al. (1973), Cheung and Parker (1974), Lynch (1994); chicken—Kinney (1969); pigs and Japanese quail—Sheridan (1988).

phological traits (means = 0.40 and 0.48 for life history and morphology, respectively). As these data do not include experiments showing symmetrical response (in which case the absolute proportional asymmetry would be zero) and these are expected to be more frequent for morphological than life history traits, the means may be strongly biased.

4.7.2 Long-Term Selection

A survey of experiments in which a plateau in response has been observed indicates that a failure to respond as predicted from the initial heritability estimate

Table 4.2 Heritability Estimates from Experiments in Which Asymmetrical Responses to Selection Have Been Reported

Species	Trait	h^2 Estimates Increase	h^2 Estimates Decrease	Reference
		Morphological Traits		
Tribolium castaneun	Pupal weight	0.17	0.33	Meyer and Enfield (1975)
	Pupal weight	0.40	0.26	Gall (1971)
Cochliomyia hominivorax	Pupal weight	0.60	0.15	McInnis et al. (1983)
Oncopeltus fasciatus	Wing length	0.53	0.31	Palmer and Dingle (1986)
Gryllus firmus	% Macroptery	0.25	0.95	Roff (1990a)
Drosophila melanogaster	Wing length	0.08	0.38	Tantawy (1956a)
	Abdominal bristles	0.53	0.37	Clayton et al. (1957a)
	Body size[a]	0.14	0.20	Baptist and Robertson (1976)
	Adult weight	0.12	0.23	Martin and Bell (1960)
	Sternopleural bristles	0.30	0.17	Barker and Cummins (1969)
Drosophila subobscura	Wing length	0.31	0.44	Prevosti (1967)
Chicken	Body weight	0.34	0.52	Festing and Nordskog (1967)
		Life History Traits		
Tribolium castaneun	Development time	0.26	0.38	Englert and Bell (1970)
	Development time	0.32	0.22	Soliman (1982)
	Development time	0.11	0.32	Dawson (1965)
Drosophila subobscura	Development time	0.19	0.06	Clarke et al. (1961)
Drosophila pseudoobscura	Development time	0	>0	Marien (1958)
Drosophila melanogaster	Fecundity	0.01	0.50	Richardson et al. (1968)
	Development time	0.70	0.17	Hunter (1959)
	Development time	0.25	0.00	Sang and Clayton (1957)
	Development time	0.20	0.19	Moriwaki and Fuyama (1963)
	Age at maturity	0.17	0.08	Hudak and Gromko (1989)
	Survival at 40°C	0.00	0.10	Morrison and Milkman (1978)
Gambusia affinis	Age at maturity	0.67	0.15	Campton and Gall (1988)
Cyprinus carpio	Growth rate	0.00	0.30	Moav and Wohlfarth (1976)

(continued)

Table 4.2 Continued

Species	Trait	h^2 Estimates Increase	h^2 Estimates Decrease	Reference
		Life History Traits (continued)		
Coturnix coturnix	Egg production	0.06	0.11	Lambio (1981)
Chicken	Hatching time	0.19	0.25	Smith and Bohren (1974)
Mus musculus	Growth rate	0.26	0.42	Falconer (1960)
	Growth rate (normal diet)	0.33	0.23	Nielsen and Anderson (1987)
	Growth rate (reduced protein diet)	0.41	0.12	Nielsen and Anderson (1987)
	6-wk body weight[b]	0.20	0.50	Falconer (1953)
	5-wk body weight	0.39	0.24	McCarthy and Doolittle (1977)
	10-wk body weight	0.32	0.35	McCarthy and Doolittle (1977)
	Litter size	0.08	0.25	Falconer (1971)
	Litter size	0.22	0.26	Joakimsen and Baker (1977)
	Age at maturity	0.45	0.52	Drickamer (1981)
		Behavioral Traits		
Drosophila melanogaster	Walking behavior	0.07	0.26	Choo (1975)
	Larval feeding rate	0.11	0.21	Sewell et al. (1975)
	Anemotaxis	0.13	0.01	Johnston (1982)
Drosophila persimilis	Geotaxis	0.07	0.05	Polivanov (1975)
	Phototaxis	0.06	0.07	Polivanov (1975)
Drosophila pseudoobscura	Mating speed	0.19	0.06	Kessler (1969)
	Mating speed	0.01	0.02	Spuhler et al. (1978)
Drosophila mercatorum	Pulse rate	0.14	0.32	Ikeda and Maruo (1982)
Coturnix coturnix	Mating ability[c]	0.00	0.22	Siegal (1980)
Chicken	Mating ability[c]	0.18	0.31	Siegal (1965)
Mus musculus	Attack latency	0.00[d]	0.30	Oortmerssen and Baker (1981)
	Maternal aggression	0.13	0.40	Hyde and Sawyer (1980)

[a]Response symmetrical but because of different phenotypic variances heritability estimates different.

[b]Weight at an age below adult is taken as a growth rate and, hence, is a life history trait.

[c]Included as a fitness-related trait in the comparison between morphological and life history traits.

[d]All four lines selected failed due to reproductive failure. Authors believe this failure to be directly related to selection for low aggressiveness.

can become evident within about 20 generations (Table 4.3). There are some notable exceptions in which continued response has occurred for many generations; for example, (1) 76 generations of selection for oil content in maize (Dudley, 1977), (2) 75 generations of selection for pupal weight in *Tribolium* (Enfield, 1980), (3) 50 generations of selection for abdominal bristle number in *Drosophila melanogaster* (Jones et al., 1968), (4) 75 generations of selection for abdominal bristle number in *D. melanogaster* [(Yoo, 1980a); not all lines continued to respond to selection], (5) 60 generations of selection for ethanol resistance in *D. melanogaster* (Weber and Diggins, 1990), and (6) 55 generations of selection for wing-tip height in *D. melanogaster* (Weber, 1990).

Population sizes used in selection experiments are typically quite small (Table 4.3); hence, the decline in heritability might result from the joint effect of selection and drift as predicted by the model of Robertson (see Section 4.4). This is illustrated in Fig. 4.13 where the ratio of the initial heritability to the heritability at generation t is plotted against t. For population sizes greater than 50, there is only a modest decline in heritability by generation 30, and, given that the number of parents used in the selection experiments is probably an overestimate of the effective population size, it is likely that some of the observed reduction in heritability results from drift. To test this hypothesis, I computed the expected ratio of heritabilities under the Robertsonian model and compared these with the observed ratios. Because the effective population size is probably overestimated by the number of parents, I estimated predicted values using both the number of parents and one-half this number; the effect on the predicted values is typically small (Fig. 4.14). Although some values fall close to the 1:1 line, the majority of observed ratios are substantially less than predicted. These results qualitatively support those obtained via simulation by Hospital and Chevalet (1993) (Fig. 4.6).

Attempts to match long-term responses with theoretical predictions have been largely unsuccessful; for example, selection for abdominal bristles in *D. melanogaster* have produced long-term responses, but "the long-term behavior of these lines is bewilderingly complex" (Clayton and Robertson, 1957, p. 166), "in general, agreement with these models was poor" (Jones et al., 1968, p. 265), and "the pattern of long-term response was diverse and unpredictable" (Yoo, 1980a, p. 1). Possible reasons for the failure are the presence of lethal genes (Clayton and Robertson, 1957; Yoo, 1980b), infertility of extreme females and heterozygosity (Clayton and Robertson, 1957), presence of a few genes of large effect (Jones et al, 1968), and an overestimation of the effective population size (Frankham, 1977b).

Weber and Diggins (1990) compared the predicted and observed relative response after 50 generations of selection using both their own data and those obtained from the literature (Fig. 4.5). For population sizes less than 10, there is fair agreement with prediction, but, thereafter, the observed values fall below, frequently substantially, the predicted value. Nevertheless, there is the same overall sigmoidal increase in response that is predicted by theory.

Table 4.3   Review of Selection Experiments in Which a Plateau in Selection Response Has Been Observed; Realized Heritabilities Calculated over Two Time Periods (T1, T2)

Species	Trait	N^a	Generations T1	Generations T2	h^2 Estimates T1	h^2 Estimates T2	Ref.[e]
Tribolium castaneum	Increased pupal wt.	16^b	1–10	1–23	0.40	0.26	1
	Decreased pupal wt.	16^b	1–10	0–25	0.26	0.23	1
	Fecundity	4–40	1–16[a]	16–32	0.26	0.15	2
Drosophila melanogaster	Wing length	6	0–10	10–20	0.30	0.18	3
	Development time	700	0–8	9–14	0.20	0.00	4
	Mating speed	20	0–7	8–25	0.30	≈0.00	5
	Copulation duration	80	0–3	0–10	0.23	0.11	6
	Geotaxis	60	0–5	0–10	0.13	0.10	7
	Walking	40	0–10	10–15	0.17	≈0.00	8
Drosophila pseudoobscura	Phototaxis	50	1–10	11–20	0.10	0.05	9
	Geotaxis	50	1–10	11–20	0.07	0.02	9
	Body wt.	30	1	28	0.24	0.04	10
	Mating speed	20	0–5	5–24	0.13	≈0.00	11
Drosophila subobscura	Light preference (6500 lux)	V^c	0–9	0–19	0.04	0.02	12
Mouse	Growth rate	F^d	0–4	4–13	0.39	0.17	13
	Body weight	108^b	0–10	11–20	0.32	0.19	14
	Litter weight	F^d	0–10	11–22	0.20	≈0.00	15
	Litter size index	40^a	1–6	19–33	0.39	0.17	16

[a]Number of males and females. Where these are not equal, approximate effective population size estimated using Eq. (8.2).

[b]Effective population size.

[c]Number of parents variable and not specified precisely.

[d]Family selection used.

[e]References:

1—Gall (1971); 2—Ruano et al. (1975); 3—Robertson and Reeve (1952); 4—Sang (1962); 5—Manning (1961), see also Manning (1963); 6—Gromko et al. (1991; data averaged over four lines); 7—Watanabe and Anderson (1976); 8—Choo (1975); 9—Dobzhansky et al. (1969); 10—Frahm and Kojima (1966); 11—Kessler (1969); 12—Kekic and Marinkovic (1974); 13—Falconer (1960), see also Roberts (1966a, 1966b) and Baker et al. (1984); 14—Wilson et al. (1971); 15—Eisen (1972); 16—Schuler (1985).

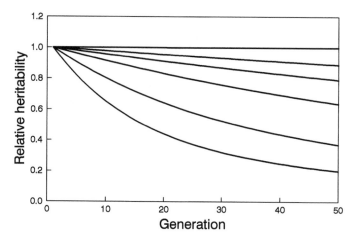

Figure 4.13 Predicted ratio of $h^2(t)/h^2(0)$ versus generation, t, for population sizes (from bottom to top) of 5, 10, 25, 50, 100, and 1000. Heritability computed using Eq. (4.35), with $\mu = 0$, and the relationship $R(t) = h^2 t S$, where S is the per generation selection differential. The predicted ratio is then given by

$$\frac{h^2(t)}{h^2(1)} = \frac{1 - C^t}{t(1 - C)}$$

where $C = 1 - 1/(2N)$.

In reviewing their theoretical analysis (Fig. 4.5), Webber and Diggins (1990, p. 593) note that, "Populations that have actually been selected to the point of near-exhaustion of response (e.g., Reeve and Robertson, 1953; Roberts, 1966a, b; Enfield, 1980; Yoo, 1980a) indicate that Robertson's upper limit of $2NR_1$ cannot be attained except in very small populations. For populations of 10 it is realistic; for populations of 50 it is not remotely possible."

A significant challenge to the theory of the long term response is that reversed or relaxed selection frequently demonstrates the presence of additive genetic variance even when little or no further progress is being made in the original direction of selection (Reeve and Robertson, 1953; F. W. Robertson, 1955; Dawson, 1965; Clayton and Robertson, 1957; Rathie and Barker, 1968; Wilson et al., 1971; Eisen, 1972; Lerner and Dempster, 1951; Dickerson, 1955; Kaufman et al., 1977; Yoo et al., 1980). Such a result could occur because (1) artificial selection in one direction is opposed by natural selection, (2) the alleles favoring the change in the direction of selection are dominant, inbreeding then causing a depression of the response (see Chapter 8), (3) overdominance [the last considered unlikely by Falconer (1989)]. The question of the maintenance of genetic variation in the face of selection is discussed in more detail in Chapter 9. The brief review given here highlights the simplicity of the present theoretical foundation. Therefore, appli-

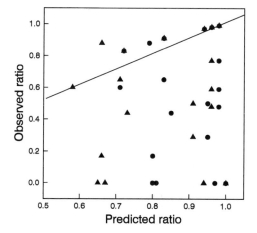

Figure 4.14 Observed versus predicted ratio of heritabilities presented in Table 4.3. Dots show predictions based on the reported population size; triangles give the predictions assuming the effective population size is one-half of that reported. Predicted values were computed using

$$\frac{h^2(t_1)}{h^2(t_2)} = \frac{t_1(1 - C^{t_2})}{t_2(1 - C^{t_1})}$$

where $C = 1 - 1/(2N)$, t_1 is the first time period (generation 1 to t_1), and t_2 is the second time period (generation 1 to t_2). Because the heritability for the second time period is estimated by averaging the response over all generations, the above ratio will underestimate those values given in Table 4.3 in which the heritability for the second time period was estimated over only the latter portion of the selection period. Ruano et al. (1975) used five selection intensities; all five are presented in the above figure.

cation of such a theory to natural populations in which one or all of the complicating factors are likely to be found must be done with considerable caution, or skepticism.

4.8   Predicting Responses in Nature

From an extensive review of estimates of directional selection in natural populations (Fig. 4.15), Endler (1986, p. 210) concluded that the range of i values "extensively overlaps the values found in animal and artificial selection experiments. For example, the i values found in Falconer (1981 [= 1989]) and in the papers cited by Robertson (1980) range from 0.15 to 1.39, with a geometric mean of 0.71. For comparison, the geometric mean of significant i in [Fig. 4.15] is 0.59. This suggests that natural selection is as often as strong as artificial selection." This observation is of extreme importance because it means that long-term pre-

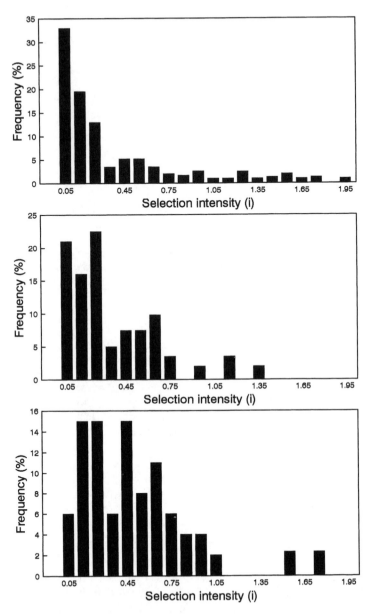

Figure 4.15 The distribution of directional selection intensities in natural populations. The top panel shows data for species in undisturbed populations (262 estimates from 25 species), the middle panel shows data from perturbed populations, field cages, or stressful environments (62 estimates from 5 species), and the bottom panel shows fossil and sub-fossil data (46 estimates from 5 species). [Redrawn from Endler (1986).]

dictions for natural populations cannot, without supporting evidence, make the assumption that the heritability value remains constant. In the rest of this section, I present three examples in which quantitative genetics has made an important contribution to the understanding of the evolution of the trait, at least in the short term.

4.8.1   Evolutionary Changes in K_P in the Lepidopteran Hyphantria cunea

The lepidopteran *Hyphantria cunea* inhabits forests throughout North America. In Canada, it is univoltine, the moths emerging in June and July, the larvae maturing in late summer and fall, and the pupae diapausing until the following spring. Diapause is broken after the pupae have been chilled and subjected to temperatures above 10.6°C. The total number of degree days [= (Actual temperature − 10.6) × Number of days] accumulated from the time of chilling to adult eclosion is defined as K_P (Morris and Fulton, 1970). There is considerable variation in K_P, both across years and among different geographical sites, and mean offspring on mid-parent regression indicates that this variation is largely additive genetic in origin (Fig. 4.16), the heritability of the trait being 0.60 [SE = 0.22 (Morris, 1971)]. In cold years, the progeny from late-emerging (high K_P) moths fail to attain the pupal stage by fall and hence die, whereas in warm years, the progeny of early-emerging (low K_P) moths suffer high mortality because they pupate early and use up their fat reserves before the onset of chilling temperatures. Because temperature fluctuates from year to year, there is directional selection each year, but its sign fluctuates.

From extensive laboratory studies, Morris (1971) constructed a simulation model predicting the change in K_P based on three equations.

Equation 1: Mean offspring on mid-parent regression

$$\ln Y = 2.456 + 0.60 \ln X_t \tag{4.52a}$$

where Y is the mean offspring value and X_t is the mean parental value in year t.

Equation 2: Proportion reaching the pupal stage

$$\text{Probit(survival)} = P(S) = 5 + (146 - 23 \ln X_t)[\ln(C - X_t) - 6.9] \tag{4.52b}$$

where C is the observed cumulative heat units for the particular year.

Equation 3: Mean K_P of the survivors

$$X_{t+1} = Y - 1030 + \exp\left(\frac{P(S/2) - 5}{146 - 23 \ln Y} + 7\right) \tag{4.52c}$$

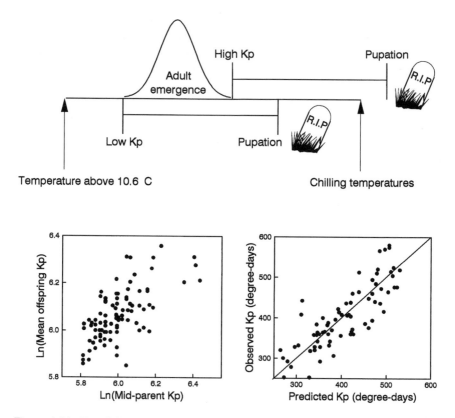

Figure 4.16 *Top*: Schematic illustration of selection on *Hyphantria cunea*. Moths that eclose too early (low K_P) deplete their fat reserves before chilling temperatures induce diapause. Moths that emerge too late (high K_P) fail to reach the pupal stage before chilling temperatures. Both types are elimated from the population. Changes in the average annual temperature favor shifts in the mean K_P. *Bottom left*: Regression of mean offsprig K_P on the mid-parent K_P. *Bottom right*: Observed K_P on value predicted from the simulation model incorporating the genetic relationship illustrated on the left. [Model and analysis from Morris (1971).]

The approach taken by Morris is not a strictly quantitative genetic approach but derives from a regression perspective. Note, in particular, the use of the offspring–parent regression as if it were fixed. Correctly, the above model should be cast in the framework of selection acting on a threshold trait. Morris used the foregoing model to predict changes in K_P for eight sites between the years 1958 and 1968, the prediction for each year being based on the predicted value from the previous year (omnibus prediction), not the observed value (stepwise prediction). This method of analysis is the most rigorous. After adjustment for a hypothesized effect due to temperature range, the match between prediction and observation was excellent (Fig. 4.16). Although the model can be faulted for its failure to incorporate an appropriate quantitative genetic model, it is possibly the first attempt to take quantitative genetics into the real world and, for that, it deserves recognition. Unfortunately, insufficient data are given in the various publications to rework the analysis using the appropriate assumptions.

4.8.2 *Morphological Change in Darwin's Finch,* Geospiza fortis

The Galapagos finches have been the subject of a long-term research program seeking to measure the forces of selection acting on morphological traits (Grant and Grant, 1989). Monitoring of climate and floristic conditions has shown that changes in rainfall are accompanied by changes in the type of seeds available, from large, hard seeds to small, soft seeds. These changes affect the survival of the Galapagos finches because the ability of an individual to handle each type of seed depends on the characteristics of a bird's bill. On the island of Daphne Major, a significant shift from large, hard seeds to small, soft seeds occurred following the El Niño event of 1982–1983. Grant and Grant (1993) measured the change in morphology of G. fortis before and after this climatic event. The results of their measurements along with predicted and observed response to selection for bill length and bill width are shown in Table 4.4. The selection differential shows that there was selection for a reduction in both measures but that on bill length was very small. The observed and predicted responses agree very well, there being no significant response in bill length but a highly significant decrease in bill width. Grant and Grant (1993, 1995) actually analyzed their data using a multivariate approach; this analysis is presented in Chapter 5, where direct and correlated responses are discussed.

4.8.3 *Can Microtine Cycles Be Explained by Genetic Changes?*

This example is not one in which quantitative genetics was used to predict a phenomenon in a natural population but rather to predict that a proposed hypothesis for a particular phenomenon is unlikely to be correct. Population fluctuations that follow regular cycles have been observed in many microtine species (Krebs and Myers, 1974). A search for the cause of such cycles has occupied an inordinate number of ecologists for an inordinately long time (Chitty, 1996). Having disposed of the simple possibilities such as predation, starvation, or disease, Chitty (1967) put forward a genetic–behavioral hypothesis. He proposed that during the bottom part of the cycle, interactions between individuals are low and selection favors nonaggressive, highly fecund genotypes. But during the peak phase of the cycle, the course of selection is reversed and aggressive genotypes are favored. He supposed that there was a negative correlation between reproductive potential and aggressiveness. There are two components required to test the validity of this hypothesis: First, it is necessary to determine the magnitude of the heritability

Table 4.4   Observed and Predicted Changes in Bill Traits in the Galapagos Finch, Geospiza fortis

Trait	h^2	S	Predicted R	Observed R
Bill length	0.65	−0.03	−0.02	0.06 (.05<P<.1)
Bill width	0.90	−0.17	−0.15	−0.12 (P<.001)

Note: Measurements are in millimeters and probabilities in parentheses show the result of t-tests on the hypothesis observed $R = 0$.

Source: Data are from Grant and Grant (1993).

required to produce the types of changes observed in the time frame observed, and second, to determine if the heritabilities in microtines are sufficiently large. Demonstration that the heritabilities are as high as required does not prove that the hypothesis is correct, only that it is plausible. However, demonstration that observed heritabilities do not correspond with the minimum requirements is sufficient to eliminate the hypothesis as a candidate for further consideration.

Anderson (1975) measured the heritabilities of several traits in *Microtus townsendii* using outside enclosures and also determined the required heritabilities for Chitty's hypothesis to work using simulation modeling. As there were clearly large maternal effects, sire estimates from a half-sib design were used. Juvenile growth rate and age at puberty had very low heritabilities [0.032 [SE = 0.019] and 0.073 [0.093], respectively]. Repeatabilities for nine behavioral traits ranged from −0.001 to 0.47, with a mean of 0.16, indicating that heritabilities for behavioral characters was probably also very low (see the Glossary for a brief discussion on the use of repeatabilities). Anderson's simulations (1975, p. 191) showed that "when h^2 is less than 0.8, population cycles cease, and the system exhibits unstable behavior."

Boonstra and Boag (1987) repeated Anderson's genetic analysis, using another vole, *M. pennsylvanicus*. Their results were essentially the same: insignificant heritabilities for a wide range of morphological and reproductive traits (Table 4.5). To further test the Chitty hypothesis, they estimated the heritabilities required to change adult weight and growth rate by the amount observed in the peak and trough portions of the *M. pennsylvanicus* population cycle within the time span of a cycle. This analysis showed that it is highly unlikely that the observed changes are being brought about by a heritable response to selection. Thus, the Chitty hypothesis can be consigned to the same bin as previously suggested hypotheses (and the search continues). More importantly, the two analyses of Anderson (1975) and Boonstra and Boag (1987) demonstrate that it is not necessary to always obtain highly accurate estimates of genetic parameters to test a quantitative genetic hypothesis.

Table 4.5 *Heritabilities Required to Produce the Observed Response in* Microtus
pennsylvanicus *Under the Chitty Hypothesis in Two, Four, or Six Generations Using
Different Selection Intensities; Heritabilities Greater Than 1 Are Not Possible*

Trait	Intensity of Selection (%)	Heritability Required for the Change to Occur in		
		2 Generations	4 Generations	6 Generations
Adult body size	0.3 (38)	>1	>1	>1
	0.6 (27)	>1	0.82	0.55
	1.2 (11)	0.82	0.41	0.27
Growth rate	0.3 (38)	>1	0.93	0.62
	0.6 (27)	0.93	0.46	0.31
	1.2 (11)	0.46	0.23	0.16

Trait	Observed heritabilities[a]:
Maximum weight	0.00 (0.34), 0.00 (0.32), 0.37 (0.31), 0.04 (0.39)
Growth rate	0.27 (0.30), 0.33 (0.38), 0.11 (0.16), 0.54 (0.39)

[a]Estimates (SE) for 1982 females, 1982 males, 1983 females, and 1983 males.

Source: Data from Boonstra and Boag (1987).

4.8.4 The Evolution of a Novel Migratory Route in the Blackcap

The blackcap, *Sylvia atricapilla*, is a widely distributed warbler, ranging across much of Europe. Most populations migrate to overwintering grounds in Africa, although a few show either partial migration or no migration (Fig. 4.17). Until the 1950s, the species rarely overwintered in Britain (Stafford, 1956), but the number of recorded overwintering individuals has risen dramatically since the 1970s [1978–1979, 2000 recorded; 1981–1984, 3000 estimated to be overwintering (Berthold, 1995)]. These birds are not birds that have taken up residence in Britain but are birds from Germany and Austria that are migrating in an entirely novel direction and for a shorter distance than is typical of the southwestern flying migrants from these countries (Langslow, 1979; Berthold, 1995). Hand-rearing of birds has shown that migratory orientation is not learned (Helbig, 1992) and, hence, the change in migration pattern represents a remarkably fast evolutionary change.

The advantages of overwintering in Britain are (Berthold, 1995) (1) lower intraspecific competition, (2) shorter migration distance, (3) earlier gonadal development and return migration (a physiological response due to differences in photoperiod; because we can assume that this response is under genetic control, this cannot be assumed to adaptive unless conditions in Europe now favor such a change), (4) earlier occupation of territories in spring, and (5) physiological acclimatization due to potentially harsh conditions initially experienced on the

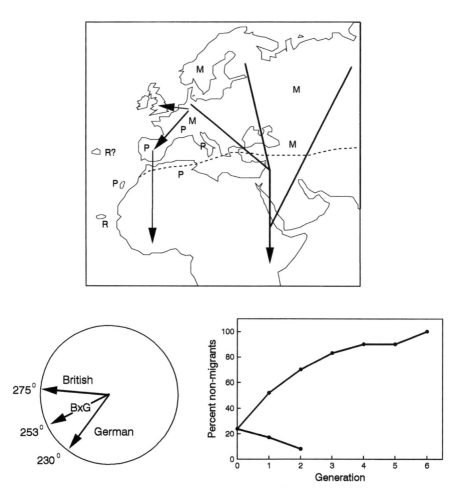

Figure 4.17 Top: The migration pattern of the blackcap (*Sylvia atricapilla*). The thick arrows indicate the main migration routes, the thin arrows the by-routes to west Africa and northern wintering grounds. Letters indicate populations that are migratory (M), partially migratory (P), or nonmigratory (R = resident). The dotted line indicates the southern border of the continental breeding grounds. [Redrawn from Berthold (1988).] *Bottom left*: Migratory orientation of offspring from "British" (adults collected in Britain during the winter), German, and hybrids (B×G). [Data from Berthold et al. (1992) and Berthold (1995).] *Bottom right*: Response to selection for reduced migratory activity. [Data from Berthold (1988).]

breeding grounds. These advantages were presumably offset by the costs before 1960; the apparent reason why blackcaps can now overwinter successfully in Britain is climatic amelioration and the presence of a large number of bird feeders, on which the blackcaps rely almost entirely (Berthold, 1995). Droughts in the Sahel zone may have also selected against birds overwintering in Africa (Sutherland, 1988).

The rapid evolution of the new migratory pattern requires high additive genetic variance for migratory orientation and migration distance. The offspring of birds collected on the overwintering grounds in Britain orient westward, the offspring of birds collected in Germany orient southwestward, and the offspring of crosses between "British" and German birds orient in a direction almost midway between these two directions (Fig. 4.17; although the "British" birds probably originate in Germany or regions thereabouts, the population of birds that migrate to Britain is very small relative to that in which birds follow the traditional route). These data suggest a high heritability for orientation, but direct estimation through pedigree analysis is clearly required. The evidence for high additive genetic variation for migration distance is very strong. First, comparison among *Sylvia* species indicates that migratory restlessness as measured in a laboratory assay coincides with migratory distance (Berthold, 1973). The presence of varying degrees of migration among blackcap populations indicates that migratory propensity can evolve, and this is supported by intraspecific crosses between partially migrant and fully migrant populations (Berthold et al., 1992). More definitive evidence comes from selection for migration propensity (Fig. 4.17) and offspring–parent regression which gives a heritability of 0.45 (Fig. 2.6). A similar heritability (0.52) was obtained for the European robin (Biebach, 1983). With such high heritabilities, migratory activity can be halved in approximately 10 generations with an 80% selection rate (Berthold and Pulido, 1994). The experimental data are, thus, in accord with the rapid evolutionary response: Actually predicting the time course of the change will prove very difficult, given the number of selective factors that must be measured (i.e., relative success of the different migration patterns). Nevertheless, the work of Berthold and his colleagues has demonstrated that the rapid evolutionary change is explicable on the basis of quantitative genetic variation.

4.9 Summary

From the mean offspring on mid-parent regression it is a simple matter to predict the response to selection. The response to a single generation of selection requires no assumptions of genetics, but prediction beyond this can only be justified using the principles of Mendelian genetics. In order to compare different traits, Houle (1992) suggested the use of the coefficient of variation, which he termed evolvability. Unfortunately, this measure is not scale independent and, hence, can only be used to compare traits that have a common scale. It cannot be used to compare disparate traits such as fecundity and body size. Unless population size is very small, simple theory predicts that truncation selection will have little effect on the heritability of a trait. However, if the population size and number of loci are finite (e.g., population size <100, number of loci <20), there may be considerable erosion of additive genetic variance, although this is greatly reduced in the case of threshold traits. Again, although simple theory predicts that the response to selection does not depend on the direction of selection, there are numerous reasons (grouped under measurement artifacts, genetic asymmetry, and finite population size) why asymmetry of response is to be frequently expected. Heritability may be estimated from the response to selection using the regression of response on the cumulative selection differential. Realized heritabilities estimated from the first 10 generations of selection are typically reasonable approximations of the heritability estimated from a breeding design (half-sib, offspring-parent, etc.). The response to selection commonly declines from that predicted from the previous heritability by about generation 20. As yet, there is no satisfactory theory to quantitatively predict the course of long-term selection. Quantitative genetic analysis has been successfully applied in at least four cases to understand the process of short-term evolution in a wild population.

5

Directional Selection and the Correlated Response

Animal and plant breeders are frequently concerned with response in a single character only. However, in natural populations, selection will operate on that suite of traits which determines overall fitness. Further, even when there is direct selection on only one trait, there may be significant effects on other traits because of a genetic or phenotypic correlation between the selected and unselected traits. In this chapter, I review the theory of correlated responses to selection on one or several traits and assess the extent to which experimental evidence validates the theory.

5.1 Derivation of the Correlated Response to Selection

Consider the offspring–parent relationships for two traits X and Y, in which X is under directional selection (Fig. 5.1). For simplicity, and without loss of generality, population means are set at zero. From standard linear regression theory, the predicted value of Y (i.e., the correlated response of Y to selection on X, CR_Y) in the next generation is

$$CR_Y = \frac{Cov_{XY}}{V_X} S_X \qquad (5.1)$$

where Cov_{XY} is the covariance between the mid-parent value of trait X and the mean offspring value of trait Y, V_X is the variance of the mid-parent values, and S_X is the selection differential applied to trait X. Recalling that $h^2 = Cov_{XX}/V_X$, where Cov_{XX} refers to the covariance between mean offspring and mid-parent values, we can rewrite Eq. (5.1) as

$$CR_Y = \frac{Cov_{XY}}{V_X} S_X = \frac{Cov_{XY}}{V_X} S_X h_X h_Y \sqrt{\frac{V_X V_Y}{Cov_{XX} Cov_{YY}}} \qquad (5.2)$$

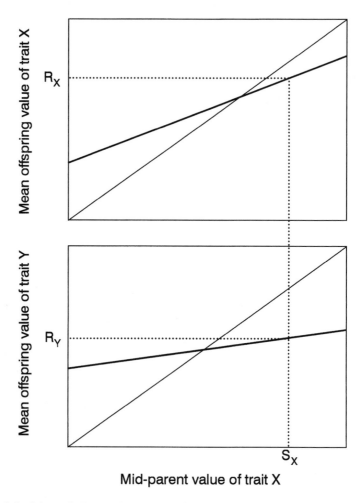

Figure 5.1 Schematic illustration of the offspring–parent relationship for two traits (*X*, *Y*), and the predicted change in these two traits as a result of selection on trait *X*. For simplicity, and without loss of generality, the population means have been set to zero.

Note that the product of the additional terms is unity. Rearranging the above, we have

$$CR_Y = \sqrt{\frac{Cov_{XY}}{Cov_{XX}Cov_{YY}}}\, h_X h_Y \sqrt{\frac{V_Y}{V_X}}\, S_X \tag{5.3}$$

The first term is simply the genetic correlation between *X* and *Y* and, thus,

$$\mathrm{CR}_Y = r_A h_X h_Y \sqrt{\frac{V_Y}{V_X}} S_X = i r_A h_X h_Y \sqrt{V_Y} \tag{5.4}$$

As can be seen from the above, the correlated response to selection depends on the heritabilities of the two traits, their phenotypic variances, and the genetic correlation between them. It is important to note that the above derivation is based on the assumption that there is no covariance between the genetic and environmental effects. An interesting feature of Eq. (5.4) is that it allows the correlated response to exceed the direct response. This may be most clearly seen by considering the ratio of correlated response to direct response:

$$\frac{\mathrm{CR}_Y}{R_Y} = \frac{i_X h_X h_Y r_A \sqrt{V_Y}}{i_Y h_Y^2 \sqrt{V_Y}} = \frac{i_X h_X}{i_Y h_Y} r_A \tag{5.5}$$

where i_X is the selection intensity on trait X and i_Y is the selection intensity applied directly to trait Y. Assuming the same selection intensities, the correlated response will be greater than that achieved by direct selection when the product of the genetic correlation and the heritability of trait X is greater than the heritability of trait Y.

If trait X is selected and the responses of both X and Y are measured, then the genetic correlation can be estimated by rearrangement of Eq. (5.4):

$$r_A = \frac{\mathrm{CR}_Y}{i h_X h_Y \sqrt{V_Y}} \tag{5.6}$$

If separate experiments are done, one in which X is selected and the other in which Y is selected, an estimate of the genetic correlation can be obtained as

$$r_A = \frac{\mathrm{CR}_Y}{R_X} \sqrt{\frac{V_{AX}}{V_{AY}}} = \frac{\mathrm{CR}_X}{R_Y} \sqrt{\frac{V_{AY}}{V_{AX}}} = \frac{\mathrm{CR}_Y}{R_Y} \frac{\mathrm{CR}_X}{R_X} \tag{5.7}$$

which does not require the separate estimation of heritabilities. So far as I know, the statistical properties of the above two methods of estimating the genetic correlation have not been investigated.

5.2 Correlated Response with Selection on One Trait

5.2.1 Theory

For the infinitesimal model with an infinite population size, continued selection on a trait has little effect on the heritability of the trait (Section 4.3). Thus, we would expect that in this case there would also be little effect on the genetic correlation. At equilibrium, the heritability of the correlated trait (Y) and the genetic correlation between X and Y are (Villanueva and Kennedy, 1990)

$$r_A(\infty) = \frac{r_A(0)}{\{1 + h_X^2(\infty)k[1 - r_A^2(0)]\}^{1/2}} \tag{5.8a}$$

$$h_Y^2(\infty) = \frac{h_Y^2(0)}{1 + h_X^2(\infty)r_A^2(\infty)k[1 - h_Y^2(0)]} \tag{5.8b}$$

where $k = i(i - z)$, i being the intensity of selection and z the standardized deviation of the point of truncation from the population mean for the trait under direct selection (X). Neither the heritability of Y nor the genetic correlation between X and Y change by more than a few percent, regardless of the intensity of selection or the initial genetical parameters.

The above result does not hold if the genetic correlation is determined by a few genes of large effect. This is shown clearly in the model described in Chapter 3 and illustrated in Fig. 3.2: As the allele frequency shifts, there is a significant change in r_A. A somewhat more complex model was analyzed by Bohren et al. (1966). They assumed four loci which contributed differently to the two traits (Table 5.1), two loci showing no pleiotropic effects (locus 1 and locus 2), one locus having a positive pleiotropic effect (locus 4), and one locus showing antagonistic pleiotropy (locus 3). Heritability of both traits was set initially at 0.5. Depending on the initial allelic frequencies, a variety of responses to selection on trait 1 or trait 2 were observed, but in all cases, there was a significant change in the genetic correlation (Fig. 5.2). Contrary to the prediction of the infinitesimal model, but in accord with the single locus approach, asymmetrical responses were easily obtained. Villanueva and Kennedy (1992) have shown that such asymmetrical responses can also result from linkage disequilibrium.

With respect to the relative change in heritabilities and genetic correlations, Bohren et al. (1966, p. 55) argued as follows:

> The additive genetic variance of any character will be made up of contributions from the separate loci. These contributions will change as the gene frequencies are altered by selection or by random drift and they will not all change in the same way, depending on the gene frequencies at the loci concerned. But the genetic covariance (if the genetic correlation is not close to 1) will either be made up of a much smaller number of terms, if all loci contribute to the covariance with the same sign, or will be made up of positive and negative contributions from different loci. In either case the *proportional* (authors' italics) change in the genetic covariance is likely to be greater than in the genetic variances themselves. *It must be therefore expected that the static description of a population in terms of additive genetic variances and covariances will be valid in prediction over a much shorter period for correlated responses than it will be for direct responses* (my italics).

Given that directional selection appears to changes heritabilities over a time period

Table 5.1 Description of Models Employed by Bohren et al. (1966)

	Independent Effects		Pleiotropic Effects	
	Locus 1	Locus 2	Locus 3	Locus 4
Trait 1	a	0	c_1	b_1
Trait 2	0	d	$-c_2$	b_2
		Values used in Fig 5.2		
Model 1	0.5	0.5	0.5	0.5
Model 2	0.5	0.5	0.2	0.5
Model 3	0.5	0.5	0.2	0.2

Notes: A general description of the contribution by the individual loci to the two trait values is shown in the first two rows. The initial allelic frequencies for the three models illustrated in Fig. 5.2 are shown in the lower three rows.

of 10–20 generations (Table 4.3), the foregoing conclusion suggests that the statistical approach to evolutionary quantitative genetics is at best a very crude approximation, and at worst, highly misleading if selection intensities are not extremely weak.

The above conclusion is reinforced by the simulation results of Parker et al. (1969, 1970a, 1970b). Two models were analyzed: the additive model and the complete dominance model. In both cases, there were 48 diallelic loci per trait, the alleles taking the values 0 or 1, no linkage, no epistasis, no genotype $\times$ environment interaction, and an initial allele frequency of 0.5, obtained by setting all individuals heterozygous. A genetic correlation was created between the two traits by having n loci in common; the genetic correlation is then simply $n/48$. Three levels of genetic correlation were examined: 0.25, 0.5, and 0.75 (i.e., $n =$ 12, 24, and 36, respectively). For the additive model, the phenotypic value of a trait was equal to the sum of allelic values plus an environmental variance set such as to produce heritabilities of 0.1, 0.4, or 0.7. The dominance model assumed that the contribution from each locus was 2, 2, or 0 (= values for $++$, $+-$, or $-$, respectively). The environmental variance was set as for the additive model (i.e., all individuals in the initial population heterozygous), but because of random mating producing homozygous individuals in later generations, with a consequent change in additive genetic variance, the actual heritabilities were 0.095, 0.33, and 0.52. Each generation consisted of 24 pairs of parents, the number of offspring produced being determined from the specified level of selection intensity, 20%, 50%, or 80% (60, 96, or 240 offspring). Pairs were chosen at random with replacement and each pair produced one offspring; thus, although the number of parents was somewhat limited, the method of pairing produced a very large number of combinations. Truncation selection was practiced on one trait only: in one direction in the case of the additive model but in both directions for the dominance

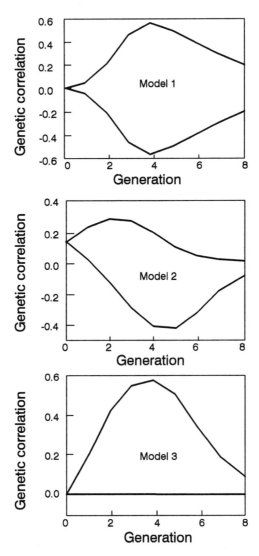

Figure 5.2 Changes in the genetic correlation as a result of selection on one trait that is genetically correlated with another trait. The upper curves show the results of selecting on trait 1, lower curves the result of selecting on trait 2. Parameter values for the three models are given in Table 5.1. The asymmetry (or, in the case of model 1, symmetry) is reflected in the genetic correlations, the genetic covariances, and the responses to selection. [Redrawn from Bohren et al. (1966).]

model in which differences are expected due to different changes in allelic frequencies (see Fig. 2.3). Each selection experiment was run for 30 generations, with two replicates per parameter combination.

Linearity of direct response to selection was observed only for the weakest selection intensity (80%); in all other cases, fixation of some alleles caused a slowing of the response (Parker et al., 1970b). As expected from the single-locus modeling approach, there was asymmetry of response in the two directions for the dominance model. When selection intensity was low (80%) or heritability small (0.1), there was little change in the genetic correlation (Fig. 5.3). At the highest selection intensity (20%) and moderate to high heritabilities (0.4 and 0.7), there was a drastic decline in the genetic correlation, the effect being different (as expected) for the two different directions of selection in the dominance model (Fig. 5.3). The data shown in Fig. 5.3 were calculated by averaging across the three values of r_A; in general, there was a trend for the relative change to be inversely related to r_A. It would be interesting to know the relative contributions of fixed versus unfixed loci to the change in r_A; this could easily have been calculated during the simulations but cannot be estimated from the published data. Nevertheless, these results do demonstrate that if there is a marked plateauing of a trait under selection, then we may also expect significant changes in genetic correlations.

Recently, Slatkin and Frank (1990) have considered further the change in the genetic correlation under directional selection—in their case, incorporating mutation. They assumed the following: (1) an infinite population size; (2) a single haploid locus; (3) phenotypic values of traits X and Y to be given by $c_X x$ and $c_Y y$, where the c's are constants and the lowercase x and y denote the contribution of the allele x or y (x and y take values of $0, \pm 1, \pm 2, \ldots$); (4) mutation can change an allele into only one of the eight adjacent allelic classes; (5) mutations come in two types: a proportion v of nonpleiotropic mutations which affect only one trait and $1 - v$ pleiotropic mutations which affect both traits. To assess the importance of directional selection on the genetic correlation, they first ran the model to its equilibrium value under the joint effects of stabilizing selection and mutation, and then continued the model for 20 generations of directional selection, the mutation rate (10^{-4}) being much less than the intensity of selection. When $c = 1$, there is a dramatic increase in r_A, but the effect is considerably reduced for $c = 0.1$, and in both cases, the increase is greatest when the proportion of pleiotropic mutations, v, is greatest (Fig. 5.4). Overall, these results reinforce the caution suggested by the analyses of Bohren et al. (1966) and Parker et al. (1969, 1970a, 1970b).

5.2.2 Experimental Findings

Given the theoretical reasons not to expect that r_A will remain constant, we must enquire from the outset, "What do we expect of the genetic correlation estimate?"

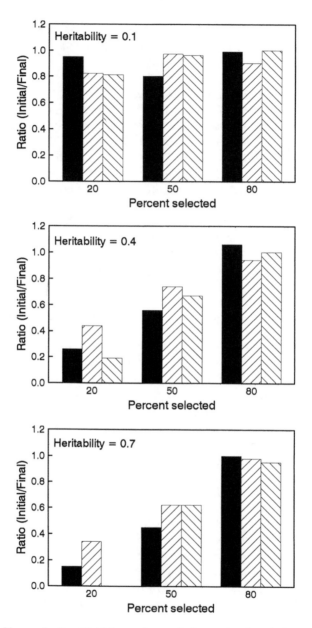

Figure 5.3 Observed ratio of initial genetic correlation to that found after 30 generations of selection in the simulation models of Parker et al. (1969, 1970b). The results for the three initial genetic correlations of 0.25, 0.5, and 0.75 have been averaged. Solid bars show results for the additive model; hatched bars for the complete dominance model (because of asymmetry in response to up or down selection, the data for the dominance model have been plotted separately).

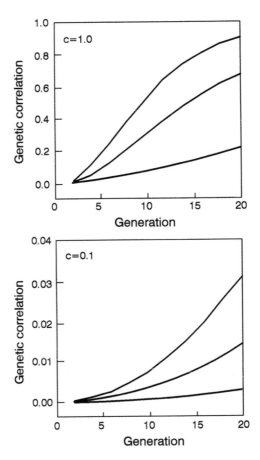

Figure 5.4 Changes in the genetic correlation resulting from directional selection in the simulation model of Slatkin and Frank (1990). For details, see text. From top to bottom within each graph, the curves show predictions for $v = 0.0$, 0.5, and 0.9, respectively.

If we wish estimates that will give very accurate predictions of response to selection for, say, 20 generations, then we are likely to be disappointed. From the empirical observation of a frequent decline in heritability after about 10 generations, we cannot expect the genetic correlation to perform any better and, for the theoretical reasons presented above, it will likely perform worse. At the other extreme, we may simply wish that the direction of the correlated response be correctly predicted by the sign of the estimated genetic correlation. If, however, the sign of the genetic correlation or the direction of response can change in as little as 10 generations, then the estimate has very little utility. I suggest that, from an evolutionary perspective, the important properties of the genetic correlation are to (1) indicate the presence or absence of pleiotropy (positive or negative) and (2) indicate the strength of the pleiotropy between traits—that is, a genetic correlation of 0.9 really does mean a stronger genetical relationship between two traits than a genetic correlation of 0.1. It is unrealistic to expect more of the parameter—it should serve as a statistical tool to suggest experiments into the mechanistic basis of correlations between traits. The validity of theory of the genetical correlation has been assessed by three methods:

1. *Comparison of the genetic correlation in the base population with that after selection (either realized or by further sib analysis).* Results from this method are summarized in Table 5.2. There is excellent agreement for wing and thorax length in *D. melanogaster* over a wide range of generations in the experiments of Reeve and Robertson (1953) and Tantawy and his colleagues. Tantawy and El-Helw (1966) found no change over nine generations of selection in the wing–thorax correlation, the wing–longevity correlation, or the thorax–longevity correlation (Table 5.3). After 23 generations of selection on thorax length, Wilkinson et al. (1990, p. 1990) measured the covariance relationships between thorax length and wing length, and three other morphological traits, and found "significant changes in covariance structure among the selected lines." It is instructive to consider this experiment in more detail. In addition to the selected lines, a series of control lines was also maintained. Initial estimates of r_A were obtained from a "base" population which consisted of a large randomly mating stock population from which the experimental lines were originally derived. In deciding whether significant changes had occurred in the selected lines, Wilkinson et al. compared the selected lines with the base population, not the controls. As it is quite evident from the work of Wilkinson et al. that the genetic correlations in the control lines differed from the base population, the comparison between the selected and base populations cannot distinguish among the effects of the intentional selection from other factors that might have caused deviation. The appropriate comparisons are between the control and the base population, and between the controls and the appropriate selected lines. Graphical displays of these comparisons are shown in Fig. 5.5. It is evident that genetic correlations involving bristle number are very small in the base population, but in both the control and selected populations, they increased in some lines but not others. The genetic correlations between the other metric traits (thorax length, wing length, wing width, and tibia length) are considerably higher and there is as much variation between the base and control populations as between the control and selected populations. For the metric traits, there is an indication that populations selected for large size are less deviant from the controls than the lines selected for small size, a point also noted by Wilkinson et al. Overall, it appears that changes not due to selection were at least as important as changes due to selection. It is unfortunate that genetic correlations were not estimated at an earlier stage, say, generations 5 and 10. Given that there were, not unexpectedly, significant changes in heritability over the 23 generations, it is not surprising that there were changes in some genetic correlations.

Table 5.2 Survey of Experiments Comparing Genetic Correlations Estimated from Pedigree Analysis with the Results from Selection Experiments

Species	Traits (Selected, Correlated)	G^a	r_A Base	r_A Sel	Reference
D. melanogaster	Wing, thorax	50	0.74	0.71	Reeve and Robertson (1953)
D. melanogaster	Wing, thorax	15	0.81	0.80	Tantawy (1956b)
D. melanogaster	Wing, thorax	5	0.80	0.81	Tantawy and Tayel (1970)
D. melanogaster	Wing, thorax	9	0.79	0.95	Tantawy and El-Helw (1966)
D. melanogaster	Thorax, 4 other morphological traits	23	See text		Wilkinson et al. (1990)
D. melanogaster	Body wt, egg size	15	0.13	0.01	Martin and Bell (1960)
D. melanogaster	Body wt, fecundity	15	0.14	0.26	Martin and Bell (1960)
D. melanogaster	Body wt, adult emergence	15	0.24	−0.10	Martin and Bell (1960)
D. pseudo-obscura	Weight at 2 densities	2	0.51	0.57	Frahm and Kojima (1966)
Mouse	Tail, weight	14	0.50	0.41	Cheung and Parker (1974)
Mouse	Tail, weight	7	0.29	0.34	Rutledge et al. (1973)
Mouse	Litter wt, 8-wk wt	22	0.44	0.46	Eisen (1972)
Mouse	Litter wt, 6-wk wt	22	0.49	0.81	Eisen (1972)
Mouse	Litter wt, number born	22	0.19	0.40	Eisen (1972)

[a]Number of generations selection run.

An apparently poor correspondence was obtained by Martin and Bell (1960) for the genetic correlation of body weight in *D. melanogaster* with egg size, fecundity, and adult emergence (Table 5.4). Martin and Bell (1960, p. 187) concluded that, "Genetic correlations of the unselected traits with body weight were highly transitory both before and during selection and were of little value for predicting association with the selected trait or changes in the means of unselected traits." This statement is correct, but Martin and Bell's work cannot, as has been done, be used as evidence against the utility of the genetic correlation. The genetic correlations estimated from sib analysis are based on two different experiments which gave very different estimates of the genetic correlation (Table 5.4: the means of these do not correspond to the sib values published by Martin and Bell and I presume that their estimates are based on some weighted average, but no details are given). Given the variability in the initial estimates, one would conclude that something was decidedly amiss in the base population, and on these grounds,

Table 5.3   Genetic Correlation Estimates Between Traits in Drosophila melanogaster Before and During Truncation Selection on Wing Length

Generation	Wing–Thorax		Wing–Longevity		Thorax–Longevity	
	Up	Down	Up	Down	Up	Down
Base	0.79 (0.08)[a]		0.43 (0.11)		0.45 (0.09)	
1	0.94	0.93	0.41	0.40	0.38	0.41
3	0.89	0.95	0.43	0.41	0.40	0.38
6	0.95	0.94	0.43	0.41	0.43	0.40
9	0.94	0.96	0.48	0.41	0.44	0.48

[a]Standard errors in parentheses. Standard errors for genetic correlations during selection are of an equivalent magnitude but for clarity of presentation are not shown.

Source: Data from Tantawy and El-Helw (1966).

prediction of a correlated response is not reasonable (for example, in two cases the pairs of genetic correlations are large but opposite in sign!).

No change in the genetic correlation was observed following two generations of selection for body weight under two larval densities in Drosophila pseudo-obscura (Frahm and Kojima, 1966) (Table 5.2). Continued selection produced erratic correlated responses, but, unfortunately, genetic correlations in later generations were not estimated.

Cheung and Parker (1974) selected either for tail length or body weight in mice and measured the genetic correlation over generations $1 - t$, where t ranged from 2 to 14 generations. In neither type of selection was there a significant change in realized r_A over the 14 generations (for selection on body weight, $r = -0.46$, $P > .05$; for selection on tail length, $r = 0.51$, $P > .05$), and the mean value (0.41) is in reasonable agreement with the estimates from the base populations (0.50, Table 5.2). Rutledge et al. (1973) also obtained good correspondence after seven generations of selection.

Eisen (1972) selected on 12-day litter weight in mice and measured the genetic correlations with three other traits after 22 generations of selection. In one case, r_A remained constant, whereas in the other two cases, the realized genetic correlations were twice as large as the initial estimates (Table 5.2). Unfortunately, Eisen did not estimate the genetic correlations using only the first 10 generations, which might have been expected to give better correspondence.

2. Comparison of the two r_A estimates from divergent selection. Genetic correlations estimated from three generations of selection for either increased weight or increased tail length in mice were 0.62 and 0.57, respectively (Falconer, 1953), providing support for the theory. However, Falconer cautions that the confidence limits are large but does not provide them. In another experiment on mice, Falconer (1960) selected for growth rate on two different diets (H, L). Good agreement was observed during the first four generations ($r_A = 0.67$ and 0.65 for the

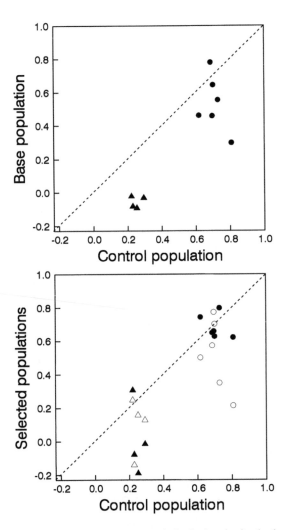

Figure 5.5 Genetic correlations between morphological traits in the base, control, and selected lines of *D. melanogaster* from the study of Wilkinson et al. (1990). Lines selected for large (closed symbols) or small (open symbols) thorax length. Triangles: Correlations between bristle number and another morphological trait. Dots: Correlations between other morphological traits (wing length, wing width, tibia length, thorax length).

Table 5.4   Genetic Correlations Between Body Weight and Other Traits Estimated from Sib Analysis and 15 Generations of Selection in Drosophila melanogaster

Trait	Heritability	Sib r_A	Realized r_A	Individual values of r_A ($\bar{x}$)
Body weight	0.21	—	—	—
Egg size	0.58	0.13	0.01	−0.47, 0.50 (0.02)
Fecundity	0.07	0.14	0.26	−0.63, 0.75 (0.06)
Adult emergence	0.51	0.24	−0.10	0.16, 0.71 (0.44)

Source: Data from Martin and Bell (1960).

H and L experiments, respectively), but continued selection caused significant divergence in the estimates (generations 4–13, r_A = 1.25 and −0.02 for H and L, respectively).

Realized genetic correlations from three types of selection experiments (selection on pupation time, larval weight, and pupal weight) in *Tribolium casteneum* are presented in Table 5.5. Among lines within experiments there is inconsistency in the sign of the genetic correlation in only one case. The standard error estimated from the individual estimates is remarkably small. However, there are significant differences between experiments in two of the three correlations; whether this reflects differences in the founding population or changes due to selection is not clear. Of the two significantly different comparisons, that involving the genetic correlation between pupation time and pupal weight is the most striking, the magnitude of the correlation remaining more or less the same but the sign changing. In reviewing these results, Englert and Bell (1969, p. 905) noted that, "While observed changes among the various components can be described in terms of selection differentials, heritabilities and genetic correlations, it is time to seek a more basic understanding of the underlying physiological processes through which the genetic systems are operating." Almost 30 years later, such a program of research is still wanting.

3. *Comparison of observed and predicted correlated responses.*

(a) Abdominal bristles and sternopleural bristles in *Drosophila melanogaster* (Clayton et al., 1957b). In the base population, the estimated r_A was approximately 0.05–0.10, based largely on the argument that the phenotypic correlations were better approximations than the genetic correlation estimates themselves (the phenotypic correlation estimates ranged from −0.04 to 0.11, whereas the genetic correlation estimates ranged from −0.58 to 0.24, the pooled genetic correlation estimate being 0.08). After five generations of selection, the lines showed a correlated response that was "consistent with a very low genetic correlation" (Clayton et al., 1957b, p. 172). Selection over 20 generations produced

Table 5.5 Estimates of Realized Genetic Correlations Between the Three Traits, 13-Day Larval Weight (D), *Pupation Time* (P), *and Pupal Weight* (W) *in Populations of* Tribolium casteneum *Subjected to Different Selection Histories*

Trait Selected	G^a	$D \times P$	$D \times W$	$P \times W$
Early pupation	6	− 1.96	—	− 0.14
Early pupation	6	− 0.76	—	− 0.06
Early pupation	6	− 0.95	—	− 0.13
Late pupation	6	− 0.95	—	− 0.12
Late pupation	6	− 1.29	—	− 0.30
Late pupation	6	− 1.46	—	− 0.27
Average (SE)		**− 1.23** **(0.18)**	—	**− 0.17** **(0.04)**
Increased 13-day larval wt (good diet)	13	− 0.47	0.58	—
Increased 13-day larval wt (poor diet)	13	− 0.13	0.43	—
Increased 13-day larval wt (poor diet)	13	− 0.32	0.38	—
Decreased 13-day larval wt (good diet)	13	− 0.27	0.66	—
Decreased 13-day larval wt (poor diet)	13	− 0.80	7.55[b]	—
Decreased 13-day larval wt (poor diet)	13	− 0.47	1.54	—
Average (SE)		**− 0.41** **(0.09)**	**0.72** **(0.21)**	—
Large pupal wt	42	—	0.24	0.28
Large pupal wt	62	—	0.19	0.23
Large pupal wt	18	—	0.46	0.63
Small pupal wt (wet environment)	21	—	0.45	0.59
Small pupal wt (dry environment)	21	—	0.68	0.17
Small pupal wt (alternating envir.)	8	—	0.52	− 0.23
Average (SE)		—	**0.42** **(0.07)**	**0.28** **(0.13)**
Probability from ANOVA		**0.002**	**0.190**	**0.007**

[a]Number of generations of selection.

[b]Not used in calculation of mean.

Source: Data from Englert and Bell (1969).

different results, the high lines showing correlated responses but the low lines showing no predictable changes.

(b) Postweaning growth rate and litter size in mice (Rahnefeld et al., 1966). Sib analysis gave a genetic correlation of 0.89, but the observed response after 30 generations was only 64% of that predicted. However, as noted by the authors, the standard error on the covariance estimate cannot preclude the observed response (the estimated covariance is 0.20 ± 0.22).

(c) Wing length and two other traits in *Oncopeltus fasciatus* (Palmer and Dingle, 1986; Dingle et al., 1988). Selection for increased and decreased wing length was undertaken in two populations of the milkweed bug, *O. fasciatus*. These populations differ in their migratory behavior, those from Puerto Rico being nonmigratory and those from Iowa showing a significant propensity to fly for long periods (Dingle et al., 1988; Dingle and Evans, 1987). Sib analysis of these two populations revealed a positive genetic correlation between wing length and head width in both populations, nonsignificant correlations with age at first reproduction, and possibly disparate correlations with fecundity (Table 5.6). The last comparison is uncertain from the sib estimates because the standard errors are very large. Correlated responses to selection on wing length were measured but the sib estimates not used to predict the magnitude of the response; therefore, results are only qualitative. Nevertheless, the observed responses are reasonably concordant with the estimates; in those cases where no correlated responses were obtained, the sib r_A's are not significantly different from zero (Table 5.6).

(d) Copulation duration and two other traits in *Drosophila melanogaster* (Gromko et al., 1991). From sib analysis, genetic correlations between copulation duration and courtship vigor, and copulation duration and fertility were estimated as -0.41 and 0.27, respectively. A total of eight lines were selected: four for increased copulation duration and four for decreased. Correlated responses were measured at generations 8 and 10 for one series of replicates, and at generation 10 only for the other series; a total of 6 comparisons are, therefore, possible. For vigor, a response in the predicted direction was found in two comparisons, no significant difference between lines in three, and, in one, the response was in the opposite direction. For fertility, the numbers are 2, 4, and 0, respectively. Thus, the results are very inconsistent. It is notable that the heritability of copulation duration decreased by one-half during the 10 generations of selection. This suggests that changes in the genetic correlations may also have occurred. If pleiotropic effects resulting from a few major loci are involved in the control of these three traits, erratic correlated responses are quite likely (see Section 5.2.1).

Table 5.6 Genetic Correlations Between Traits in Two Populations of Milkweed Bug, Oncopeltus fasciatus *and the Observed Correlated Response to Selection on Wing Length*

Trait	Genetic Correlations (SE)		Correlated Response[a]	
	Puerto Rico	Iowa	Puerto Rico	Iowa
Head width	0.68 (0.10)	0.50 (0.09)	As predicted	As predicted
Age at 1st reproduction	−0.15 (0.19)	0.06 (0.11)	No significant response	No significant response
Fecundity	−0.25 (0.46)	0.24 (0.10)	No significant response	As predicted

[a]"As predicted," trait increased in lines selected for long wing length and decreased in lines selected for short wing length.

Source: Data from Palmer and Dingle (1986) and Dingle et al. (1988).

The original paper by Gromko (1987) describing the estimation of the genetic correlations raises a troubling question concerning the estimated standard errors. In his paper, Gromko estimates the heritabilities and genetic correlations between seven courtship traits. Of the 21 possible genetic correlations, only four had standard errors smaller than the estimate itself, and of these four, only two (those monitored in the selection experiment) had standard errors less than one-half of the estimate (i.e., estimate $-2SE > 0$). The question arises as to whether these two correlations can be considered significant on this basis or should be judged on the significance level after a Bonferroni correction. The significance of these correlations is also suspect because neither vigor nor fertility have significant heritabilities (0.09 and 0.15, respectively). In the light of these two problems, the hypothesis that the genetic correlations may, in fact, be close to zero seems tenable. A second estimation should have been undertaken immediately prior to the selection experiments to confirm that the results from the first experiments were not chance events.

(e) Wing muscle histolysis and wing dimorphism in *Gryllus firmus* (Roff, 1994b). The genetic correlation between these two threshold traits was estimated from full-sib analysis and the response of wing muscle histolysis predicted after 14 generations of selection for increased and decreased proportion macropterous (=long-winged, flight-capable morph). The correlation between predicted and observed response was 0.98 ($n = 6$, $P < .0001$), with no significant deviation from the 1:1 line. The same data set was also used to predict changes in fecundity, also with considerable accuracy (<5% error).

From the above survey, I conclude that the empirical evidence is not in accord with theoretical expectations from the infinitesimal model but are in accord with predictions when finite numbers of loci, asymmetric gene frequencies, and so forth (Section 5.2.1) are considered. As with heritability, predictions seem to be reasonably accurate for about 10 generations but are likely to deviate from the observed trajectories beyond this point. With the exception of the *Tribolium* data (Table 5.5), there are no qualitatively surprising results such as a correlated response in the opposite direction to that predicted by an accurate estimate of the genetic correlation. The importance of the genetic correlation is that it indicates the presence and approximate strength of genetical relationships between traits. Such correlations can have important effects on response in the short term, but long-term predictions will need to be based on a more detailed understanding of the processes (genetical, physiological, ecological) that create the correlation. The observation of relatively high genetic correlations in nondomestic populations raises the issue of the maintenance of genetic variation, a subject tackled in Chapter 9.

5.3 Correlated Response to Selection on Several Traits

5.3.1 *Theory*

Thus far we have assumed only that selection operates on one of the two traits. In reality, natural selection is likely to operate simultaneously on both traits. To determine the response to joint selection on two traits, we must take into account the combined effects of direct and correlated responses. The expected response in phenotypic standard deviation units is (Young and Weiler, 1960)

$$R_X = \frac{(i_X - r_P i_Y)h_X^2 + (i_Y - r_P i_X)r_A h_X h_Y}{1 - r_P^2} \tag{5.9}$$

with a similar term for R_Y. Substituting the appropriate variances, covariances, and selection differentials, the response in measurement units is

$$R_X = \frac{V_{AX}(V_{PY}S_X - \text{Cov}_P S_Y) + \text{Cov}_A(V_{PX}S_Y - \text{Cov}_P S_X)}{V_{PX}V_{PY} - \text{Cov}_P^2} \tag{5.10}$$

The pair of equations for R_X and R_Y can be rewritten in matrix form that makes extension to more than two traits immediate:

$$\begin{pmatrix} R_1 \\ R_2 \end{pmatrix} = \begin{pmatrix} V_{A11} & V_{A12} \\ V_{A21} & V_{A22} \end{pmatrix} \begin{pmatrix} \beta_1 \\ \beta_2 \end{pmatrix} \tag{5.11a}$$

where

$$\begin{pmatrix} \beta_1 \\ \beta_2 \end{pmatrix} = \begin{pmatrix} V_{P11} & V_{P12} \\ V_{P21} & V_{P22} \end{pmatrix}^{-1} \begin{pmatrix} S_1 \\ S_2 \end{pmatrix}$$

$$= \frac{1}{V_{P11} V_{P22} - V_{P12} V_{P21}} \begin{pmatrix} V_{P22} & -V_{P12} \\ -V_{P21} & V_{P11} \end{pmatrix} \begin{pmatrix} S_1 \\ S_2 \end{pmatrix} \qquad \textbf{(5.11b)}$$

where X and Y have been replaced by the numerals 1 and 2 (hence, the additive genetic variances are V_{A11} for trait 1, V_{A22} for trait 2, and $V_{A12} = V_{A21} =$ covariance between trait 1 and 2; phenotypic variances are written similarly). In shorthand, the above can be written using bold type as

$$\mathbf{R} = \mathbf{V_A}\boldsymbol{\beta} = \mathbf{V_A}\,\mathbf{V_P}^{-1}\mathbf{S} \qquad \textbf{(5.12)}$$

where $\mathbf{V_P}^{-1}$ is the inverse of the matrix of phenotypic variances and $\mathbf{S}$ is the vector of selection differentials. The vector $\boldsymbol{\beta}$ is called the **selection gradient** vector. For the two-trait case, the response to selection can also be written as

$$R_X = V_{AX}\beta_X + \mathrm{Cov}_A\beta_Y \qquad \textbf{(5.13a)}$$

$$R_Y = V_{AY}\beta_Y + \mathrm{Cov}_A\beta_X \qquad \textbf{(5.13b)}$$

where V_{AI} is the additive genetic variance in trait I, Cov_A is the additive genetic covariance between X and Y, and the selection gradients are

$$\beta_X = \frac{S_X V_{PX} - S_Y \mathrm{Cov}_P}{V_{PX} V_{PY} - \mathrm{Cov}_P^2}$$

$$\beta_Y = \frac{S_Y V_{PY} - S_X \mathrm{Cov}_P}{V_{PX} V_{PY} - \mathrm{Cov}_P^2} \qquad \textbf{(5.13c)}$$

If the data are in standardized units (raw data divided by the phenotypic standard deviation), Eqs. (5.13) can be converted to the standardized response

$$R_X = \beta_X h_X^2 + \beta_Y h_X h_Y r_A \qquad \textbf{(5.14)}$$

$$R_Y = \beta_Y h_Y^2 + \beta_X h_X h_Y r_A$$

where R and β are in standardized units. Inspection of the above equations [(5.9)–(5.14)] reveals the following:

1. Changes in the phenotypic correlation change the response [Eq. (5.9)]. This arises because if the genetic correlation is kept constant, a change

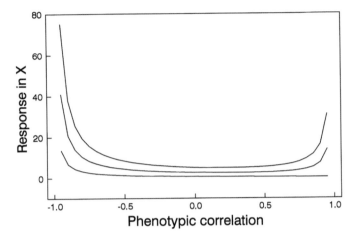

Figure 5.6 Response in trait X as a function of the phenotypic correlation between traits X and Y. For simplicity, the selection intensity on X is set equal to c times the selection intensity on Y, and the response is given in terms of units of the applied selection intensity, that is,

$$R_X = i_Y \frac{(1 - r_P c)h_Y^2 + (c - r_P)r_A h_X h_Y}{1 - r_P^2}$$

Parameter values are $h_X^2 = 0.5$, $h_Y^2 = 0.25$, and $r_A = 0.5$, $c = 1, 5,$ and 10, reading curves from bottom to top.

in the phenotypic correlation implies a change in the environmental correlation. In particular, as the phenotypic correlation approaches ± 1, the response to selection increases dramatically (Fig. 5.6: when $r_P = \pm 1$, the denominator equals zero and the response goes to $\pm \infty$).

2. If the genetic correlation is exactly ± 1, the genetic variance–covariance matrix is singular (i.e., $V_{AX}V_{AY} = \mathrm{Cov}_A$) and evolutionary response in constrained (Lande, 1979a; Maynard Smith et al., 1985; Via and Lande, 1985). The reason for this can be visualized by considering the regression of the additive genetic values of traits X and Y (Fig. 5.7). According to the infinitesimal model, we have $Y = c + bX +$ error, where c and b are constants and the error term is normally distributed with mean zero. Provided the variance in the error term is not zero, any combination of Y and X is possible; hence, evolution can always move the traits to such a combination. However, if the variance of the error term is zero, which occurs when the (genetic) correlation between Y and X is ± 1, then the traits are constrained to lie on the regression line. The number of cases in which r_A is exactly equal to unity is probably very small. One possible circumstance is as follows: consider a trait, X, determined

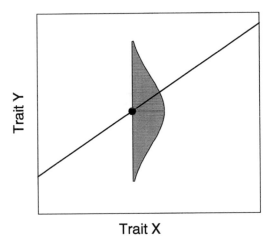

Trait X

Figure 5.7 The assumed statistical relationship between the additive genetic values of traits X and Y. According to standard statistical theory, values of Y are normally distributed about the regression line, and, in principle, any combination of trait values can occur.

by a biochemical reaction which is itself determined by a particular set of genes and an environmental component, say temperature; thus, we have X = function(genes, temperature). Changes in temperature will change X, but as the same set of genes is involved, the genetic correlation between two temperatures will be unity. Although evolutionary change may not, in principle, be limited when the correlations are not equal to unity, the rate of evolutionary change can be radically altered, particularly if the correlated response is opposing the direct response. This issue is dealt with in more detail in Chapter 6.

3. Even if there is no direct selection on trait X, there will be a correlated response to selection on trait Y if either the genetic or phenotypic covariances are not equal to zero. Further, the correlated response may be positive or negative, depending on the value of the environmental correlation; this is illustrated in Fig. 5.8, along with the corresponding responses for the six other categories of selection [see also Deng and Kibota (1995)].

4. When the numerator [Eqs. (5.9) and (5.10)] is zero, there is no response to selection in spite of both genetic variance for the trait and a genetic correlation with another trait under selection. The second observation is particularly important because it implies that a trait may be under selection but show no response. This can be most easily seen by relating the selection differential as a function of the selection gradient and phenotypic variances and covariances:

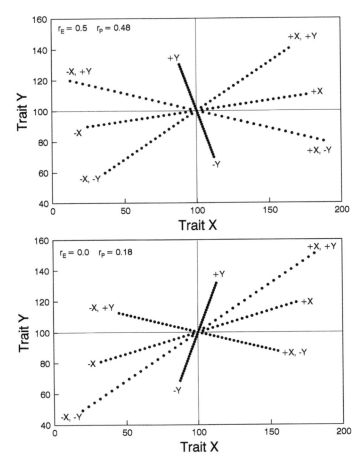

Figure 5.8 The predicted direct and correlated responses for two different combinations of genetic and environmental correlations. Each dot represents the response after each generation of selection (20 generations in total), and the symbols at the end of each trajectory denote the type of selection applied. In all cases, $h_X^2 = 0.5$, $h_Y^2 = 0.25$, $r_A = 0.5$. Upper panel shows results assuming $r_E = 0.5$, and lower panel, $r_E = 0$.

$$S_X = \beta_X V_{PX} + \beta_Y \operatorname{Cov}_P \qquad (5.15)$$

Now suppose that (1) there is no additive genetic variance for trait Y, (2) trait Y is phenotypically correlated with X but genetically uncorrelated ($\operatorname{Cov}_A = 0$), (3) there is no direct selection on trait X (i.e., $\beta_X = 0$), but there is selection on trait Y (i.e., $\beta_Y > 0$). From Eq. (5.15) it can be seen that a positive selection differential is generated ($S_X > 0$), but from Eq. (5.13) ($R_X = V_{AX}\beta_X + \text{Cov}_A\beta_Y$), it is apparent that there will be no response to this selection, even if there is additive genetic variance for X. A likely case in which this situation can arise is when trait X is a heritable trait such as fecundity and Y is a trait such as nutritional status which might have zero heritability: Thus, well-nourished individuals have high fecundity which gives them an apparent selective advantage, but this is not realized because selection is acting only on the environmentally determined component of fecundity. The mechanism described above, or ones conceptually similar, have been proposed for the observation of directional selection without evolutionary response in breeding date in birds (Price et al., 1988), clutch size in birds (Price and Liou, 1989), and tarsus length in birds (Alatalo et al., 1990; Thessing and Ekman, 1994). The prediction of genetic responses in a natural situation clearly depends on including all relevant parameters in the quantitative genetic model: Tienderen and de Jong (1994) have suggested that discrepancies between prediction and observation may be used to discover such "missing" factors. Although in principle this is possible, the large sampling errors attached to the various parameters may make such a procedure very difficult.

The theory described above assumes that the genetic covariance (i.e., the genetic correlation) remains constant. As demonstrated for the simple case of selection acting on only a single trait, this is unlikely to be the case. With selection acting on both traits, we might expect that the genetic correlation would change even faster, or certainly show more diverse responses. According to the "simple" pleiotropy model, selection for two traits in the same direction will generate a negative genetic correlation (Lush, 1948; Lerner, 1950; see Chapter 3). However, if the genetic correlation is due to a partitioning of resources, different effects may be found. This was first discussed by Rendel (1963, 1967) and explicitly modeled by James (1974) using the "partition of resources" model of Sheridan and Barker (1974). Two traits X and Y are supplied with resources from a common pool of size T. Trait X receives a fraction f of the resources, the remaining fraction, $1 - f$, going to trait Y. The heritabilities of T and f are assumed to be both 1 and uncorrelated. The phenotypic values of X and Y are

$$X = fT + E_X \tag{5.16a}$$
$$Y = (1 - f)T + E_Y \tag{5.16b}$$

where E_X and E_Y are the environmental components of X and Y, respectively. The environmental effects are assumed to be uncorrelated with each other and any other component. The covariance between X and Y is, therefore, entirely genetic. Because the covariance between f and T is zero, the additive genetic covariances are

Table 5.7 *Effects of Various Types of Selection on Genetic Correlation Between Two Traits in the "Partition of Resources" Model*

	Effect on the Genetic Correlation, r_A	
Selection	$\mu_f < 0.5$	$\mu_f > 0.5$
For X	May increase	Decreases
For $X + Y$	Decreases	Decreases
For $X - Y$	Increases	Decreases
Against X	Uncertain	Increases
Against $X + Y$	Increases	Increases
Against $X - Y$	Decreases	Increases

$$\text{Cov}_A(X, Y) = \mu_f(1 - \mu_f)V_{AT} - \mu_T^2 V_{Af}$$
$$\text{Cov}_A(X, f) = \mu_T V_{Af}$$
$$\text{Cov}_A(Y, f) = -\mu_T V_{Af}$$
$$\text{Cov}_A(X, T) = \mu_f V_{AT}$$
$$\text{Cov}_A(Y, T) = (1 - \mu_f)V_{AT}$$

where μ_f is the mean value of f, μ_T is the mean value of T, and V_{AT} and V_{Af} are the additive genetic variances of T and f, respectively. Assuming that the genetic variances V_{AT} and V_{Af} remain constant, the change in the covariances is a function of μ_f and μ_T, and its sign is readily obtained; depending on the type of selection, the genetic correlation may increase, decrease, or remain the same (Table 5.7). The actual magnitude and rate of change will depend on the values of the parameters but obviously could be very large and very rapid. The partition of resources model has been explored in more detail by Riska (1986), and for a general analysis of resource allocation models, see Houle (1991) and Jong and van Noordwijk (1992).

5.3.2 Experimental Findings

Sen and Robertson (1964) used two types of joint selection on the abdominal and sternopleural bristles of *D. melanogaster*: (1) index selection, in which the highest 10 of 40 individuals of each sex were selected using the index, abdominal score + 1.5 × sternopleural score, and (2) independent culling selection, in which the highest 20 of 40 individuals of each sex were first selected according to their abdominal score, and then from these, the highest 10 of each sex according to their sternopleural score. There appears to have been a decrease in the heritability of abdominal bristle number but not in that of sternopleural bristles (Table 5.8). The estimate of r_A in the base population is quite uncertain, but despite this, the results from the selection experiments indicate no significant decline after 12 generations; if anything, there may have been an increase (Table 5.8).

Table 5.8 Genetic Correlations Between Abdominal and Sternopleural Bristles in
Drosophila melanogaster *Before and After 12 Generations of Joint Selection*

	h^2 (SE)		
Population	Abdominal	Sternopleural	r_A (SE)
Base	0.42 (0.09)	0.27 (0.07)	$\approx$0.1
Index selected	0.24 (0.03)	0.25 (0.08)	0.21 (0.06)
Independently culled	0.30 (0.17)	0.32 (0.10)	0.21 (0.09)

Source: Data from Sen and Robertson (1964).

Sheridan and Barker (1974) also used joint selection on bristle number in *D. melangaster*—in this case, the coxals and sternopleurals. They used all four possible combinations to test the hypothesis that selection for two characters in the same direction will cause a negative change in the genetic correlation, whereas selection in opposing directions will increase the genetic correlation (see Chapter 3 for the rationale of this hypothesis). This hypothesis is not supported by their results, there being a general trend for the genetic correlation to increase regardless of the type of selection (Table 5.9). These results parallel those of Sen and Robertson (1964), and suggest that the "simple" view of the effect of joint selection is incorrect. Note, however, that the hypothesis was supported in a comparative analysis of genetic correlations between life history traits versus morphological traits (Chapter 3). The observed results also do not fit the partition of resources model. It is assumed in this model that the variance remain unchanged: There is evidence of a slight decline in the heritabilities (Sheridan and Barker, 1974) (Table 5.9), but whether this is sufficient to "rescue" the model is unclear.

Bell and Burris (1973) selected over eight generations for both 13-day larval weight and pupal weight in *Tribolium casteneum*. The genetic correlation shows no obvious change during the course of selection (Table 5.10). However, the quantitative predictions of change in larval and pupal weight are very poor (Fig. 5.9), which could be a consequence of poor estimates or changes in these during the course of selection (Bell and Burris, 1973). Both the predicted and observed changes were greatest when selection operated in a **reinforcing** sense (i.e., for both increased or both decreased weights). When selection was **antagonistic** (i.e., against the genetic correlation), the observed responses were very erratic, particularly when selection was for increased pupal weight but decreased 13-day larval weight (Fig. 5.9). A similar pattern has been observed in selection for egg and body weight in chickens (Fig. 5.10). Nordskog (1977, p. 576) called selection into the $-X$, $+Y$ (large egg weight, small body size) quadrant **incompatible antagonistic selection** because not only is it contrary to the sign of the genetic correlation but also because "very small chickens don't naturally lay very large eggs." Nordskog (1977, p. 576) further noted that "current quantitative genetics theory seems not to be of help in this case."

Table 5.9   Genetic Correlations Between Coxal and Sternopleural Bristles in Drosophila melanogaster *After 10 and 22 Generations of Joint Selection*

	h^2			Sign of Response	
Population	Coxals	Sternopleurals	r_A (SE)	Predicted	Observed
Base	0.09	0.15	0.24 (0.08)		
Generation 10					
UU[a]	0.07	0.14	0.15 (0.06)	−	−
DD	0.08	0.13	0.45 (0.12)	−	+
UD	0.06	0.11	0.37 (0.09)	+	+
DU	0.06	0.13	0.39 (0.07)	+	+
Generation 22					
UU	0.06	0.12	0.54 (0.13)	−	+
DD	0.07	0.07	0.53 (0.21)	−	+
UD	0.07	0.09	0.23 (0.09)	+	0
DU	0.04	0.13	0.40 (.014)	+	+

[a]First letter indicates direction of selection on coxals and second letter the direction of selection on the sternopleurals (U = "up," D = "down").

Source: Data from Sheridan and Barker (1974).

Table 5.10   Genetic Correlations Between 13-Day Larval Weight and Pupal Weight in Tribolium casteneum *After 8 Generations of Joint Selection*

		r_A	
Population	Replicates	Mean[a]	
Base	0.49, 0.55	0.51	
Control	0.62, 0.72	0.67	
UU[b]	0.76, 0.68	0.72	
DD	0.53, 0.91	0.75	
UD	0.82, 0.62	0.72	
DU	0.82, 0.32	0.61	

[a]Mean based on pooled variance components.

[b]First letter indicates direction of selection on larval weight and second letter the direction of selection on pupal weight (U = "up," D = "down").

Source: Data from Bell and Burris (1973).

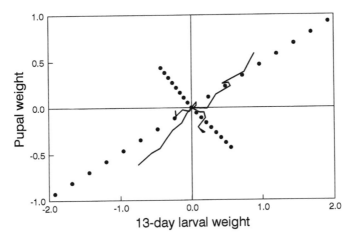

Figure 5.9 Predicted (dots show values per generation) and observed (solid lines) responses to joint selection on larval weight and pupal weight in *Tribolium casteneum*. For clarity, the responses have been individually centered at zero. [Data from Bell and Burris (1973).]

A similar conclusion was reached by Rutledge et al. (1973) from their analysis of selection for body weight and tail length in mice. Genetic correlations estimated from single-trait selection gave excellent agreement with the half-sib estimate [half-sib $r_A = 0.29$ (SE $= 0.09$); from selection on weight, $r_A = 0.31$ (0.09); from selection on tail, $r_A = 0.38$ (0.06)]. Two-trait selection produced consistent results in that the trait values were driven into the appropriate quadrant (Fig. 5.11), but was quantitatively very different, the overall estimate of r_A being 1.00 (0.19). The actual aggregate response to selection varied between 50% and 75% of what was predicted, leading Rutledge et al. (1973, p. 724) to conclude, "In contrast to single-trait-selection responses, the responses to index selection were not consistent with current theory. . . . Our results indicate that in the dynamic situation of antagonistic selection, the genetic correlation may be more powerful in impeding component responses than predicted from presently available theory."

As has been previously suggested, the only way to resolve this problem is to dissect in greater detail the underlying mechanisms that generate the phenotypic and genetic correlations (Riska, 1989). It is evident from the mechanistic models described by Sheridan and Barker (1974), Houle (1991), and Jong and van Noordwijk (1992) that a variety of responses are possible even if we do not permit the genetic correlation to change. Presently needed is a stronger theoretical base for the level of selection at which genetic correlations may be expected to remain unchanged over extended periods, and a stronger empirical base on which to base models of the phenotypic and genetic relationship between traits.

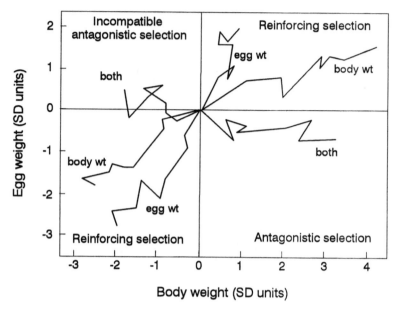

Figure 5.10 Observed responses to single or joint selection on egg weight and body weight in chickens. [Data from Nordskog (1977).]

5.3.3 Observations on Natural Populations

I am aware of only one examination of multiple-trait evolution in a natural population that actually made predictions of response to selection—the evolution of morphological traits in Darwin's medium ground finch, *Geospiza fortis* (Grant and Grant, 1995). Over two periods, 1976–1977 and 1984–1986, the Grants measured the survival of banded birds on Daphne Major, a small island within the Galapagos archipelago. Both periods were characterized by a severe drought, although the direction of selection differed between episodes (Table 5.11). Six morphological traits were measured and the variance–covariance matrices estimated from offspring on mid-parent regression of banded, free-ranging birds. The complete data set is, unfortunately, not given, but heritabilities ranged from 0.48 to 0.97 and genetic correlations from 0.67 to 0.94 (Grant and Grant, 1995). When both episodes are considered together, there is a highly significant correlation between predicted and observed evolutionary responses, although it is evident that the predictions for the second period do not fit as well as for the first period (Fig. 5.12). This analysis clearly demonstrates that it is possible to take quantitative genetic analysis into the real world. It is likely to be more difficult to make accurate predictions of changes in life history traits, because the heritabilities and

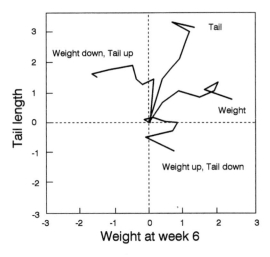

Figure 5.11 Observed responses to single or joint selection on tail length and body weight in mice. For clarity, each line has been individually centered on zero. [Data from Rutledge et al. (1973).]

Table 5.11 *Directional Selection on Morphological Traits in Darwin's Medium Ground Finch Measured over Two Time Periods*

| | Standardized Selection Coefficients | | | |
| | Selection Differential (S) | | Selection Gradient (β) | |
Trait	1976–1977	1984–1986	1976–1977	1984–1986
Weight	**+0.74**	−0.11	**+0.477**	−0.040
Wing length	**+0.72**	−0.08	**+0.436**	−0.015
Tarsus length	**+0.43**	−0.09	+0.001	−0.047
Bill length	**+0.54**	−0.03	−0.144	**+0.245**
Bill depth	**+0.63**	**−0.16**	**+0.528**	−0.135
Bill width	**+0.53**	**−0.17**	**−0.450**	−0.152
Sample size	634	556	632	549
Survival	15%	32%	15%	32%

Note: Coefficients in boldface are significantly different ($P < .05$) from zero.

Source: Modified from Grant and Grant (1995).

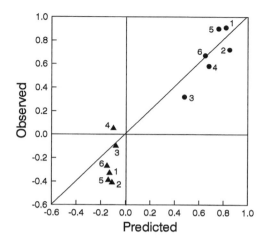

Figure 5.12 Comparison of observed and predicted response (in standard deviation units) to natural selection in Darwin's medium ground finch. ● 1976–1977, ▲ 1984–1986. The actual values are those predicted for and observed in the generations born in 1978 and 1987, respectively; 1 = weight, 2 = wing length, 3 = tarsus length, 4 = bill length, 5 = bill depth, 6 = bill width. [Data from Grant and Grant (1995).]

genetic correlations will generally be lower than with morphological traits and, hence, estimation more difficult.

5.4 Summary

Because of genetic correlation between traits, selection on one trait will result in a correlated response in other traits. Like heritability, the genetic correlation may be quickly eroded with strong directional selection, the rate of erosion probably being faster than with heritability. Given the predicted "fragility" of the genetic correlation, it is unreasonable to expect accurate long-term predictions. However, the genetic correlation should be at least capable of predicting the direction of the response. Experimental evaluation of correlated response to selection on a single trait suggest that the infinitesimal model is inadequate and that a model incorporating a finite number of loci is required. Quantitative prediction of response has been erratic, some experiments giving excellent results, whereas others have been very inaccurate. Nevertheless, with one exception (Table 5.5), the correlated response was in the direction predicted. When selection acts on several traits, the responses can be predicted using a multivariate, matrix approach. Responses to selection may be highly variable, depending on the relative values of phenotypic and genetic variances and covariances. A correlated response to selection may occur not only when there is nonzero genetic correlation but also when there is a nonzero phenotypic correlation but a zero genetic correlation. Further, even when there is genetic variance for a trait and selection on it, there may still be no response (e.g., selection on breeding date in birds). Such a situation arises when selection acts only on the environmental component of the trait. Experimental analyses of response to selection on two traits have generally shown qualitative agreement but poor quantitative correspondence with predicted responses.

6

Phenotypic Plasticity and Reaction Norms

"Almost every conceivable life-history trait is known to respond to almost every conceivable environmental factor in at least one species of plant or animal" (Travis, 1994a, p. 181). This observation can be readily extended to cover not just life history traits but practically every conceivable trait. The observation is so commonplace that it hardly needs experimental verification: It is because of the interaction between environment and phenotype that we are able to increase crop yields with fertilizers and particular watering regimes, and increase the rate of growth and reproduction of domestic animals by altering diets. Given that the natural world is a very heterogeneous place, it is to be expected that selection would favor interactions with the environment that increase fitness. For example, suppose there are two types of habitats characterized by a particular environmental variable E, the values for the two habitats being E_1 and E_2. Further, suppose that the optimal trait value in habitat 1 is X_1^*, and in habitat 2, it is X_2^*. Clearly, the most fit genotype is that which is able to perceive the environmental value and react in such a manner that its trait values in habitats 1 and 2 are X_1^* and X_2^*, respectively. In other words, selection will favor the evolution of some response $f(E)$ such that $X_i = f(E_i) = X_i^*$. From a quantitative genetic perspective, the important observation is that there is genetic variability in how individuals respond to their environment. Such variation has been termed **phenotypic plasticity** and can be formally defined as "a change in the average phenotype expressed by a genotype in different macro-environments" (Via, 1987, p. 47). The presence of such variation has been demonstrated in innumerable studies using clones, inbred lines, reciprocal transplants, or common garden experiments, some examples of which are listed in Table 6.1. The "flip side" to phenotypic plasticity is **canalization**, which refers to the production of the same phenotype regardless of the environment *or* the same phenotype by several different genotypes (Waddington, 1942). Closely akin to phenotypic plasticity is the concept of the **reaction norm**, defined as follows: "A reaction norm as coded for by a genotype is the systematic change in mean expression of a phenotypic character that occurs in response to a systematic change in an environmental variable" (Jong, 1990a, p. 448). This

Table 6.1 Examples of Genetic Variation in Phenotypic Plasticity

Species	Trait(s)	Environment(s)	Reference
Clones			
Polygonum spp. (3)	Morphological	Light, temperature	Zangerl and Bazzaz (1983), Sultan and Bazzaz (1993)
Acyrthosiphon pisum	Development time, survival	Plants	Via (1991), Sandstrom (1994)
Alsophila pometria	Weight	Plants	Futuyma and Philippi (1987)
Daphnia magna	Length, clutch size, maturity	Food	Ebert et al. (1993a, 1993b)
Different Strains			
Drosophila melanogaster	Body weight	Temperature	Fondevila (1973)
Drosophila melanogaster	Development time, weight	Temperature, food	Gebhardt and Stearns (1993a, 1993b)
Drosophila pseudoobscura	Development time, % emergence, wing length	Temperature, food	Taylor and Condra (1978)
Drosophila pseudoobscura	Bristles, viability, development time	Temperature, density	Gupta and Lewontin (1982)
Muscidifurax raptor	Sex ratio	Temperature	Antolin (1992a, 1992b)
Arabidopsis thaliana	Flowering, time, height, leaf number, siliqua number	Temperature	Westerman (1970)
Arabidopsis thaliana	Fecundity-size allometry	Light, nutrient, pot volume	Claus and Aarssen (1994)
Different Geographic Populations			
Drosophila melanogaster, D. simulans	Wing length, emergence	Temperature	Tantawy and Mallah (1961)
Drosophila serrata	Adult survival, fecundity	Temperature	Birch et al. (1963)
Allonemobius socius	Diapause	Photoperiod, temperature	Bradford and Roff (1995)
Jadera haemotoloma	Mating behavior	Social environment	Carrol and Corneli (1995)
Poecilia formosa	Size at maturity	Temperature	Travis (1994a)

(*continued*)

Table 6.1   Continued

Species	Trait(s)	Environment(s)	Reference
Different Geographic Populations (*continued*)			
Phlox drummondii	Morphological, reproductive traits (10)	Water regime, nutrients	Schlichting and Levin (1990)
Arabis serrata	Morphology	Nutrients	Oyama (1994a, 1994b)
Betula papyrifera	Seed germination	Temperature	Bevington (1986)
Selection Changes Reaction Norm			
Drosophila melanogaster	Morphology	Temperature	Waddington (1960), Waddington and Robertson (1966), Kindred (1965), Scharloo et al. (1972), Thompson and Rook (1988)
Drosophila melanogaster	Body size, development time	Diet	F.W. Robertson (1960a, 1990b, 1963), Hillesheim and Stearns (1991)
Drosophila pseudoobscura	Body size	Temperature	Druger (1962)
Laodelphax striatellus	Wing dimorphism	Density	Mori and Nakasuji (1990)
Dianemobius fascipes	Wing dimorphism	Photoperiod	Masaki and Seno (1990)
Tribolium casteneum	Larval weight	Diet	Yamada and Bell (1969)
Tribolium casteneum	Fecundity	Temperature	Orozco (1976)
Bicyclus anynana	Wing eye spots	Temperature	Holloway and Brakefield (1994)
Menidia menidia	Sex ratio	Temperature	Conover et al. (1992)
Mice	Size	Diet	Falconer (1990)
Schizophyllum commune	Growth rate	Temperature	Jinks and Connolly (1973)
Nicotiana rustica	Final height	Sowing date	Brumpton and Jinks (1977)

Note: For other examples see Bradshaw (1965), Schlichting (1986), Kuiper and Kuiper (1988), Scheiner (1993a), Travis (1994a).

definition does not exclude discrete environments because they can be subsumed under the definition by the statistical approach of dummy variables.

Phenotypic plasticity has been divided into two categories. (1) Graded responses to the environment. Examples include changes in photosynthetic rate with temperature and light level, changes in flowering time and flowering height (Fig. 6.1), and variation in life history traits with density or morphology with temperature (Fig. 6.2). Such responses were called **dependent development** by Schmalhausen (1949) and **phenotypic modulation** by Smith-Gill (1983). (2) Discrete variation produced in different environments. Examples include cyclomorphosis, paedomorphosis, wing dimorphism, diapause, and so forth (Table 6.2). This type of variation was termed **autoregulatory morphogenesis** by Schmalhausen (1949) and **developmental conversion** by Smith-Gill (1983). The reaction norms measured by Windig (1994a, 1994b) are interesting because in the wild *Bicyclus anaynana* displays basically only two morphs: a "wet season" morph and a "dry season" morph (those produced at the two extreme temperatures). The production of intermediate forms in the laboratory suggests that the "dimorphism" observed in the field is a response to discrete environmental conditions and not a consequence of the trait being a threshold trait. This serves warning that field data can be unreliable indicators of the mode of inheritance.

The above twofold classification of phenotypic plasticity is not particularly meaningful, as the second type can be subsumed under the first using the threshold model of quantitative genetics. The underlying continuously distributed trait varies in a graded manner with the environment, but the phenotypic shift between morphs occurs only when the trait value exceeds the threshold (Fig. 6.2). At the population level there is a graded response in the proportion of the morph with the environment (Fig. 6.3). I shall refer to both types of variation simply as reaction norms.

There is no reason to suppose that all reaction norms are adaptive (Scharloo, 1984; Schlichting, 1986; Sultan, 1987; Stearns, 1989), but the ubiquity of their occurrence and the obvious advantages of such responses argues very strongly that selection has molded many, if not most, of them. The interest in reaction norms goes back to the beginning of this century with the work of Woltereck (1909) on cyclomorphosis in *Daphnia* and was emphasized as an important factor in evolution by Schmalhausen (1949), but it has been only comparatively recently that a concerted effort has been made to understand how reaction norms evolve. Much of the attempt has centered on phenotypic models, addressing the question, "Given a particular set of environmental variables and variation, what is the optimal reaction norm?" Examples of this approach include the work of Bradshaw (1965), Hairston and Munns (1984), Lloyd (1984), Lively (1986), Stearns and Koella (1986), and numerous others [see Roff (1992) and Stearns (1992) for reviews of this approach]. These models presume that there exists sufficient genetic variation to achieve whatever reaction norm is optimal. Another perspective is that of quantitative genetics: How do we describe and measure the genetic

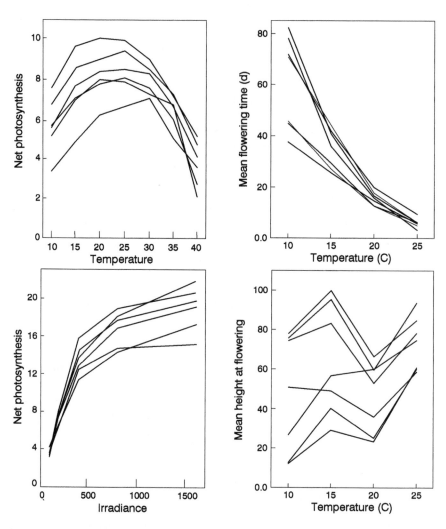

Figure 6.1 Examples of reaction norms that show a graded response. Data on the right show responses of clones in the plant *Polygonum pennsylvanicum* [redrawn from Zangerl and Bazzaz (1983)]. Data on the left show response of inbred lines in the plant *Arabidopsis thaliana* [data from Westerman (1970)].

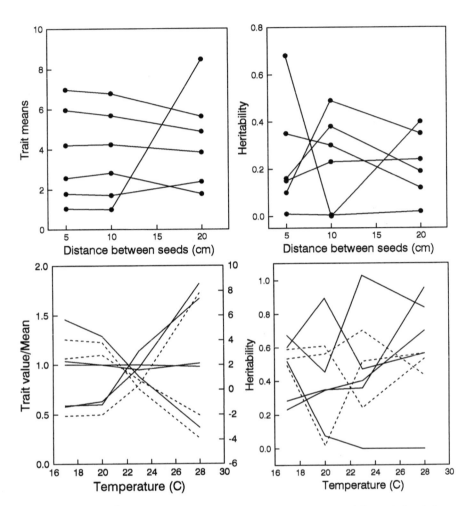

Figure 6.2 Examples of population reaction norms (left panels) and the variation in heritabilities of the traits across environments (right panels). The upper panels show data on reproductive traits from wild radish, *Raphanus sativus* [data from Mazer and Schick (1991)]. To enable all traits to be plotted using a common axis, trait means have been divided by 10^c, where c is a value such that the resulting value lies between 0 and 10. The lower panels show morphological data (wing size and characteristics of the wing pattern) for the butterfly *Bicyclus anynana* [data from Windig (1994a)]. Mean trait values for each environment have been divided by the mean across environments, solid lines being plotted using the left axes and dotted lines plotted using the right.

Table 6.2 *Review of Environmental Factors That Induce the Development of a Particular Morph in Species in Which Phenotypic Variation is Discrete*

Taxon	Morph Induced	Morph Induced by
Protective Polymorphisms		
Daphnia (4 spp.)	Helmeted	Substance released by invertebrate and vertebrate predators
Daphnia pulex	Neck toothed	Substance released by invertebrate predators
Brachionous calyciflorus, *Keratella* (3 spp.)	Spined	Substance released by invertebrate predators
Asplanchna (2 spp.)	(1) Cruciform	(1) Substance (α-tocopherol) released by algal cells
	(2) Giant	(2) Dietary α-tocopherol + large prey
Onychodromus quadricornutus	Spined	Substance produced by the Giant morph
Euplotes (6 spp.)	Spined	Substance released by predatory ciliates
Membranipora membranacea	Spined	Grazing by nudibranch
Chthamalus anisopoma	Bent	Substance released by predator
Corals and sea anemones	Catch tentacles	Proximity of competitors
Papilionidae	Pupal color	Photoperiod, substrate color, food plant odor
Trophic Polymorphisms		
Ambystoma tigrinum	Cannibal	High density of conspecifics
Scaphiopus multiplicatus	Carnivores	If fed on macroscopic prey (shrimp, tadpoles)
Salvelinus alpinus	Both	Environmental induction inferred from common garden expt.

(continued)

variation underlying phenotypic plasticity, how do we model it, and under what circumstances might evolutionary response be limited? These questions are the subject of the present chapter.

6.1 Two Perspectives: Character State Versus Reaction Norm

The evolution of phenotypic plasticity can be addressed using two apparently different mathematical perspectives: the character state approach and the reaction

Table 6.2   Continued

Taxon	Morph Induced	Morph Induced by
Life-Cycle Polymorphisms		
Notophthalmus viridescens	Paedomorph	Low density of conspecifics
Ambystoma talpoideum	Paedomorph	Lack of fish predators, drying regime of pond, density (complex interaction with drying regime)
Cecidomyiidae	Paedogenetic larva	Uncrowded larval conditions or abundant food
Insects	Volant	Density, photoperiod, temperature (effects depend on species)
Various invertebrates	Diapause	Photoperiod, temperature, density
Some invertebrates and vertebrates	Sex	Temperature, crowding, nutrition
Mating Polymorphisms		
Euterpina acutifrons	Small ♂	Increasing temperature
Pachypygus gibber	Small ♂	Age of female host
Mites (2 spp.)	Fighter	Good diet. Decreased density in *Caloglyphus berlesei* but not *Rhizoglyphus robinii*
Forficula auricularia	Large cerci	Good diet

Source: After Roff (1996a).

norm approach. Both approaches are actually interchangeable and each has advantages and disadvantages.

6.1.1   The Character State Approach

We have already encountered this approach in the discussion of the genetic correlation (Chapter 3). Falconer (1952) pointed out that, conceptually, the same trait measured in two environments can be considered as two traits that are genetically correlated. This is illustrated in Fig. 6.4, where each line represents a separate genotype. If the lines joining the phenotypic values in the two environments (E_1 and E_2) intersect at a common point between E_1 and E_2, the genetic correlation between the two traits is -1 (Fig. 6.4, panel D). If the lines intersect outside the range E_1 to E_2, the genetic correlation is $+1$ (Fig. 6.4, panel A; note that parallel lines also produce a genetic correlation of $+1$, because, mathematically, their point of intersection is at the point of infinity point beyond E_1 or E_2). The genetic correlation will differ from ± 1 if there is no common point of intersection (Fig.

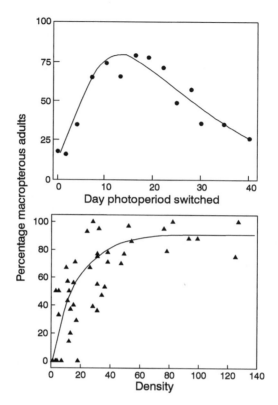

Figure 6.3 Examples of population reaction norms for a dimorphic trait. Top panel shows change in percentage of macropterous (= long winged) adults of the cricket *Pteronemobius taprobanensis* as a function of the day the photoperiod was shifted from a long day (16L:8D) to a short day (12L:12D) [Redrawn from Tanaka et al. (1976)]. Lower panel shows the percentage of macropterous adults of the leafhopper *Prokelisia dolus* as a function of the density of nymphs [data from Denno et al. (1991)].

6.4, panels B and C). The above description is based on clones; with a sexually reproducing organism, genetic variation can be visualized by using family means, and the variance–covariance matrix estimated using the mixed-model analysis of variance (ANOVA) described in Chapter 3.

As an alternative to the above ANOVA, Via (1984) suggested arbitrarily pairing individuals within families from different environments, which then permits the standard method of estimation of genetic variances and covariances (Chapter 3). In her analysis of development time and pupal weight of *Liriomyza sativae* on two different hosts (cowpea and tomato), Via (1984) obtained genetic correlations from the mixed-model ANOVA of 0.07 and 0.75, respectively [dam data from reanalysis by Fry (1992)], and estimates from the arbitrary pairing method of 0.38

and 0.71, respectively. The considerable differences between the estimates for development time might reflect non-Mendelian effects because the estimate from the sires gave 0.69 (Fry, 1992) and the mean of the sire and dam is 0.38, in accord with the arbitrary pairing estimate. The statistical behavior of such an approach needs to be explored before its use can be recommended, although it appears promising.

From the definition of plasticity given in the introduction, an obvious way to define phenotypic plasticity for a genotype is the difference in the phenotypic value between the two environments. Note that genetic variation in phenotypic plasticity (i.e., differences among genotypes in plasticity) does not necessarily mean that selection is not constrained: Genetic variation for plasticity exists in all cases shown in Fig. 6.4 but is unconstrained in only two of the cases.

The above approach can readily be extended to a finite number of cases, the variance–covariance matrix simply increasing in size. However, in many cases, such as variation in, say, temperature or photoperiod, the number of character states is infinite. It is obviously not possible to measure variances and covariances across all possible values. To solve this problem, Kirkpatrick and Heckman (1989) introduced their **infinite-dimensional model**. This approach is appropriate for any continuous reaction norm, such as growth trajectories (Kirkpatrick et al., 1990) or variation across environments (Gomulkiewicz and Kirkpatrick, 1992). Because the mathematics are rather complex, I present here only a verbal description of the approach. The approach is essentially twofold: First, a variance–covariance matrix is estimated for a fixed number of environments (ages, etc.), and, second, the values for the intervening points obtained by interpolation, leading to two matrices, one defining the polynomials used for the interpolation and a second giving the coefficients of the polynomials. An example is presented in Fig. 6.5, using variation in node number in *Abutilon theophrasti* as a function of variation in light level. The variance–covariance matrix was estimated under five light levels, generating a 5 × 5 matrix. There is considerable complexity in the interpolated surface (Fig. 6.5), although there is considerable statistical uncertainty in the parameter values. Given the difficulty of measuring variances and covariances between just two traits (see Chapter 3), the problems of accurately estimating an entire surface seem vast indeed.

6.1.2 *Norm of Reaction Approach*

This approach was implicitly used by Scheiner and Goodnight (1984) and Scheiner and Lyman (1989a) in their approach to partitioning of phenotypic variance into plastic and nonplastic components. Whereas the character state approach sees the phenotype as two points in state space, the norm of reaction approach sees a line (Fig. 6.6; Jong, 1990b; Jong and Stearns, 1991). Thus, for two environments, phenotypic variation can be described as

$$X(E) = c_0 + c_1 E + e \qquad \textbf{(6.1)}$$

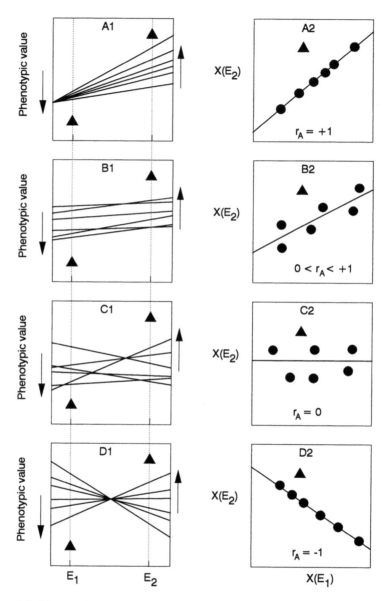

Figure 6.4 Phenotypic plasticity in two environments as viewed from the character state perspective. The panels on the left show the character states in the two environments E_1 and E_2, each line joining the trait values of a single genotype (the reaction norms of the genotype). The panels on the right show the regression of the trait value from the second environment, $X(E_2)$, on the trait value from the first environment, $X(E_1)$.

Caption continues on facing page

◀ (A) The reaction norms meet at a single point beyond the range of the two environments. The correlation between trait values is $+1$. Note that parallel reaction norms also give a genetic correlation of $+1$ (mathematically this is because the lines meet at infinity).

(B) and (C) The reaction norms cross at several points within the range of the two environments. Depending on the distribution of intersections the genetic correlation will be positive but less than $+1$ (B), zero (C), or negative but greater than -1 (not shown).

(D) The norms of reaction intersect at a single point between the two environments. In this case the genetic correlation is -1.

The triangles show a hypothetical optimal combination of trait values. Because in cases A and D all the points lie on a single line ($r = \pm 1$) this combination might not be achievable. In all other cases, because in principle the distribution about the regression line is normal (i.e., no value is excluded), selection can move the population to the joint optimum. [Figure modified from Via (1987).]

where the two parameters c_0 and c_1 are viewed as traits, and e is an error term, which is normally distributed with mean 0 and variance V_e. The coefficient c_0 is the intercept (elevation) of the reaction norm; if the environment is scaled such that $E = 0$, then c_0 is the mean value of the trait in environment E. The coefficient c_1 is the slope of the reaction norm. Evolutionary change in X depends on the heritabilities of c_0 and c_1 and the genetic correlation between them. If the genetic correlation is ± 1, then the line is genetically fixed and corresponds, as it should, to cases A and D in Fig. 6.4. This description is merely an alternate formulation of the character state approach (Tienderen and Koelwijn, 1994; Jong, 1995) and, in fact, has a longer history, as Woltereck (1909), who introduced the concept of reaction norms, emphasized that it is actually the reaction norm that is inherited. The conceptual advantage of this approach is that it extends quite naturally to continuously distributed environmental variables such as temperature. The above model can clearly apply to any linear reaction norm; if the relationship between trait and environment is more complex, a more complex function can readily be substituted [e.g., the quadratic $X(E) = c_0 + c_1 E + c_2 E^2 + e$].

For the linear reaction norm, an obvious measure of phenotypic plasticity is the slope c_1, but there is no clear definition for nonlinear reaction norms. Jong (1995) suggested the first derivative of the function, but this describes only part of the reaction norm shape and, hence, could be misleading. For any given environment, the use of the first derivative could be useful, but it must be remembered that plasticity will vary with the environment. Because the potential for considerable change in shape is possible with nonlinear reaction norms, it is probably better to dispense with any formal quantitative definition of phenotypic plasticity in such cases.

An alternative statistical definition of plasticity, suggested by Scheiner and Goodnight (1984), is

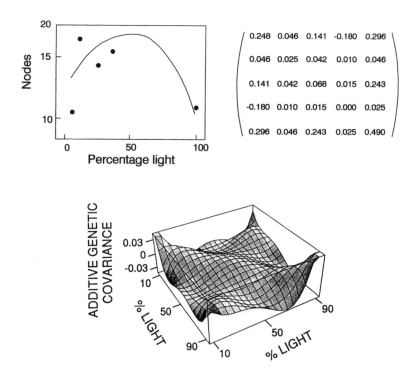

Figure 6.5 An example of the use of the infinite-dimensional model to describe the genetic variance–covariance matrix across an indeterminate number of environments. The upper left plot shows the population reaction norm of node number versus light level in *Abutilon theophrasti* (solid line is fitted polynomial curve). On the right is the genetic variance–covariance matrix estimated from these data. The bottom panel shows the interpolated three-dimensional relationship between the additive genetic variances/covariances and light levels. [After Gomulkiewicz and Kirkpatrick (1992).]

$$V_{\text{PL}} = V_E + V_{G \times E} \tag{6.2}$$

where V_{PL} is phenotypic plasticity, V_E is the contribution from the environment, and $V_{G \times E}$ is the genotype by environment variance. The remaining genotypic variance V_G was assumed by Scheiner and Goodnight (1984) to be independent of the environment. For a linear reaction norm, this is equivalent to the separation of c_0 and c_1 and is a valid partition, but if the reaction norm is nonlinear, there are, of course, additional coefficients and V_G is no longer independent of the environment [i.e., not the genetic variance of c_0; Jong (1990b), Muir et al. (1992), and Scheiner (1993a)].

Assuming that the environmental covariance is zero and that the environmental variance remains constant, for the linear reaction norm the genetic and phenotypic

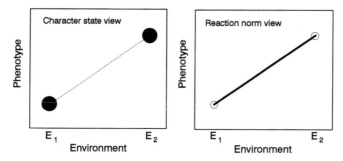

Figure 6.6 A pictorial representation of the two viewpoints of phenotypic plasticity. The character state approach (upper panel) focuses attention on the trait values in the two environments, whereas the reaction norm approach (bottom panel) focuses on the line joining the trait values. [Redrawn from Jong (1990b).]

variances, $V_A\{X(E)\}$ and $V_P\{X(E)\}$ in a given environment E are (Gavrilets and Scheiner, 1993)

$$V_A\{X(E)\} = V_A(c_0) + 2\,\text{Cov}_A E + V_A(c_1)E^2 \qquad \textbf{(6.3a)}$$

$$V_P\{X(E)\} = V_A\{X(E)\} + V_e \qquad \textbf{(6.3b)}$$

where $\text{Cov}_A E$ is the additive genetic covariance between c_0 and c_1. From the above two equations, it follows that, in general, heritability will be a quadratic function of the environment (Fig. 6.4, panels C and D; i.e., whenever the genetic correlation between environments is not exactly unity). Thus, the fundamental assumption underlying the use of the mixed-model ANOVA for the testing of genotype by environment interaction that the variances remain constant (Fry, 1992) is violated. Analysis of variance is fairly robust to the assumption of heteroscedasticity, but work is needed to assess just how much variation can be allowed before estimation becomes unreliable. The assumption of equal variances is not an assumption of the genetic model; therefore, what is really required is a method of testing and estimation that does not depend on equality of variances. Maximum likelihood offers such an avenue, and the method needs to be explored in more detail. For two environments, the data can be rescaled as described in Chapter 3.

As expected from the above theory, variation in heritabilities across environments is very common, two examples of which are shown in Fig 6.2 [for other examples, see Gebhardt-Henrich and Noordwijk (1991), Gebhardt and Stearns (1993a, 1993b), Carrière and Roitberg (1994, 1995), and Simons and Roff (1995); for an example in which heritabilities remained constant between two environments, see Scheiner et al. (1991)]. Variation is likely to be general although many (most?) studies lack the statistical power to detect even large changes in variances.

The magnitude of the differences in heritabilities will depend to some extent on the difference in the environment; however, there is no general theory that defines what is meant by "small" or "large." In the two examples shown in Fig. 6.2, heritabilities can range from 0.0 to greater than 0.40, but because of large sampling errors, how much of this variation is "real" is not clear. It is also evident that the heritabilities show no simple pattern with environmental variation. Again, this may reflect sampling error or an underlying complexity in the reaction norms and variance–covariance matrices.

An important feature of the covariance between c_0 and c_1 is that it is not independent of how the environment is scaled. Consider, for example, the following two alternate scenarios for the two-environment case: (1) the environments are labeled 0 and 1, giving $c_0 = X_1$ and $c_1 = X_2 - X_1$, where X_i is the genotypic trait value in environment i and (2) the environments are labeled -0.5 and 0.5, in which case $c_0 = (X_1 + X_2)/2$ and $c_1 = X_2 - X_1$. In the first case, $\text{Cov}_A = \text{Cov}_A(X_1, X_2) - V_A(X_1)$, and in the second, $\text{Cov}_A = [V_A(X_2) - V_A(X_1)]/2$. From the point of view of statistical estimation, the second definition is preferable because it requires only the separate variances. However, there is no theoretical reason why one should be preferred over the other. Comparison of the genetic correlation between the two coefficients can only be made using a common definition of the environmental scale.

For a linear reaction norm and two environments, a simple procedure based on a half-sib mating scheme for the estimation of the genetic parameters was developed by Gavrilets and Scheiner (1993). Each family is divided between the two environments and there are an equal number of offspring per full-sib family per environment (n). Let $m_{ij} = \Sigma X_{ij}/n$ be the mean value in full-sib family i of trait X measured in environment j ($j = 1$ or 2). We now construct two variables, the overall mean for family i, $\bar{m}_i = (m_{i1} + m_{i2})/2$, and the difference between the two means $d_i = m_{i1} - m_{i2}$. The former is the intercept, c_0, when the environments have been scaled (without loss of generality) such that $E_2 = -E_1, = \frac{1}{2}$ and the latter is the slope, c_1. By using the mean values, the design can now be viewed as a half-sib design in which each sire is mated to several dams and each dam produces *one* offspring with two traits $\bar{m}_1$ and d_i. The analysis of variance is then exactly the same as for the full-sib with multiple offspring per family (Table 2.2), replacing the term "among families" with "among sires," and "among progeny within families" with "among progeny within sires." There are two such analyses, one for each "trait." The additive genetic variance for each is estimated as

$$V_A = \frac{4(\text{MS}_{\text{AF}} - \text{MS}_{\text{AP}})}{k} \tag{6.4}$$

where for simplicity I have retained the subscripts as defined in Table 2.3 ($k =$ number of dams per sire). Note that the multiplier is 4, not 2, as these are half-sib families. The heritability of the means ($\bar{m}_i$) is equivalent to the heritability of

c_0, whereas the heritability of the difference (d_i) is equivalent to the heritability of plasticity, c_1. The heritability estimates are biased due to the use of the full-sib means; hence, a correction factor is required for the phenotypic variances (Gavrilets and Scheiner, 1993):

$$V_P(c_0) = V_P(c_0)^* + \frac{V_{e1} + V_{e2}}{4}\left(1 - \frac{1}{n}\right) \tag{6.5a}$$

$$V_P(c_1) = V_P(c_1)^* + (V_{e1} + V_{e2})\left(1 - \frac{1}{n}\right) \tag{6.5b}$$

where the values superscripted with an asterisk are the unadjusted phenotypic variances and V_{ei} is the error variance from an ANOVA within environment i. If the number per family differs between the environments, the term $1/n$ is replaced by $2/(n_1 + n_2)$, where n_i is the number of offspring per full-sib family per environment. The covariance between c_0 and c_1 is obtained in the same manner as above by using ANOVA Table 3.3 for $\bar{m}_i$ and d_i. As described in Chapters 2 and 3, standard errors can be estimated using the jackknife, although the robustness of the technique needs to be checked with simulation.

A similar procedure was used by Ebert et al. (1993b) to estimate broad sense H^2 estimates for clones of *Daphnia magna* raised at two food levels. Instead of using differences between clonal means, they used the arbitrary pairing method; an individual from the low food environment was randomly paired with an individual from the same clone raised in the high food environment. As there were three individuals per clone per environment, this method resulted in three estimates per clone of the difference between environments. Two analyses of variance were performed: a one-way ANOVA using the differences between clones, and a two-way ANOVA using clone and environment as main effects. The broad sense heritability of plasticity, H^2, can be estimated from each method as

$$H^2 = \frac{V_d}{V_d + V_{e1}} = \frac{V_{\text{int}}}{V_{\text{int}} + V_{e2}} \tag{6.6}$$

where V_d is the expected variance of the difference between environments, V_{e1} is the error variance obtained from the one-way ANOVA, V_{int} is the expected variance of the clone by environment interaction, and V_{e2} is the error variance obtained from the two-way ANOVA. The expected variance due to clone estimated from the two-way ANOVA estimates the genetic covariance between environments, Cov_G. The broad-sense genetic correlation between environments, r_G, is

$$r_G = \frac{\text{Cov}_G}{\sqrt{V_h V_l}} \tag{6.7}$$

where V_h and V_l are the genetic variances in the high and low food environment, respectively. There is no simple proportional relationship between H^2 and r_G. This can be seen most clearly by considering the case $V_h = V_l = V$. Noting that $V_d = V_h + V_l - 2\,\text{Cov}_G$ and, hence, $V_d = 2(V - \text{Cov}_G)$, we have

$$H^2 = \frac{2(V - \text{Cov}_G)}{2(V - \text{Cov}_G) + V_{e1}} \tag{6.8a}$$

$$r_G = \frac{\text{Cov}_G}{V} \tag{6.8b}$$

Substituting $\text{Cov}_G = Vr_G$ in Eq. (6.8a) and rearranging gives

$$\frac{1}{H^2} = 1 + \frac{V_{e1}}{2V(1 - r_G)} \tag{6.9}$$

Under the above assumption, the two heritabilities for the trait measured in the separate environments are the same ($H_h^2 = H_l^2$) and equal to $V/(V + V_{e1})$. Rearranging to obtain V as a function of H_h^2, substituting in Eq. (6.8) and then further rearranging gives

$$H^2 = \frac{2[H_h^2(1 + \text{Cov}_G) - \text{Cov}_G/V_{e1}]}{2\,[H_h^2(1 + \text{Cov}_G) - \text{Cov}_G/V_{e1}] + 1} \tag{6.10}$$

Equations (6.8) and (6.10) suggest that there will be a positive correlation between the heritability of plasticity and the heritabilities for the separate environments, but no strong relationship between the heritability of plasticity and the genetic correlation. This is confirmed by the *Daphnia* data (Fig. 6.7): H_h^2 is positively correlated with H_l^2, both are positively correlated with plasticity, but there is no significant correlation with r_G.

Tienderen and Koelewijn (1994) present a method of estimating the heritabilities of the coefficients of a reaction norm that is applicable to multiple environments and any type of reaction norm. Unfortunately, it appears to be as hard to implement in the case of sexually reproducing organisms as that of Kirkpatrick and Heckman (1989) for the character state approach. One could proceed by using mean values as done by Gavrilets and Scheiner (1993), but correcting for the error introduced by the use of means is difficult, if not impossible [see appendix 2 of Tienderen and Koelewijn (1994) for a discussion]. If the reaction norm is clearly nonlinear, the best approach is probably to seek a transformation that linearizes the relationship.

The arbitrary pairing method could be expanded to more environments (in which case, it should be called the arbitrary assignment method), but interpretation

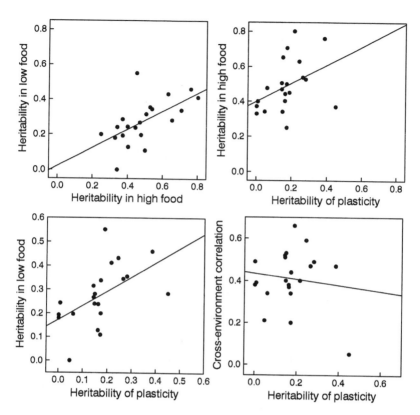

Figure 6.7 Observed relationships in the cladoceran *Daphnia magna* between heritabilities of morphological and life history traits measured in two environments, the heritability of plasticity, and the cross-environment genetic correlation. [Data from Ebert et al. (1993b).]

of the estimates would have to be very cautious in the absence of more information on its statistical properties. The arbitrary assignment method was used by Windig (1994b) in the estimation of reaction norm parameters of morphological traits in the butterfly *Bicyclus anynana* as a function of temperature measured at three values. A full-sib breeding design was used, and the parameter values and associated confidence intervals were estimated by the following bootstrap procedure. For each family, three individuals, one from each temperature, were selected at random. A linear regression was fitted to these three points, the intercept and slope then being the trait values for this "individual." The procedure was repeated for each family until the number reached the lowest sample size per family (e.g., if the family sizes were 10, 7, and 15, then 7 random samples were drawn, giving 7 estimates for this particular family). Sampling was with replacement. Variance components were then estimated for this bootstrap sample, and the whole procedure repeated 1000 times. The final estimates were taken as the mean of all individual estimates and the 95% confidence intervals determined using the observations ranked 26th and 975th.

Weis and Gorman (1990) adopted the following procedure for full-sib families across a continuous gradient (see Section 6.2.7 for a detailed discussion of the experiment). First, they computed separate quadratic fits for each full-sib family and tested for the significance of the quadratic term after correcting for experimentwise error with the Bonferroni method. They found no significant overall effect of the quadratic term and so proceeded using the linear model

$$X_{ij} = \mu + \beta + F_i + \beta_i + \varepsilon_{ij} \qquad (6.11)$$

where X_{ij} is the trait value for individual j in family i, μ and β are the overall mean and slope, respectively, F_i is the deviation from the overall mean of family i, β_i is the overall deviation in the slope due to family i, and ε_{ij} is the error term. Statistically significant variation in F and the family-specific slope was taken to indicate significant additive genetic variation in the intercept and slope, respectively. The heritabilities of these components were taken as twice the proportion of phenotypic variance accounted for by each component. Weis and Gorman (1990) did not estimate the genetic correlation between the two coefficients, nor did they discuss how this might be done. They computed the correlation between slope and elevation, but this was not the genetic correlation because it included some of the within-family variance.

The genetic correlation between environments indicates the extent to which two traits can evolve independently, but by itself, it gives little information on how plasticity will evolve. Similarly, the heritability of plasticity gives information on how the slope of the reaction norm can evolve, but interpretation cannot be unambiguously made unless one also has estimates of the heritability of the intercept and its genetic correlation with the slope. What is the relationship between the reaction norm parameters and the genetic correlation? To answer this question, we first rescale the two environments so that the first has a value of 0 and the second a value of 1: As noted earlier, the covariance between c_0 and c_1 is dependent on the environmental scaling and thus the following argument is restricted to this particular circumstance, although the general message is not affected. For a particular genotype i, the value in the first environment is c_{0i} and the value in the second is $c_{0i} + c_{1i}$. The genetic correlation between the two traits thus depends on the correlation between c_0 and c_1. The genetic correlation, r_A, is equal to $Cov(c_0, c_0 + c_1)/\{V_A(c_0)V_A(c_0 + c_1)\}^{1/2}$, from which we can infer, as before, that r_A will not necessarily be correlated with c_1. The genetic correlation between environments is a part–whole correlation of c_0 and $c_0 + c_1$ and can be written as (Sokal and Rohlf, 1995)

$$r_A = \frac{\sqrt{V_A(c_0)} + r_{01}\sqrt{V_A(c_1)}}{[V_A(c_0) + 2r_{01}\sqrt{V_A(c_0)V_A(c_1)} + V_A(c_1)]^{1/2}} \tag{6.12}$$

where r_{01} is the correlation between c_0 and c_1. If there is a positive or zero correlation between c_0 and c_1, then there will be a positive genetic correlation between environments. If c_0 is negatively correlated with c_1, then there is a complex relationship between r_A and r_{01}, which is dependent on the value of r_{01} and the ratio $V_A(c_1)/V_A(c_0)$ (Fig. 6.8). In general, it is likely that the variances of the trait in the two environments will be approximately equal, implying that $V(c_0) > V(c_1)$, in which case r_A will be positive. The lack of a trade-off between the coefficients of the reaction norm cannot be inferred from a positive genetic correlation between environments, but a negative correlation does imply a negative correlation between c_0 and c_1 (Fig. 6.8).

The foregoing analysis indicates that one cannot readily interpret parameters from an analysis from a particular perspective (e.g., character state) in the context of the alternate perspective (e.g., reaction norm). For this reason, it is useful to estimate both sets of parameters.

6.2 Evolution of Plastic Traits

Although there is abundant evidence that traits are genetically correlated between environments, some examples of which are presented in Table 6.3, estimates of the heritability of plasticity per se are far fewer (Table 6.4), largely because this perspective has been explicitly explored only recently. The few available estimates suggest that there is considerable potential for the evolution of plasticity. This conclusion is aptly summarized by Jinks and Pooni (1988, p. 521): "Genetic variation for environmental sensitivity is as ubiquitous as that for mean performance and is at least in part independent of it. As we learn more about the genetic variation for environmental sensitivity and its specificity in respect of character and environmental variable, it becomes clear that it is possible to select a pattern of response to environmental variation to meet almost any requirements."

6.2.1 Directional Selection

From the character state perspective, the formula for the change in trait values is the same as for any multitrait system:

$$R = V_A V_P^{-1} S \tag{6.13}$$

The approach is the same for the reaction norm perspective. First, we must convert the equations into matrix format. For simplicity, consider the linear reaction norm. The changes in c_0 and c_1, designated as Δc_0 and Δc_1, respectively, are given by (Gavrilets and Scheiner, 1993)

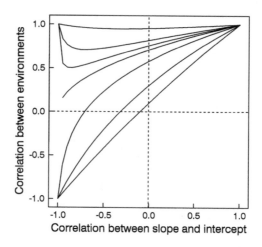

Figure 6.8 The relationship between the genetic correlation between environments and the genetic correlation between intercept and slope of the reaction norm when the environments are scaled as 0 and 1. Each line shows a different value of the ratio $V_A(c_1)/V_A(c_0)$. Reading from top to bottom the values are 0.1, 0.5, 0.75, 1, 2, 10, and 100.

$$\begin{pmatrix} \Delta c_0 \\ \Delta c_1 \end{pmatrix} = \begin{pmatrix} V_A(c_0) & \mathrm{Cov}_A \\ \mathrm{Cov}_A & V_A(c_1) \end{pmatrix} \begin{pmatrix} 1 \\ E \end{pmatrix} P^{-1}S \qquad (6.14)$$

Notice that this is identical in format to the character state formula except for the addition of the vector $(1, E)^T$. Letting this vector be denoted by $\mathbf{B}^T$ and the genetic variance–covariance matrix by $\mathbf{V}_{Ac}$, we have

$$\Delta \mathbf{c} = \mathbf{V}_{Ac}\mathbf{B}^T\mathbf{P}^{-1}\mathbf{S} \qquad (6.15)$$

The phenotypic variance matrix also depends on the environment:

$$\mathbf{P} = \mathbf{B}\mathbf{V}_{Ac}\mathbf{B}^T + e \qquad (6.16)$$

Extension to nonlinear reaction norms is immediate, requiring only obvious modification of the matrix $\mathbf{B}$. The response of the elevation (c_0) and slope (c_1) to selection differential S on trait X in environment E for a linear reaction norm is given by the very simple formulas

$$\Delta c_0 = \frac{[V_A(c_0) + \mathrm{Cov}_A\{E\}]S}{V_P\{X(E)\}} \qquad (6.17a)$$

$$\Delta c_1 = \frac{[V_A(c_1) + \mathrm{Cov}_A\{E\}]S}{V_P\{X(E)\}} \qquad (6.17b)$$

Table 6.3   Some Estimates of Genetic Correlations Between Environments

Species	Traits (n)	Environment	r_A	Reference
Oncopeltus fasciatus	Fecundity (Pop. 1)	Photoperiod	−0.42	Groeters and
	Fecundity (Pop. 2)		0.11	Dingle
	Age at maturity (2)		0.36–0.45	(1987)
Oncopeltus fasciatus	Life history (5)	Temperature	0.16–0.44	Groeters and
	Wing length		0.43 (0.08)	Dingle (1988)
Liriomyza	Pupal weight	Plant species	0.75	Fry (1992)
	Development time		0.07	
Alsophila pometaria	Weight		1.0	
Colias philodice	Life history (2)		0.48–0.51	
	Physiological (2)		−1.0–1.0	
	Pupal weight		0.91	
Choristoneura rosaceana	Pupal weight (2)	Plant species	0.76–1.54	Carrière and
	Life history (4)		1.01–1.25	Roitberg (1995)
Nemophila menziesii	Weight	Density, competitor	≈1	Shaw and Platenkamp (1993)
Gryllus firmus	Wing dimorphism	Photoperiod	1.0	Roff (1994c)
	Wing dimorphism (4)	Photoperiod, temperature	0.49–0.65	
Gryllus pennsylvanicus	Morphology (11)	Lab versus field	0.43–0.68	Simons and Roff (1996)
	Development time (2)		0.17–0.33	
Dysdercus fasciatus	Weight	Moisture	0.28	Kasule (1992)
	Reproductive allocation		0.34	
Bicyclus anynana	Morphology (48)	Temperature	−0.78–0.97	Windig (1994b)
	Development time (24)		−0.64–0.77	
	Pupal weight (24)		−0.24–1.1	
Senecio integrifolius	Life history (4)	Light	0.27–0.47	Andersson and Widen (1993)

where $\text{Cov}_A\{E\}$ is the additive genetic covariance between c_0 and c_1 in environment E. Gavrilets and Scheiner (1993) refer to c_0 in the above equations as the mean value of the trait. Although it is true that the environment can be scaled such that this is the case, it gives a false impression because such scaling sets $E = 0$ and, hence, the genetic variation in X is due only to the genetic variance of

Table 6.4   Estimates of the Heritability of Plasticity

Species	Trait	Environment	h^2 (SE)	Reference
Drosophila melanogaster	Thorax length	Temperature	0.11 (0.03)	Scheiner et al. (1991)
	Wing length		0.29 (0.06)	
	Bristle number		0.05 (0.02)	
Eurosta solidaginis	Gall size	Lag time	0.54 (0.25)	Weis and Gorman (1990)
Bicyclus anynana	Morphology (14)	Temperature	0.40–0.76	Windig (1994a)

c_0. In some arbitrary environment $E \neq 0$, selection on X will result in changes in both the intercept and the slope of the reaction norm.

There has been some discussion in the literature as to whether plasticity is itself a trait that can evolve independently of the mean value of the observed trait (Scheiner, 1993b; Schlichting and Pigliucci, 1993; Via, 1993a, 1993b; Via et al., 1995). The temporary confusion over this issue arose because of the two separate perspectives. For a linear reaction norm, it is, in principle, possible to independently select for a change in slope (plasticity, c_1) and intercept (c_0), but selection on either will necessarily lead to a change in the mean value over two environments. If the reaction norm is a higher-order polynomial, then it is possible to select for a change in the coefficients such that the reaction norm shape changes but the mean value in the two "end" environments does not change. In general, however, the mean value as measured across several environments will change.

6.2.2   The Jinks–Connolly Rule

From a series of selection experiments on the basidiomycete *Schizophyllum commune*, Jinks and Connolly (1973, p. 40) drew the following conclusion:

> When the direction of selection and the effect of the environment of selection are opposed in their effects on the deviation of the resulting phenotype from the mean phenotype, e.g. selection for a low rate of growth at a temperature that leads to a relatively high rate of growth, the resulting selection lines are less sensitive to environmental variation than when the direction of selection and the effect of the environment of selection are reinforcing, e.g. selection for a low rate of growth at a temperature that leads to relatively low rate of growth.

An example is shown in Fig. 6.9 for the case of selection on growth rate of mice on two diets. This conclusion was restated in a later paper (Jinks and Connolly, 1975), and again with respect to selection experiments on *Nicotiana rustica* and

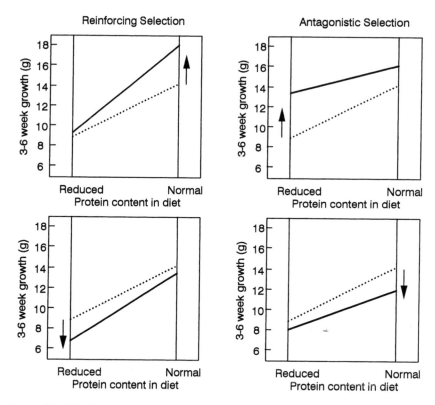

Figure 6.9 The Jinks–Connolly rule illustrated with data from selection for growth rate in mice on two diets. Dotted line shows the growth rates of the control mice, the arrow the direction and environment of selection, and the solid line the direct and correlated responses after five generations of selection. [Data from Nielsen and Andersen (1987), as reported in Falconer (1990).]

N. tabacum (Jinks and Pooni, 1988). Falconer (1989, p. 325) also drew the same conclusion from experiments on mice, naming the phenomenon the "the Jinks–Connolly rule." In a later article, Falconer (1990) showed that there is no theoretical justification for the rule. It is instructive to consider this problem from the two perspectives.

Consider two environments, E_1 and E_2, in which the mean trait value in E_1 is greater than E_2 [i.e., $X(E_1) > X(E_2)$; e.g., Fig. 6.6]. Let the reaction norm be described as $X(E) = c_0 + c_1 E + e$, where, without loss of generality, the environmental variable is scaled such that the $E_1 = 0$ and $E_2 = 1$, in which case the intercept, c_0, is then the mean value of the trait in the first environment, $X(0) = c_0$, and the mean value in the second environment is $X(1) = c_0 + c_1$. Using the character state approach, the Jinks–Connolly rule holds only if the correlated response in the second environment is greater than the direct response in the first environment. The direct response in environment 1, R_0, and the correlated response in environment 2, CR_1, are

$$R_0 = ih_0 V_{A0} \qquad \textbf{(6.18a)}$$

$$CR_1 = ih_0 r_A V_{A1} \qquad \textbf{(6.18b)}$$

where, for notational simplicity, the parameters for the traits in each environment have been subscripted with the environmental value [e.g., $V_{A1} = V_A\{X(1)\}$]. The change in slope, Δc_1, is then the difference in the direct and correlated responses,

$$\Delta c_1 = ih_0(V_{A0} - r_A V_{A1}) \qquad \textbf{(6.19)}$$

The Jinks–Connolly rule will be broken if $V_{A0} - r_A V_{A1} > 0$ (i.e., when $r_A < V_{A0}/V_{A1}$). This situation is not unlikely.

Consider now the situation using the reaction norm approach. Selection on the trait in the first environment is equivalent to selection on c_0 and it is clear that the response in the second environment, and the response of plasticity ($=$ sensitivity $= c_1$) will depend on the genetic correlation between c_0 and c_1. The change in plasticity thus depends on the sign of the genetic correlation, a viewpoint implicitly, but not explicitly, acknowledged by Brumpton and Jinks (1977) and Jinks et al. (1977) in their analyses of the genetic correlation between mean performance and sensitivity in *Nicotia rustica*. Let the absolute value of the correlated response in c_1 be R_1. The expected changes in plasticity are shown in Table 6.5. The Jinks–Connolly rule requires that the genetic correlation between the reaction norm coefficients be negative. Although, for reasons given below, this might be likely, it is certainly not inevitable.

Although there is no reason, on the basis of the mathematical theory, to suppose that the Jinks–Connolly rule is correct, it does find empirical support (Falconer, 1990). Data testing the rule are presented in Table 6.6; the change in plasticity is in the direction predicted 48 times and in the opposite direction 18 times, a difference that is statistically different from random ($\chi^2 = 13.6$, df $= 1$, $P < .001$). Because the number of data points varies between traits, this test must be viewed cautiously. With the exception of the pig data, the change in plasticity is more often in the predicted than the opposite direction in all cases (Table 6.6), a deviation that is highly significant ($\chi^2 = 6.0$, df $= 1$, $P < .01$, the pig data omitted because the number is the same in each category). If the pattern is correct, then it indicates that, in general, there is a negative genetic correlation between c_0 and c_1. If the character (e.g., body size, fecundity, etc.) is directly connected with fitness, then selection will act to fix alleles that produce a positive correlation in trait values between environments, leaving alleles which show negative pleiotropy (for reasons given in Chapter 3, this process might reflect the rate of loss

Table 6.5 Predicted Change in Plasticity for the Two Types of Selection Done in Environment 1 (E = 0); Changes Corresponding to the Jinks–Connolly Rule are Shown in Bold Type

Type of Selection	Correlated Response in Plasticity		
	$-R_1$	0	$+R_1$
Antagonistic	$c_1 - R_1$	c_1	$c_1 + R_1$
Synergistic	$c_1 + R_1$	c_1	$c_1 - R_1$

and addition of the particular types of alleles and not the indefinite preservation of antagonistic alleles). The traits examined in the experiments cited in Table 6.6 are indeed ones that might be expected to contribute to fitness; hence, the hypothesis of a negative correlation seems plausible in these cases.

6.2.3 Spatial Variation: Continuous Model

Via and Lande (1985) considered the evolution of phenotypic plasticity in a two-patch universe using the character state approach. Patches are assumed to vary spatially but not temporally. At each generation, the two populations are mixed, mated at random, and their offspring redistributed to the two patches. Selection is assumed to be sufficiently weak and population size sufficiently large that genetic variation depleted by selection is restored by mutation (i.e., the variance–covariance matrix remains constant). From the perspective of the character state approach, selection acts on the trait in each patch, plasticity then being considered a by-product of the two regimes of selection (Via, 1993a). As already discussed, evolution to the joint optima is possible in such a universe, provided the genetic correlation between environments is not ±1. Consider selection acting to move the two traits to the points indicated in Fig. 6.4 by triangles: If the genetic correlation is ±1, then there is no variation about the regression of $X(E_2)$ on $X(E_1)$ and hence selection cannot, in general, achieve the optimal combination. This is equivalent to the condition noted in Chapter 5 that response to selection is limited if the variance–covariance matrix is singular. When the genetic correlation is not ±1, there is no restriction to evolutionary response. If the genetic correlation between environments is zero, the traits will follow a trajectory of the steepest (fastest) ascent up the fitness surface (Fig. 6.10). If the genetic correlation is greater in magnitude than 0 but less than 1, the path will be curved, as the independent evolution is impeded (curved paths in Fig. 6.10). Further, the rate at which the optimum is approached will decrease as the magnitude of the genetic correlation increases (note the increase number of generations on the curved paths in Fig. 6.10). Finally, if the genetic correlation is ±1, the trajectory cannot attain the joint optimum.

Thus, in the Via–Lande model, there are no limits to plasticity under any but the most unusual circumstances. However, as the number of patches increases,

Table 6.6   Evidence For and Against the Jinks–Connolly Rule; Except Where Indicated, Data Summarized from Falconer (1990)

Organism	Trait	Number of Cases in Which Change in Plasticity Was in Direction	
		Predicted	Not Predicted
Schizophyllum	Growth rate	5	3
Mouse	Weight	15	4
Pig	Weight	1	1
Tribolium	Larval weight	5	2[a]
	4-Day fecundity	4	2
Drosophila	Thorax length[b]	6	2
	Adult weight[c]	12	4

[a]In one line there was no change: this has been ignored.
[b]Data from Scheiner and Lyman (1991).
[c]Data from Hillesheim and Stearns (1991).

and hence also the number of genetic correlations, the rate of progress to the optima can become extremely slow, as the requirement for no response to selection on n characters is that the average correlation be $-1/(n-1)$ (Dickerson, 1955); thus, even low correlations can impede progress. If the reaction norm is linear, then, in general, an optimal reaction norm can evolve only in a two-patch environment, because the reaction norm cannot be made to pass through all points in a multipatch universe. For a habitat consisting of three-patch types, a quadratic relationship is required. As the number of patch types increase so also will the general complexity of the reaction norm. This is implicit in the character state approach and can be captured in the reaction norm approach by assuming a Taylor series approximation (Jong, 1995).

In the Via–Lande model, the attainment of the joint equilibrium does not depend on whether selection is hard or soft. However, a different pattern is found if there is a "cost of adaptation." Via and Lande (1985) assumed that the fitness function for environment E, $W(E)$, followed a Gaussian curve,

$$W(E) = \exp\left(-\frac{[X(E) - \theta_E]^2}{2\gamma}\right) \tag{6.20}$$

where θ_E is the optimum value in environment E (i.e., fitness is at its maximum) and γ is a constant which measures the strength of stabilizing selection. Tienderen (1991) also considered such selection acting in a two-patch universe but he added a second component, $W_c(E)$, which incorporated a cost to being a generalist in the sense of being adapted to both habitats:

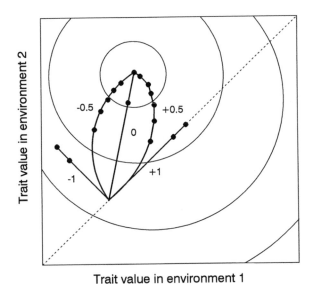

Figure 6.10 The Via–Lande model for the evolution of phenotypic plasticity in a spatially variable environment. Contours indicate combinations of mean fitness as a function of the average phenotypic value in each environment. The solid lines show the trajectories given the genetic correlations indicated beside the lines, each dot representing the change after about 50 generations. [Modified from Arnold (1992, after Via and Lande, 1985).]

$$W_c(E) = \exp\left(- \frac{[X(E_2) - X(E_1) - \theta_{2-1}]^2}{g(E)^2}\right) \qquad (6.21)$$

where $X(E_i)$ is the mean trait value in ith environment, θ_{2-1} is the cost-free reaction (a function of the difference between habitats in average response), and $g(E)$ is a parameter that is inversely related to the strength of selection against deviating from the cost-free reaction in habitat E. The fitness surfaces of Eqs. (6.20) and (6.21) for soft and hard selection are shown in Fig. 6.11. As already observed by Via and Lande (1985), there is a single peak when there is no cost to adaptation. However, the fitness surface generated by the cost function is a ridge so that now there is no single optima but rather a line. Under soft selection, the two selective forces acting together produce a single peak that is shifted away from the peak when there is no cost to adaptation (Fig. 6.11). Under hard selection, the dual action is capable of producing several peaks (Fig. 6.11); as a consequence, the combination to which selection will drive a population depends on the initial conditions.

6.2.4 Spatial Variation: Threshold Traits

Hazel et al. (1990) analyzed a quantitative genetic model for the evolution of a threshold trait in a variable environment. In contrast to the usual threshold model,

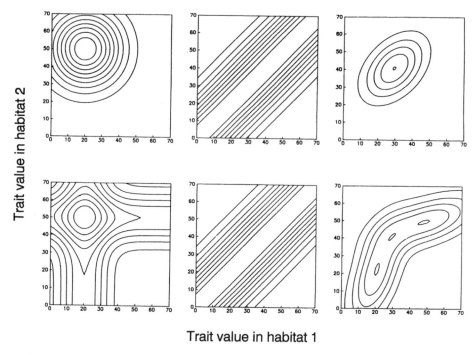

Trait value in habitat 2

Trait value in habitat 1

Figure 6.11 Adaptive landscapes under different types of selection and costs to adaptation. Top panels show soft selection, bottom panels hard selection. Panels on the left show the landscape when there is no cost to adaptation and fitness follows a Gaussian function in each habitat. Notice that in both cases there is a single joint optimum. Middle panels show the situation in which the fitness function is Gaussian with a cost to adaptation. Note that the maximal fitness occurs along a ridge. Right panels show the effect of the joint action of the two fitness functions shown to the left. When selection is soft, there is still only a single optimum, but it is shifted relative to the case when there is no cost to adaptation. Hard selection can result in multiple peaks, in which case initial conditions determine the evolutionary equilibrium. [Modified from Tienderen (1991).]

they assumed that the underlying trait distribution remains constant and the threshold of response is heritable (Fig. 6.12). Thus, the proportion of individuals that develop into phenotype A in environment E, $p(E)$, is given by

$$p(E) = \frac{1}{\sqrt{V_P \, 2\pi}} \int_E^\infty \exp\left(-\frac{(x - \mu)^2}{2V_P}\right) dx = \int_E^\infty \phi(x - \mu) \, dx \quad \textbf{(6.22)}$$

where μ is the mean threshold value in the population and V_P is the phenotypic variance of thresholds. The proportion that develop into phenotype B is $1 - p(E)$.

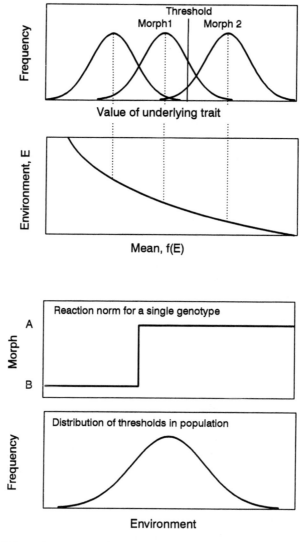

Figure 6.12 Schematic representation of the effect of environment on a threshold trait. Bottom two panels show the environmental threshold model of Hazel et al. (1990). For each genotype there is a fixed threshold of the environmental variable (e.g., temperature) that switches the phenotype from morph 1 to morph 2. These temperature-sensitive thresholds are inherited as a typical polygenic trait, generating the normal distribution of thresholds shown in the bottom panel. Top two panels show the interpretation using the standard threshold model of quantitative genetics: There is a fixed threshold and the underlying trait value changes with the environment, three examples of which are shown in the top panel. The relationship between the mean value of the underlying trait and the environmental variable, E, can be described by some function, $f(E)$, as shown in the lower panel. By using this transformation, the model can be made to be mathematically equivalent to the environmental threshold model. [From Roff (1996a).]

This model is mathematically equivalent to the usual threshold model if one simply assumes that the mean value of the underlying trait is a monotonic function of the environment (Fig. 6.12). As in the previous models, Hazel et al. (1990) assume that individuals mate in a common pool and then distribute themselves at random among the habitats. At equilibrium, the selection differential is zero, which for a two-patch model occurs when

$$(W_{A1} - W_{B1})\,f\phi(E_1 - \mu^*) + (W_{A2} - W_{B2})(1 - f)\phi(E_2 - \mu^*) = 0 \quad \textbf{(6.23)}$$

where μ^* is the mean threshold value at equilibrium, W_{Ai} is the fitness of phenotype A in patch i, W_{Bi} is the fitness of phenotype B in patch i, f is the frequency of patch 1, and E_i is the value of the environment in patch i. After some algebraic manipulation, the equilibrium condition is

$$\frac{\phi(E_2 - \mu^*)}{\phi(E_1 - \mu^*)} = \left(\frac{W_{A1} - W_{B1}}{W_{B2} - W_{A2}}\right)\frac{f}{1 - f} \quad \textbf{(6.24)}$$

After further rearrangement, we have

$$\mu^* = \frac{\ln C}{E_2 - E_1} + \frac{E_2 + E_1}{2} + \frac{\ln[f/(1 - f)]}{E_2 - E_1} \quad \textbf{(6.25)}$$

where $C = (W_{A1} - W_{B1})/(W_{B2} - W_{A2})$, and μ^* is given in units of standard deviation. For fixed fitnesses, the equilibrium mean value, μ^*, is approximately a linear function of f over much of the range of f (Fig 6.13). Hazel et al. (1990) attempted to use data from two species of swallowtail butterflies to test their model. Both species, *Battus philenor* and *Papilio polyxenes*, produce two types of pupae, green or brown. In nature, *B. philenor* choose green pupation sites 3% of the time, whereas *P. polyxenes* select green sites 94% of the time. These data do not permit a direct estimation of μ^* in the two species, but one can predict the difference in the means, *assuming that the parameter values in the model, other than f, are approximately the same for both species and one has an estimate of* $E_2 - E_1$. These assumptions appear to have been overlooked in the article and the prediction is based on arbitrary values of E_2 and E_1 (6 and 2, respectively) that happen to produce a reasonable fit between prediction ($\mu^* = 1.6$) and observation ($\mu^* = 2.04$). This test does not, therefore, provide any support for the model.

An important feature of the model of Hazel et al. (1990) is that there cannot be evolution to separate optima, such as μ_1^* and μ_2^*. From the character state perspective, this can be seen as due to a genetic correlation of 1 between environments (Roff, 1994c). Evidence from four separate experiments on wing dimorphism in insects and one on sex ratio in a turtle support this assumption (Roff,

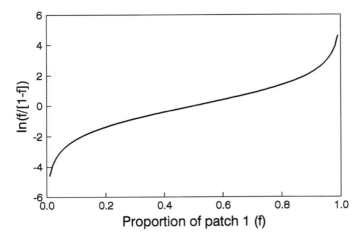

Figure 6.13 Relationship between $\ln[f/(1-f)]$ and f, where f is the proportion of patch 1 in the model of Hazel et al. (1990).

1994c; Matsumar, 1996). Thus, in this model, given variation across a single environmental gradient, such as temperature or photoperiod, there cannot be independent evolution. The conditions for a genetic correlation of 1 when more than one environmental factor varies are quite stringent and unlikely to be generally true; therefore, if both photoperiod and temperature vary, it is possible to evolve the optimal response in two environments that differ in both environmental factors. An alternative model that does permit unrestricted evolution when only a single environmental factor varies is $\mu = f(E)$, where μ is now the mean value of the underlying continuously distributed trait and the threshold is fixed, and $f(E)$ is some function of the environment in which the coefficients are genetically variable.

6.2.5 Spatial Variation: Predicting the Sign of the Genetic Correlation Between Hosts

In the above models, no prediction was made on the sign of the genetic correlation between environments. Joshi and Thompson (1995) argue that with respect to performance of insects on different hosts there should exist a negative genetic correlation. The argument for this is precisely that given earlier, namely that selection for alleles conferring a positive correlation will become fixed, leaving only alleles which have opposite effects in the different environments. As already noted in Chapter 3, even if this hypothesis is not true in a particular instance, a weaker version, that the rate of a loss and influx via mutation should be greater for antagonistic alleles, might be true. A problem with the analysis of genetic correlations in a laboratory environment or by using novel hosts is that the genetic correlation so obtained might be positive and not reflect that which would be found under natural conditions (Joshi and Thompson, 1995). Thus, of the four possible combinations (types of host, types of environment), Joshi and Thompson predict that three will give positive or no genetic correlation (+ /0) between hosts, and one (normal hosts and environment) will give a negative correlation (−). However, as discussed previously (see Section 3.4), the evidence is far from conclusive that a novel environment will change the sign of the correlation, and, therefore, no prediction is really possible for the first three cases. Joshi and Thompson reviewed data from 11 experiments on genetic correlations in performance on different hosts for 10 insect species. In eight cases they (but not I for the reasons given above) predict and observe + /0 correlations, and in three cases, they predict negative correlations and observe one case in which all three types (+ / − /0) were observed, one in which two types (− /0) were observed, and one in which a negative correlation was observed. One of the negative predictions is questionable because the stock was assumed to have adapted to the laboratory environment after only five generations. Although the data might be "consistent with the more general predictions from our hypothesis" (Joshi and Thompson, 1995, p. 88), the lack of experiments in which a negative correlation was predicted and observed is so small that the strongest conclusion that can be drawn is that, in general, the experiments are unsatisfactory for predicting evolution under natural conditions. As heritability estimates from laboratory studies do not appear to greatly overestimate estimates for wild populations, it is quite likely that the use of novel environments will not so adversely affect the genetic correlation as to change its sign. In this case, the published data refute the hypothesis of negative genetic correlations between hosts.

6.2.6 Temporal Variation: Plasticity and Predictability

It is obvious that if there are cues that an organism can utilize to predict what phenotype will be the most fit in a particular environment, then selection will favor a plastic response. But suppose that no such cues exist; under what circumstance will an increase in plasticity (c_1) evolve? Orzack (1985) considered the problem using a simple single-locus, two-allele model in an environment that varied temporally between two states. In an uncorrelated environment in which the frequency of each is close to 0.5, selection favors the least plastic (i.e., most homeostatic) genotype. Autocorrelation between environments can alter this conclusion, depending on the details of the model.

Gavrilets and Scheiner (1993), using a reaction norm model, reached the same conclusion as Orzack (1985), that in a temporally uncorrelated but variable environment, selection will lead to a decrease in plasticity. This result is unsurprising: If there is no correlation between the response to the environment and fitness, then the variance in fitness will be increased and the variance in geometric mean fitness decreased. The complexity of the genetical analysis required to reach this conclusion argues strongly for the simpler phenotypic approaches of optimality modeling (e.g., Stearns and Koella, 1986; Lynch and Gabriel, 1987; Perrin and Rubin, 1990; Gabriel and Lynch, 1992; Kawecki and Stearns, 1993; Bradford and Roff, 1996).

Gavrilets and Scheiner (1993) attempted to test the prediction that plasticity will be negatively correlated with predictability of the environment using four studies but found no conclusive support. However, the quality of the data is highly questionable. The first test considered was the study of Etges (1989) on the developmental characteristics of two populations of *Drosophila mojavensis*. There are no quantitative data measuring relative degrees of predictability and, more significantly, the major host plant differs between the two populations. Given the lack of quantification and the major qualitative differences in habitat irrespective of relative predictabilities, the suggested support for the hypothesis cannot be taken seriously. The second test used data from a study by Rabinowitz et al. (1989) on eight different plant species. Comparisons among species are fraught with difficulties because species differ in a large number of features (Roff, 1992).

The third study compared two allopatric species of the desert annual *Eriogonum abertianum*. Fox (1990, p. 140) noted that the Sonoran desert site "is significantly hotter, drier and has more variable rainfall, especially during the summer" than the Chihuahuan site. According to Gavrilets and Scheiner (1993, p. 41), Fox "found that flowering time initiation was more plastic for the Sonoran Desert populations in agreement with the prediction (Fox, 1990)." Unfortunately, the relevant data are not presented. The two regimes studied by Fox were "watered" versus "drought"; plasticity is then measured as the difference in time to initiate flowering in each regime. This information is not given in the article, but extracting the modal dates from their Fig. 2, I obtain differences of approximately 30 days for Sonoran plants and 15 days for Chihuahuan plants, which is in the direction predicted. However, as a test of the hypothesis for the evolution of plasticity, this experiment seems somewhat lacking in quantification. A further problem with the experiment from the perspective of the present hypothesis is that natural populations have a bimodal distribution of flowering times, but under both experimental regimes, only unimodal distributions were observed (Fox, 1990). Thus, the environmental regimes might bear no relationship to what the plants actually experience in the field; consequently, the responses may be a laboratory artifact. An example of this can be seen in the production of diapause eggs by females of the cricket *Allonemobius socius*. Northern populations are strictly univoltine, and under natural photoperiod/temperature conditions, they produce only diapausing eggs, but under more "southerly" conditions, they produce some nondiapausing eggs (Bradford and Roff, 1995: see also the example of *Bicyclus anynana* given in the introduction to this chapter).

The fourth and final "test" examines paedomorphosis in six populations of *Ambystoma talpoideum*. The original data come from a study by Semlitsch et al.

(1990) who estimated the annual probability of each natal pond drying based on their recent history, which covered from 3 to 20 years. The actual probabilities of drying (with number of years in parentheses) were 1.00 (10 yr), 1.00 (3 yr), 0.90 (10 yr), 0.80 (10 yr), 0.27 (12 yr), and 0.15 (20 yr). Semlitsch et al. (1990) hypothesized that individuals from permanent ponds (the last two) would show a greater propensity to become paedomorphic than individuals from temporary ponds (the first four). There was significant variation among populations in the propensity to metamorphose under different conditions, but this was not correlated with the type of pond [permanent versus temporary; Semlitsch et al. (1990)]. There was no correlation between propensity to metamorphose and probability of the pond drying (Gavrilets and Scheiner, 1993). In all populations, a very high proportion of individuals became paedomorphic ($>20\%$), which given the supposedly high probability of drying in some ponds seems very odd. There are several possible explanations: (1) because of the relatively few years of observation, the probabilities of pond drying are inaccurate; (2) there is migration between ponds and hence the evolutionary response is impeded; Semlitsch et al. observed no migration between ponds during the course of their study; (3) the probability of pond drying may have been altered by recent agricultural practices (suggested by the authors); (4) the propensity to metamorphose might be influenced by the type of habitat surrounding the pond (suggested by the authors); (5) the experimental conditions did not accurately mimic the conditions experienced by the natural populations. Because of these problems, it is not possible to use this study as a reasonable test.

My conclusion from the above four data sets is that none are sufficient to provide support either for or against the hypothesis. It remains to be tested. The best tests require that fitness functions be developed and quantitative predictions made. Again, phenotypic models, at present, offer the best approach [see, for example, the evolution of diapause initiation in the cricket *Allonemobius socius* detailed by Bradford and Roff (1997)].

6.2.7 Temporal Variation: Selection on Plasticity in the Eurosta–Solidago *System*

The study by Weis and Gorman (1990) on the evolution of gall size in the insect *Eurosta solidaginis* is one of the few studies that has examined selection on a reaction norm in a natural population. *Eurosta solidaginis* attacks the goldenrod *Solidago altissima*, laying its eggs in the stems, and the larvae causing the formation of a gall. From field data, Weis and Gorman determined that survival was a complex function of gall diameter, with an optimum at approximately 24 mm (Fig. 6.14). On the basis of analysis of 16 full-sib families (described in Section 6.1.2), they obtained a linear relationship between final gall diameter and the time lag between oviposition and gall initiation (Fig. 6.14). Because the time lag showed variation associated with plant sibship but not among insect sibship, Weis

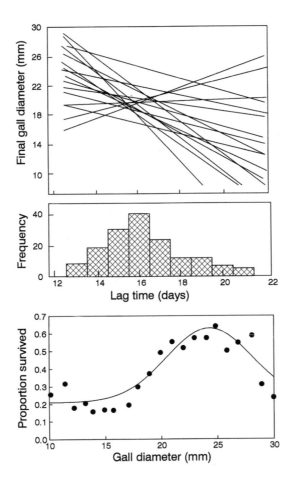

Figure 6.14 Bottom panel: Proportion of survivors as a function of gall size (*x*) in *Eurosta solidaginis.* Dots show observed value, the solid line the fitted curve

$$\text{Survival} = 0.21 + 0.42 \exp\left[-\left(\frac{x-24.3}{3.84}\right)^2\right]$$

Middle panel: Frequency distribution of lag time between oviposition and gall initiation. *Top panel:* Reaction norms from the 16 full-sib families. [Graphs redrawn from Weis and Gorman (1990).]

and Gorman (1990) argued that time lag is a trait of the plants and not the insects. Thus time lag is, from the insect's perspective, an environmental gradient associated with the plant. The heritability of the gall size/lag time intercept was determined as 0.21 (SE = 0.18) and the slope as 0.54 (SE = 0.25). Although the former estimate is not significantly different from zero, the ANOVA indicated significant family effects. The correlation between slope and intercept computed for each full-sib family was −0.27, and not significantly different from zero ($P = .3$).

Maximum survival occurs when gall diameter is approximately 24.3 mm; hence, we would expect selection to favor reaction norms that produced a gall of this size, regardless of lag time. In this regard, it is significant that the greatest intersection of the reaction norms occurs at the modal lag time but gives a gall diameter of approximately 19 mm (Fig. 6.14). Thus, the fitnesses of the different families is approximately equal in the most frequent environment but is lower than the maximum fitness. Selection should act to shift this area of intersection upward, which could be done by increasing the intercepts or changing the slopes. In the former case, an overall increase of approximately 4 mm would achieve the required maximization of fitness within the modal environment, but in the latter case, it appears that a much greater range in slopes must be achieved. Therefore, we might expect that selection will act most strongly in a directional sense on the intercept, as the same change in all families is likely to have a greater impact on fitness than a similar change in the slope. This prediction can be tested by estimating the relative strengths of selection on the two parameters.

The relative fitness of family i, W_i, is

$$W_i = \frac{\sum_{j=0}^{n} S_{ij} f_j}{W} \tag{6.26}$$

where S_{ij} is the survival of family i in environment j, f_j is the frequency of environment j, and W is the summed fitness of all families. Selection on the reaction norm coefficients was evaluated using the general approach outlined by Lande and Arnold (1983). Directional selection was determined using the model

$$W_i = \text{constant} + b_0 c_{0,i} + b_1 c_{1,i} \tag{6.27}$$

where $c_{0,i}$, and $c_{1,i}$ are the intercept and slope of the reaction norm for family i, respectively, and b_0 and b_1 are the partial regression coefficients of the reaction norm intercept and slope, respectively. Stabilizing selection was evaluated using a quadratic model. As predicted, directional selection was stronger on the intercept than the slope (intercept/slope = 4.4), but stabilizing selection acted approximately equally on the two components (intercept/slope = 1.4).

The confidence region about the heritability of the intercept is sufficiently large that little can be said about the expected response to selection. The analysis of variance suggests that genetic variance is present and, hence, that the reaction norms could evolve. The relative fitnesses would change if either the survival function or the distribution of lag times changed. Interannual variation could conceivably produce such variation. This study reinforces the point made in the previous section; namely the presence of genetic variation is necessary for reaction norms to evolve to their optima, but at the same time the continued presence of genetic variation remains a phenomenon that itself must be explained.

6.2.8 Canalization and Developmental Stability

Waddington (Waddington, 1940, 1942, 1957; Waddington and Robertson, 1966) and Lerner (1954) put forward the hypothesis that selection is capable of acting to buffer developmental events from the action of the environment. Consider two environments: Waddington's hypothesis states that selection can act such that the slope of the reaction norm between the two environments is zero. Viewed in this perspective, there is nothing exceptional about the hypothesis and it falls within the general framework of the evolution of phenotypic plasticity. Two types of experimental investigations have been used to test this hypothesis: stabilizing selection and fluctuating asymmetry. The first type of experiment consists of selecting as parents those individuals which are closest to the mean value, the argument being that such a selection will favor genotypes which show minimal deviation. There are, however, two mechanisms by which such stability could be achieved—selection for developmental canalization genes or selection for homozygosity. The results of stabilization experiments are discussed in detail in Chapter 9; suffice it to note here that the evidence suggests that stabilization selection experiments have selected for homozygosity rather than developmental canalization genes. A stronger selection experiment with respect to Waddington's hypothesis would be to select for the same response in two environments; the results already outlined in this chapter suggest that such an experiment would be successful.

A second approach to the study of canalization is the study of developmental variability, particularly **fluctuating asymmetry** (FA), which is defined as nondirectional deviations from bilateral symmetry (Palmer and Strobeck, 1986). The hypothesis is that selection in bisymmetrical organisms will favor a minimal amount of fluctuating asymmetry, a hypothesis that can (as so many can) be traced back to Darwin: "It might be anticipated that deviations from the law of symmetry would not have been inherited" (Darwin, 1868, p. 456). In modern terms, this hypothesis states that the heritability of FA should be close to zero. Heritabilities of FA show wide variation and in several cases are quite high [Table 6.7 and review by Moller and Thornhill (1997)], refuting Darwin's hypothesis.

Fluctuating asymmetry can actually be viewed as consisting of two traits, the

Table 6.7 Estimates of the Heritability of Asymmetry in Morphological Traits

Species	Trait	h_X^2	h_{FA}^2	SE	n	Min h_{FA}^2	Max h_{FA}^2	Ref.[f]
D. melanogaster	Wing length	0.67	−0.01[a]	0.02	2	−0.03	0.005	3
Rainbow trout	Meristic	0.52	0.02[b]	ns	1	—	—	8
Mouse	Molars	0.6	0.02[c]	0.02	6	−0.01	0.11	5
D. melanogaster	Sternopleural bristles	~0.1	0.06[d]	0.02	7	0.02	0.17	1
Human	Finger-ridge counts	0.95	0.27[d]	0.02	2	0.25	0.29	6
Rhesus Macaque	Cranial features	0.55	0.35[a]	0.14	12	−0.39	1.14	2
Barn swallow	Feather and skeletal	ng	0.41	—	8	—	—	1
Stickleback	Lateral plates	0.79	0.63	0.16	1	—	—	4
Scorpionfly	Wing length	ng	1.07	0.44	1	—	—	7

Note: Data ranked according to h_{FA}^2. Also shown is the average heritability, h_X^2, of the two traits showing the asymmetry.

[a] Separate traits, same study.

[b] Asymmetry measured as the number of characters (1–5) for which an individual was asymmetric. The heritability, h_X^2, is the arithmetic mean of the five separate traits.

[c] Same trait, different lines.

[d] Single trait, separate studies.

[e] Heritability estimate reported in Moller and Thornhill (1997) based on unpublished data and other statistics not reported. However, the data in Moller (1994) reporting on four of the traits gives one significant result ($h^2 = 1.34$, based on regression of sons on fathers and daughters on mothers) and three nonsignificant (values unreported).

[f] References: (1) Moller and Thornhill (1997); (2) McGrath et al. (1984); (3) Scheiner et al. (1991); (4) Hagen (1973); (5) Leamy (1986); (6) Singh (1970); Martin et al. (1982), h_T^2 from Holt (1968); (7) Thornhill and Sauer (1992); (8) Leary et al. (1985).

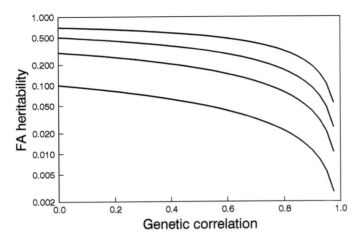

Figure 6.15 Heritability of fluctuating asymmetry as a function the genetic correlation between the two traits assuming that the phenotypic and additive genetic variances of the separate traits are the same (from bottom to top, the curves show the results for $h^2 = 0.1$, 0.3, 0.5, and 0.7) and that $r_E = 0$.

trait (e.g., wing length) on the left side and the trait on the right side plus a genetic correlation between them. The phenotypic and additive genetic variances of the two traits are likely to be very similar; for illustrative simplicity, assume that they are equal at V_A and V_P, respectively. The additive genetic variance of FA (defined as |trait value on one side − trait value on other side|) is equal to $2V_A - 2\text{Cov}_A = 2V_A (1 - r_A)$. Similarly, the phenotypic variance is $2V_P (1 - r_P)$. Hence, the heritability of FA, h_{FA}^2, is

$$h_{FA}^2 = h^2 \left(\frac{1 - r_A}{1 - r_P} \right) \tag{6.28}$$

where h^2 is the heritability of the separate traits. If the asymmetry is due to developmental noise, then the environmental correlation, r_E, should be close to, or equal to, zero. Assuming $r_E = 0$ and making use of the relationship between the three correlations [Eq. (3.7)], we arrive at

$$h_{FA}^2 = h^2 \left(\frac{1 - r_A}{1 - h^2 r_A} \right) \tag{6.29}$$

The heritability of FA will be approximately equal to that of the separate traits unless the genetic correlation is greater than 0.8, at which point there is a marked decrease in heritability with r_A (Fig. 6.15). The latter is demonstrated by the

Table 6.8 Heritabilities and Genetic Correlations in Two Morphological Traits in Drosophila melanogaster

Trait	Heritabilities			Correlations	
	Left	Right	FA	r_A	r_P
Wing length 19°C	0.693	0.686	0.005	1.001	0.985
Wing length 25°C	0.648	0.655	−0.026	1.001	0.975
Bristles 19°C	0.038	0.069	0.039	1.198	0.183
Bristles 25°C	0.122	0.084	0.027	1.026	0.420

Source: Data from Scheiner et al. (1991).

analysis of Scheiner et al. (1991) on fluctuating asymmetry in wing length and bristle number of *D. melanogaster*. The heritabilities of right and left wing lengths are large, whereas those of bristle number are very small; however, in both cases, the genetic correlation is approximately 1, giving an expected heritability of FA close to zero, which is what is actually observed (Table 6.8). Because of the low heritability, there will be little response to selection on FA, although selection directly on wing length should produce an immediate response, as has frequently been demonstrated.

We can arrange developmental asymmetry along a gradient: at one extreme, under strong selection, and at the other, under, weak or even disruptive selection. In the former category are placed morphological structures such as wings that have be symmetrical in order to function correctly. A priori, the heritability of FA in this category will be low, as selection will have eliminated additive genetic variance (i.e., selection for a genetic correlation of 1 between the two sides). This hypothesis is supported by the aforementioned data on FA of wing length in *D. melanogaster* but not by the heritability of FA in wing length in the scorpion fly, *Panorpa vulgaris* [$h^2 = 1.07$, SE $= 0.44$; Thornhill and Sauer (1992); Table 6.8).

The second extreme class of traits is those in which asymmetry is not critical or might even be detrimental. Consider, for example, the right and left color pattern on an amphibian such as the leopard frog: If this patterning serves as camouflage, then symmetry will be exactly the opposite to what is needed. Thus, we might expect that the genetical correlation between right and left sides to be zero or even negative, and the heritability of FA to be relatively large. If symmetry in the trait on the right and left halves serves no purpose, the heritability of FA is also expected to be large, because it is under weak or no selection; the FA of fingerprint patterns might fall into this group ($h^2 = 0.27$, Table 6.8). Unfortunately, the traits given in Table 6.8 cannot be readily assigned one of the two classes, either a priori or a posteriori. What is evident is that canalization through genes that regulate development can clearly occur because the heritabilities of the individual traits can be very high but the heritability of asymmetry very low.

Traits that are closely related to fitness should be more strongly buffered against random environmental noise or differences in genetic background than traits less related to fitness (Stearns and Kawecki, 1994). Although this hypothesis is appealing, it is not clear how it can be adequately tested. Stearns and Kaweki (1994) and Stearns et al. (1995) suggested the use of the coefficient of variation (CV). They showed that the CV of a trait decreased as its impact on fitness increased. Unfortunately, as discussed in the section on evolvability (Section 4.2), the CV cannot be used to compare traits which are measured on different scales. There were, however, two traits, early and late fecundity, which might reasonably be compared, as they are traits which have a common scale. Visually, at least the analyses of early and late fecundity support the hypothesis.

6.3 The Genetic Basis of Plasticity

The different ways of viewing reaction norms have generated considerable confusion and controversy over the nature of the genetic basis of plasticity. Much of this confusion stems from a failure to recognize the difference between a mathematical model and a mechanistic model. Biometrical genetics is basically a mathematical interpretation of the process of the inheritance of quantitative traits. Thus, for example, the mechanistic processes of dominance and epistasis are decomposed in the biometrical approach into the mathematical components of linear and nonlinear effects (see Chapter 2). The confusion over the separation of these two aspects is well illustrated by two of the three models of plasticity proposed by Scheiner and Lyman (1991). These two models are termed the pleiotropy model and the epistatic model. According to Scheiner and Lyman (1991, p. 25), in the pleiotropy model, "plasticity is a function of differential expression of the same gene in different environments (Falconer, 1981 [=1989]; Via and Lande, 1985; Via, 1987)", whereas in the epistatic model, "plasticity is due to genes that determine the magnitude of response to environmental effects which interact with genes that determine the average expression of the character. . . . (Lynch and Gabriel, 1987; Jinks and Pooni, 1988; Scheiner and Lyman, 1989)." In fact, none of the cited articles makes any reference to the mode of action of genes; they are all mathematical descriptions of the process. With the exception of the model of Lynch and Gabriel (1987), all of the models are mathematically equivalent, the pleiotropy model being the character state perspective and the epistatic model being the reaction norm perspective. It is true that Jinks and his collaborators have analyzed the pattern of inheritance using diallel crosses to separate additive, dominance, and epistatic effects, but this is no different from separating such effects for other types of quantitative characters. For *Schizophyllum commune*, Jinks and Connolly (1975) found that the interaction between growth and temperature was largely due to additive and dominance gene action, but they did not determine the relative contribution of each to the overall genetic variance. In contrast, significant epistatic effects have been observed in *Nicotiana tabacum* and *N. rustica*

(Jinks et al., 1973; Perkins and Jinks, 1973; Pooni et al., 1987; Jinks and Pooni, 1988), although, again, the impact on heritability estimates has not been determined. Interestingly, epistasis is found only in the extreme environments, a phenomenon which Jinks et al. (1973) attribute to stabilizing selection at the intermediate environments and directional selection at the extremes. At this time, the genetical architecture of plasticity is unexplored in nondomestic species.

The model of Lynch and Gabriel (1987) is very different from those previously discussed: The reaction norm is assumed to be described by a Gaussian function

$$X(E) = \frac{1}{\sqrt{2\pi C_1}} \exp\left(- \frac{(C_0 - E)^2}{2C_1} \right) \qquad (6.30)$$

where the trait X is directly associated with fitness, and C_0 and C_1 are constants characteristic of a particular genotype. The highest fitness occurs at $E = C_0$, and C_1 defines the tolerance of the genotype to deviation from this optimum; an increase in C_1 leads to a flattening of the curve and a more tolerant genotype, although with a loss of maximum fitness. Thus, there is a trade-off between maximum fitness and tolerance. As with the previous models, this model makes no assumption about the specific genetic architecture; indeed, the model is analyzed on the basis of clonal variation. It is, therefore, a phenotypic model rather than a quantitative genetic model.

A further source of confusion introduced by Scheiman and Lyman (1991, p. 25) is the use of the terms "pleiotropy" and "epistasis" in an idiosyncratic manner, "in the present context the terms refer to effects that are manifest across environments only, not within a single environment." Scheiner and Lyman (1991, p. 25) conclude that, "The pleiotropy model predicts a weak response to selection on plasticity because there is no directional selection on the character itself and the distribution of allelic expressions in different environments is independent. The epistasis model predicts a response to selection proportional to the heritability of plasticity because there exist independent plasticity genes." Opposing predictions such as these cannot be derived simply by virtue of variation in the mathematical description of the reaction norm. How they can be derived by a consideration of the supposed gene action is unclear.

Another model considered by Scheiner and Lyman (1991) is one they call the overdominance model, in which plasticity is a function of heterozygosity. This model posits that heterozygotes are buffered against environmental variation and so can maintain the same phenotype in the face of environmental fluctuation (Lerner, 1954; Dobzhansky and Levene, 1955; Marshall and Jain, 1968; Gillespie and Turelli, 1989). This model is based on the premise that response to an environment reduces fitness, but there is no a priori reason to suppose that this it true. For example, suppose that in environment 1, the time available for growth and reproduction is t_1 and in environment 2, it is t_2, with $t_2 > t_1$, and that there is some environmental cue *E* from which the duration of the season can be predicted. Further, for illustrative simplicity, assume that the organism is semelparous and that in both environments, the optimal phenology is one generation [for an analysis of this type of model, see Roff (1980, 1983) and Iwasa et al. (1994)]. Given that fecundity increases with body size, which increases with development time, the optimal development time in environment 2 is greater than environment 1, and we would expect a reaction norm to evolve in which development time was conditional on the environmental cue. In this case, genotypes which were not phenotypically plastic would have a lower fitness. It is, of course, possible that plasticity is somehow an unavoidable consequence of heterozygosity. However, interspecific comparisons in plants species show no clear pattern between heterozygosity and plasticity [see reviews in Schlichting (1986) and Sultan (1987)]. For a discussion of fluctuating asymmetry and heterozygosity, see Chapter 9.

Schlichting and Pigliucci (1993, 1995) recognized two distinct forms of genetic control of plasticity: **allelic sensitivity** and **regulatory plasticity**. In the former, the expression of genes is altered by changes in external conditions, whereas in the latter, "genes detect the change in external conditions (through appropriate receptors) and alter the expression of other genes (an indirect response)" (Schlichting and Pigliucci, 1995, p. 156). According to Schlichting and Pigliucci, these two types of genetic control are the basis for discrete and continuous responses. There is, however, no logical reason for such a distinction (Via et al., 1995): Discontinuous variation can be understood using a threshold model, and the underlying continuously distributed trait can show the same graded response as, say, body size. The type of genetic regulation cannot be ascertained from the phenotypic expression of the trait.

6.4 Summary

Phenotypic plasticity is ubiquitous and no evolutionary analysis can, in the long run, ignore it. The phenomenon can be viewed from two perspectives: the character state model and reaction norm model. The former approach focuses on the character values in the different environments, connecting them via the genetic correlation, whereas the latter approach sees phenotypic plasticity as a function that relates character values with environmental values, the traits inherited being the coefficients of the function. Both approaches have particular advantages and it is worthwhile to analyze data in both ways. Whereas the statistical analysis of variation between two environments is relatively well worked out, the extension to multiple environments still needs much work. Given the complexities and difficulties of estimating the statistics for just two environments, present experimental work is probably best restricted to such situations. From either perspective, it is clear that phenotypic plasticity will respond to directional selection, as has been observed in many different species. Theoretical analyses of the evolution of phenotypic plasticity in spatially and/or temporally variable environments has shown that the attainment of joint optima is not generally constrained, but the time taken to achieve these values may be extremely long because of the inertia created by the genetic variance–covariance relationships. Finally, the connection between the statistical models of phenotypic plasticity and the mechanistic models has generated considerable confusion and we decidedly need a general theoretical framework to connect the two perspectives.

7

Sex-Related Effects on Quantitative Variation

Thus far, it has been assumed that inheritance is autosomal and that environmental effects can be defined without regard to the parents. Both of these assumptions can be wrong in particular instances. Sex-determining chromosomes carry not only genes that determine gender but also other traits; thus, the contribution by one sex to the genetic value of a trait may be different from the other sex. Because the different contributions are simply numerical, the problems introduced by sex-linked inheritance are relatively minor and easily resolved (Bohidar, 1964; Griffing, 1965; James, 1973; Bulmer, 1985, p. 98). However, problems introduced by non-Mendelian parental effects are another matter altogether. The phenotype of an off-spring may depend not only upon its own genetic constitution but also upon the actions or phenotype of the parents, which may themselves be inherited. This effect is most frequently found via the mother; hence, the importance and consequence of maternal effects has received considerable attention first from breeders and more recently from evolutionary biologists. A maternal effect in animals can result from three sources: first, the mother may pass factors to its offspring through the cyto-plasm of the egg; second, mothers may differ genetically in how they nourish their offspring; third, environmental conditions experienced by the mother may effect her contribution to the offspring phenotype. The last two factors can be illustrated by consideration of lactation or nursing behavior in mammals: Females may differ genetically in the amount of milk that they produce or their proclivity to nurse their offspring; similarly, conditions experienced during ontogeny could generate dif-ferences in milk production between females due to, say, differences in body size, these differences being environmentally determined. In plants, there is an addi-tional source of maternal contribution via the endosperm, which always contains more doses of female genes than male genes (Roach and Wulff, 1987).

7.1 Influence of Loci Located on the Sex Chromosomes

Gender is determined by particular chromosomes in many organisms, the hetero-gametic sex being either male or female (Table 7.1). In principle, the homoga-

Table 7.1   Distribution of Heterogamy in Plants and Animals

Phylum	Class[a]	Male XY	Female XY	Male XO
Angiosperms	Angiosperms (F)	11	2	
Platyhelminthes	Trematoda (F)		1	
Nematoda	Phasmida (O)			4
Arthropoda	Arachnida (SC)			2
	Crustacea (SC, O)	5	3	Some
	Chilopoda	1		
	Insecta (SC, O)	3	3	11
Vertebrata	Osteicthys (F)	21	8	
	Amphibia (F)	5	6	
	Reptilia (F)	2	1	
	Aves		1	
	Mammalia	1		

[a]Symbols in parentheses denote the taxonomic level examined: SC = subclass; O = order; F = family.

Source: Adapted from Bull (1983).

metic sex produces twice the gene product that the heterogametic sex produces. Failure to produce such a difference is caused by **dosage compensation**, by which the products in the two sexes remain equal. In mammals, this is caused by inactivation of one of the X chromosomes, which one being due to chance, leading to female mammals being a mosaic (Bulmer, 1985; Graves, 1987). Different molecular mechanisms underlying dosage compensation may occur in *Drosophila* and *Caenorhabditis* (Hodgin, 1987; Baker et al., 1994; Parkhurst and Meneely, 1994). Dosage compensation can alter the response to selection (Frankham, 1977a), although its possible importance in this regard appears not to have been well studied.

Similar in effect to dosage compensation, although not due to linkage of genes on the sex chromosomes, is **genomic imprinting**, in which chromosomes function differently according to their parental origin (Crouse, 1960; Chandra and Brown, 1975; Peters and Ball, 1990). Genomic imprinting can be defined and described as "an epigenetic gamete-of-origin dependent modification of the genome.... When an allele at an imprintable locus passes through gametogenesis of one sex, its ability to be expressed is unaffected. However, when this allele passes through gametogenesis of the opposite sex, it becomes inactivated and is unable to be expressed" (Peterson and Sapienza, 1993, p8). Imprinting has been observed in *Chlamydomonas* (Sager and Ramanis, 1973), maize (Birchler and Hart, 1987; Kermicle and Alleman, 1990), wheat (Flavell and O'Dell, 1990), *Sciara* (Crouse, 1960), coccids (Brown and Nur, 1964), *Drosophila* (Spofford, 1976), *Nasonia* (Werren, 1991), and mammals (Searle and Beechey, 1990 Thomas and Rothstein, 1991; Varmuza and Mann, 1994).

Genes that are carried on the sex-determining chromosomes might show differential expression because in one sex, they might be represented by a single allele, whereas in the other, there might be two alleles; well-known examples are red–green color blindness and hemophilia in humans. Crosses between closely related species of Lepidoptera (in which females are XY and males are XX) have demonstrated the involvement of sex-linked genes in such quantitative traits as development rate, body size, diapause, color, egg weight, and pheromone composition [reviewed by Sperling (1994)]. Similar crosses in *Drosophila* species and two species of mosquito have revealed large effects of the sex chromosomes on male and female viability, fertility, and mating ability (reviewed in Coyne and Orr, 1989). Crosses between strains have also demonstrated sex-linkage effects on growth in chickens (Barbato, 1991; Barbato and Vasilatos-Younken, 1991).

Because of the differential expression, it is necessary to determine the extent to which a quantitative trait is influenced by genes carried on the sex-determining chromosomes. Consider an organism in which one sex, say the males, is heterogametic (XY) and the other, the females, is homogametic (XX). If the genes determining some trait are carried on the Y chromosome, then the slope of offspring value on parental value will depend on the sex of the offspring and the parent used: only the male offspring on male parent will give a slope greater than zero. An example in which a trait appears to be very largely determined by Y-linked genes is body size in the sailfin molly, *Poecilia latipinna* (Travis, 1994b, Fig. 7.1). Patriclinal inheritance of traits such as body size, age at maturity, and color pattern has been demonstrated in many species of poecilid (Kallman, 1989; Houde, 1992). In the case of color pattern, the analysis is made somewhat more difficult because the color is expressed only in the males. In the guppy, *Poecilia reticulata*, son on father regressions, the sire estimates from the half-sib analysis and realized heritabilities are generally greater than one if the trait is assumed to be autosomally inherited, but take sensible values if the trait is assumed to be entirely Y-linked (Table 7.2). Further, the dam estimates from the half-sib analyses are negligible, which also argues for primarily sex-linked inheritance.

A particular trait may be sex-linked in one species but not in another, an example of which is to be found in the inheritance pattern of wing morphology in the crickets *Gryllus firmus* and *G. rubens*. In both of these species, two wing morphs occur, a macropterous (long-winged) morph that is capable of flight and a micropterous (short-winged) morph in which the wings do not extend beyond the tegmina, lacks wing muscles, and is incapable of flight. The genetic basis of this dimorphism can be understood using the threshold model (see Chapter 2). Selection for increased and decreased percentage macroptery produced rapid responses in both species, producing after approximately 10 generations, lines that were close to monomorphic (Roff, 1990; Walker, 1987; Table 7.3). In the case of *G. firmus*, crosses between the high and low selected lines more closely match the predictions based on autosomal inheritance than sex linkage (females are XX and males XO; Table 7.3). However, the reverse is true for *G. rubens* (Table 7.3).

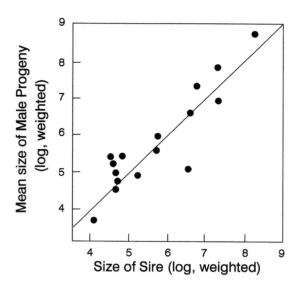

Figure 7.1 Son-on-sire regression of standard length (mm) in the sailfin molly, *Poecilia latipinna*. Each point is the weighted mean from a paternal half-sib family. [Redrawn from Travis (1994b).]

In unselected wing dimorphic orthoptera, the percentage of macropterous males is typically less than the percentage of macropterous females (Roff and Fairbairn, 1993), indicating that genes controlling the expression of gender also determine in part the expression of wing morphology. The foregoing data suggest that in *G. firmus*, such genes on the sex-determining chromosomes are largely fixed, whereas in *G. rubens*, it is autosomal genes that influence wing morphology that are fixed.

7.2 Sexual Dimorphism

Many quantitative traits are differentially expressed in males and females; for example, in animals, female body size is typically greater than that of males (Fairbairn, 1990). Such differences could result from a genetically invariant difference between the sexes but, more likely, whatever factors (e.g., hormonal) that determine the differences also show genetic variability. To analyze the evolution of these differences, we can take the approach suggested by Falconer (1952) and regard the sexes as two separate environments connected genetically by the additive genetic correlation (Frankham, 1968). From this it follows that the response to selection is given as (Leutenegger and Cheverud, 1985)

$$R_m = \frac{1}{2} \left[h_m^2 S_m + h_m h_f r_A \left(\frac{V_{Pm}}{V_{Pf}} \right)^{1/2} \right] \tag{7.1a}$$

Table 7.2 Heritability Estimates of the Relative Amount of Orange in Male Guppies;
Estimates from Separate Experiments

Method of Estimation	Assuming No Sex Linkage	Assuming Complete Y Linkage
Son on father	1.08	0.54
Son on father	1.42	0.71
Son on father	0.76	0.38
Half-sib, sire estimate	1.76	0.88
Half-sib, dam estimate	0.00	0.00
Half-sib, sire estimate	1.78	0.89
Half-sib, dam estimate	0.10	0.05
Selection, high line	0.20	0.10
Selection, high line	1.50	0.75
Selection, low line	1.08	0.54
Selection, low line	1.12	0.56

Source: Data from Houde (1992).

$$R_f = \tfrac{1}{2}\left[h_f^2 S_f + h_m h_f r_A \left(\frac{V_{Pf}}{V_{Pm}}\right)^{1/2}\right] \tag{7.1b}$$

where R is the response to selection, the subscript denoting the sex (m = male, f = female), h^2 is the heritability, S is the selection differential, r_A is the genetic correlation between the sexes, and V_P is the phenotypic variance. The correlated response, and hence the change in sexual dimorphism, depends on all the foregoing parameters, but particularly the size of the genetic correlation; if it is zero or very low, then the sexes are relatively free to evolve independently. Although the data are frequently gathered, from which the genetic correlation between the sexes can be calculated, it is surprisingly infrequently reported. The available data suggest that the genetic correlation between the sexes is typically very high (>0.8), at least for morphological traits (Table 7.4). Thus, although at equilibrium it may be possible for both sexes to attain their respective optima (Lande, 1980b), the time taken to achieve this depends critically on parameter values (Cheverud et al., 1985; Reeve and Fairbairn, 1996).

A direct test of the change in size dimorphism in *Drosophila melanogaster* resulting from selection on one or both sexes was performed by Reeve and Fairbairn (1996). Whereas the realized genetic correlation between the sexes (0.95) was very similar to that obtained from a preliminary half-sib experiment (0.93), the realized heritabilities were considerably lower (0.13 versus 0.66 for females, and 0.13 versus 0.43 for males). More importantly, it was not possible to obtain reasonable predictions of the observed responses in sexual size dimorphism: The ratio of female to male size actually showed a concave relationship with female

Table 7.3   Predicted and Observed Percentages of Macropterous Individuals of the Crickets Gryllus firmus *and* G. rubens *in Crosses of Lines Selected for Increased and Decreased Proportion Macropterous*

Female Parent	Male Parent		Gryllus firmus		Gryllus rubens	
			Female	Male	Female	Male
High	High	Observed	95	88	99[a]	99[a]
Low	Low	Observed	2	3	2[a]	2[a]
High	Low	Predicted NS[b]	66[c]	76	39	39
		Predicted S[b]	50	100	50	100
		Observed	61	64	51	99
Low	High	Predicted NS[b]	66	76	39	39
		Predicted S[b]	50	0	50	0
		Observed	67	48	47	11

[a]No significant difference between the sexes and data combined by Walker (1987).

[b]NS = Not sex-linked; S = Sex-linked; males are XO and females are XX.

[c]Predicted value obtained by converting proportions to underlying scale.

Source: Data from Roff (1990) and Walker (1987).

size. This phenomenon can be accounted for if selection is actually acting on the growth trajectories: Because of nonlinearity in the growth curves, those of males and females are not parallel, and selection on body size within a particular sex causes a correlated termination of the growth in the opposite sex, which results in a change in dimorphism that cannot be predicted without knowledge of the growth curve (Reeve and Fairbairn, 1996). These results argue that viewing organisms as a black box, the output of which can be simply understood using the statistical framework of quantitative genetics, can be misleading and that a fuller understanding of processes of development may be necessary.

7.3   Maternal Effects: A Theoretical Framework

Although science is clearly built on the experimental demonstration of phenomena, it is possible to argue the existence of a particular phenomenon fairly convincingly prior to such experimental verification; maternal effects is such a case. Consider, for example, the Mammalia: A defining characteristic of mammals is lactation; this is done only by the female (with the possible exception of a bat species) and is the only means of nourishment for the early neonates. Likewise, the developing embryos within the mother receive their nourishment from the body of the mother. Consequently, it is obvious that there must be a maternal contribution to the initial condition of the offspring that generally outweighs that of the male. It is plausible that this maternal effect also has a paternal component if, for example, the male parent feeds the female during gestation and lactation

Table 7.4   *Survey of Genetic Correlations Between the Sexes*

Species	Trait	Mean	Min	Max	N[a]	Reference
Chicken	Body weight	0.80			1	Becker et al. (1964)
Turkey	Body weight	0.91			1	Becker et al. (1964)
Quail	Body weight	1.00			1	Becker et al. (1964)
Geospiza fortis	Morphology	0.92	0.75	1.03	4	Price (1984)
Taeniopygia guttata	Bill color	0.91			1	Price and Burley (1993)
Mouse	Morphology	0.89	0.87	0.90	2	Eisen and Legates (1966), Eisen and Hanrahan (1972)
Drosophila melanogaster	Bristles	0.69	0.41	0.91	4	Sheridan et al. (1968)
	Bristles	0.92	0.90	0.93	2	Frankham (1968)
	Morphology	0.89	0.25	1.27	28	Cowley et al. (1986), Cowley and Atchley (1988)
	Body size	0.93			1	Reeve and Fairbairn (1996)
Dung fly	Body size	0.39			1	Simmons and Ward (1991)
Gryllus firmus	Wing dimorphism	0.86			1	Roff and Fairbairn (1993)
Gryllus pennsylvanicus	Morphology	0.88	0.79	0.95	4	Simons and Roff (1995)
	Gonad weight	−0.15			1	Simons and Roff (1995)
	Development time	0.77			1	Simons and Roff (1995)

[a]Number of estimates.

and hence indirectly provides the nourishment to the young. The same argument can be advanced for other organisms—egg or seed construction being under the control of the female more than the male. Direct maternal or paternal influences will generally decline rapidly with age, as few organisms have protracted parental care. Indirect effects resulting from conditions experienced during early ontogeny may have long-term effects and determine such things as time to maturity or adult body size. The foregoing argument suggests that maternal effects should be common, but it is still necessary to experimentally demonstrate their existence. Further, it is necessary to examine how important such effects are in the life history of the organism. Finally, if such effects are important, how can they be incorporated into a quantitative genetic model? The design of experiments to detect maternal effects depends in part on the quantitative model assumed, two alternative theoretical frameworks having been suggested: the Willham model and the Kirk-patrick–Lande model. The latter model is more general than the former, but as much of the statistical analysis of maternal effects is based on the Willham model, it is necessary that both approaches be described.

7.3.1 The Willham Model

The basis of this model was introduced by Dickerson (1947) and developed by Willham (1963, 1972). When there is an effect of the maternal phenotype, X_M, on the phenotypic value of her offspring, X_O, the latter can be decomposed into

$$X_O = A_O + A_M + C + E + \text{nonadditive effects} \qquad (7.2)$$

where A_O is the additive direct genetic effect (i.e., genetic effects resulting from the offspring's own genes), A_M is the additive maternal genetic effect (i.e., additive change in the offspring's phenotypic value resulting from the genotype of the mother, also called the indirect maternal genetic effect), C is the environmental effect resulting from a common environment (e.g., the same nest), and E is the environmental effect peculiar to each individual. For simplicity, nonadditive effects will be assumed to be negligible. The additive (direct) maternal effect is exerted by the mother, but the offspring also inherit the genes that determine this effect; this effect is known as the direct maternal genetic effect and will be designated A_{OM} (= maternal effect genes of the offspring).

The phenotypic variance of the offspring is thus

$$V_{PO} = V_{AO} + V_{AM} + \text{Cov}_A(A_O, A_{OM}) + V_C + V_E \qquad (7.3)$$

where

V_{PO} = Phenotypic variance in the offspring

V_{AO} = Direct additive genetic variance (i.e., the variation in the offspring due to the additive contribution of its own genes)

V_{AM} = Maternal additive genetic variance (also called the indirect additive genetic variance) [i.e., variance in the trait value of the offspring due to the additive genetic variance of the mother (e.g., genetic variation in lactation, egg provisioning, maternal care)]

$\text{Cov}(A_O, A_{OM})$ = Direct maternal additive genetic covariance (i.e., the additive genetic covariance between the offspring's trait value as determined by its own genes and the genes that determine the maternal performance, which are also transmitted to the offspring)

V_C = Maternal environmental variance

V_E = Environmental variance (i.e., that which remains after above sources have been accounted for)

As shown in Chapter 4, the response to selection is equal to

$$R = \text{Cov}_{BX}\beta \tag{7.4}$$

where Cov_{BX} is the covariance between breeding value B and the phenotypic value X, and β is the selection gradient (regression of relative fitness on phenotype). Because of maternal effects, the more usual formulation $R = h^2 S$ does not have the same meaning because the heritability term now includes a maternal component. The breeding value is equal to the sum of the direct additive genetic effect (A_O) and the genes determining the direct maternal effect (A_{OM}). Assuming no correlation between environmental effects, the covariance between breeding value and phenotypic value is

$$
\begin{aligned}
\text{Cov}_{BX} &= \text{Cov}\{(A_O + A_{OM}), (A_O + A_M)\} \\
&= \text{Cov}(A_O, A_O) + \text{Cov}(A_O, A_M) \\
&\quad + \text{Cov}(A_{OM}, A_O) + \text{Cov}(A_{OM}, A_M)
\end{aligned}
\tag{7.5}
$$

The maternal effect on the phenotype of the offspring comes only from the mother but the genes determining this effect come from both parents; therefore,

$$\text{Cov}(A_{OM}, A_M) = \tfrac{1}{2}\,\text{Cov}(A_M, A_M) = \tfrac{1}{2}\,V_{AM} \tag{7.6a}$$

and

$$\text{Cov}(A_O, A_{OM}) = \tfrac{1}{2}\,\text{Cov}(A_O, A_M) \tag{7.6b}$$

Substituting the above into Eq. (7.4), we arrive at

$$\text{Cov}_{BX} = V_{AO} + \tfrac{3}{2}\,\text{Cov}(A_O, A_M) + \tfrac{1}{2}\,V_{AM} \tag{7.7}$$

The above derivation is taken from Hanrahan (1976) and Cheverud and Moore (1994); for a derivation based on path analysis, see Willham (1963) or Riska et al. (1985). Equation (7.7) is valid for mother–offspring, father–offspring, or full-sibs. For any arbitrary coefficient of relationship r, the equation becomes (Willham, 1972)

$$\text{Cov}_{BX} = V_{AO} + (1 + r)\,\text{Cov}(A_O, A_{\text{Rel}}) + rV_{A\text{Rel}} \tag{7.8}$$

where the subscript Rel distinguishes the type of relationship and r is the coefficient of relationship. From Eq. (7.7) the response to selection, R, is (Dickerson, 1947)

$$R = \text{Cov}_{BX} \beta = [V_A + \tfrac{3}{2} \text{Cov}(A_O, A_M) + \tfrac{1}{2} V_{AM}]\beta \qquad (7.9)$$

Dividing throughout by the phenotypic variance gives the relationship in terms of heritabilities and the correlation between the direct and indirect effects (r_{OM})

$$R = (h_O^2 + \tfrac{3}{2} r_{OM} h_O h_M + \tfrac{1}{2} h_M^2) i \qquad (7.10)$$

where i is the intensity of selection. The important observation from Eq. (7.9) is that the response to selection may be impeded or even reversed if $\text{Cov}(A_O, A_M)$ is negative. Negative covariances are common in domestic mammals (Table 7.5) and have been observed in some invertebrates [body weight in *Drosophila* (DeFries and Touchberry, 1961); pupal weight and family size in *Tribolium* (Bondari et al., 1978); body size in springtails (Janssen et al., 1988)]. The effect in *Drosophila* occurs because large females produce many larvae which then suffer a decreased body size because of the crowded conditions. The possible importance of maternal effects in natural populations remains to be demonstrated, although the work on springtails (Janssen et al., 1988) and that on clutch size in birds (described in Section 7.4.7) suggests that the phenomenon warrants serious study.

7.3.2 The Kirkpatrick–Lande Model

The Willham model has been further developed and explored by Hanrahan (1976), Van Vleck et al. (1977), Cheverud (1984), and Mueller and James (1985). All of these can be subsumed within the general model of Kirkpatrick and Lande (1989), which is based on the conceptual framework suggested by Falconer (1965b). This model is best understood by considering three of the most frequent types of maternal inheritance encountered (Fig. 7.2).

7.3.2.1 Case 1: A Single Character, X, Maternally Affecting Itself

An example of this type of character is early offspring size; the size of hatchlings, seedlings, or neonates is determined in part by the genes inherited from the mother and, in part, by the phenotype of the mother. The phenotypic value of the trait in generation $t + 1$, $X(t + 1)$, is given by

$$X(t + 1) = A(t + 1) + E(t + 1) + mX_M(t) \qquad (7.11)$$

where $A(t + 1)$ is the additive genetic component, $E(t + 1)$ is the environmental component (including nonadditive effects such as dominance and epistasis), m is the **maternal-effect coefficient**, and $X_M(t)$ is the phenotypic value of the mother.

Table 7.5   *Sign of the Covariance Between the Direct Genetic Effects and the Indirect Genetic Effects Through the Maternal Effect,* $Cov(A_O, A_{OM})$, *for Various Traits in Mammals*

Species	Trait(s)	Sign	Reference
Mouse	Wt at 12 weeks	+	Eisen et al. (1970)
	Wt at 12 weeks	−	Young and Legates (1965)
	Wt at 3, 6, 8 weeks	−	Hanrahan and Eisen (1973), Eisen and Durrant (1980)
	Wt at weeks 2,3..10	+	Riska et al. (1985)
	Litter size	−	Hanrahan and Eisen (1974)
Swine	Wt at 1, 2, 3, 4 weeks	+	Ahlschwede and Robison (1971)
	Wt at 5, 6, 7, 8 weeks	−	Ahlschwede and Robison (1971)
	Birth wt	+	Kuhlers et al. (1977)
	Wt at 3, 8, 12, 16, 20 weeks	−	Kuhlers et al. (1977)
	Embryonic survival	−	Gama et al. (1991)
Sheep	Wt at weaning, postweaning[a]	−	Ch'ang and Rae (1972)
	Live wt gains, Nos. of oestruses	+	Ch'ang and Rae (1972)
	Wt at 6 weeks, weaning wt	+	Hanrahan (1976), Nasholm and Danell (1994)
	Weaner greasy fleet wt.	−	Swan and Hickson (1994)
	Weaning weight	+	Swan and Hickson (1994)
	Survival to weaning	+	Matos et al. (1994)
Cattle	Calving ease, birth wt, gestation length	−	Philipsson (1976), Burfening et al. (1981)
	Weaning wt	−, 0[b]	Koch and Clark (1955), Deese and Koger (1962), Mangus and Brinks (1970), Hohenboken and Binks (1971), Robinson (1994; review of 6 studies)
	Birth wt, weaning wt, yearling wt.	−	Arthur et al. (1994)
	Birth wt, weaning wt, growth rate	−	Pang et al. (1994)
	Calf survival	−	Martinez et al. (1983)

[a]Magnitude of covariance declined with age.

[b]Covariance zero in one herd.

Mother   Offspring

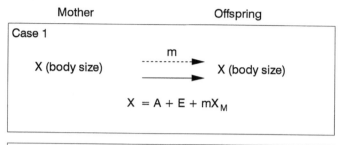

Case 1

X (body size) $\xrightarrow{\quad m \quad}$ X (body size)

$$X = A + E + mX_M$$

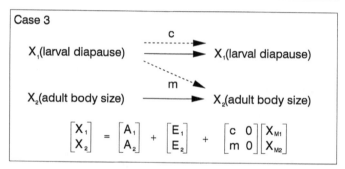

Case 2

X_1(egg provisioning) $\longrightarrow$ X_1(egg provisioning)

m

X_2(hatchling survival) $\longrightarrow$ X_2(hatchling survival)

$$\begin{bmatrix} X_1 \\ X_2 \end{bmatrix} = \begin{bmatrix} A_1 \\ A_2 \end{bmatrix} + \begin{bmatrix} E_1 \\ E_2 \end{bmatrix} + \begin{bmatrix} 0 & 0 \\ m & 0 \end{bmatrix} \begin{bmatrix} X_{M1} \\ X_{M2} \end{bmatrix}$$

Case 3

c

X_1(larval diapause) $\longrightarrow$ X_1(larval diapause)

m

X_2(adult body size) $\longrightarrow$ X_2(adult body size)

$$\begin{bmatrix} X_1 \\ X_2 \end{bmatrix} = \begin{bmatrix} A_1 \\ A_2 \end{bmatrix} + \begin{bmatrix} E_1 \\ E_2 \end{bmatrix} + \begin{bmatrix} c & 0 \\ m & 0 \end{bmatrix} \begin{bmatrix} X_{M1} \\ X_{M2} \end{bmatrix}$$

Figure 7.2 Three possible maternal-effects models using the theoretical framework of Kirkpatrick and Lande (1989).

Case 1: The trait, such as offspring size, is determined by the genes inherited by the offspring and by the mother's phenotype, which itself is genetically determined. For example, body size may be determined both by the genes of the offspring and the environment in which the offspring develop. Large mothers have high fecundity, which causes density-dependent reduction in growth, leading to small offspring which then have low fecundities and, hence, large offspring both because of the environment (low density) and the genetic constitution of the offspring. Note that the effect functionally involves more than one trait. It is difficult to erect a reasonable scenario in which the effect is functionally direct, although it can be written mathematically in such a fashion.

Case 2: Two traits, one maternally affecting the other. Egg provisioning affects the survival of hatchlings but does not itself directly affect the amount of egg provisioning by the offspring. *Figure 7.2 caption continues on facing page.*

Case 3: Two traits one of which maternally affects both itself and the other trait. In the example shown, larval diapause is determined by the environmental conditions experienced both by the mother and the offspring. An offspring that enters diapause might show a different growth pattern upon emergence from the diapause than a larva that undergoes direct development, leading to different adult body sizes. Thus, body size is a function of the genes controlling body size, the genes controlling larval diapause, and the maternal effect that comes through the diapause pathway.

The above equation is a very simple linear regression relationship between offspring and parent, and it reduces to the standard equation when $m = 0$. The coefficient m measures the relative effect of the phenotypic value of the mother; it is defined as the partial regression of the offspring's phenotype on its mother's phenotype, holding genetic sources of variation constant. The maternal effect can be either positive or negative: in the latter case, large mothers may produce genetically large offspring (i.e., large A), but their phenotypic effect (m) is to reduce phenotypic size, so that if m is sufficiently large in magnitude, small offspring are produced by large mothers. This is equivalent to the negative covariance in the Willham model.

To obtain the mean value of offspring, we take the expectations of Eq. (7.11), taking into account selection:

$$
\begin{aligned}
E\{X(t + 1)\} &= E\{A(t + 1) + mX_M(t)\} \\
&= A^*(t) + \text{Cov}_{AX}S(t)/V_P + m[X^*(t) + S(t)] \qquad \textbf{(7.12)} \\
&= A^*(t) + \text{Cov}_{AX}\beta(t) + m[X^*(t) + V_P\beta(t)]
\end{aligned}
$$

where the asterisk indicates mean values, Cov_{AX} is the covariance between the additive genetic value and the phenotypic value, $S(t)$ is the selection differential, V_P is the phenotypic variance (assumed to remain constant), and $\beta(t)$ is the selection gradient. In the absence of maternal effects ($m = 0$), $\text{Cov}_{AX}\beta(t)$ is equal to the usual $h^2 S(t)$ (heritability $\times$ selection differential). At equilibrium, Cov_{AX} is equal to [see Appendix 1 of Kirkpatrick and Lande (1989)]

$$
\text{Cov}_{AX} = \frac{V_A}{2 - m} \qquad \textbf{(7.13)}
$$

and the phenotypic variance is

$$
V_P = \frac{(2 + m)V_A + (2 - m)V_E}{(2 - m)(1 - m^2)} \qquad \textbf{(7.14)}
$$

The covariances between mother and offspring, Cov_{MO}, and father and offspring, Cov_{FO}, is

$$\text{Cov}_{MO} = \frac{V_A}{2 - m} + mV_P \tag{7.15a}$$

$$\text{Cov}_{FO} = \frac{V_A}{2 - m} \tag{7.15b}$$

The response to selection, $R(t) = X^*(t + 1) - X^*(t)$, is

$$R(t) = (\text{Cov}_{AX} + mV_P)\, \beta(t) + mR(t - 1) - mV_P\beta(t - 1) \tag{7.16}$$

Response to selection is the sum of the change in the present generation due to (1) the change in the genetic component (Cov_{AX}) plus the maternal effect (m), (2) the change resulting from the maternal effect as determined by the change in the phenotype in the previous generation, and (3) the loss in response due to purely phenotypic effects resulting from maternal influence in the previous generation. Assuming the population to be initially at equilibrium, the response to one generation of selection is approximately

$$R(1) \approx \left(\frac{2V_A}{2 - m} + mV_P \right) \beta(1) \tag{7.17}$$

An important property of maternal traits, evident from Eqs. (7.16) and (7.17), is that there is a time lag in the evolutionary response. Two consequences follow from this: First, there can be, initially, a reversed response to selection (Fig. 7.3), which occurs when $\text{Cov}_{AX} < -mV_P$, and, second, the response to selection changes each generation even under constant directional selection, only asymptotically reaching a constant value. In the present case, the asymptotic response is approximately

$$R(\infty) \approx \frac{2V_A\beta}{(2 - m)(1 - m)} \tag{7.18}$$

Once selection ceases, the response to selection does not initially cease but declines exponentially (Fig. 7.3); this continuation is called the **evolutionary momentum** by Kirkpatrick and Lande (1989). Maternal effects do not prevent the population from attaining an optimal value, only the time and trajectory to this point (Naylor, 1964; Kirkpatrick and Lande, 1989).

7.3.2.2 Case 2: A Maternal Trait, X_1, That Affects Another Trait, X_2 (Fig. 7.2)

Possible candidates for this case are litter size, milk production, or egg provisioning as the maternal trait, and body size or offspring survival as the offspring

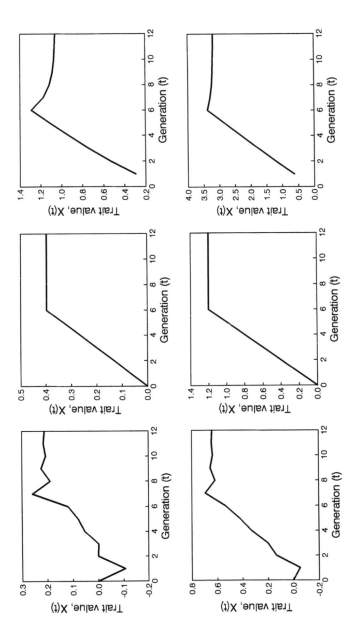

Figure 7.3 The evolution of the mean of a trait under a negative maternal effect (far right panels, $m = -0.5$), no maternal influence (middle panels, $m = 0.0$), and a positive maternal influence (far left panels, $m = 0.5$). In all cases, there is directional selection until generation 6, the response to selection being determined from Eq. (7.16), with $\beta = 0.2$, $V_E = 1.0$, $V_A = 0.3$ (top panels), and $V_A = 1.0$ (bottom panels).

trait. (e.g., egg provisioning phenotypically affects offspring survival but does not affect the egg provisioning of the offspring). This is the model most frequently considered. Letting 1 denote the maternal trait and 2 the offspring trait, the two equations describing this case are

$$X_1(t + 1) = A_1(t + 1) + E_1(t + 1) \tag{7.19a}$$

$$X_2(t + 1) = A_2(t + 1) + E_2(t + 1) + mX_{M1} \tag{7.19b}$$

where X_{M1} is the phenotypic value of the maternal trait and $|m| < 1$. These two equations can be conveniently represented by the set of matrices

$$\begin{pmatrix} X_1(t + 1) \\ X_2(t + 1) \end{pmatrix} = \begin{pmatrix} A_1(t + 1) \\ A_2(t + 1) \end{pmatrix} + \begin{pmatrix} E_1(t + 1) \\ E_2(t + 1) \end{pmatrix} + \begin{pmatrix} 0 & 0 \\ m & 0 \end{pmatrix} \begin{pmatrix} X_{M1}(t) \\ 0 \end{pmatrix} \tag{7.20}$$

There are four covariances between parent and offspring; those for the mother–offspring combination are

$$\text{Cov}_{MO} = \begin{pmatrix} \dfrac{V_{A11}}{2} & \dfrac{V_{A12}}{2} + \dfrac{mV_{A11}}{4} \\ \dfrac{V_{A12}}{2} + mV_{P11} & \dfrac{V_{A22}}{2} + \dfrac{mV_{A12}}{4} + mV_{P11} \end{pmatrix} \tag{7.21a}$$

and those for the father–offspring are

$$\text{Cov}_{FO} = \begin{pmatrix} \dfrac{V_{A11}}{2} & \dfrac{V_{A12}}{2} + \dfrac{mV_{A11}}{4} \\ \dfrac{V_{A12}}{2} & \dfrac{V_{A22}}{2} + \dfrac{mV_{A12}}{4} \end{pmatrix} \tag{7.21b}$$

where V_{ij} is the covariance between characters i and j, with A standing for the additive genetic value and P the phenotypic value. The phenotypic variances and covariances are

$$V_{P11} = V_{A11} + V_{E11}$$
$$V_{P12} = V_{P21} = V_{A12} + V_{E12} + \frac{mV_{A11}}{2} \tag{7.22}$$
$$V_{P22} = V_{A22} + V_{E22} + mV_{A12} + m^2V_{P11}$$

The response to selection on the maternal trait (X_1) follows the usual formulation, as it is not itself maternally affected. For the second (offspring) trait (X_2), the response in the first generation and at equilibrium [see appendix 1 in Kirkpatrick and Lande (1989)], assuming that there is no environmental correlation between traits 1 and 2, is approximately

$$R_2(1) \approx \left(V_{A22} + \frac{3m}{2} V_{A12} + \frac{m^2}{2} V_{A11} \right) \beta_2(1) \qquad (7.23)$$

where the subscripts denote the position in the additive genetic variance matrix; hence, V_{A22} is the additive genetic variance of trait 2 (offspring trait), V_{A12} is the additive genetic covariance between the two traits ($= V_{A21}$), and V_{A11} is the additive genetic variance of trait 1 (maternal trait). The term $\beta_1(1)$ represents the selection gradient on trait 2 in the first generation of selection.

7.3.2.3   Case 3: A Trait, X_1, That Maternally Affects Itself and Another Trait, X_2

The two types of traits discussed above can be combined as indicated in Fig. 7.2; for example, a trait such as litter size can both affect the body size of the offspring and their own future litter size. This phenomenon can easily be accommodated by the addition of another coefficient, c, into the maternal-effects matrix,

$$\begin{pmatrix} X_1(t+1) \\ X_2(t+1) \end{pmatrix} = \begin{pmatrix} A_1(t+1) \\ A_2(t+1) \end{pmatrix} + \begin{pmatrix} E_1(t+1) \\ E_2(t+1) \end{pmatrix} + \begin{pmatrix} c & 0 \\ m & 0 \end{pmatrix} \begin{pmatrix} X_{M1}(t) \\ X_{M2}(t) \end{pmatrix} \qquad (7.24)$$

where $|c| < 1$ (but not necessarily m). This process can be extended to any sort of pathway of maternal effects. The equations can also be modified to include common family environment, and maternal and paternal effects separately (Kirkpatrick and Lande, 1989; Lande and Kirkpatrick, 1990). The phenotypic covariances become quite complex, as also does the complete description of response to selection and the reader is referred to appendix 1 of Kirkpatrick and Lande (1989) for further details. The complexity of the response can be seen by the equation for the asymptotic response of trait 2 (offspring trait):

$$R_2(\infty) = \left(V_{A22} + \frac{m^2 V_{A11}}{(2-c)(1-c)} + \frac{m(3-2c)V_{A12}}{(2-c)(1-c)} \right) \beta_2 \qquad (7.25)$$

The impact of maternal effects can cause the population to evolve in the opposite direction for an indefinitely long time (Cheverud, 1984; Kirkpatrick and Lande, 1989), although we lack sufficient empirical data to say how important such effects might be in natural populations.

7.4 Measuring Maternal Effects

7.4.1 Use of Single-Locus Variants

A single-locus variant can be used in two ways. The first is exemplified by the study of maternal effects on ethanol tolerance in relation to the alcohol dehydrogenase locus in *Drosophila melanogaster* (Kerver and Rotman, 1987). Four different genotypes were initially established; two ADH-positive and two ADH-negative strains. The two ADH-positive strains are designated FF and SS (fast and slow electrophoretic alleles). The two ADH-negative strains, ff and ss, were created by introducing the Adh^{n1} allele into the same genetic backgrounds as the F and S strains. These strains were then crossed as indicated in Fig. 7.4 to produce and F_1 generation that was either heterozygous or homozygous for the negative alleles (the mating scheme is female × male). Reciprocal crosses were then made and the two types of offspring tested for their survival to 4% ethanol added to their food. The survival of ADH-positive flies did not depend on the type of cross (open histograms in Fig. 7.4), but the survival of the ADH-negative flies was greater when their mother was herself ADH-positive (black histograms in Fig. 7.4). The reason for this difference is that ADH-positive females can transfer ADH or mRNA for synthesizing ADH to the egg cytoplasm, but the ADH-negative females cannot (Kerver and Rotman, 1987). For more examples of this type of reciprocal crossing design utilizing mutants, see Barnes (1984).

The second way in which maternal effects can be established using variation at a single locus can be illustrated by the experiment of Weigensberg et al. (1997) on the maternal control of egg and hatchling size in the cricket *Gryllus firmus*. In the laboratory stock, a single-locus eye mutant was isolated; whereas the usual eye color is red in the embryo and black in the nymph and adult, the homozygous mutant eye color is white in the embryo and orange in the nymph and adult. The mutant is recessive and does not appear to be deleterious. Triplets were set up consisting of a female homozygous for white eye and two males, one wild type and one homozygous for white eyes. Eggs laid on the same day were collected from the female and allowed to develop on moist filter paper. Measurements were made of egg length on the day they were laid and of head width of the newly hatched nymphs. Differences due to the genotype of the offspring were assessed using a nested ANOVA with sire nested within dam. If egg size is a trait of the mother, then there should be no difference in egg size between sires. At the time of laying, egg size was independent of sire, indicating the maternal control of initial egg size. However, at hatching there was a significant difference in hatchling head width attributable to sire, and hence the genetic constitution of the offspring. Thus, by the time of hatching, the genotype of the offspring had asserted itself.

7.4.2 Full-sib and Half-sib Designs

The expectations using the Willham model for these two standard mating designs assuming no dominance or epistasis are shown in Table 7.6. The maternal effect

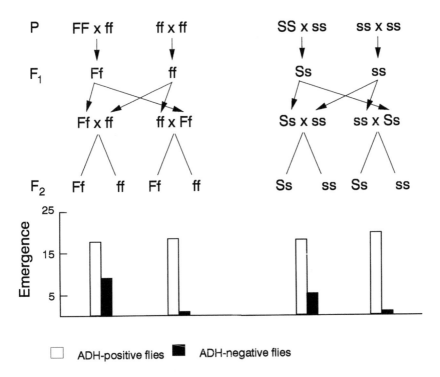

Figure 7.4 Use of a single-locus variant to establish the existence of a maternal effect for ethanol tolerance in *Drosophila melanogaster.* In the P and F_1 crosses, females are shown on the left, whereas the line giving the F_2 displays the type of offspring without regard to sex. Capital letters (F and S) designate ADH-positive alleles and lower case letters (f and s) designate ADH-negative alleles. The survival of ADH-positive flies is independent of the type of cross (open histograms), but the survival of ADH-negative flies is greatest when the female of the F_1 generation carries the ADH-positive allele (solid histograms in the leftmost groups). This increased survival arises because ADH or mRNA for synthesizing ADH is passed through the cytoplasm.

enters in three manners: the maternal additive variance, the direct maternal additive covariance, and the maternal environmental variance. The half-sib design gives an estimate of the additive genetic variance, but the three maternal effects cannot be separated (V_C can be eliminated by using a split-family nested design). This design can, therefore, be used to demonstrate the existence of maternal effects, *provided nonadditive genetics effects are assumed to be negligible.*

7.4.3 Offspring–Parent Regression

If a trait such as propagule size is determined by the mother, then it is more rightly regarded as a trait of the mother rather than a trait of the propagule. In this case,

Table 7.6   Expectations of Covariances Obtained from the Three "Standard" Breeding Designs; Dominance Variance Assumed Zero

Relationship	Expected Relationship
Paternal half-sibs	$\frac{1}{4} V_{AO}$
Full-sibs within sire	$\frac{1}{4} V_{AO} + V_{AM} + \text{Cov}(A_O, A_{OM}) + V_C$
Individuals within full-sib family	$\frac{1}{2} V_{AO} + V_E$
Offspring–sire	$\frac{1}{2} V_{AO} + \frac{1}{4}\text{Cov}(A_O, A_{OM})$
Offspring–dam	$\frac{1}{2} V_{AO} + \frac{1}{2}V_{AM} + \frac{5}{4}\text{Cov}(A_O, A_{OM})$

Note:

V_{AO} = Direct additive genetic variance (i.e., the variation in the offspring due to the additive contribution of its own genes)

V_{AM} = Maternal additive genetic variance (also called the indirect additive genetic variance) [i.e., variance in the trait value of the offspring due to the additive genetic variance of the mother (e.g., genetic variation in lactation, egg provisioning, maternal care)]

$\text{Cov}(A_O, A_{OM})$ = Direct maternal additive genetic covariance (i.e., the additive genetic covariance between the offspring's trait value as determined by its own genes and the genes that determine the maternal performance, which are also transmitted to the offspring)

V_C = Maternal environmental variance (i.e., effect of environment provided by the mother on the offspring trait value that is not due to additive or dominant genetic effects in the mother); for example, provisioning of propagules can be a function of the mother's size, which is itself a function of genetic and environmental factors.

V_E = Environmental variance (i.e., that which remains after above sources have been accounted for)

Source: From Hanrahan and Eisen (1973).

the heritability of the trait is obtained by regressing the mean trait value of the offspring on the mean trait value in the mother (e.g., mean size of offsprings' eggs on mean size of the mothers' eggs). Alternatively, some other design such as a full- or half-sib could be used, again using the mean value of the character as the trait value. Estimates of the heritability of egg size using this approach have typically produced values around 0.5 (Table 7.7).

If the trait is regarded strictly as a trait of the offspring, the usual offspring–parent regression is used (e.g., the mean egg size produced by the offspring on the mid-parent egg size). In this design, if egg size is actually a maternal trait, the offspring–dam regression will be significant, but the offspring–father regression will not. This arises because if egg size were entirely maternally derived, then the egg size from which the mother hatched is an estimate of the mean egg size produced by her mother (the grandmother of the offspring measured in this experiment). On the other hand, in this case, the egg size of the male parent makes no contribution to the egg size from which his offspring hatch; therefore, there should be no correlation between sire egg size and the mean egg size from which his offspring hatch. There will, however, be a correlation between the sire's egg size and the mean egg size produced by his offspring (the grandparent effect, discussed in Section 7.4.4).

Table 7.7   Heritability Estimates of Propagule Size in Various Animal and Plant Species

Species	Common Name	h^2 (SE)	Method of Analysis	Reference
Sorghum vulgare	Sorghum	0.63	Cross between two lines	Voigt et al. (1966)
Glycine max	Soybean	0.93	Not clear	Fehr and Weber (1968)
Lupinus texensis	Lupine	0.10 (0.004)	Offspring–mother	Schaal (1980)
Raphanus raphanistrum	Wild radish	1.26 (0.61)	Offspring–mother[b]	Stanton (1984)
Anthoxanthum odoratum	Grass	0.04, 0.17[a]	Diallel cross	Antonovics and Schmitt (1986)
	Strawberry (fruit size)	0.31 (0.10)	Offspring–mother	Shaw (1989)
Phlox drummondii	Phlox	0.0, 0.56[a]	Diallel cross	Schwaegerle and Levin (1990)
Callosobruchus maculatus	Seed beetle	0.59 (0.10)	Offspring–mother	Fox (1993)
Cyprinus carpio	Carp	0.24		Kirpichnikov (1981)
Gasterosteus aculeatus	Stickleback	0.38 (0.29)	Full-sib, mother's trait[c]	Snyder (1991)
Hyla crucifer	Tree frog (hatchling size)	0.28, 0.13	Diallel cross	Travis et al. (1987)
Scaphiopus couchii	Spadefoot toad (hatchling length)	0.0, 0.69	Diallel cross	Newman (1988)
	Chicken	0.49	Half-sib, mother's trait	Kinney (1969)
	Turkey	0.42 (0.14)	Full-sib, mother's trait[b]	Nestor et al. (1972)

[a]For diallel crosses, the analysis separates two components: that due to genes in the offspring and that due to the mother. Genetic and environmental components cannot be separated in the maternal component. Seed size might thus be due to genetic variation in resource allocation by the mother. Heritability estimates from this method of analysis thus cannot be readily compared with estimates obtained by offspring–parent regression. Therefore, I have presented the proportion of variance accounted for by the two components.

[b]The trait is regarded as an entirely maternal trait. The value for each individual based on 10 eggs/female.

[c]The trait is regarded as an entirely maternal trait. Mean of 26 estimates; where estimated, SE varied from 0.02 to 0.13. Mean of 6 offspring–mother estimates = 0.32.

(continued)

Table 7.7   Continued

Species	Common Name	h^2 (SE)	Method of Analysis	Reference
Anas platyrynchus	Mallard	0.55	Offspring–mother	Prince et al. (1970)
Lagopus lagopus	Red grouse	0.66 (0.14)	Offspring–mother	Moss and Watson (1982)
Anser caerulescens	Snow goose	0.53 (0.27)	Offspring–mother	Lessels et al. (1989)
Parus major	Great tit	0.61	Offspring–mother	Jones (1973)
		0.86	Offspring–mother	Ojanen et al.
		0.61 (0.25)	Offspring–mother	(1979)
				Noordwijk et al. (1980)
Branta leucopsis	Barnacle goose	0.67 (0.27)	Offspring–mother	Larsson and Forslund (1992)
Ficedula hypoleuca	Pied flycatcher	0.90 (0.28)[d]	Offspring–mother	Potti (1993)

[d]Egg length. Estimates for egg width and egg volume were not significant (-0.06 and 0.09, respectively).

An offspring on parent design can be used to estimate the parameters of the Willham model, providing that V_E is assumed to be zero or is estimated by using a nested design, *and that dominance and epistatic effects are negligible.* The additive genetic variance is estimated from the full-sib analysis and then subtracted from the offspring–sire component to give $\frac{1}{4}\text{Cov}(A_O, A_{OM})$. The remaining component, V_{AM}, is then estimated from the offspring–dam regression. If the dominance variance cannot be assumed to be zero, the foregoing estimates will be biased.

If the covariance term is positive, the slope of the mean offspring value on the mother's phenotypic value should then be higher than that on the sire (Table 7.6). In particular, the "heritability" of the trait ($= 2 \times$ slope) will appear higher using the offspring on dam regression than the offspring on sire regression. The standard errors associated with offspring on one parent regressions are typically very high (see Chapter 2); hence, large sample sizes are required to detect a significant difference. The absence of a difference between the two regressions does not necessarily demonstrate lack of a maternal effect because if the covariance is negative, the two slopes could be very similar.

Lande and Price (1989), using the general maternal effects model proposed by Kirkpatrick and Lande (1989), developed an offspring–parent method for the estimation of the maternal components, "*providing that all of the maternal characters influencing the characters of interest are measured and included in the analysis*" (Lande and Price, 1989, p. 918; authors' italics). The inclusion of all relevant interactions might not be as difficult as might appear at first glance. A hypothetical matrix including all the likely effects is given in Table 7.8. Three observations are noteworthy: (1) Development time, although possibly dependent on hatchling size, does not enter into the matrix because the relevant relationship is between mother and offspring; (2) in no case does a trait influence itself (all effects act via interactions with other traits); and (3) there are comparatively few interactions.

To estimate the maternal-effects matrix, all that is required are the two matrices of partial regression coefficients of offspring on mothers and offspring on fathers. Letting the covariance matrix between offspring and mother be $\mathbf{C}_M$ and that between offspring and father be $\mathbf{C}_F$, then the two matrices of partial regression coefficients are $\mathbf{C}_M\mathbf{P}^{-1}$ and $\mathbf{C}_F\mathbf{P}^{-1}$, respectively. The coefficient of the maternal-effects matrix is the difference between the coefficients of the two aforementioned matrices

$$\mathbf{M} = \mathbf{C}_M\mathbf{P}^{-1} - \mathbf{C}_F\mathbf{P}^{-1} \tag{7.26}$$

The additive genetic variance–covariance matrix is estimated from the relationship

$$\mathbf{G} = 2\mathbf{C}_F(\mathbf{I} - 0.5\mathbf{M}^T) \tag{7.27}$$

where $\mathbf{I}$ is the identity matrix and $\mathbf{T}$ stands for transpose. Considerable difficulty can be encountered if the trait is sex limited as will often be the case (e.g., lactation, litter size). In this case, it may be necessary to assume that specific elements in the $\mathbf{M}$ matrix are zero. For example, suppose we have reason to believe that case 2 applies; from the offspring–mother covariance matrix [Eq. (7.20)], we can estimate the additive genetic variance of trait 1

$$V_{A11} = 2C_{M11} \tag{7.28a}$$

and m and V_{A12} from

$$m = \frac{4(C_{M21} - C_{M12})}{4V_{P11} - V_{A11}} \tag{7.28b}$$

$$V_{A12} = \frac{8V_{P11}C_{M12} - 2V_{A11}C_{M21}}{4V_{P11} - V_{A11}} \tag{7.28c}$$

and, finally,

$$V_{A22} = 2C_{M22} - 0.5mV_{A12} - 2mV_{P12} \tag{7.28d}$$

Table 7.8 Hypothetical Matrix of Maternal Effects

Offspring	Fecundity (F_M)	Hatchling Size (H_M)	Mother Adult Size (B_M)	Development Time	Survival (S_M)
Fecundity, F	0	0	0	0	0
Hatchling size, H	$-$ [a]	0	$+$ [b]	0	0
Adult size, B	0	0	0	0	0
Development time	0	0	0	0	0
Survival, S	$-, 0, +$ [c]	0	$+$ [d]	0	0

[a] As fecundity increases, hatchling size will decrease assuming that the same amount of reproductive biomass is available.

[b] Hatchling size will increase by the same argument as above.

[c] Survival can depend on fecundity if offspring are deposited in a single location and survival is density dependent.

[d] Survival may increase with mother's body size if she is able to better protect her offspring from sources of mortality such as predators.

Below is the equation necessary to compute the offspring values. Note that development time has been omitted because it has no direct effect through maternal influence.

$$\begin{pmatrix} F \\ H \\ B \\ S \end{pmatrix} = A + E + \begin{pmatrix} 0 & 0 & 0 & 0 \\ - & 0 & + & 0 \\ 0 & 0 & 0 & 0 \\ [-,0,+] & 0 & + & 0 \end{pmatrix} \begin{pmatrix} F_M \\ H_M \\ B_M \\ S_M \end{pmatrix}$$

7.4.4 The Grandparent Effect

A maternal influence that has a heritable basis will be determined by both the genes the females received from her mother and also the genes she received from her father. Therefore, the traits so influenced by the maternal effect will be visibly correlated with the phenotype of their grandfather rather than their father. Suppose, for example, initial propagule size is maternally determined but that this determination is genetically based. Let the two parents be designated $P_♀$ and $P_♂$. The propagule size produced by this mating is a function of the genotype of the mother, $G(P_♀)$, and is unrelated to that of the father. However, the phenotype of the propagules produced by the F_1 is a function of $P_♀$ and $P_♂$. With respect to these propagules, $P_♂$ is their grandfather. This provides an experimental approach to both demonstrating the presence of a maternal effect and also that it is heritable, as illustrated by Reznick (1981, 1982) for offspring weight in the mosquito fish, *Gambusia affinis*, and the guppy, *Poecilia reticulata*. Reznick crossed mosquito fish whose grandparents had been collected from two widely separated localities in North America, North Carolina (NC), and Illinois (Ill): crosses were made in all possible combinations giving four combinations of parents. The several generations of laboratory rearing removed any possible environmental effects arising directly from the natal locations. From each combination, Reznick measured six offspring from six females. Analysis of variance showed that offspring weight was highly correlated with the female but not the male (Fig. 7.5). This experiment demonstrates the presence of a maternal effect but cannot separate environmental from genetic sources. To do this, Reznick examined three further crosses involving wild caught males from Illinois (I_δ) and North Carolina (N_δ). In one cross, I_δ were mated with female offspring from the Illinois × Illinois (II) cross. For the second cross, I_δ were mated with female offspring from the Illinois$_\text{\female}$ × North Carolina$_\delta$ (IN) cross. Finally, N_δ were mated with female offspring from the Illinois$_\text{\female}$ × North Carolina$_\delta$ (IN). If offspring weight is due to a heritable maternal effect, then offspring from the latter two matings should not differ, as they share the same grandfather stock (North Carolina). On the other hand, offspring from the first cross have different grandfathers (Illinois and North Carolina, respectively) and hence should be different. This pattern was observed (Fig. 7.5), confirming that the maternal effect did depend on nuclear genetic variation.

A simple variant on this scheme which allows at least the demonstration of maternal (or paternal) effects in natural populations is to collect females from the field and, from these, raise two generations under constant conditions. Assuming that parents in the field experience conditions different from those in the laboratory (either in mean or variance), then a maternal effect in, say, early development will be generated in the offspring produced by the field-collected females but will not be present in subsequent generations (grandparental effects could be present in the first laboratory generation but would be lost by the second generation). This is illustrated by a study on the diapause propensity of the lepidoperan *Choristoneura rosaceana*: The mean proportion of larval diapausing in the first generation (i.e., from field-collected females) was 0.70 (SE = 0.24), but in the second and third generations, only 0.17 (0.14) and 0.17 (0.13), respectively, of the larvae entered diapause (Carrière, 1994). Note that the proportions in the second and third generations were identical, indicating the absence of any effects further back than the mother.

Jenkins and Hoffman (1994) used regressions across three generations to demonstrate the presence of genetic and maternal effects in heat resistance of *Drosophila melanogaster*. The first generation of females was collected from the field and then two subsequent generations reared in the laboratory. Assuming that (1) maternal additive and maternal environmental effects are passed only through one generation (i.e., no maternal effects via the grandparent that cannot be accounted for in the mother), (2) no nonadditive genetic effects, and (3) no genotype by environment interactions, leading to differences in variance components between lab and field, then the slopes of the relationship between relatives are as shown in Table 7.9. Jenkins and Hoffman included effects that might be passed

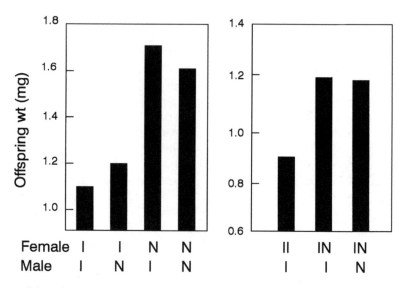

Figure 7.5 Offspring sizes of mosquito fish from different crosses. Left panel shows offspring size from the four crosses of the stocks from Illinois and North Carolina. Note that offspring size depends on the female but not the male. Right panel shows crosses made from female offspring in the left panel and wild-caught males. Females that had grandfathers from different localities (II and IN) produce offspring that differ in size but females that had grandfathers from the same locality (IN and IN) produce the same-sized offspring irrespective of the source of the father (I and N). [Modified from Reznick (1981).]

through the cytoplasm (e.g., mitochondrial DNA). Because males and females with the same heat resistance in the F_1 generation were mated to produce the F_2 generation, the coefficients of V_A differ from that expected under random mating. Heritabilities were estimated assuming no maternal effects. The heritabilities estimated from the dam–F_2 regressions and the F_1–F_2 regressions are very similar, suggesting that cytoplasmic inheritance is negligible (this is not unexpected). The higher heritabilities obtained from the dam–F_1 regressions compared to the sire–F_1 regressions suggest the presence of maternal effects. This is further indicated by the decline in heritability when estimated using dam–F_2 regressions. The heritability estimated from the sire–F_1 regressions is approximately twice as high as that obtained from the dam–F_2 regressions or the F_1–F_2 regressions: Jenkins and Hoffman suggest that this indicates that r_1 is close to 1 and that $V_{AF} > V_{AL}$. This does not seem to be an entirely satisfactory explanation because this does not account for the relatively low value of the dam–F_2 estimate, which should equal that of the sire–F_1, regardless of the G × E terms (Table 7.9). In summary, although the three-generation regression method allows the detection of maternal effects, it is not able to separate the two important components, V_{AM} and V_C.

Table 7.9   Expectations of Variance Components in Offspring on Parent and Offspring on Grandparent Regressions; Dams Collected from the Field, the F_1 and F_2 Progeny Raised in the Laboratory

Comparison	No G×E	G×E (Assuming No Cytoplasmic Effects)	Estimated h^2 Assuming No Maternal Effects	
			Males	Females
Dam–F_1	$\frac{1}{2}V_A + \frac{1}{2}V_{AM} + V_{CM} + V_C$	$\frac{1}{2}r_1 (V_{AL}V_{AF})^{1/2} + \frac{1}{2}r_2 (V_{AML}V_{AMF})^{1/2} + V_C$	0.80	0.70
Dam–F_2[a]	$\frac{1}{2}V_A + V_{CM}$	$\frac{1}{2}r_1 (V_{AL}V_{AF})^{1/2}$	0.28	0.30
F_1–F_2	$V_A + V_{CM}$	V_{AL}	0.22, 0.29[b]	0.18, 0.22[c]
Sire–F_1	$\frac{1}{2}V_A$	$\frac{1}{2}r_1 (V_{AL}V_{AF})^{1/2}$	0.48	0.46

Note: V_A = direct additive genetic variance; V_{AM} = maternal additive genetic variance; V_{CM} = maternal cytoplasmic variance; V_C = maternal environmental variance. Terms with subscripts L and F refer to the lab and field, respectively (in the absence of G×E these variances are equal). r_1 = correlation between direct additive genetic effects in lab and field; r_2 = correlation between maternal additive effects in lab and field.

[a]The coefficient used by Jenkins and Hoffman is 0.5 rather than 0.25 because the F_1 matings used individuals with the same heat resistance (i.e., assortative mating). This also accounts for the differences in coefficients in the dam–F_1 and the continued presence of V_{CM}.

[b]Father–son, mother–son.

[c]Father–daughter, mother–daughter.

Source: Modified from Jenkins and Hoffman (1994).

7.4.5   Reciprocal Crosses

The method discussed in Section 7.4.1. made use of reciprocal crosses distinguishable on the basis of variation at a single locus. The same principle can be applied using inbred lines, strains, or individuals (e.g., Chandraratna and Sakai, 1960; Jinks and Broadhurst, 1963; Smith and Fitzsimmons, 1965; Fleming, 1975; Corey et al., 1976; Millet and Pinthus, 1980; Cadieu, 1983; Garbutt and Whitcombe, 1986). The statistical analysis of such data is discussed by Cockerham and Weir (1977), from whom the following description has been extracted. The general model for reciprocal crosses is

$$X_{ijk} = \mu + G_{ij} + \text{error} \qquad (7.29)$$

where X_{ijk} is the trait value of the kth offspring from the ith maternal parent (or line) mated to the jth paternal parent (or line), μ is the overall mean, and G_{ij} is the effects attributable to the two parents (lines). Cockerham and Weir propose three statistical models for the decomposition of G_{ij}; here, I consider that designated by them as the **bio model,** using a diallel design for purposes of illustration. For this model, we have

$$G_{ij} = \text{Nuc}_i + \text{Nuc}_j + I_{ij} + \text{Mat}_i + \text{Pat}_j + II_{ij} \qquad (7.30)$$

where Nuc_i and Nuc_j represent the nuclear contributions of the parents, I_{ij} is the nuclear $\times$ nuclear interaction, Mat_i is the extranuclear maternal effect, Pat_j is the extranuclear paternal effect, and II_{ij} are all other interactions. The mating design for a diallel cross is shown in Table 7.10, along with the table of expectations and the formulas necessary to calculate the components. It is assumed that the number of observations per cross is constant (n). Definitions of the components are (1) general, V_G, the additive genetic effect arising from the nuclear contributions, (2) specific, V_S, represents nonadditive nuclear effects, (3) reciprocal general, V_{RG}, representing maternal and paternal non-nuclear effects, and (4) reciprocal specific, V_{RS}, includes all other interactions. The maternal variance component, V_{Mat}, is estimated from

$$V_{\text{Mat}} = \frac{V_{RG} - V_{RS}}{nN} - \frac{T_M - T_P}{2} \qquad (7.31)$$

where

$$T_M = \frac{\sum_i x_{i..}^2 - \sum_{i=j} x_{ij.}^2}{n^2 N(N-1)(N-2)}, \qquad T_P = \frac{\sum_i x_{.i.}^2 - \sum_{t=j} x_{ij.}^2}{n^2 N(N-1)(N-2)}$$

The paternal component, V_{Pat}, can then be estimated from Table 7.10 by subtraction. Tests of hypotheses can be constructed as follows:

Hypothesis	F-Test
$V_{II} = 0$	V_E/V_{RS}
$V_{\text{Mat}} + V_{\text{Pat}} = 0$	V_{RS}/V_{RG}
$V_I = 0$	V_{RS}/V_S
$V_{\text{Nuc}} = 0$	$V_G + (N-2)V_{RS})/N / \{V_S + (N-2)V_{RG}/N\}$

Note that there is no separate test for the maternal and paternal components. However, if $V_{\text{Mat}} + V_{\text{Pat}}$ is significant, the hypothesis $V_{\text{Mat}} = V_{\text{Pat}}$ can be tested against the hypothesis that one is larger by choosing the larger of V_{Mat} and V_{Pat} as the numerator in the F-ratio. Because this is a two-tailed test, the probability level is doubled (Cockerham and Weir, 1977). If the crosses consist of individuals rather than lines, then $V_A = 4V_{\text{Nuc}}$, $V_D = 4V_I$, and $V_E = \frac{1}{2}V_A + \frac{3}{4}V_D + V_e$ (where V_e is the environmental variance). For examples of the application of this model to nondomestic plant species, see Hayward and Nsowah (1969), Antonovics and Schmitt (1986), and Montalvo and Shaw (1994). The results for the last analysis demonstrate the important general finding that maternal effects dissipate with age, early traits (seed mass) being determined primarily by maternal effects and later traits (leaf width) by the genotype of the offspring (Table 7.11).

Table 7.10   Mean Squares and Their Expectations for the Diallel Cross

CROSSING SCHEME

Maternal Parent	Paternal Parent			Marginals
	1	j	N	
1	None			$x_{1..}$
i		$x_{i,j,1}, . \, x_{i,j,k..} \cdots x_{i,j,n}$		$x_{i..}$
N			None	$x_{N..}$
Marginals	$x_{.1.}$	$x_{.j.}$	$x_{.N.}$	$x_{...}$

EXPECTATIONS

Source	df	MS	Expectation
General	$N-1$	V_G	$V_E + nV_{II} + 2nV_I +$ $n(N-2)(V_{Mat} + V_{Pat})/2 +$ $2n(N-2)V_{Nuc}$
Specific	$N(N-3)/2$	V_S	$V_E + nV_{II} + 2nV_I$
Reciprocal general	$N-1$	V_{RG}	$V_E + nV_{II} + nN(V_{Mat} + V_{Pat})/2$
Reciprocal specific	$(N-1)(N-2)/2$	V_{RS}	$V_E + nV_{II}$
Error	$N(N-1)(n-1)$	V_E	V_E

Formulas

$$V_G = \left(\frac{\sum_i (x_{i..} + x_{.i.})^2}{2n(N-2)} - \frac{2x_{...}^2}{nN(N-2)} \right) \bigg/ (N-1)$$

$$V_S = \left(\frac{\sum_{i<j} (x_{ij.} + x_{ji.})^2}{2n} - \frac{x_{...}^2}{nN(N-1)} - (N-1)V_G \right) \bigg/ [0.5N(N-3)]$$

$$V_{RG} = [\sum_i (x_{i..} - x_{.i.})^2]/[2nN(N-1)]$$

$$V_{RS} = \left(\frac{\sum_{i<j} (x_{ij.} - x_{ji.})^2}{2n} - (N-1)V_{RG} \right) \bigg/ [0.5(N-1)(N-2)]$$

$$V_E = \left(\sum_{i \ne j,k} x_{ijk}^2 - \frac{\sum_{i \ne j} x_{ij.}}{n} \right) \bigg/ [N(N-1)(n-1)]$$

Source: From Cockerham and Weir (1977).

Table 7.11   Estimates of Variance Components (%) in the Plant Aquilegia caerulea

Trait	V_M	V_A	V_D
Seed mass	37.4[a]	3.6	0.0
Emergence time	5.5[a]	3.4	14.6
Initial leaf width	0.5	21.0[a]	3.1
Leaf width, year 1	0.0	23.2[a]	9.5
Leaf width, year 2	0.0	18.6[a]	0.1
Leaf width, year 3	0.0	4.5	16.3

[a]Significantly greater than 0 ($P < .05$).

Source: Data from Montalvo and Shaw (1994).

Cockerham and Weir (1977) present the necessary formulas for the North Carolina 2 design. For examples of its use, see Edwards and Emara (1970) and Mazer (1987a, 1987b). It is generally not feasible to do a complete diallel or North Carolina 2 design with animals, and various modified partial diallels have been used (see Section 2.2.6.2 for references).

7.4.6   Eisen's Pedigree Analysis

For many organisms, reciprocal crosses are not easily obtained, but several generations of crossing can be achieved. Most insects, such as *Drosophila* and *Tribolium*, and some vertebrates, such as mice, fall into this category. Eisen (1967) suggested three possible breeding designs requiring three generations of breeding to separate additive and dominance genetic variances from maternal effects in the Willham model with dominance effects (Fig. 7.6). In design I, full-sib males from family 1 are mated to females from families 2 and 3. Design II is more complex, with both males and females being used from families 1 and 2. The mating scheme of design III is similar to design II except that family 3 consists of half-sib females. The data are analyzed according to the statistical model

$$V_P = V_{AO} + V_{DO} + V_{AM} + V_{DM} + \text{Cov}(A_O, A_{OM})$$
$$+ \text{Cov}(D_O, D_{OM}) + V_C + V_E \quad (7.32)$$

where

V_P = Phenotypic variance in the offspring

V_{AO} = Direct additive genetic variance (i.e., the variation in the offspring due to the additive contribution of its own genes)

V_{DO} = Direct dominance genetic variance (i.e., the dominance deviation due to the genetic constitution of the offspring)

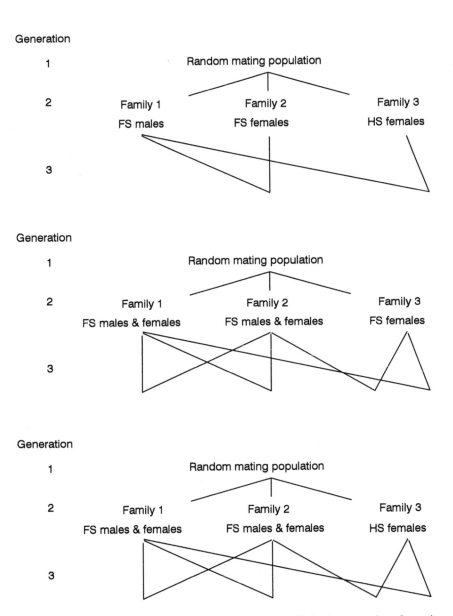

Figure 7.6 Three breeding designs suggested by Eisen (1967) for the separation of genetic and maternal effects. FS = full-sib, HS = half-sib. [Modified from Thompson (1976).]

V_{AM} = Maternal additive genetic variance

V_{DM} = Maternal dominance genetic variance (= indirect dominance genetic variance)

$Cov(A_O, A_{OM})$ = Direct maternal additive genetic covariance

$Cov(D_O, D_{OM})$ = Direct maternal dominance covariance

V_C = Maternal environmental variance

V_E = Environmental variance (i.e., that which remains after above sources have been accounted for)

This model does not take into account possible effects that may come through the grandparent or some more distant source. Such effects can be incorporated, but the model becomes very unwieldy (Willham, 1963, 1972). The crosses from the three mating schemes produce a wide range of covariances (Table 7.12), and from these, the variance components can be separated either by ANOVA (Eisen, 1967) or maximum likelihood (Thompson, 1976). These designs are logistically difficult to arrange and I have not found an example of their use.

7.4.7 Environmental Manipulation

Consider a random sample of individuals drawn from a single population; one half is exposed to one set of environmental conditions during development and the other half to an alternate environmental regime. If the offspring from the two groups, when raised in a common environment, display different phenotypes, then there must be maternal and/or paternal effects. This approach has been used in insects to demonstrate the importance of maternal influence on diapause, wing dimorphism, body size, survival rate, growth rate, lipid content, and sex ratio [reviewed in King (1987) and Mousseau and Dingle (1991)]. Similar experiments have shown that maternal influences are important in plants, particularly with respect to seed characteristics [reviewed by Roach and Wulff (1987)]. Alternatively, a female can be successively exposed to differing conditions; for example, suppose the environment is rotated through a sequence A, B, A, B, and so forth and the female produces eggs of size $E1$, $E2$, $E1$, $E2$, and so forth, then this is evidence for a maternal effect on egg size. However, if the sequence were $E1$, $E2$, $E3$, $E4$, then this indicates variation in egg size, but it is not possible to determine whether the variation is a result of the female or due to the offspring genotypes (for example, there may be physiological changes in the mother that cause selective mortality of genotypes). Several environments might be provided simultaneously. For example, King (1988) provided females of the parasitic wasp *Spalangia cameroni* with different sizes of hosts and found that the sex ratio of offspring varied according to host size; further analyses indicated that this was not due to differential mortality. In some cases, the effect of the mother may be indirect, as in *Sarcophaga crassipalpis*, in which diapause is determined by the

Table 7.12   Relationships Obtained in the Three Designs Proposed by Eisen (1967). An asterisk indicates that the design gives the specified relationship.

Relationship	I	II	III
Paternal half-sibs	*	*	*
Single first cousins (sires full-sibs)	*	*	*
Paternal half-sibs plus single first cousins (dams full-sibs)	*	*	*
3/4 Sibs (dams paternal half-sibs)	*		*
Double first cousins (sires full-sibs and dams full-sibs)	*	*	*
Single first cousins (sires full-sibs) plus half first cousins (dams paternal half-sibs)	*		*
Full-sibs	*	*	*
Within full-sibs	*	*	*
Dam–offspring	*	*	*
Sire–offspring	*	*	*
Maternal uncle (or aunt) with nephew (or niece)	*	*	*
Maternal half uncle (or aunt) (paternal half-sibs) with nephew (or niece) or single first cousins (opposite sexes full-sibs)	*		*
Paternal uncle (or aunt) with nephew (or niece)	*	*	*
Double first cousins (opposite sexes full-sibs)		*	*
Single first cousins (dams full-sibs)		*	
Half first cousins (dams paternal half-sibs)			*

Source: Modified from Thompson (1976).

larva as a consequence of a photoperiodic signal received through the integument of the female (Denlinger, 1970, 1971, 1972).

The above experiments can demonstrate the presence of a maternal effect but do not allow the separation of genetic from environmental effects. Variation among females in the maternal effect they have on their offspring could be due at least in part to genetic variation between females (Falconer, 1965b; Willham, 1972). Schluter and Gustafsson (1993) combined the experimental manipulation approach with a pedigree analysis to dissect these two components in the determination of clutch size in the collared flycatcher, *Ficedula albicollis*. The experimental manipulation consisted of transferring one egg from a clutch to the clutch of a neighboring female. The pedigree analysis used information on mothers and their daughters from the first-year and second-year clutches. Clutch size of the daughters was assumed to be determined by the genotype of the mother and two maternal components acting through the daughter's condition, one due to the condition of the mother (M) and one due to the clutch size which the daughter experienced (m, Fig. 7.7). The situation was further complicated by the fact that clutch size of a female depends on the size of the previous clutch. To take this

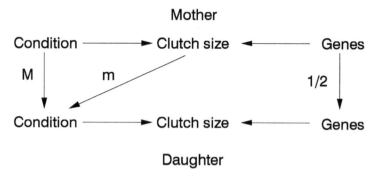

Figure 7.7 Maternal effects model suggested by Schluter and Gustafsson (1993) for clutch size in the collared flycatcher. Arrows connect dependent variables, symbols showing the partial regression coefficients (= 1 if not indicated). [Redrawn from Schluter and Gustafsson (1993).]

into account, Schluter and Gustafsson proposed the set of relationships given in Table 7.13. The clutch size of the mother in her first year is the sum of an additive genetic value (A_0) and a condition value (C_0). In her second year, her clutch size is a linear function of her additive genetic value (A_0), the clutch size she produced in her first year (bX_0), and an environmental deviation (e_0). The clutch size of her first-year daughters producing in their first year is equal to their additive genetic value (A_1) plus a condition value that is made up of three components: a factor that depends on the condition of the mother (MC_0), a factor that depends on the clutch size from which the daughter came (mX_0), and an environmental deviation (e_1). The clutch size in the second year is the sum of the additive value (A_1), the previous clutch size (bX_1), and an environmental deviation (e_3). Daughters from the second clutch produce clutch sizes that are similarly determined, the condition parameters ($bX_0 + e_0$, mY_0) being set by the condition of the mother as before.

From the above equations, it is possible to determine the expected covariance between offspring and parents. In addition to the three parameters m, M, and b, there is a parameter h_a^2, defined as the ratio of the additive genetic variance to the phenotypic variance. In the absence of maternal effects, h_a^2 is simply the heritability of the trait; however, in the presence of maternal effects, genetic and nongenetic components of clutch size are correlated (Kirkpatrick and Lande, 1989); hence, the phenotypic variance is the sum of the separate variances plus a covariance. With maternal effects, the covariance between the phenotypic and additive genetic values is equal to the additive genetic variance ($h_a^2 V_P$) plus the covariance between the additive genetic component and the condition component (Table 7.14). The covariances between mother and daughter are similarly augmented (Table 7.14). Note that in the case of second clutches, the covariance is influenced by the cost (b) of the first clutch. These equations illustrate the considerable complexity that can be introduced with just a relatively simple maternal effect.

Table 7.13 Equations Relating Clutch Size Between Mother and Daughters in the Collared Flycatcher

Type	Clutch Size in First Year	Clutch Size in Second Year
Mother	$X_0 = A_0 + C_0$	$Y_0 = A_0 + bX_0 + e_0$
First daughters	$X_1 = A_1 + MC_0 + mX_0 + e_1$	$Y_1 = A_1 + bX_1 + e_3$
Second daughters	$X_2 = A_2 + M(bX_0 + e_0) + mY_0 + e_4$	$Y_2 = A_2 + bX_2 + e_5$

Note: Clutch size is made up of two components: an additive genetic component and a condition component. The condition of the daughter in her first year is a function of the mother's genotype and condition; in the second year, the condition of a female depends on its genotype and the clutch size it produced in the previous year.

A_i: Additive genetic component

C_i: Effect due to condition

M, m: Direct maternal effects

b: Coefficient relating clutch size in second year to clutch size in first year

e_i: Environmental effect

The two parameters m and b were estimated from the clutch manipulation experiment. To obtain the remaining two parameters (M and h_a^2), the estimated values of m and b were substituted in the covariance equations (Table 7.14; also the covariances between daughters) and the values of M and h_a^2 found that maximized the probability of obtaining the observed clutch sizes. This is a good example of a situation in which maximum likelihood is appropriate.

The clutch size of a daughter was altered by approximately 0.25 eggs, giving an estimate of $m = -0.25$. The cost of the first clutch with respect to the second was found to be similarly high ($b = -0.25$). From the maximum likelihood analysis, values of $M = 0.43$ (95% limits of 0.25 to 0.58) and $h_a^2 = 0.33$ (-0.10 to 0.72) were obtained. Thus, if mothers in good condition produce large clutches, the negative maternal effect coming via the clutch size (m) is offset in part by the positive maternal effect coming from the condition of the mother (M). The apparent heritability (h_a^2) is also quite high, although the confidence region includes zero.

Estimation of parameters by experimental manipulation is a very valuable tool and greatly simplifies the statistical analysis. Schluter and Gustaffson suggest the following protocol by which the second maternal parameter M might be estimated by experiment. First, increase the condition of some mothers by providing additional food. Second, remove eggs from the experimental birds to reduce clutch sizes to those of control birds (leaving mother's condition unaffected): M is then estimated by the ratio of the mean increase in daughter clutch size of experimental birds, to the increase in clutch size experienced by their mothers.

The clutch size model analyzed by Schluter and Gustaffson (1993) demonstrates that heritability estimates from offspring–parent regression can be con-

Table 7.14 Covariances Between Phenotypic and Additive Genetic Components in the Maternal-Effects Model of Schluter and Gustafsson (1993)

Traits	Equation
	Covariance Between Phenotypic and Additive Genetic Values
X_0, A_0 for type 1 mothers[a]	$h_a^2 V_P + (mh_a^2\,V_P)/(2 - [m+M])$
X_0, A_0 for type 2 mothers	$h_a^2 V_P + \{h_a^2 V_P[m + b(m+M)]\}/[2 - b(m+M)]$
	Offspring Clutch Size on Mother's First Clutch Size
X_0, X_1	$(M+m)V_P + (\tfrac{1}{2} - M)\,\mathrm{Cov}(X_0, A_0)$
X_0, X_2	$(M+m)\,\mathrm{Cov}(1, 2) + (\tfrac{1}{2} - M)\,\mathrm{Cov}(X_0, A_0)$
X_0, Y_1	$b\,\mathrm{Cov}(X_0, X_1) + \tfrac{1}{2}\,\mathrm{Cov}(X_0, A_0)$
X_0, Y_2	$b\,\mathrm{Cov}(X_0, X_2) + \tfrac{1}{2}\,\mathrm{Cov}(X_0, A_0)$
	Offspring Clutch Size on Mother's Second Clutch Size
Y_0, X_1	$(M+m)\,\mathrm{Cov}(1, 2) + (\tfrac{1}{2} - M)[h_a^2 V_P + b\,\mathrm{Cov}(X_0, A_0)]$
Y_0, X_2	$(M+m)V_P + (\tfrac{1}{2} - M)[h_a^2 V_P + b\,\mathrm{Cov}(X_0, A_0)]$
Y_0, Y_1	$b\,\mathrm{Cov}(Y_0, X_1) + \tfrac{1}{2}[h_a^2 V_P + b\,\mathrm{Cov}(X_0, A_0)]$
Y_0, Y_2	$b\,\mathrm{Cov}(Y_0, X_2) + \tfrac{1}{2}[h_a^2 V_P + b\,\mathrm{Cov}(X_0, A_0)]$

Note: Variances of the six clutch sizes are assumed to be equal (V_P). Covariances between first and second clutches are also assumed to be equal [$\mathrm{Cov}(1,2)$]. For parameter definitions, see text.

[a]Type 1 mothers are those which are themselves first-clutch daughters. Type 2 mothers are those that are themselves second-clutch daughters.

founded by maternal effects. In their model, at equilibrium the slope of the daughter's first clutch size on her mother's first clutch size is

$$\text{Slope} = \tfrac{1}{2}\,h_a^2\,\frac{(M - 2)(2M - 1)}{2 - m - M} + (m + M) \tag{7.33}$$

In the absence of maternal effects, the offspring–mother regression estimates the heritability of the trait. Using the estimates obtained for the collared flycatcher gives an estimated slope of 0.2 and an apparent heritability of 0.4, which is not far removed from the value of 0.32 (SE = 0.15) actually estimated by daughter–mother regression (Gustaffson, 1986). This correspondence is largely fortuitous due to the particular values of M and m. In the extreme cases where either m or M equals zero (recall that m is negative and M positive), we have

$$\text{if } M = 0, \text{ then slope} = \tfrac{1}{2}\,h_a^2\,\frac{2}{2 - m} + m$$

$$\text{if } m = 0, \text{ then slope} = \tfrac{1}{2}\,h_a^2\,(1 - 2M) + M \tag{7.34}$$

When $m = 0$ and $M = 0.5$, the resemblance between mother and daughter is determined solely by M (i.e., the apparent heritability is 1.0). Estimates of the heritability of clutch size using dam on daughter regressions have typically given significant values, as also has a single selection experiment (Table 7.15). In the light of the Schluter and Gustafsson analysis, the former estimates must be viewed with caution. In one case, the great tit studied by Noordwijk et al. (1980), regressions based on three generations support the daughter–dam estimate (Fig. 7.8). All the heritabilities based on the assumption that the trait is maternally derived give similarly high estimates, whereas those based on the assumption of a significant paternal contribution give uniformly small and nonsignificant values. It must be stressed that this does not imply that the male does not contribute clutch-determining genes to his daughter; this is demonstrated by the daughter on paternal grandparent regression, which although statistically nonsignificant (0.68, SE $= 0.38$) is of the same order as that of daughter on dam (0.37, SE $= 0.12$). As shown by Fig. 7.8, the "male" regressions amount to regression on the randomly drawn female which are predicted to give insignificant heritability estimates if the trait is a maternally expressed character. The similarity of the estimates across two generations argues for a heritable basis to the trait. Similar results have been obtained for other populations, confirming the conclusion that approximately 40% of the phenotypic variation in clutch size is due to additive genetic variation (Noordwijk et al., 1981).

7.4.8 Cross-fostering

If parental effects have no influence on the phenotype of the offspring, then switching offspring between parents should not influence, on average, the phenotypic value of the offspring. This approach has been used in the detection of maternal/paternal effects in mice and birds.

Brumby (1960) did two types of cross-fostering in mice. First, he transferred ova between mothers, thereby providing a prenatal fostering environment, and, second, he switched babies at birth to examine the effect of the postnatal environment. Initial experiments showed that the act of transplantation or movement of young among mothers did not itself have any effects on the subsequent growth of young. Brumby used three strains of mice—one selected for high body weight at 6 weeks (large strain), one selected for low body weight (small strain), and a control (control strain). As expected from their previous selection regime, the growth rates of large (L) and small (S) strain offspring differed markedly from each other (Fig. 7.9). This difference was evident even when the young were transplanted and suckled by control (C) females, but the growth rate of such fostered offspring was greater than that of offspring reared by females of their own strain. It appears that C mothers provide a better environment than either type of selected female. Although the statistical analysis of these differences is not adequate, the substantial difference in means leaves no doubt that the effect

Table 7.15 Heritability Estimates of Clutch Size in Birds

Species	h^2	SE	Method[a]	Reference
Mallard	0.46	0.57	D–D	Prince et al. (1970)
Great tit	0.48	0.05	D–D	Perrins and Jones (1974)
Great tit	0.37	0.12	D–D	Noordwijk et al. (1980)
Starling	0.33	0.02	S	Flux and Flux (1982)
Collared flycatcher	0.32	0.15	D–D	Gustaffson (1986)
Darwin's ground finch	−0.17	0.12	D–D	Gibbs (1988)
Lesser snow goose	0.20	0.08	D–D	Findlay and Cooke (1987)

[a]D–D = daughter on dam regression; S = selection.

is significant. To examine the relative importance of the two environments (prenatal versus postnatal), Brumby used a hybrid line (H) produced by crossing the L and S strains and transplanted embryos either into an S or C female, and then used either an S or C female as a postnatal foster mother. Thus, three combinations of rearing were generated: C/C, S/C, and S/S, where the first letter denotes the prenatal type of mother and the second letter the postnatal type of mother. The differences among the growth rate of the three types of environment appear to be negligible (Fig. 7.9), suggesting that the prenatal environment plays no important role in generating the differences observed in the first experiment. [For further evidence, see Moore et al. (1970) and Riska et al. (1984)].

Statistically, more satisfactory experiments have since been carried out by Rutledge et al. (1972), Atchley and Rutledge (1980), Cheverud et al. (1983), Cheverud (1984), Leamy and Cheverud (1984), and Riska et al. (1984, 1985) using single strains of random-bred mice. I shall use the study by Riska et al. (1984, 1985) to illustrate the approach. At birth, litters were standardized to eight pups and four (two of each sex if possible) were exchanged with four pups from another mother. There are two types of genetic effects that can be disentangled from this experiment: First, there is the usual direct genetic contribution made by both the mother and the father; second, there is an indirect contribution made by the female by virtue of the environment she provides, which can itself have a heritable component. Females can differ genetically in the type of care they provide (e.g., different physiologies or behaviors) and this will be passed on as a genetic heritage to the offspring although they experience it as an environmental effect. These effects were separated using the Willham model.

Two types of analysis are possible with the experimental design of Riska et al.

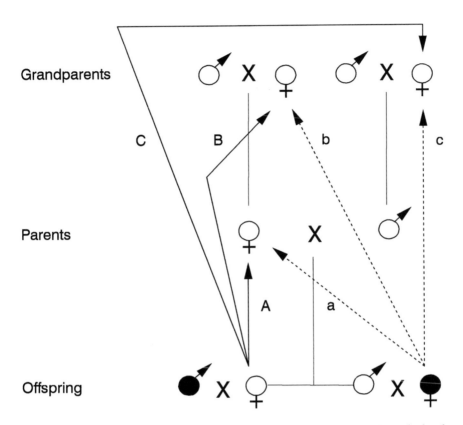

Figure 7.8 Heritability estimates of clutch size in the great tit, *Parus major*, calculated from descendant on ancestor regressions [data from Noordwijk et al. (1980)]. Solid lines show pedigree relationships; closed symbols represent male and females drawn at random from the population. Arrows show the various regressions used: Solid arrows are those using the female descendants; dotted lines are those using the male descendants. Because clutch size is a sex-limited trait, it is calculated in all cases from the female, the hypothesis for the male descendants being that the male makes a contribution in some manner to the female, which causes his genes to be expressed in her clutch size.

Symbol	Relationship	Heritability (SE)
A	Daughter on dam	0.37 (0.12)
B	Granddaughter on maternal grandam	0.38 (0.56)
C	Granddaughter on paternal grandam	0.68 (0.38)
a	Son on sire	0.05 (0.11)
b	Grandson on maternal grandsire	−0.04 (0.36)
c	Grandson on paternal grandsire	0.05 (0.31)

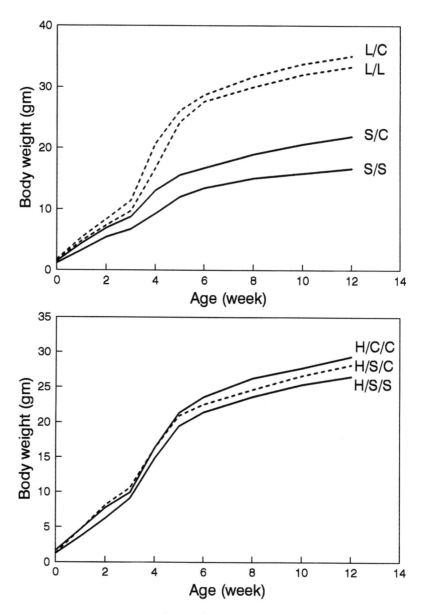

Figure 7.9 Growth rate of mice under different rearing regimes. *Upper panel:* Mice reared by their own mother (S/S and L/L, where S = small strain and L = large strain) or transplanted as ova into a female of the control (C) strain of mice (L/C and S/C). *Lower panel:* Growth rate of hybrid mice (H) when transplanted as ova into females of the S strain (H/S/S), when transplanted as ova into females of the S strain but raised from birth by control females (H/S/C), and when transplanted as ova into females of the control strain (H/C/C). [Data from Brumby (1960).]

(1984, 1985), one utilizing only the cross-fostered pups and one using all the data. The linear model for the first analysis is

Offspring weight $= c_0 + c_1$ Pair $+ c_2$ Nursing mother
$$+ c_3 \text{ Genetic mother} + \text{error} \quad \textbf{(7.35)}$$

where the c's are coefficients. Assuming no cage effects and no dominance, the variance component due to the genetic mother is equal to one-half the direct-additive genetic variance (because the pups are full-sibs), and the variance due to the nursing mother is equal to the indirect-additive variance (maternal additive genetic variance) plus the common environmental variance (maternal or cage if not zero). The ratio of the maternal variance ($=$ nursing mother) to the direct-additive genetic variance was very large at an early age and even though it declined with age, it was still a significant ratio at age 10 weeks (Fig. 7.10).

The data from both pups reared by their natural mothers and the pups reared by foster mothers can be analyzed using the variance decomposition previously given:

$$V_P = V_{AO} + V_{AM} + \text{Cov}(A_O, A_{OM}) + V_C + V_E \quad \textbf{(7.36)}$$

where V_P is the phenotypic variance in offspring weight, V_{AO} is the direct-additive genetic variance, V_{AM} is the indirect-additive (maternal additive) genetic variance, $\text{Cov}(A_O, A_{OM})$ is the covariance between the additive direct- and indirect-genetic effects, V_C is the common environmental variance, and V_E is the residual environmental variance. For simplicity, I have dropped the dominance components, which Riska et al. (1985) assumed to be zero. Unfortunately, although the covariance can be estimated, this design does not provide a reliable estimate of V_{AM} [modification of the design such that the dams are themselves related does permit such an estimation; see Cheverud and Moore (1994) for further details]. Because the cross-fostered mice are reared in an environment produced by genes that they do not share, the phenotypic variance of these mice is equal to $V_P - \text{Cov}(A_O, A_{OM})$. Therefore, by two analyses of variance, one estimating the phenotypic variance of pups raised by their own mothers and one estimating the phenotypic variance of cross-fostered pups, the covariance can be estimated by subtraction. Riska et al. (1985) obtained positive values of Cov (A_O, A_{OM}) although negative values are common (Table 7.5). Negative covariances could arise because of variation in litter size, an effect eliminated in the present experiment by standardization of litter size at birth. However, negative covariances have been found in experiments in which litter size either does not vary substantially (e.g., cattle) or in which litter size standardized (e.g., cattle) or in which litter size was standardized (e.g., Eisen et al., 1970).

The cross-fostering of mammals in natural populations is generally not feasible,

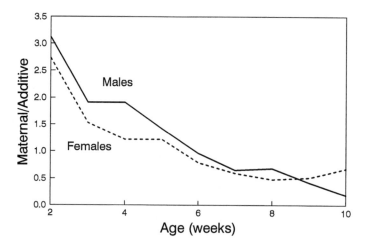

Figure 7.10 Ratio of direct-additive genetic variance to maternal variance in mice. [Data from Riska et al. (1984).]

although there are some species, such as squirrels, where it may be possible. In birds, however, it is relatively easy to switch eggs between nests and thus this group is ideal for the study of maternal/paternal influences on offspring traits. Nine experiments of this kind have all shown that the offspring resemble their biological parents and not their foster parents and that the heritabilities are independent of the rearing environment (Table 7.16). Thus, at least with respect to morphological traits, parental effects appear to be of minor importance in birds. But, as discussed above, this is not the case for clutch size.

7.5 Summary

Sex-related effects may arise because of physical linkage of genes on the sex chromosomes, because the phenotypic manifestation of a trait is sex related or because a particular sex has a phenotypic effect on their offspring. In the first category are such traits as red–green color blindness in humans, although many quantitative traits also appear to have loci located on the sex chromosomes. Predictions of the consequences of such sex-linked traits are relatively easily analyzed using simple Mendelian genetics. An example of a sex-related effect that is not due to physical linkage on the sex chromosome is sexual dimorphism in body size. The evolution of this category of trait can be understood using the concept of a genetic correlation between the sexes. As in the case of other genetically correlated traits, the evolution to different equilibria for the sexes might not generally be prevented but might take an extremely long time. The third category of

Table 7.16   Comparison of Heritabilities of Morphological Traits in Birds Estimated from Mean Offspring on "Mid-parent" Regressions Using Unswitched or Switched Families, Two Estimates Being Obtained from the Latter; One from the True Parents and One from the Foster Parents [Except Where Indicated the Trait Measured Was Tarsus Length]

Species	Unswitched (SE)	True (SE)	Foster (SE)	Ref.[a]
Melospiza melodia	0.79 (0.26)	0.67 (0.22)	−0.10 (0.33)	1
Tachycineta bicolor	0.04 (0.04)	−0.1 (0.05)	0.01 (0.01)	2
	0.75 (0.20)	0.50 (0.16)	0.16 (0.16)	2
Parus caeruleus	0.61 (0.13)	0.64 (0.18)	−0.32 (0.20)	3
Parus montanus	0.61 (0.24)	0.67 (ng)	ns	4
Ficedula hypoleuca	0.53 (0.10)	0.50 (0.22)	0.04 (0.23)	5
Ficedula albicollis	0.46 (0.05)	0.45 (0.18)	−0.19 (0.19)	6
Sturnus vulgaris	0.43 (0.17)	0.49 (0.12)	0.18 (0.18)	7
Experiments measuring differences but not estimating heritabilities				
Parus ater	Tarsus length (deviation from parents)	−0.07 (0.13) −0.06 (0.16)	−0.96 (0.12) 0.76 (0.16)	8
Passerina cyanea	Wing arc	0.89 (0.04)	1.33 (0.08)	9

[a]References and notes: (1) Smith and Dohndt (1980); mean of four heritability estimates (beak length, beak depth, beak width, tarsus length). (2) Wiggins (1989); body mass, tarsus length. (3) Dhondt (1982). (4) Thessing and Ekman (1994); SE approximately 0.24, no significant correlation between foster parents and offspring ($r = -0.03$). (5) Alatalo and Lundberg (1986); offspring on female parent. Based on mother–offspring versus father–offspring regressions, Potti and Merino (1994) suggested that in their population maternal effects may be significant. (6) Gustafsson and Merila (1994); beak width, beak depth, beak length, tarsus length, wing length, tail length, primary length. (7) H.G. Smith (1993), Smith and Wettermark (1995). (8) Alatalo and Gustafsson (1988); results from two experiments. Offspring did not differ significantly from their true parents but did differ from their foster parents. (9) Payne and Payne (1989); body size of offspring did not differ from true parents. Comparison with foster parents not made.

traits "classically" includes maternal effects. Such effects are particularly important in organisms that show parental care, which might include maternal provisioning of the egg or seed. Two quantitative genetic frameworks have been proposed to measure and predict the consequences of maternal effects (including paternal effects, but the former are more frequent): the Willham model and the Kirkpatrick–Lande model. The former framework is derived from an ANOVA approach, whereas the second is based on an extension of the variance–covariance matrix approach. Analysis of either model shows that evolutionary trajectories can be significantly altered by maternal effects. The statistical detection of maternal/paternal effects is possible by a variety of experimental designs, but the full elucidation of the genetic parameters requires considerable effort and, to date, we have no complete analysis for a single nondomestic species (and possibly not even for a domestic species), although several experiments with wild bird species demonstrate that a complete analysis is feasible.

8

Bottlenecks, Finite Populations, and Inbreeding

Although not explicitly stated, an assumption made throughout most of the analyses of the foregoing chapters is that population size, N, is an appropriate index of the number of individuals making a genetic contribution to the next generation. Specifically, it has typically been implicitly assumed that mating is at random, family size follows a Poisson distribution, N is constant, and the sex ratio is 1:1. Perhaps more significantly, it has generally been assumed that population size is sufficiently large that effects due to sampling variation can be ignored. In nature there will be many circumstances in which one or more of the preceding assumptions are violated. The consequences of such violations is the subject of this chapter. Three important issues are dealt with: (1) Effective population size. Analysis of the relationship between the census population size (the total number of organisms counted in the population) and the effective population size, which is the population size that is relevant for the discussion and analysis of genetic changes in a population. (2) Effect of a finite population size on genetic variance. This analysis is divided into two components: first, the consequences of a population passing through a single generation in which population size is restricted and, second, the effect of a population being maintained over many generations at a finite size. (3) Inbreeding. One consequence of a finite population size is that mating may occur between close relatives, a phenomenon that can also occur in a very large or infinite population due to characteristics of the life history of the organism. Inbreeding frequently leads to inbreeding depression, the subject of the third section of this chapter. Also considered is the converse of inbreeding depression, outbreeding depression, a phenomenon for which there is much less clear evidence of its general occurrence.

8.1 Effective Population Size

8.1.1 Theory

Thus far, we have assumed that the number of individuals in a population is a suitable measure of the size of the breeding population. Specifically, this assumption requires that (1) individuals mate at random (including selfing), (2) the number of progeny per parent follows a binomial (or its limiting case, the Poisson) distribution, and (3) population size remains constant through time (Fisher, 1930; Wright, 1931). In fact, in natural populations, all of these assumptions are likely to be violated to a greater or lesser degree. Violation of the assumptions will lead to an **effective population size** that can differ substantially from the actual number of individuals in the population. Consider a population of $\frac{1}{2}N$ males and $\frac{1}{2}N$ females that mate at random with selfing permitted: the probability that a particular gene in two different individuals comes from the same individual of the parental generation is $1/N$. We can, therefore, define the effective population size as that population size which, given the three assumptions previously stated, will give the same probability of two genes coming from a common parent as actually observed. Thus, suppose the number of males and females is N_m and N_f, respectively, and that mating is at random; the probability that the same gene in two individuals comes from the same male parent is $1/4N_m$, and from the same female parent it is $1/4N_f$. Therefore, the overall probability is

$$\frac{1}{4N_m} + \frac{1}{4N_f} = \frac{1}{N_e} \tag{8.1}$$

Hence,

$$N_e = \frac{4N_mN_f}{N_m + N_f}$$
$$= 4p_m (1 - p_m)N \tag{8.2}$$

where p_m is the proportion of males and $N = N_m + N_f$. The above definition is called the **inbreeding effective population size**. An alternative definition is the **variance effective population size** defined as follows. In an ideal population, the sampling variance of allele frequency, V, is $V = pq/2N$; the variance effective population size is then $N_e = pq/2V_{obs}$, where V_{obs} is the observed variance in the actual population. For detailed discussion of these two measures, see Crow and Kimura (1970, pp. 345–365) or Crow and Denniston (1988); it is sufficient to note here that in most natural situations the two will be similar. Examples of the relationship between N_e and N, some of which are discussed in more detail below, are given in Table 8.1.

A formal definition of inbreeding effective population size is "the size of an ideally behaving population that would have the same homozygosity increase as in the observed population" (Crow and Kimura, 1970, p. 103). From this, we can readily derive the equation for the effective population size in a population that fluctuates in size. The heterozygosity at time t for a population of constant size is (Bulmer, 1985, p. 220)

Table 8.1 Equations Relating Population Parameters to the Effective Population Size Under a Variety of Conditions; the Idealized Population Is One in Which There Is Random Mating, Poisson (or Binomial) Distribution of Family Size, Constant N, Equal Sex Ratio, and No Selection

Condition	Equation
1. No self-fertilization	$N + 0.5$
2. Different number of males and females	$\left(\dfrac{1}{4N_m} + \dfrac{1}{4N_f}\right)^{-1} = \dfrac{4N_m N_f}{N_m + N_f}$
3. Variable N	$\left(\dfrac{1}{t}\sum \dfrac{1}{N_i}\right)^{-1}$
4. Variable N, $m \neq f$	$\left[\dfrac{1}{t}\sum\left(\dfrac{1}{4N_m} + \dfrac{1}{4N_f}\right)\right]^{-1}$
5. Nonrandom contribution by parents	
5.1 Haploid species	$\dfrac{N - 1}{V_k}$
5.2 Monoecious (+ selfing)	$\dfrac{4N - 2}{2 + V_k}$
5.3 Separate sexes	$\dfrac{4N - 4}{2 + V_k}$
5.4 X linkage (male is XY), large N	$\dfrac{9N_m N_f}{4N_m + 2N_f}$
5.5 Lottery polygyny	$\dfrac{4N_m N_f n}{Nn + N_m}$
6. Truncation selection	
6.1 One generation, trait not fitness	$\dfrac{N}{1 + i^2 h^2/2}$
6.2 Fitness (= fertility)	$\dfrac{4N - 2}{2 + (3h^2 + 1)V_k}$

(continued)

Table 8.1 Continued

Condition	Equation
7. Overlapping generations	In general, replace N by $N_c T$ in above formulas
7.1 Equal sex ratio	$\dfrac{N}{2 - 1/T}$
8. Subdivided population	$\dfrac{4s}{[s - (s-1)(d_m + d_f - d_m d_f)][2\phi_f + \phi_m(1 - \phi_f)]}$

Notes:

1. Avoidance of other types of matings between relatives also has little effect on N_e (Caballero and Hill, 1992; Caballero, 1994).
2. N_e is largely determined by the value of the less numerous sex.
3. N_e is largely determined by the smallest value of N_i. Thus bottlenecks can have important effects on N_e.
4. For the derivation, see Chia and Pollack (1974).
5. V_k is the variance in offspring number. In the idealized population, the distribution of progeny is Poisson.
   - 5.1 When the distribution of progeny is Poisson ($V_k = 1$) and N is large, then $N_e = N$. For a formula for an n-ploid population, see Caballero (1994, p. 666).
   - 5.2 As in the case of 1, there is little difference between the monoecious case and that of separate sexes.
   - 5.4 For a thorough treatment, see Caballero (1995).
   - 5.5 In lottery polygyny, each female mates at least once and each male attempts to mate with every female he meets. Each female mates with n males. For other systems of mating, see Crow and Denniston (1988) and Nunney (1993).
6. N_e declines with the intensity of selection (i) and the heritability. For further discussion and references, see Caballero (1994) and Santiago and Caballero (1995).
7. N_c is the number of individuals in a cohort; T is the generation time. N_e approaches $N/2$ as T increases. See Nunney (1991) and Caballero (1994) for derivations and other cases.
8. s is the number of breeding groups in the population, d_m and d_f are the migration rates of males and females, respectively, and the functions designated by ϕ are the probabilities that random progeny are the offspring of a particular adult male or female

$$\phi_m = \frac{n_m[V_c + c(c-1)]}{n_f(n_f - 1)}, \qquad \phi_f = \frac{V_k + k(k-1)}{k(kn_f - 1)}$$

where n_m is the number of males per breeding group, n_f is the number of breeding females, $c = n_f/n_m$, V_c is the variance in c, k is the mean offspring per female, and V_k is the variance in progeny number. For the derivation, see Chesser et al. (1993).

$$H_t = H_0 \left(1 - \frac{1}{2N}\right)^t \tag{8.3}$$

and so for a fluctuating population, we have

$$
\begin{aligned}
H_t &= H_0 \left(1 - \frac{1}{2N_0}\right)\left(1 - \frac{1}{2N_1}\right) \cdots \left(1 - \frac{1}{2N_{t-1}}\right) \\
&= H_0 \prod_{i=0}^{t-1} \left(1 - \frac{1}{2N_i}\right) \\
&= H_0 \left(1 - \frac{1}{2N_e}\right)^t
\end{aligned}
\tag{8.4}
$$

If population sizes (N_i) are fairly large, the effective population size is approximately equal to the harmonic mean of the population sizes:

$$\frac{1}{N_e} \approx \frac{1}{t} \sum \frac{1}{N_i} \tag{8.5}$$

When the number of offspring per parent varies in a non-binomial fashion but population size remains constant the effective population size is (Wright, 1938; Crow and Kimura, 1970)

$$N_e = \frac{4N - 2}{V_k + 2} \tag{8.6}$$

where V_k is the variance in the number of progeny per parent. If selfing is not permitted, the 2 in the numerator is replaced by 4. Suppose the population is stationary and the distribution of family sizes follows a binomial distribution; in this case, the mean number per family is 2 with binomial variance $2(N - 1)/N$. Substitution in Eq. (8.6) gives the expected $N_e = N$. A decrease or increase in the variance will lead to a concomitant increase or decrease in the effective population size. For example, consider the extreme case in which all females have exactly the same number of offspring ($V_k = 0$); in this case, $N_e = 2N - 1$. Thus, in many laboratory experiments, in which the number of progeny per parent is kept constant, the effective population size is actually far larger than the census number. This result is potentially important for the conservation of genetic variation in endangered species; if feasible, it would be better to equalize family size rather than permitting family size to vary randomly (Loebel et al., 1992; Borlase et al., 1993). An increasing tendency for a few individuals to monopolize reproduction will increase V_k and lead to a drastic reduction in N_e, in some cases possibly leading to a reduction of several orders of magnitude. Crow and Denniston (1988) and Nunney (1993) give formulas for the effective population size under various types of breeding system.

When generations overlap, an approximate formula is (Hill, 1979)

$$N_e \approx \frac{4N_cT}{V_k + 2} \tag{8.7}$$

where T is the generation interval and N_c is the size of each cohort. Nunney (1991, 1993) considered the more complex case of overlapping generations and different mating systems. These equations are complex and the reader is referred to Table 1 in Nunney (1993) for a description. Two important conclusions were reached from this analysis: first, that "special circumstances are required for N_e to be close to N," and, second, "that special circumstances are required for N_e/N to be much less than 0.5. These conclusions indicate that, in general, if the number of adults in a population is known, then the effective size of the population will probably not be far from half that number" (Nunney, 1993, p. 1338). From their review of data on 12 species of terrestrial birds, Grant and Grant (1992) suggested that for this group, N_e will typically be one-quarter of the census population size. These general conclusions echo that arrived at by Crow and Morton (1955, p. 213), "We conclude that only under rather unusual conditions would the effective number be of a different order of magnitude than the actual number." To illustrate the point, Table 8.2 gives some examples of the required parameter values that produce the ratio $N_e/N = 0.1$ under different models. The most likely case in which the effective population size is much less than that of the census population size is when there is great variation in fecundity and/or there is considerable fluctuation in the census population size. Such a situation is far more likely to occur in poikilotherms and plants than in homeotherms (see Section 8.1.3).

8.1.2 Estimating the Effective Population Size

The effective population size can be estimated using either census information or allelic frequencies at several loci. The latter methods can be grouped as follows:

1. **Allelism of lethals**: A method developed by Wright, which uses a comparison of the rate of allelism within and between populations [used primarily with *Drosophila* spp.: see Dobzhansky and Wright (1941, 1943), Prout (1954), and Begon et al. (1980)].

2. **Linkage disequilibrium method**: Developed by Hill (1981) and utilizing the relationship between population size and linkage disequilibrium. It does not appear to have been used very much.

3. **Allelic distribution method**: Developed by Chakaborty and Neel (1989), uses the predicted relationship between allelic variation and population

Table 8.2   *Parameter Values Required Such That the Effective Population Size is One Tenth the Value of the Census Population* ($N_e = 0.1N$). *Unless Otherwise Stated, There Is Random Mating, a Poisson Distribution of Family Size, Constant N, Equal Sex Ratio, and No Selection*

Condition	Equation when $N_e = 0.1N$ $N_e =$	Notes and Parameter Values
1. Different number of males and females	$\approx \dfrac{c}{40}$	$c = \dfrac{N_m}{N_f}$, i.e., $\dfrac{N_m}{N_f} = 40$
2. Variable N	$\left(\dfrac{1}{2}\left[\dfrac{1}{N_1} + \dfrac{1}{cN_1} \right] \right)^{-1}$	Assuming N takes two values with equal frequency, $N_2 = cN_1$, with $c = 0.05$.
3. Nonrandom contribution by parents		
3.1 Haploid species	$\dfrac{N-1}{V_k}$	$N_e = N$ when $V_k = 1$ ($N \gg 1$). The required value in the present case is $V_k \sim 10$
3.2 Separate sexes	$\dfrac{4}{38 - V_k}$	In a random mating population, $V_k = 2$. Provided N_e is reasonably large, $V_k \approx 38$.
3.3 Lottery polygyny	$\dfrac{n40cN_e}{(c+1)[n(c+1)+1]}$	n is the number of matings per female, $c = N_f/N_m$. In the present case, $38 < c < 40$, regardless of the value of n.
4. Truncation selection		
4.1 One generation, trait not fitness	$\dfrac{20N_e}{2 + i^2h^2}$	Rearranging gives $i^2h^2 = 18$. Assuming $h^2 = 0.5$, $i = 6$, which is equivalent to selecting considerably less than 0.01% of the population.
4.2 Fitness	$\approx \dfrac{40N_e}{2 + (3h^2+1)V_k}$	Rearranging gives $(3h^2+1)V_k = 38$. Assuming $h^2 = 0.5$, then $V_k = 15.2$.
5. Overlapping generations, equal sex ratio		Not possible, $N_e \rightarrow N/2$, as $T \rightarrow \infty$, where T is generation length.

size at mutation–drift balance. The method estimates the product of population size times the mutation rate and, thus, is of limited value unless independent estimates of mutation rate are available [for an example of its use in estimating N_e of various plant species, see Schoen and Brown (1991)].

4. **Temporal method**: First used by Krimbas and Tsakas (1971) and probably the most widely used, is based on Eq. (8.5), from which one can derive the approximation

$$N_e \approx \frac{t}{2V_s} \tag{8.8}$$

where t is the time interval over which the survey is conducted, and V_s is a standardized variance used to compensate for differences in initial allele frequencies (Waples, 1989). For examples of its use, see Krimbas and Tsakas (1971), Begon et al. (1980), and Husband and Barrett (1992; see Nunney [1995] and Husband and Barrett [1995] for further discussion of the problems in applying this method to the data set of Husband and Barrett). The method is very sensitive to allele frequencies: for example, the estimate of N_e for a Greek population of *D. subobscura* ranged from 268 to infinity, depending on which alleles were used. In this case, Begon et al. (1980) suggested that the low estimate was a consequence of two loci being under selection rather than neutral as required.

The use of census information has to take into account the natural history of the organism and, therefore, might differ from species to species (Harris and Allendorf, 1989; Nunney and Elam, 1994; Rockwell and Barrowclough, 1995). Factors that must be considered are the area of the panmictic circle, the rate of migration between populations, the numbers of breeding males and females, the distribution of family sizes, and the breeding system (Begon et al., 1980). The estimation of effective population size in the wood frog, *Rana sylvatica*, illustrates the general approach [Easteal (1985) and Easteal and Floyd (1986) provide a very detailed, and sobering, analysis using both gene frequency and ecological methods for the marine toad, *Bufo marinus*. See also the analysis by Johnson and Black (1995) on neighborhood size in the intertidal snail, *Bembicium vittatum*]. Berven and Grudzien (1990) derived three estimates: (1) Using Eq. (8.1) with the counts of the males and females. (2) By taking into account the breeding behavior. Breeding is explosive with all reproduction occurring often only in one night. Therefore, when the sex ratio was male biased, N_e was estimated as $2N_f$, and when female biased, as $N_m + N_f$. (3) Harmonic means. Population size varied greatly between years and, therefore, the harmonic means of the previous two estimates were calculated. Methods 1 and 2 gave very similar results and differed little from the total population count (Fig. 8.1). In contrast, the effective population size in the angiosperm *Eichhornia paniculata*, estimated using allelic frequencies, differed

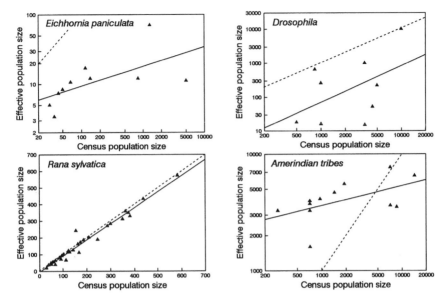

Figure 8.1 A comparison of estimates of effective population size with the census population size in a plant *Eichhornia paniculata* [data from Husband and Barrett (1992); $r = 0.62$, $n = 10$, $P < .05$], captive populations of *Drosophila* [data from Laurie-Ahlberg and Weir (1979) and Briscoe et al. (1992); $r = 0.49$, $n = 9$, $P > .1$], the wood frog, *Rana sylvatica* [data from Berven and Grudzien (1990); $r = 0.99$, $n = 32$, $P < .001$], and Amerindian tribes [data from Chakraborty and Neel (1989); $r = 0.53$, $n = 12$, $P < .05$]. The dotted line shows the 1:1 line, the solid line the fitted regression. All statistical tests are one tailed.

markedly from the census population size (Fig. 8.1). Likewise, very low effective population sizes despite large census population sizes have been consistently found in captive populations of *Drosophila* (Fig. 8.1; Frankham, 1995a). On the other hand, Chakraborty and Neel (1989) consistently obtained estimates of effective population size in Amerindian tribes in excess of the census size (Fig. 8.1).

8.1.3 Effective Population Size in Different Taxa

Most of the data on effective population size comes from studies on vertebrates (Table 8.3), and even in this group, it is clear that the species studied are not a representative sample of the taxa. Of the species studied, a very high proportion (50%) have effective population sizes less than 100 (Fig. 8.2). In these cases, drift is likely to play an important role in evolutionary change, whereas in those species in which N_e exceeds 1000 (16.7% of cases), drift is likely not to be of consequence. Shields (1993) proposed the following formula to estimate the approxi-

Table 8.3   *Estimates of Effective Population Size in Different Taxa of Animals; Values in Parentheses Are Estimates Using Eq. (8.9) with K = 3*

Species	Common Name	N_e	Reference
Invertebrates			
Drosophila pseudoobscura	Fruit fly	500–1,000	Dobzhansky and Wright (1941, 1943)
Drosophila subobscura	Fruit fly	>4,000	Begon et al. (1980)
Drosophila subobscura	Fruit fly	250	Pollack (1983)
Dacus olea	Fly	189–722	Nei and Tajima (1981)
Aquarius remigis	Waterstrider	170	Preziosi and Fairbairn (1992)
Vertebrates—Poikilotherms			
Oncorhynchus nerka	Sockeye salmon	206	Altukhov (1981)
Cynoscion nebulosus	Sea trout	22,900	Ramsey and Wakeman (1987)
Sciaenops ocellatus	Red drum	1,800	Ramsey and Wakeman (1987)
Acris crepitans	Cricket frog	4–111	Gray (1984)
Rana sylvatica	Wood frog	38–152	Berven and Grudzien (1990)
Rana lessonae	Pool frog	35	Sjogren (1991)
Rana pipiens	Leopard frog	2–112	Merrel (1968)
Bufo woodhousei	Fowler's toad	38–152	Breden (1987)
Bufo marinus	Marine toad	390–460	Easteal (1985), Easteal and Floyd (1986)
Notophthalmus viridescens	Red-spotted newt	25	Gill (1978)
Anolis grahami	Graham's anole	641	Taylor and Gorman (1975)
Uta stansburiana	Iguanid lizard	17	Tinkle (1965)
Sceloporus olivaceus	Rusty lizard	225–270	Kerster (1964)

Vertebrates—Homeotherms

Species	Common name	Number	Reference
Odocoileus virginianus	White-tailed deer	45–800	Chepko-Sade et al. (1987)
Equus cabullus	Horse	50 (77)	Chepko-Sade et al. (1987)
Ursus americanus	Black bear	552	Chepko-Sade et al. (1987)
Helogale parvaula	Dwarf mongoose	1–12 (52)	Chepko-Sade et al. (1987)
Canus lupus	Wolf	804–1,661	Chepko-Sade et al. (1987)
Cynomys ludovicianus	Black-tailed prairie dog	23–31 (120)	Chepko-Sade et al. (1987)
Dipodomys spectabilis	Kangaroo rat	7–16	Chepko-Sade et al. (1987)
Ochotona princeps	Pika	2–59 (6)	Chepko-Sade et al. (1987)
Homo sapiens	Japanese popn	1,993	Nei and Imaizumi (1966)
	Amerindians	3,744	Chakraborty and Neel (1989)
	New Guinea	651	Wood (1987)
Mus musculus	House mouse	5–80	Petras (1967a, 1967b)
Geospiza scandens	Darwin's ground finch	38	Grant and Grant (1992)
Geospiza fortis	Darwin's medium ground finch	60 (3,000)	Grant and Grant (1992)
Geospiza conirostris	Large cactus finch	75–190	Grant and Grant (1989)
	Florida scrub jay	~300 (750)	Barrowclough (1980, 1983), Woolfenden and Fitzpatrick (1978)
Malurus splendens	Splendid fairy-wren	12–84 (15)	Rowley et al. (1993)
	Spotted owl	220	Barrowclough and Coats (1985)
	Acorn woodpecker	81–961 (94)	Koenig and Mumme (1987)
	White-crowned sparrow	36	Baker (1981)

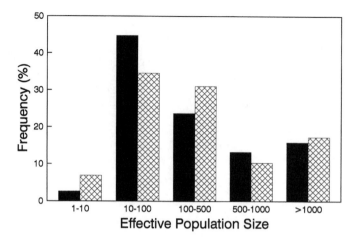

Figure 8.2. Frequency distributions of N_e for a variety of animal species. Solid bars show data presented in Table 8.3. Hatched bars show values estimated using the approximation proposed by Shields (1993), $N_e = 3/P_I$, where P_I is the frequency of close inbreeding. Data from Rowley et al. (1993), Shields (1993), and A.T. Smith (1993).

mate effective population size from the frequency of **close inbreeding** [= matings between first-degree kin (e.g., brother–sister or father–daughter)]:

$$N_e = \frac{K}{P_I} \qquad (8.9)$$

where K is the number of first-degree kin and P_I is the probability of close inbreeding. In a numerically stable population, $K = 3$ (i.e., each male has three first-degree kin—his mother, daughter, and sister—with which he can mate). The frequency distribution obtained using this approximation is remarkably similar to that obtained by other more precise methods (Fig. 8.2). Only eight species are represented in both data sets, and within this small set, there is modest agreement between the "precise" and estimated values of N_e (a notable exception is *G. fortis*). In the absence of direct estimates of N_e but sufficient pedigree data to estimate P_I, the Shield formula appears to give a reasonable working value.

Two potential major sources of error in the estimation of effective population size are an insufficient time span over which variation in N_e is measured and an underestimation of migration rates [for detailed discussion of the latter, see Chesser et al. (1993)]. The former will lead to an overestimation, whereas the latter will give an underestimate; consequently, any estimate should be viewed as, at best, only an order-of-magnitude estimate. From an analysis of chromosomal rearrangements, Lande (1979b, p. 247) concluded, assuming approximate neutrality of such rearrangements, that "during the evolution of many animal taxa effective deme sizes have been in the range of a few tens to a few hundreds of individuals." This is consistent with the estimates shown in Table 8.3, particularly given that these estimates were made over a very limited time span. Even in those populations which typically number in the thousands, bottlenecks probably occur at least on rare occasions and these will have significant impact on the effective population size.

In some cases, data can be used to estimate the ratio N_e/N, even if neither parameter can be estimated separately. This is illustrated by the study of Heywood (1986) on the variation in the above ratio in populations of annual plants. Heywood used the fact that the ratio of the variance effective number to the census number is

$$\frac{N_e}{N} = \frac{1}{(1 + F)(1 + V_f/m_f^2)} \qquad (8.10)$$

where F is the inbreeding coefficient (discussed in detail in Section 8.4), V_f is the variance in fecundity, and m_f is the mean fecundity. Note that if outbreeding is assumed ($F = 0$) and population size is assumed to be constant ($m_f = 2$), the above formula reduces to $4/(4 + V_f)$, which is approximately equal to the inbreeding effective number (Table 8.1). Because seed production is approximately a linear function of biomass, Heywood was able to estimate the above ratio using data on variation in plant size in a population. Lacking estimates of F, Heywood computed the values for the two extremes, $F = 0$ and 1: Estimates using the lower value will be larger than those obtained using the upper value. Most estimated ratios are less than 0.5, reflecting the large variation in potential fecundity among annual plants (Fig. 8.3). In contrast, estimates for birds and mammals, derived using a variety of methods, are more frequently greater than 0.5 (Fig. 8.3). The difference is consistent with the observation that variation in potential fecundity is not likely to be as large in homeotherms as plants. More variability in fecundity might be expected in vertebrate ectotherms and, thus, N_e/N should be typically lower than in homeotherms, which is what is indeed observed (Fig. 8.3). The data for invertebrates show a bimodal distribution which might be an artifact introduced by the majority of the populations being laboratory stocks (see legend to Fig. 8.3).

In a review of the ratio N_e/N, Frankham (1995a, p. 95) concluded that, "Comprehensive estimates of N_e/N (that included the effects of fluctuation in population size, variance in family size and unequal sex-ratio) averaged only 0.1–0.11. Wildlife populations have much smaller effective population sizes than previously recognized." This conclusion must be tempered by the following considerations: (1) A significant number of the estimates come from *Drosophila* and (2) a low

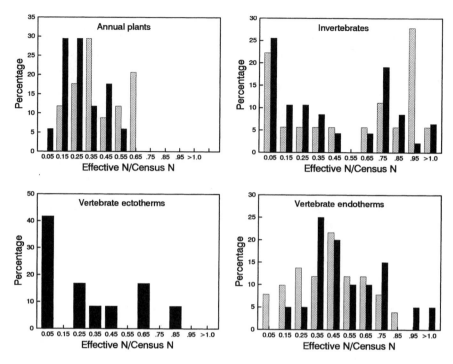

Figure 8.3 Distributions of N_e/N for different taxa. Plant data from Heywood (1986); animal data from Frankham (1995a).

Annual plants ($n = 33$): Solid bars based on estimation assuming $F = 1$, hatched bars based on estimation assuming $F = 0$.

Invertebrates: Solid bars show data only for *Drosophila*, primarily *D. melanogaster* ($n = 47$). Hatched bars show data for other invertebrates ($n = 18$). The very high ratios (>0.8) are all from laboratory populations of *Tribolium*.

Vertebrate ectotherms: Fish, $n = 5$; amphibians, $n = 5$; reptiles, $n = 2$.

Vertebrate endotherms: Solid bars show data for nonhuman homeotherms ($n = 51$), hatched bars show data for human populations ($n = 20$). Barrowclough (1980) also presents data for 15 bird species (range in ratio $= 0.84 - 1.15$), but because these may be less precise and are not given by Frankham (1995a), they are not included in the present plot.

ratio does not mean a small effective population size, because if the ratio is 0.1 and the census population size is 10^6, the effective population size is still 10^5. The relevant data are presented in Table 8.4: What is readily apparent is that as predicted, the low ratio is more characteristic of ectotherms than endotherms (the wood duck data is for a captive inbred population and so its utility is doubtful).

Table 8.4   Estimates of N_e/N *Which Take into Account Fluctuating Population Size, Variance in Family Size, and Unequal Sex Ratio*

Species	Common Name	N_e/N (mean)
Insects		
Coelopa frigida	Seaweed fly	0.0047, 0.0009 (0.0028)
Dacus oleae	Olive fruit fly	0.18
Drosophila melanogaster	Fruit fly	0.03–0.14 (0.13,[a] $n = 5$)
Drosophila pseudoobscura	Fruit fly	0.012, 0.036 (0.024)
Molluscs		
Crassostrea gigas	Pacific oyster	$<10^{-6}$
Fish		
Atractoscion nobilis		0.27–0.40 (0.335[b])
Oncorhynchus mykiss	Rainbow trout	0.90
Oncorhynchus tshawytscha	Chinook salmon	0.013, 0.043 (0.028)
Amphibians		
Bufo marinus	Cane toad	0.016–0.088 (0.052[b])
Notophthalamus viridescens	Red-spotted newt	0.073
Birds		
Cairina scutulata	White-winged wood duck	0.052, 0.094 (0.073)
Mammals		
Bison bison	Bison	0.069
Cerocebus galeritus	Tana River crested mangabey	0.19–0.29 (0.24[b])
Lasiorhinus krefftii	Hairy-nosed wombat	0.18, 0.59 (0.385)
Pteropus rodricensis	Rodrigues fruit bat	0.18–0.43 (0.305)

[a]To avoid bias, the estimates for each study were averaged and then the mean estimates across studies averaged.

[b]Only range given, value in parentheses is midpoint.

Source: Data from Frankham (1995a).

8.2   The Influence of Population Bottlenecks on Quantitative Genetic Variation

8.2.1   Predicted Change in Genetic Variances

We shall consider an infinitely large population that passes through a single generation in which the population size is reduced to N individuals. Further, we focus

Table 8.5   Genotypic Distributions in a Population That Passes Through a Single-Generation Bottleneck of Two Individuals

A_1A_1	A_1A_2	A_2A_2	P^a	p'^b	V_A
2	0	0	p^4	1.00	0.000
0	2	0	$(2pq)^2$	0.50	0.500
0	0	2	q^4	0.00	0.000
0	1	1	$2(2pq)(q^2)$	0.25	0.375
1	1	0	$2(p^2)(2pq)$	0.75	0.375
1	0	1	$2p^2q^2$	0.50	0.500

Note: Assuming random selection the probability of picking a particular genotype is $P(A_1A_1) = p^2$, $P(A_1A_2) = 2pq$, and $P(A_2A_2) = q^2$. In the far right column is shown the genetic variance assuming an additive model (V_A).

[a]Probability of this distribution of genotypes.

[b]Proportion of A_1 alleles in the population after the bottleneck.

on a single locus with two alleles A_1 and A_2 at frequencies p and q ($= 1 - p$), respectively. A qualitative picture of the effect of drift on the additive genetic variance and heritability can be obtained by consideration of Fig. 2.3, which shows the change in V_A and h^2 with a change in allelic frequency. Due to drift, the allelic frequency will move away from its initial position; depending on its initial position, a change in p in any particular population can generate an increase or decrease in additive genetic variance and a consequent change in heritability.

In the initial population, the heritability is

$$h^2 = \frac{V_A}{V_A + V_D + V_E} \tag{8.11}$$

where (Chapter 2)

$$V_A = 2pq[(1 + d(q - p)]^2 \tag{8.12a}$$

$$V_D = (2pqd)^2 \tag{8.12b}$$

and V_E is the environmental variance. The probability that heritability will increase varies with the degree of dominance (d) and the allele frequency (p). Suppose the population passes through a bottleneck of two individuals: There are six possible outcomes, each occurring with the probability shown in Table 8.5. Over a wide range of allele frequencies, the probability that heritability will increase can be extremely high (Fig. 8.4). The probability of an increase given a bottleneck of size N can be more generally obtained using the multinomial probability function (Sokal and Rohlf, 1995; Mitchell-Olds, 1991)

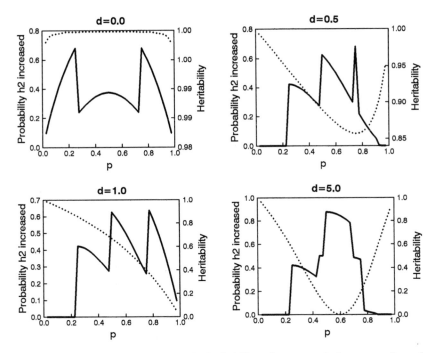

Figure 8.4 Probability of an increase in heritability when a population passes through a bottleneck of size 2 (solid line). The genetic model is a single locus with two alleles. The sudden shifts in the probabilities are due to the limited number (6) of genotypes possible (see Table 8.4). To avoid undefined values, the environmental variance is set at 0.0001. The dotted line shows the heritability prior to the bottleneck.

$$P(n_1, n_2, n_3) = \frac{N!}{n_1!\, n_2!\, n_3!}\, p_1^{n_1}\, p_2^{n_2}\, p_3^{n_3} \tag{8.13}$$

where n_i ($i = 1, 2, 3$) are the number of A_1A_1, A_1A_2, and A_2A_2, respectively, $N = n_1 + n_2 + n_3$, $p_1 = p^2$, $p_2 = 2pq$, and $p_3 = q^2$. The frequency of A_1 after the bottleneck, p', is thus $p' = (n_1 + 0.5n_2)/N$. The probability of an increase in heritability is a function of the allele frequency, the environmental variance, and the bottleneck size (Fig. 8.5). The case shown in which $V_E = 0.0001$ might give a somewhat unrealistic picture because such a low environmental variance is not typical and because the analysis considers only the probability of an increase in heritability, no matter how small (for a very large bottleneck, there is a .5 probability that p will increase, although this will be by a trivial amount). In the case of $V_E = 1.0$, the additional condition has been imposed that the increase must be greater than 10% of the initial value. Even in this case, there may be a signifi-

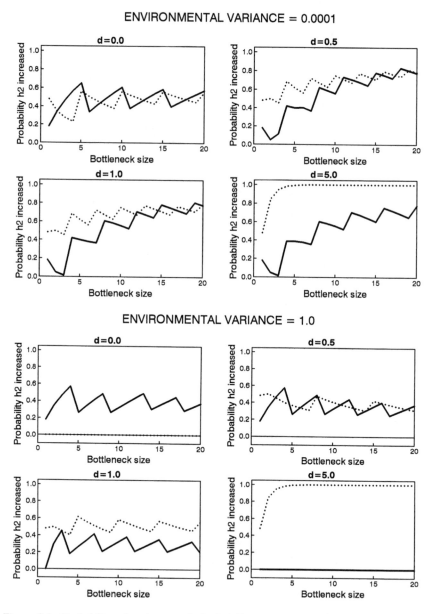

Figure 8.5 Probability of an increase in heritability when a population passes through a single bottleneck from 1 to 20 individuals. The genetic model is a single locus with two alleles. Initial allelic frequencies are 0.1 (solid line) and 0.6 (dotted line). When $V_E = 1$, only increases leading to a minimum increase of 10% in the heritability are considered.

cant probability of an increase when genetic variation is strictly additive and the allele frequency is small (10%).

The expected variances can be obtained using Eq. (8.13) or more directly by analysis of the moments of p'. From the latter, we have (Willis and Orr, 1993),

$$
\begin{aligned}
E(V_A) &= E\{2p'q' [1 + d(q' - p')]^2\} \\
&= 2\mu_1 (1 + 2d + d^2) - 2\mu_2 (1 + 6d + 5d^2) \qquad \textbf{(8.14a)} \\
&\quad + 8\mu_3 (d + 2d^2) - 8\mu_4 d^2
\end{aligned}
$$

$$
\begin{aligned}
E(V_G) &= E(V_A) + E([2p'q'd]^2) \\
&= 2\mu_1 (1 + 2d + d^2) - 2\mu_2 (1 + 6d + 3d^2) \qquad \textbf{(8.14b)} \\
&\quad + 8\mu_3 (d + d^2) - 4\mu_4 d^2
\end{aligned}
$$

where μ_i is the ith moment (i.e., the expected values of p', p'^2, p'^3, and p'^4; for their derivation, see page 335 of Crow and Kimura, 1970). These moments are functions of the bottleneck size. Populations passing through a single bottleneck of size 2 can experience large increases in heritability due to the shift in allelic frequency, but only if $d > 0$ (Fig. 8.6). These results are dependent on the assumption of a single locus: If the trait is determined by a large number of loci, then the expected additive genetic variance will always decrease.

The above analysis ignored epistatic effects, which require by definition at least two loci. As is evident from Fig. 2.5, similar increases in heritability can occur when epistasis is present. An epistatic model involving two loci and no dominance has been analyzed by Goodnight (1987, 1988; Table 8.6). The alleles in the initial population are set at 0.5, in which case the additive genetic variance is $(a^2 + b^2)/2$. If, as a consequence of random sampling through the bottleneck, the B_1 allele is fixed while the frequency of A_1 remains at 0.5, the additive genetic variance becomes $(a^2 + i^2)/2$, which, depending on the value of the epistatic component i relative to b, may lead to an increase in the additive genetic variance. This phenomenon is also seen in the two examples shown in Fig. 2.5: in the first case, heritability goes down, whereas in the second case, it is increased. Goodnight (1987) concluded, for the particular epistatic model he examined, that (1) "a single founder event, regardless of size, will lead to an increase in the additive genetic variance, provided the epistatic genetic variance in the ancestral population is at least ⅓ of the additive genetic variance" (p. 446); (2) "When founder events are of size 2, regardless of linkage, more than half of the epistatic genetic variance in the ancestral population is converted to additive genetic variance in the new colony. With free recombination, fully 75% of the epistatic genetic variance is converted to additive genetic variance each generation that the population size remains at two individuals" (p. 447); (3) the amount of conversion decreases rapidly with population size, 50% being converted when $N = 1$, approximately 10% when $N = 5$, and 6% when $N = 16$.

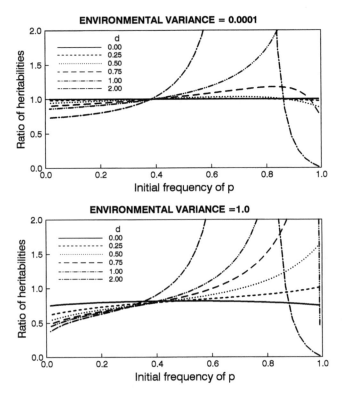

Figure 8.6 Relative increase in expected heritability when a population passes through a bottleneck of size 2. The genetic model is a single locus with two alleles.

Regardless of the model chosen, any changes in allelic frequency other than fixation need not necessarily result in permanent change. If the original allelic frequency is a product of selection, then the shift in genetic variance will only be temporary, selection eventually restoring the initial conditions. This might not occur if random drift breaks up adaptive gene complexes, leading to the establishment of new combinations [Wright's shifting-balance process, Goodnight (1995)]. A second important point to be remembered is that there must necessarily be a reduction in genetic variation in an absolute sense if the bottleneck is very small (Nei et al., 1975; Maruyama and Fuerst, 1985). For example, suppose that the bottleneck consists of a single male and female; in this case, the maximum number of alleles that can pass through the bottleneck is four. This loss of variation is seen in a comparison of the mean number of alleles per locus of source and founder populations (Fig. 8.7), in a positive correlation between population size and electrophoretic variation in the plants *Scabiosa columbaria* and *Salvia*

Table 8.6   *Epistatic Model Analyzed by Goodnight (1988); All Alleles at a Frequency of 0.5 in the Initial Population*

	A_1A_1	A_1A_2	A_2A_2
B_1B_1	$a + b + i$	b	$-a + b - i$
B_1B_2	a	0	$-a$
B_2B_2	$a - b + i$	$-b$	$-a - b + i$

pratensis (Bijlsma et al., 1994), and in the loss of electrophoretic variability following a bottleneck caused by vulcanism in wild populations of the rodent *Ctenomys maulinus* (Gallardo and Kohler, 1994; Gallardo et al., 1995). The loss of alleles could negate the first effect, because after the bottleneck, part of the genetic variability is lost and it might, therefore, not be possible to return to the original conditions (e.g., one locus might be fixed).

8.2.2   Predicted Change in Trait Value

As above, letting the genotypic values be $A_1A_1 = 1, A_1A_2 = d, A_2A_2 = -1$, the mean trait value, M, is (Chapter 2)

$$M = (p - q) + 2pqd \tag{8.15}$$

The expected value of M following the bottleneck, $E(M')$ is

$$E(M') = M - 2pqd \frac{1}{2N} \tag{8.16}$$

If there is no dominance, the expected value remains the same, whereas with dominance, there is a decline, which is an inverse function of the size of the bottleneck. The change in the value of the trait can be attributed to the effects of inbreeding due to the finite population size; when the trait value declines the phenomenon is called **inbreeding depression** (for a detailed discussion, see Section 8.4). Because the change is negative, the direction of change is toward the value of the more recessive allele (A_1 is dominant to A_2 when d is positive). The magnitude of the change is proportional to the allelic frequency, being greatest when $p = q = 0.5$. Although the mean might decline, the probability that in any particular case it will increase can be high, as illustrated in Fig. 8.8 for a population passing through a bottleneck of size 2 individuals.

The above analysis applies to a single locus; assuming independence between loci, the effect of multiple loci can be obtained by summing across loci,

$$E(M') = M - 2 \sum_i p_i q_i d_i \frac{1}{2N} \tag{8.17}$$

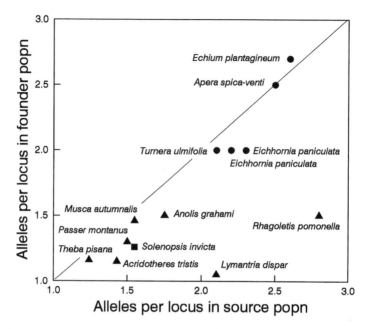

Figure 8.7 The relationship between allelic variation, as revealed by electrophoresis, in source and founder populations (the two populations are typically on different continents). A similar pattern is observed using average heterozygosity or proportion of polymorphic loci, but heterozygosity is typically a less sensitive indicator (Leberg, 1992). Data are from Howard (1993), ▲; Barrett and Husband (1990) ●; and Ross et al. (1993) ■.

The expected change in the trait value depends on the general direction of the dominance effects. If there is **directional dominance**, meaning that the dominance values tend to have the same sign, then there is more likely to be a change in the mean value of the trait. The variability in the change in trait value will decline with the number of loci.

8.2.3 *Empirical Studies*

Results of single bottleneck experiments have been undertaken on *Drosophila melanogaster* (Frankham, 1980; Lints and Bougois, 1984), the housefly, *Musca domestica*, (Bryant et al., 1986, Bryant and Meffert, 1991), and the butterfly, *Bicyclus anynana* (Brakefield and Saccheri, 1994). All report changes in additive genetic variance and heritabilities, but beyond this, the results are difficult to evaluate. The most extensively reported study is that by Bryant and his colleagues on *M. domestica*. From a large outbred base population, four replicate lines per bottleneck size of 2, 8, and 32 flies were set up and allowed to expand to approximately 2000 individuals, which took "about five generations" (Bryant et al.,

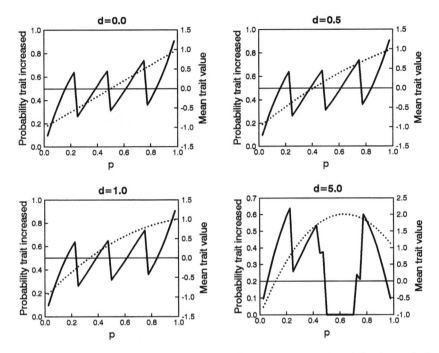

Figure 8.8 Probability of a change in the mean trait value (solid line) following a single generation bottleneck of size 2. The genetic model is a single locus with two alleles. The sudden shifts in the probabilities are due to the limited number (6) of genotypes possible (see Table 8.4). To avoid undefined values, the environmental variance is set at 0.0001. The dotted line shows the mean trait value prior to the bottleneck.

1986, p. 1193). As a consequence of the five generations of (presumed) random mating following the bottleneck, linkage disequilibrium was not likely to have influenced the estimates of genetic parameters (Lynch, 1988a; see Fig. 3.1). Because of decreased viability in some lines, these parameters were estimated by pooling replicate families, which significantly compromises the interpretation of results. The pooled estimates suggest that at least for some morphological traits, additive genetic variance and heritabilities increased at some bottleneck sizes (Fig. 8.9). However, the previous theory does not make this a surprising result. Bryant et al. (1986) found a significant decrease in body weight and viability as the bottleneck size was decreased (Fig. 8.10) and suggested that this was indicative of a genetic architecture with dominance and possibly epistatic effects. However, the traits showing such a decline were not those examined for changes in heritability (although body weight is probably a reasonable index of the linear body measures actually scored) and, further, decreases in trait values can occur in a strictly additive system (Fig. 8.8). The problem is that there were simply too few

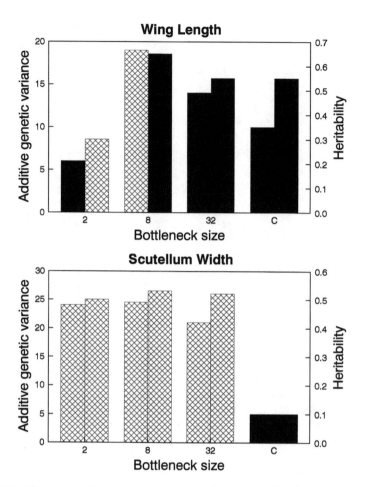

Figure 8.9 Observed additive genetic variance (left bar of each group) and heritabilities (right bar) for two morphological traits in *Musca domestica* populations that have passed through a single-population bottleneck. Values not significantly different from the control (C) are plotted in solid color, values that are significantly different from the control are cross-hatched. Data from Bryant et al. (1986).

lines in relation to the probable variation in genetic variability in the bottlenecked line to draw any firm conclusion. The coefficient of variation in the within-line genetic variance caused by variation in the genetic variation in the founder population assuming only additive effects is approximately $[2/(N_e L)]^{1/2}$, where N_e is the effective population size and L is the number of replicate lines (Lynch, 1988a). For the housefly experiment, this coefficient of variation is 0.45, 0.24, and 0.12 for the three bottleneck sizes (N_e is approximately $N + 0.5$ because self-

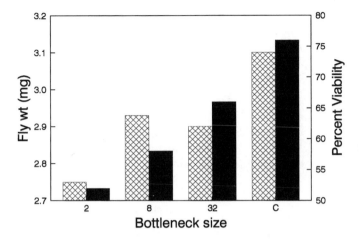

Figure 8.10 Changes in body weight (cross-hatch) and viability (solid) in *Musca domestica* populations that have passed through a single-population bottleneck (control population denoted by C). Data from Bryant et al. (1986).

fertilization is excluded; see first section of this chapter). To reduce this source of variation to an acceptable level of 10% requires approximately 100, 25, and 6 replicates lines per bottleneck size, sample sizes which would certainly be logistically difficult to handle if a minimum of 50 families per replicate line are required to adequately estimate the genetic parameters within a line.

8.3 The Influence of Finite Population Size on Quantitative Variation

8.3.1 Changes in Genetic Variance over Time

As with case of the effect of a single bottleneck, the qualitative consequences of a continued finite population size on genetic parameters can be most easily understood by consideration of the fate of two alleles at a single locus. In the absence of any selective forces that maintain an allele at some intermediate frequency, random sampling from generation to generation will cause the allele frequency to drift between the bounds 0 and 1. But these two extremes are absorbing boundaries (i.e., once the allele frequency attains 0 or 1, then it must, in the absence of recurrent mutation, remain there). Thus, over time, a population will become increasingly homozygous. If the initial allelic frequency is p, then in p proportion of cases, the allele will be lost. This can be easily seen by considering the allele in its initial state on the 0–1 line: Because drift is entirely random, the probability that the allele will travel the distance from p to 0 is equal to the relative length of the line, which is, of course, p. Therefore, if we consider a large number of populations that start from the same state, the frequency of the allele will on average stay constant at p, although in any particular population, it must eventually fix or be lost.

For the additive model, the variance within any particular line at some time t is

$$V_G = V_A = 2p_t q_t \qquad (8.18)$$

The mean variance within lines can be designated $2P_t Q_t$, where $P_t Q_t$ is the mean value of $p_t q_t$ over all lines. This mean value is the overall heterozygosity when all lines are considered a single population. The expected heterozygosity, $E(H_t)$ at time t can be shown to be (Bulmer, 1985, p. 220)

$$E(H_t) = H_0 \left(1 - \frac{1}{2N}\right)^t \qquad (8.19)$$

Now at $t = 0$, $H_0 = 2p_0 q_0$ and, hence,

$$V_A(t) = 2p_0 q_0 \left(1 - \frac{1}{2N}\right)^t \qquad (8.20)$$

where $V_A(t)$ is the additive genetic variance at time t. On average, the additive genetic variance is declining asymptotically approaching zero, as intuitively expected. Initially, the variance among lines is by definition zero, and because alleles fix in proportion to their initial value, the variance among lines must approach $p_0 q_0$. The variance among lines, $V(p_t)$, approaches this value according to the relationship (Crow and Kimura, 1970, pp. 327–329; Bulmer, 1985, pp. 218–219)

$$V(p_t) = p_0 q_0 \left[1 - \left(1 - \frac{1}{2N}\right)^t\right] \qquad (8.21)$$

The dynamics of the process thus consists of two elements: Within populations, there is a decline in variance, whereas among populations, there is an increase as populations drift apart. Initially, we have a single point on the 0–1 line corresponding to the frequency p_0; over time, the distribution of values of p develops into an approximately bell-shaped curve and then into a U-shaped curve as an increasing fraction of the lines becomes fixed (Fig. 8.11). The process of increasing variance between lines and fixation is well illustrated by the experiment of Buri (1956), utilizing a mutation in *Drosophila melanogaster* [the figure is reproduced in Falconer (1989, p. 56)], although this experiment was not run long enough for the development of the U-shaped distribution.

Extension of the model to include dominance effects is complex. Only the case

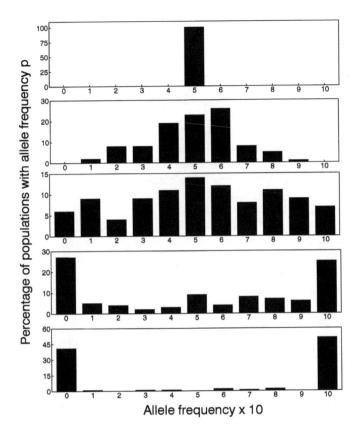

Figure 8.11 Changes over time in allele frequency due to drift (from top to bottom, generation number = 0, 1, 5, 10, 30). There are initially 100 populations of size 10 with an allele frequency of 0.5. At each generation, 10 offspring are produced by random selection of parents from the population.

of complete dominance has been investigated in detail (Robertson, 1952; Bulmer, 1985). The expected additive genetic variance, $E\{V_A(t)\}$ is

$$E\{V_A(t)\} = 4\left[\frac{3}{5}pqC_0^t + pq(q - p)C_1^t - 2pq(pq - \frac{1}{3})C_2^t\right] \quad (8.22)$$

where, for simplicity, the zero subscripts on p and q have been dropped, and

$$C_0 = 1 - \frac{1}{2N}, \qquad C_1 = 1 - \frac{3}{2N}, \qquad C_2 = 1 - \frac{3}{N}$$

The expected total genetic variance, $E\{V_G(t)\}$, is

$$E\{V_G(t)\} = 4\left[\frac{4}{5}pqC_0^t + pq(q - p)C_1^t - pq(pq - \frac{1}{3})\,C_2^t\right] \quad \textbf{(8.23)}$$

Both of the above variances asymptotically approach zero but might show an initial increase. The additive genetic variance will increase when $p_0 > 0.5$ and the total variance will increase when $p_0 > 0.59$ (Bulmer, 1985, p. 226); three examples are shown in Fig. 8.12. Note that in all cases the heritability does not approach zero but 0.75. This arises because as t increases, the first term in the above two equations dominates and, hence, $V_A(t)/V_G(t) \rightarrow (3/5)/(4/5) = 3/4$. This result depends on there being no environmental variance; in the more typical case in which $V_E > 0$, the heritability will tend to zero as the variances become progressively smaller fractions of the total variance ($V_G + V_E$).

Two further sources of variation are random departures from Hardy–Weinberg and linkage equilibrium (Bulmer, 1971a, 1976, 1985; Avery and Hill, 1977). These two factors cause variation in the total variance which are, on average, zero but which have mean square deviations of V_T^2/N and $5V_T^2/3N$, respectively, where V_T is the variance observed if the gene frequencies were in perfect Hardy–Weinberg (Bulmer, 1985, p. 231). Thus, these two factors can contribute significantly to the total variance, as demonstrated by the simulation results of Bulmer (1976). Departures from linkage equilibrium do not lead to any permanent change but must be considered in experimental analyses of the effects of drift (Lynch, 1988a).

Goodnight (1988) examined the change in additive genetic variance over time for the epistatic model described in Table 8.5. As with the case of dominance, as the genetic variance ultimately declines to zero, there may be initial increases in the additive portion relative to the total genetic component (Fig. 8.13). As with the additive model, linkage disequilibrium does not contribute to permanent shifts in allele frequencies (Tachida and Cockerham, 1989).

8.3.2 Changes in Trait Value over Time

As previously, we consider a one-locus, two-allele model. Letting the genotypic values be $A_1A_1 = 1$, $A_1A_2 = d$, and $A_2A_2 = -1$, we have (Section 8.2.2)

$$M_t = (p_t - q_t) + 2p_tq_td \quad \textbf{(8.24)}$$

In any particular line, drift will eventually fix one of the alleles leading to some lines with $M_t = 1$ and some lines with $M_t = -1$. The expected value of the trait at time t is given by

$$E(M_t) = M_0 - 2p_0q_0d\left[1 - \left(1 - \frac{1}{2N}\right)^t\right] \quad \textbf{(8.25)}$$

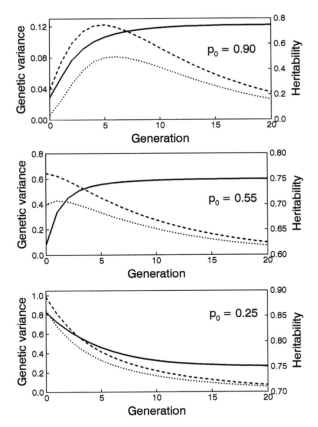

Figure 8.12 Predicted changes in additive genetic variance (dotted line), total genetic variance (dashed line), and heritability (solid line) for the case of a single locus with two alleles, one dominant to the other.

The expected value is decreased in the presence of dominance, the value asymptotically approaching $M_0 - 2p_0q_0d$. Assuming independence between loci, the effect of multiple loci can be obtained, as before, by summing across loci:

$$E(M_t) = M_0 - 2 \sum_i P_{0,i} q_{0,i} d_i \left[1 - \left(1 - \frac{1}{2N} \right)^t \right] \qquad (8.26)$$

As described above, if there is directional dominance, there will be a general decline in the trait value (= inbreeding depression). Inbreeding and its consequences are the subject of the remainder of this chapter.

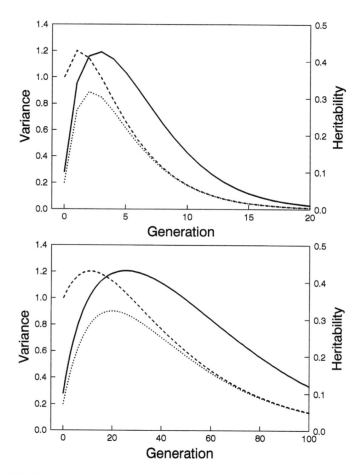

Figure 8.13 The time course of additive genetic variance (dotted line), total genetic variance (dashed line), and heritability (solid line) for the epistatic model of Goodnight (1988). In the initial population, 20% of the total variance is additive, the remainder due to epistasis.

8.4 Inbreeding

8.4.1 The Concept of Inbreeding

Thus far, it has been assumed that mating is random. **Inbreeding** is defined as the mating of two related individuals (e.g., full-sibs, cousins, etc.). Nonrandom mating could result in inbreeding (or the converse, **outbreeding**). As population size is decreased, the probability that two parents will be related even if mating is at random will clearly increase. To make use of the concept of inbreeding, we define genes as being **identical by descent** if they have originated from the replication of the **same gene**. This should not be confused with identity of effect: Two alleles at a diallelic locus can be the same in the sense that both are A_1 or both A_2 but not be identical by descent because they have originated from the replication of two different A_1's or A_2's (Fig. 8.14). The probability that the two genes at a locus are identical by descent is termed the **coefficient of inbreeding**, F. A **base population** is defined as that population in which $F = 0$ (i.e., there is no inbreeding). The **rate of inbreeding** is defined as

$$\Delta F = \frac{F_t - F_{t-1}}{1 - F_{t-1}} \tag{8.27}$$

The term $1\text{-}F$ is called the **panmictic index**, denoted by the symbol P.

The estimation of the inbreeding coefficient for autosomal loci can be calculated following a simple addition rule [for a detailed discussion, see Crow and Kimura (1970, pp. 69–73) or Falconer (1989, pp. 85–97)]. Letting two individuals produce an individual X and the inbreeding coefficient in the ancestral generation be F_A, we have

$$F_X = \sum \left(\frac{1}{2}\right)^n (1 + F_A) \tag{8.28}$$

where n is the number of individuals in any continuous path of relationship beginning with one parent of X and ending with the other (i.e., the paths connecting X to its two parents are not counted). It is usual to assume that F_A is zero (generally because it is not known). Examples of the use of Eq. (8.28) to calculate the inbreeding coefficient for full-sibs and half-sibs are shown in Fig. 8.15. To calculate the inbreeding coefficient for offspring–parent, we need simply note that any offspring will share one-half of its alleles with its parents and, hence, $F_X = \frac{1}{4}$ in this case (Fig. 8.15).

8.4.2 Changes in Means and Variances as a Result of Inbreeding

The results presented in the previous sections can be expressed in terms of the inbreeding coefficient as follows. Defining the separate lines to be our base populations (i.e., $F_0 = 0$), we consider the change in the inbreeding coefficient over time. Because mating is at random and selfing permitted, the probability that any two genes will be identical by descent is $\frac{1}{2}N$; thus, $F_1 = \frac{1}{2}N$. In the second generation there are two ways in which genes may be identical by descent: first, because they originate by replication in the previous generation and second, because they originate from the same gene in the first generation (Fig. 8.14). From this we can write

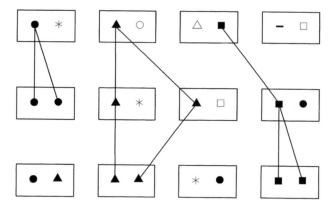

Figure 8.14 A schematic illustration of the concept of identity by descent. Each box represents an individual and each symbol an allele. In the initial population (top row), all alleles are considered unique (i.e., by definition they are unrelated by descent). In the second generation, an individual can inherit two alleles that are identical by descent only if there is self-fertilization (●). In the third generation, two alleles may be identical by descent due to a common ancestor in the first (▲) or second (■) generation.

$$F_2 = \frac{1}{2N} + \left(1 - \frac{1}{2N}\right)F_1 \qquad (8.29)$$

The rate of inbreeding, ΔF, is equal to $\frac{1}{2}N$ and arises from self-fertilization; if self-fertilization is not permitted, the increment is not much affected as the effect is to simply push the process back to the grandparental generation. Note that the rate of inbreeding is inversely proportional to population size. Applying the same argument as used to derive F_2 gives the recursive equation

$$F_t = \frac{1}{2N} + \left(1 - \frac{1}{2N}\right)F_{t-1} \qquad (8.30)$$

which can be shown to be equivalent to

$$F_t = 1 - \left(1 - \frac{1}{2N}\right)^t \qquad (8.31)$$

Hence, for a model incorporating both additive and dominance effects, the expected changes in the mean [Eq. (8.25)] and variance [Eq. (8.21)] are given by

$$E(M_t) = M_0 - 2p_0q_0dF_t \qquad (8.32)$$

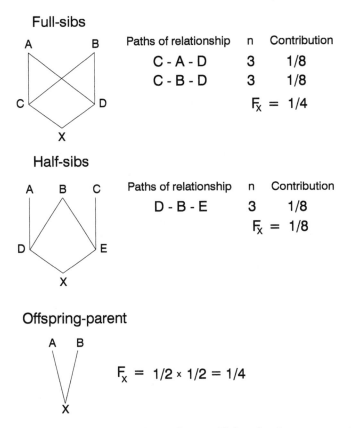

Figure 8.15 The estimation of the inbreeding coefficient for three common breeding designs.

$$V(p_t) = p_0 q_0 F_t \tag{8.33}$$

Selection experiments on wing length in *D. melanogaster* by Tantawy (1956a) using both an outbreeding and an inbreeding (full-sib) design show, very clearly, the effects of inbreeding on reducing additive genetic variance (Fig. 8.16). Under full-sib mating, the heritability of the focal trait (wing length), a trait highly genetically correlated with this trait (thorax length), and a trait modestly genetically correlated with the focal trait (longevity) all showed dramatic declines, although in the outbred stock, heritabilities might even have increased (upper panels, Fig. 8.16). Similar dramatic reductions were observed in the genetic correlations between these traits (lower panels, Fig. 8.16).

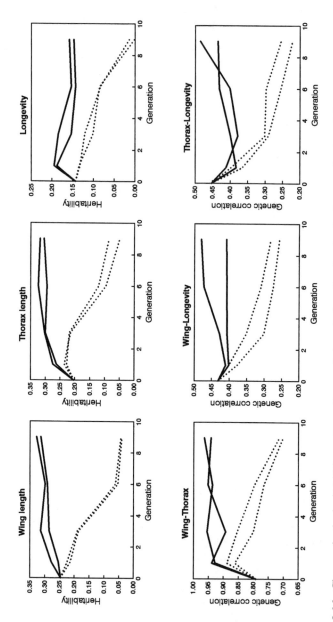

Figure 8.16 Changes in the heritabilities and genetic correlations during directional selection on wing length in *Drosophila melanogaster*. Two lines (solid) were maintained under outbreeding conditions and two lines (dotted) were maintained under brother–sister mating. Data from Tantawy (1956a).

8.4.3 Inbreeding Depression

The deleterious effects of inbreeding were reported by Darwin (1868, 1876) and had undoubtedly been realized by plant and animal breeders long before. **Inbreeding depression** can be defined as the decline in the value of a trait (typically connected with fitness) as a result of inbreeding. In the analysis of inbreeding depression, the usual statistic computed is the **coefficient of inbreeding depression**, δ,

$$\delta = 1 - \frac{w_I}{w_O} \tag{8.34}$$

where w_I is the trait value from inbred progeny and w_O is the trait value from outcrossed progeny. In plants, the comparison is typically between selfed and outbred individuals, but in animals, selfing is frequently not possible. The coefficient of inbreeding depression can be calculated using population or family means, but, because it is a ratio, these two estimates will not be the same, although the difference is not likely to be so great as to qualitatively change the conclusions. Johnston and Schoen (1994) recommend the use of the family-based method because (1) each family is an independent observation, (2) pairwise comparisons are possible, and (3) a confidence interval can be placed on the population estimate from the family data. However, in evolutionary models, it is the population means for outbred and inbreds that are useful, so that there is a strong argument for the opposite of what Johnston and Schoen recommend (Charlesworth, personal communication).

From the previous analyses, it was noted that inbreeding depression will not occur, on average, if there is only additivity. When there is dominance, the expected trait value declines linearly with the degree of inbreeding [Eq. (8.32)]. When there is epistasis but no dominance, there will be no inbreeding depression, and when both epistasis and dominance are present, the relationship between inbreeding depression and F can be linear or quadratic, although more likely the latter (Crow and Kimura, 1970, pp. 77–85).

Examples of a linear and a nonlinear decline in trait value with F are shown in Fig. 8.17: The average yield of maize declines linearly with the inbreeding coefficient, demonstrating the presence of dominance (Fig. 17a), whereas litter size in the house mouse shows a nonlinear decline with F, suggesting the presence of dominance and epistatic interactions (Fig. 17b). Nonlinear relationships must be interpreted cautiously because they could a consequence of an inappropriate scale. This is illustrated in the analysis of inbreeding effects on competitive ability in *D. melanogaster* (Latter and Robertson, 1962; Fig. 8.17). Competitive ability appears to decline in a highly nonlinear fashion with the inbreeding coefficient (Fig. 8.17c), but this nonlinearity is completely removed if a log scale is used [Fig. 8.17d; see also Latter et al. (1995)]. Thus, on the latter scale there is no

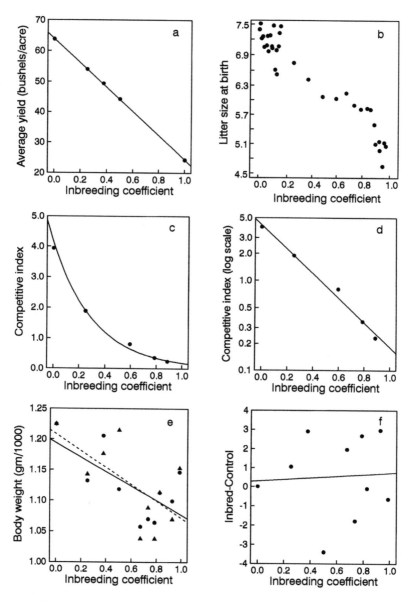

Figure 8.17 Some examples of the relationship between metric traits and the inbreeding coefficient. (a) Average yield of maize [data from Crow and Kimura (1970)]; (b) litter size in a wild strain of house mouse [data from Connor and Bellucci (1979)]; (c) an index of competitive ability in *Drosophila melanogaster* [data from Latter and Robertson (1962)]; (d) the same trait as in (c) but plotted on a log scale; (e) body weight in *Drosophila melanogaster* [circles and solid line show results for inbred lines, triangles and dashed line show results for control lines (outbred but placed on the graph in the corresponding position to the relevant inbred line), data from Kidwell and Kidwell (1966)]; (f) the difference between inbred and control lines for the data shown in panel e.

compelling evidence for epistatic effects. As in all experiments, it is essential that control lines be maintained to account for any systematic environmental effect. The need for such lines is amply demonstrated by the experiment of Kidwell and Kidwell (1966) on inbreeding and body size in *D. melanogaster*. There is a highly significant decline in body size with the inbreeding coefficient, but there is a similar decline in the control line, and the difference between the two values shows no relationship with *F* (Figs. 8.17e and 8.17f).

The linear relationship between inbreeding depression and *F* suggests the following experimental design for the detection of inbreeding depression: A series of lines are subjected to a continued program of inbreeding and the average trait value is regressed on the average inbreeding coefficient. Although this approach has been used frequently, it has the serious statistical flaw that the samples are not independent. Lynch (1988a) provides an approximate method of correcting this problem: Application of the test to data sets on mice, sheep, *D. melanogaster*, *Tribolium casteneum*, barley, and corn supports the original conclusion that inbreeding depression is typically present [see Table 2 of Lynch (1988a)]. The following alternate, conservative test based on one generation of mating was suggested by Lynch (1988a): *L* females are mated at random, whereas *L* females are mated consanguineously (self-fertilization or full-sib), and from each mating, a single random offspring is measured. From these data, compute the *t*-statistic, which has *L*-1 degrees of freedom:

$$t = (X_R - X_I) \sqrt{\frac{L}{cV_P}} \tag{8.35}$$

where X_R is the mean value in the random-bred progeny, X_I is the mean in the inbred progeny, V_P is the phenotypic variance in the random-bred progeny, and $c = 11/4$ for self-fertilization and $17/8$ for full-sib matings.

Because, as shown in Chapter 2, dominance is expected and observed to be larger in traits closely connected to fitness, we would expect inbreeding depression to be most evident in fitness-related traits such as fecundity and survival. This is demonstrated by Mackay's (1985) analysis of inbreeding depression associated with homozygosity ($F = 0.65$) of the third chromosome of *Drosophila melanogaster*: Inbreeding depression was large in viability ($\delta = 0.44$), fertility (0.81), and female productivity (0.56), but considerably smaller in male weight (0.10), abdominal bristle number (0.05), and sternopleural bristle number (0.00). Fitness, as measured by competitive ability against flies carrying a standard third chromosome, had the greatest inbreeding depression (0.87).

Empirical data from nondomestic species shows overwhelming evidence for significant inbreeding depression in gymnosperms (Table 8.7), angiosperms (Table 8.8), and animals (Table 8.9, Figs. 8.17, 8.18). An inbreeding depression of only 10% might appear small, but when applied to a series of traits, such as survival at various life-cycle stages and fecundity, that contribute to fitness, the

Table 8.7   Some Estimates of Inbreeding Depression ($\delta \times 100$) in Gymnosperms, Listed in Approximate Order of Selfing Rate

Species	Selfing Rate (%)	% Filled Seeds	Germination	Trait[a]
Pinus radiata	4	<99	22	2–12
Pinus taeda	0–4	88	14	4 (S)
Pinus elliottii	6	78–84	53	22
Pinus jeffreyi	6		9	46
Pinus ponderosa	4–19	63		36
Picea mariana	8	54	ns[b]	20
Pseudotsuga menziesii	7–10	89	11	18 (S)
Pinus banksiana	0–17	58	2	5
Pinus sylvestris	11–24	88		37
Picea abies	11–49	50	75	11
Abies procera	41	31	ns	24
Pinus attenuata	50			37 (F)
Seqoia sempervirens	59	−2		73
Picea glauca	60	80	ns	63
Latrix decidua	71			16

[a]Trait is size unless followed by S (= survival) or F (= fertility).

[b]ns = Not significant.

Source: Modified from Charlesworth and Charlesworth (1987).

combined multiplicative effect can produce an enormous depression (see, for example, *Lymnaea peregra* and *Coturnix coturnix* in Table 8.9). Estimation of inbreeding depression in plants grown under different conditions indicates that the degree of depression is increased under field or stressful conditions (Allard, 1965; Antonovics, 1968; Schemske, 1983; Mitchell-Olds and Waller, 1985; Dudash, 1990; Schmitt and Ehrhardt, 1987, 1990; Johnston, 1992; Eckert and Barrett, 1994; McCall et al., 1994). Chen (1993) also observed an increase in inbreeding depression when the gastropod *Arianta arbustorum* was raised under stressful conditions. These results demonstrate the importance of circumstance in the determination of fitness differences between inbred and outbred individuals.

8.4.4   Theories of Inbreeding Depression

There are two hypotheses to account for the existence of inbreeding depression: the overdominance hypothesis and partial dominance hypothesis. The overdominance hypothesis posits that heterozygotes have superior fitness and, thus, inbreeding, leading to homozygosity, will cause a diminution of fitness (East, 1908). The partial dominance hypothesis supposes that inbreeding depression results from the increase in frequency of deleterious recessive or partially reces-

Table 8.8   Examples of Inbreeding Depression ($\delta \times 100$) in Angiosperms. Estimates
Have Been Selected as Representing Species with Low Selfing Rate (0–10%), Moderate
Selfing Rate (30–63%), and High Selfing Rate (90–100%)

Species	Selfing Rate (%)	Seeds	Germination	Size	Fertility[a]	Survival	Ref.[d]
Costus allenii	~0	33	12	25			1
Costus laevis	0	37	−7	20			1
Costus guanaiensis	0	34	6	19			1
Phlox drummondii[b]	0	17				17	1, 2
Ipomopsis aggregata[b]	0	95					3
Gilia achilleifolia	4	ns	ns			44	1
Limnanthes douglassii	6				7		1
Sabatia angularis[c]	7	3	ns		10	22	4
Decodon verticallis[c]	30				6–39	6	5
Erythronium americanum	38	75					1
Clarkia tembloriensis	42		−1	13	44	12	6
Aquilegia caerulea	52	11	9	4		39	7
Mimulus guttatus	63		−1	36	47		8
Impatiens capensis	95		ns	41	16–27	7	1, 9
Thlaspi alpestre	95		18				1
Begonia semiovata	95	12	5	33		3	10
Begonia hirsuta	97	15	2	22		−1	10
Leavenworthia crassa	97	19	15		28	2	11
Clarkia tembloriensis	97		8	4	23	29	6
Amsinckia gloriosa	100				12	9	12
Amsinckia spectablis	100				20–40	0	12

[a]Fertility measures include differences in seed production per flower and also differences in total
fecundity.

[b]Self-incompatible.

[c]Data are for greenhouse experiment: inbreeding depression greater in field trials.

[d]References: (1) Charlesworth and Charlesworth (1987); (2) Levin (1991); (3) Waser and Price
(1989); (4) Dudash (1990), fertility measure is total fruit mass; (5) Eckert and Barrett (1994);
(6) Holtsford and Ellstrand (1990); (7) Montalvo (1994); (8) Dole and Ritland (1993), Latta and
Ritland (1994); (9) Charlesworth et al. (1990a); (10) Agren and Schemske (1993); (11) Charlesworth
et al. (1994); (12) Johnston and Schoen (1995).

Table 8.9   Estimates of Inbreeding Depression ($\delta \times 100$) in a Variety of Animal Species

Trait						

Trait	0.125	0.1875	0.25	0.375	0.50	Ref.[a]
Lymnaea peregra (snail)						
Eggs/day					49	1
Hatchability					61	
Survival at 14 days					47	
Survival at 30 days					43	
Size at hatching					3	
Size at 14 days					19	
Size at 30 days					16	
Overall fitness					94	
Salmo gairdneri (rainbow trout)						
Hatchability	5	−2	−10, 9, 14	6, 16	−9, 6	2
Fry survival	3	14	8, 11	4, 8	11, 25	
Wt at 91 days	−4	16	18	33	30	
Wt at 150 days	−9	2	12	21	22	
Brachiation rerio (zebra fish)						
Hatchability	90		89			3
% Eggs fertilized	83		91			
% Normal fry	9		8			
Survival to 30 days	40		43			
Length at 30 days	14		11			
Ictalurus punctatus (channel catfish)						
Hatchability			−6			4
Egg weight			10			
Survival to 4–12, 12–26, 12–40 wks				1, 0, −9		
Body weight at 4 wks			43			
Body weight at 12 wks			ng, 7	7		
Body weight at 16 wks				5		
Body weight at 26 wks				−27		
Body weight at 40 wks				−25		
Coturnix coturnix japonica (Japanese quail)						
Survival 0–5 wks			10	29	65	5
Survival 5–16 wks			3	7		
Fertility			37	58		
Body weight			7	9		
Egg weight			4	7		
Probability of zygote leaving offspring			66			

(continued)

Table 8.9   Continued

Trait	Inbreeding Depression ($\delta \times 100$) when $F =$					Ref.[a]
	0.125	0.1875	0.25	0.375	0.50	
Peromyscus maniculatus (deer mouse)						
Litter size			15			6
Weight on day 1, day 2			6, 5			
Age at first reproduction			− 17			
Litters per female			7			
Offspring per litter			2			
Survival to weaning			8			
Mus musculus (house mouse, wild population)						
Litter size			10	5	9	7
Body weight at 10 days			− 5	11	16	
Body weight at 53 days			− 5	− 11	− 14	
Nesting behavior score			10	5	9	

[a]References: (1) Jarne and Delay (1990); (2) Kincaid (1976a, 1976b); Gjerde et al. (1983), Aulstad and Kittelsen (1971); (3) Mrakovcic and Haley (1979); (4) Bondari and Dunham (1987); (5) Sittmann et al. (1966); (6) Hill (1974), Haigh (1983); (7) Lynch (1977).

Further examples in nondomestic animals:

Invertebrates: *Tribolium*—Pray et al. (1994), Pray and Goodnight (1995); *Drosophila*—Hollingsworth and Smith (1955), Tantawy and Reeve (1956), Tantawy (1957), Maynard Smith (1956), Sharp (1984), Mackay (1985), Ehiobu et al. (1989), Frankham et al. (1993), Garcia et al. (1994), Latter and Mulley (1995); oyster—Mallett and Haley (1983); Lepidoptera—Roush (1986), Cacoyianni et al. (1995); parasitoids—Hey and Gargiulo (1985), Hooper et al. (1993, review).

Vertebrates: Poikilotherms—Waldman and McKinnon (1993); captive mammals—Ralls et al. (1988), Lacy et al. (1993); *Parus major*—Greenwood and Harvey (1978), Noordwijk and Scharloo (1981); *Mus musculus*—Connor and Belluci (1979), Pape and Lasalle (1981); *Peromyscus*—Brewer et al. (1990), Keane (1990), Jimenez et al. (1994); *Microtus oeconomus*—dos Santos et al. (1995).

sive alleles (Davenport, 1908; Jones, 1917). The two models make different predictions concerning the mean trait value of a population undergoing continued inbreeding. Under the overdominance model, the heterozygotes are being purged from the population, leading in most cases to a continued decline in the trait value (Ziehe and Roberds, 1989; Charlesworth and Charlesworth, 1990). A temporary increase in the trait value can occur if the initial population is not in genetic equilibrium (Hill and Robertson, 1968), and Minvielle (1979) has shown that it is possible to construct a model in which the trait value in some cases can equal or exceed that of the original population. The dynamics expected in the partial dominance model are best understood by considering a population divided into a large number of lines being separately inbred. As inbreeding proceeds, deleterious alleles will be purged from some of the lines and will become fixed in other lines,

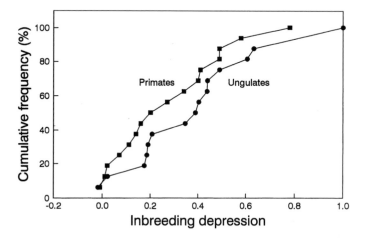

Figure 8.18 Effect of inbreeding on juvenile survival in captive populations of different ungulate (●) and primate (■) species. The median inbreeding depression in ungulates is approximately 0.4, and in primates, it is approximately 0.2. Data from Ralls et al. (1979) and Ralls and Ballou (1982).

with some lines disappearing because of the fixation of the deleterious alleles. The overall population and the trait value will first decline but then increase to a value equal to or greater than that in the starting population. The predictions of these two hypotheses are readily demonstrated by simulation models [Fig. 8.19; for the partial dominance model, see also Hedrick (1994)], although it should be remembered that, as in the case of bottlenecks, in small populations there may be considerable variation in response (Hauser et al., 1994). Crossing of inbred lines leads to restoration of fitness in the overdominant model and an increase in fitness in the partial dominance model.

Roberts (1960) produced moderate inbreeding in house mice and observed a decline in litter size and body weight, which was restored upon crossing of the inbred lines (Fig. 8.20). These results are consistent with either hypothesis. However, in a similar experiment, Bowman and Falconer (1960) inbred their lines sufficiently strongly that some lines failed to propagate themselves and, as predicted by the partial dominance model, the mean litter size increased in the remaining lines, presumably due to the purging of the deleterious alleles (Fig. 8.21). Further evidence for the purging of deleterious alleles in inbred mice is that some strains of highly inbred laboratory mice appear to have greater "vigor" than their wild progenitors (Strong, 1978, p. 56). The apparent reduction of inbreeding depression through selective breeding in Speke's gazelle, *Gazella spekei*, was attributed to the purging of deleterious alleles, although the statistical analysis, hampered by small sample size and questionable assumptions, leaves room for

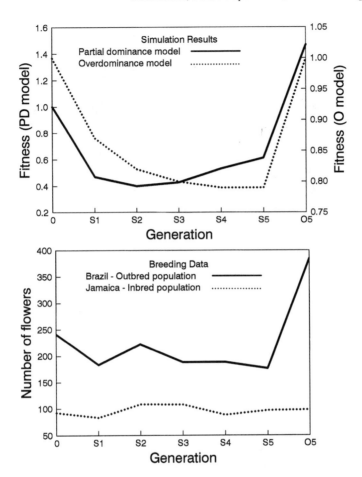

Figure 8.19 Top panel: Results from simulations based on the partial dominance and overdominance models. The populations were outbred in generation 0, selfed for four generations, and both selfed and outcrossed in the fifth generation. Fitnesses have been scaled such that the fitness of the initial outbreeding population is unity. Full details of the models are provided in Barrett and Charlesworth (1991); see caption to their Fig. 2. In the partial dominance model, there is an initial decline in fitness followed by a recovery as deleterious alleles are purged from the population. As a consequence of the purging, there is an increase in fitness when the population is outbred (O5). In contrast, due to loss of heterozygosity, there is a continual decline in fitness for the overdominance model, followed by a recovery only to the initial level upon outbreeding.

Bottom panel: Mean number of flowers in experimental populations of the water hyacinth, *Eichhornia paniculata.* The experimental protocol followed that outlined above for the simulation models, the outcrossed results for generation 5 being obtained by crossing the selfed lines. Standard errors are not shown but were 5–10% of the estimate. Two populations were examined: a population from Brazil that is naturally outcrossing (selfing rate = 0.06), and a population from Jamaica that is naturally inbreeding (selfing rate ≈ 1.00).

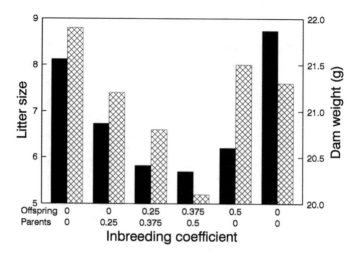

Figure 8.20 Effect of inbreeding on litter size (solid bars) and body weight (hatched bars) in a laboratory strain of house mouse. Values are the means of 10 separate lines. The last two data sets show the results of crossing all lines over two generations. Data from Roberts (1960).

doubt [Templeton and Read, 1983, 1984; for critiques of this analysis, see Hedrick (1994) and Frankham (1995b)].

Barrett and Charlesworth (1991) performed the same experiment as Bowman and Falconer using both an outbreeding and an inbreeding population of the plant *Eichhornia paniculata*. In the naturally outbreeding population, five generations of selfing produced an initial decline in flower number, followed by stabilization, and then a substantial increase upon crossing the lines (Fig. 8.19). In contrast, the naturally inbreeding population showed no change during the experiment. High-yielding strains of inbred strains of maize have been developed, which, although not as high as the best hybrids, are higher than predicted if overdominance played a major role (Crow, 1993). Although initial evidence suggested that overdominance was responsible for inbreeding depression in maize, later research showed that this was a consequence of linkage disequilibrium and "overdominance is no longer regarded as a major factor in the yield of hybrid maize" [Crow, 1993, p. 15; see also Sedcole (1981) and Geiger (1988)]. The foregoing evidence points to the partial dominance model as being the more general reason for inbreeding depression, although more data are required before any generalization can be reasonably made.

It might appear from the above description that the partial dominance model predicts that high levels of selfing, as observed in many plant species and some invertebrate species, would lead to a complete purging of deleterious alleles from a population. For the case of multiplicative fitnesses (i.e., the overall fitness is the

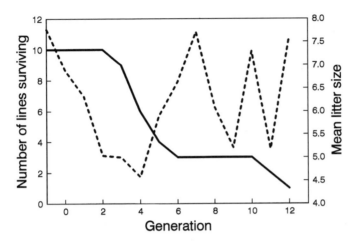

Figure 8.21 Effect of continued full-sib matings on the reproductive output of a laboratory strain of house mouse. The experiment commenced with 10 lines, some of which were lost (solid line) when they failed to produce a male and female with which to produce another generation. Initially, the litter size (dashed line) declined, but in the later generations, it rose back approximately to its original size. The inbreeding coefficients were (generation, *F*): 0, 0.125; 2, 0.438; 4, 0.633; 6, 0.760; 8, 0.843; 10, 0.897; and 12, 0.936. Data from Bowman and Falconer (1960).

product of the fitnesses at each locus), an approximate expression for the inbreeding depression in a completely selfing population is (Charlesworth et al., 1991)

$$\delta \approx 1 - \exp\left[-\left(\tfrac{1}{2} - D\right)U\right] \qquad (8.36)$$

where D is the dominance coefficient and U is the mutation rate per diploid genome per generation. With respect to a single locus, the dominance coefficient is defined as follows: Letting A_1 be the wild-type allele and A_2 be the mutant allele, the relative fitnesses of the three genotypes A_1A_1, A_1A_2, and A_2A_2 are 1, $1 - Ds$, and $1 - s$, respectively, where s is the selection coefficient. When $D = 0$, A_1 is dominant to A_2, and when $D = 0.5$, there is additivity, A_1A_2 lying midway between the two homozygotes. This definition is extended to the multilocus case simply by assuming that both D and s are the same across all loci. Equation (8.36) remains approximately true for any type of selection model (Charlesworth et al., 1991). Mutation produces deleterious alleles which are ultimately purged from the population, but there will be a dynamic balance between these two processes. The question is whether the rate of purging is much greater than the rate at which new mutations enter the population. Estimates of U vary between 0.5 and 1.5

(Charlesworth et al., 1990a; Crow, 1993). As can be seen from Fig. 8.22 when the dominance coefficient is small (implying strong directional dominance), there can be a considerable inbreeding depression maintained at mutation–selection equilibrium. Other models of how fitness is determined by the mode of interaction between loci give similar results (Lande and Schemske, 1985; Charlesworth et al., 1991).

The important message from the above analyses is that high levels of inbreeding depression are possible even in populations that are highly selfing. This is an intuitively surprising result and illustrates the importance of mathematical modeling in refining our predictions. From the foregoing, we can predict that inbreeding depression should be smaller in highly selfing populations but not necessarily negligible. This prediction is amply supported by measures of inbreeding depression in various plant species (Tables 8.8 and 8.9). Husband and Schemske (1996) showed that cumulative inbreeding depression, based on either three or four stages, decreased with selfing rate (Fig. 8.23). Using Spearman's rank correlation, Husband and Schemske obtained significant negative correlations between inbreeding depression of seed production and primary selfing rate ($r_s = -0.49$, $n = 35$, $P = .004$) and between survival inbreeding depression and primary selfing rate ($r_s = -0.41$, $n = 35$, $P = .02$) but not between primary selfing rate and germination δ ($P = .16$) or growth/reproduction δ ($P = .12$). Statistically similar results were obtained using the angiosperm data alone. However, somewhat different results are obtained using the parametric model $\delta = c_0 + c_1$Selfing rate $+ c_2$Dummy $+ c_3$Interaction, where "dummy" is a dummy variable distinguishes angiosperms and gymnosperms. Analysis using the three-stage cumulative inbreeding depression shows no significant effect due to taxon or interaction but a highly significant negative correlation with selfing ($r = 0.47$, $P < .001$). In contrast, all effects are highly significant ($P < .0001$) for the four-stage cumulative inbreeding depression. Regressions of the component δ's on selfing rate show only a significant effect between survival and taxon. The foregoing analyses indicate that there is an important contribution by taxonomic grouping (angiosperm or gymnosperm) and that the conclusion is sensitive to the type of statistical test used (parametric versus nonparametric). Whereas the overall conclusion that inbreeding depression decreases with increased selfing rate, the effect of selfing rate on the component life history stages remains uncertain.

Qualitative support for a decrease in inbreeding depression with selfing in gastropods was obtained by Doums et al. (1997) (Table 8.10). The previously described experiment by Barrett and Charlesworth (1991) on *E. paniculata* also provides support in that inbreeding within a highly selfing population produced no inbreeding depression, whereas, as noted, inbreeding within an outbreeding population led to an initial increase in inbreeding depression (Fig. 8.19). Latta and Ritland (1994) compared inbreeding depression in 15 populations of *Mimulus* with estimates of inbreeding rate within the populations. Inbreeding depression of flower number, plant height, and a composite measure of fitness all showed a

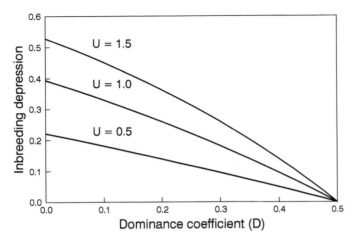

Figure 8.22 Inbreeding depression versus mutation rate (*U*) and the dominance coefficient (*D*).

negative relationship between trait value and *F*, but only that with plant height was significant (Fig. 8.24). These results are in general accord with the partial dominance model (purging of deleterious alleles), and the high inbreeding values obtained even for highly selfing populations support the prediction that high levels of inbreeding depression can be maintained at mutation–selection balance.

Self-fertilization has the advantage from the parent's perspective that the genetic contribution it makes to its offspring is 100%. On the other hand, in outcrossing, each parent makes only a 50% genetic contribution to each offspring. Assuming that outcrossing and selfing have the same fertilization efficiency, it is intuitively obvious that selfing will be selected against if the cost of inbreeding is less than the twofold advantage obtained from selfing (i.e., an allele promoting selfing will spread through the population if $\delta < 0.5$) (Maynard Smith, 1978; Lloyd, 1979; Lande and Schemske, 1985). The foregoing assumes that any seeds not outcrossed are self-fertilized; conditions for selfing to spread can be less restrictive under different assumptions (Table 8.11). Simple models predict that stable states are either entirely selfing or entirely outcrossing (Lande and Schemske, 1985). However, other models indicate that under some circumstances, intermediate levels of selfing can be maintained in a population (Charlesworth and Charlesworth, 1987; Charlesworth et al., 1990a, 1990b, 1991). Schemske and Lande (1985) reviewed data on outcrossing rates in plants and found a strong bimodal distribution, although they did not correct for possible bias due to phylogenetic effects. The issue of whether selection will generally drive a population to an extreme remains a question still in need of further empirical and theoretical study.

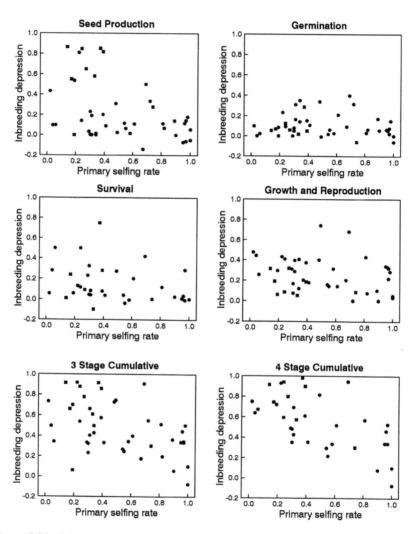

Figure 8.23 Inbreeding depression versus primary selfing rate (= proportion of selfed progeny at time of fertilization) among angiosperms (●) and gymnosperms (■). The three-stage cumulative inbreeding depression does not include survival. Data from Husband and Schemske (1996).

Table 8.10   Inbreeding Depression and Level of Selfing, as Indicated by the Average Number of Alleles per Locus in Snails

Average No. of Alleles per Locus	Inbreeding Depression		Species
	General[a]	$\delta \times 100$[b]	
2.17	High	81	*Lymnaea peregra*
2.15	High	51[c]	*Arianta arbustorum*
2.01	Intermediate	?[d]	*Physa heterostropha*
1.52	Intermediate	18, 61[e]	*Bulinus globosus*
1.39	High	99	*Triodopsis albolabris*
1.39	Not detected	− 1	*Biomphalaria straminea*
1.01	Not detected	− 27	*Bulinus truncatus*
1.00	Not detected	6	*Rumina decollata*

[a]Assessment given in Doums et al. (1997).

[b]Inbreeding depression of fecundity × hatching success.

[c]Full-sib versus outcrossed; all others are selfed versus outcrossed.

[d]I cannot find data supporting the conclusion of "intermediate inbreeding depression" in the cited reference of Wethington and Dillon (1993).

[e]First estimate from Doums et al. (1996), second from Jarne et al. (1991). The midpoint was used in the statistical analysis presented below.

Notes: Sources for estimates of δ: *L. peregra* (Jarne and Delay, 1990), *A. arbustorum* (Chen, 1993), *T. albolabris* (McCracken and Brussard, 1980), *B. straminea, B. truncatus* (Doums et al., 1996), *R. decollata* (Selander et al., 1974).

Statistical analysis: Spearman rank correlation = 0.60, $n = 7$, $P = .1$, one-tailed test.

Source: (Modified from Doums et al., 1997).

8.4.5   Outbreeding Depression

Thus far, theory and experiments have suggested that outbreeding will increase trait values. However, a plausible hypothesis is that selection at a particular geographic location will favor particular combinations of alleles. Under this scenario, the crossing of populations that have become adapted to their own particular conditions will lead to a maladaptive phenotype and a depression of trait values, a phenomenon called **outbreeding depression**. There are two types of maladaptive crosses, one due to a breakdown of the integration of the genome, leading to a reduced fitness even under optimal conditions, and one due simply to the intermediate being ecologically maladapted; I shall refer to these as physiological outbreeding depression and ecological outbreeding depression, respectively. The distinction is by no means complete and serves solely to emphasize the relative role of genetic architecture versus natural history.

Ecological outbreeding depression can be illustrated by considering the migration pattern in different populations of the blackcap, *Sylvia atricapilla*, previously

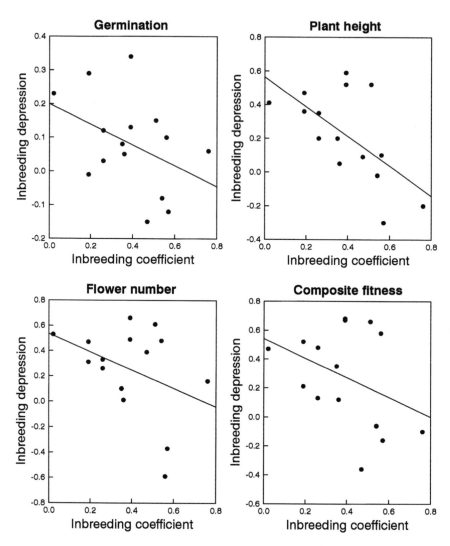

Figure 8.24 Inbreeding depression versus the inbreeding coefficient among 15 populations and species of *Mimulus*. Statistics (all tests one tailed): germination, $r = -0.41$,
$P = .06$; plant height, $r = -0.61$, $P = .007$; flower number, $r = -0.39$, $P = .08$; composite
fitness (germination $\times$ plant weight), $r = -0.38$, $P = .08$. Data from Latta and Ritland
(1994).

Table 8.11   Conditions Under Which an Allele for Selfing Will Spread in a Population

Assumptions	Selfing Increases If $\delta <$
1. Any zygote not selfed is outbred.	0.5
2. Selfing occurs prior to outcrossing and a fraction c_1 of gametes not selfed are successfully cross-fertilized.	$1 - c_1/2$
3. Outcrossing occurs prior to selfing and a fraction c_2 of gametes not outcrossed are successfully cross-fertilized.	$1 - 1/2c_2$
4. More highly selfing genotypes disperse fewer male gametes and individuals with a selfing rate r suffer a reduction c_3r of male gametes.	$(1 - c_3)/2$

Source: From Lande and Schemske (1985).

discussed in Chapter 4. A population resident in Scandinavia flies southwestward through Britain to their overwintering grounds in southern latitudes; some individuals from Germany fly westward and overwinter in Britain (Berthod et al., 1992). Migratory direction has been shown to be inherited (Berthold, 1988; Berthold et al., 1992) and, hence, crosses between German and Scandinavian populations will result in birds which, assuming only additive genetic effects for the purposes of illustration, will orient in a direction that is inappropriate for either migratory route. In captivity, there will be no obvious decline in fitness, but under natural conditions, such hybrids would be strongly selected against. This class of outbreeding depression is likely to be very common, particularly when crosses are made between populations that are adapted to very different ecological conditions. At the same time, it is one that can only be demonstrated by reference to the naturally occurring conditions.

Physiological outbreeding depression is most likely to occur in crosses between populations that are in the process of speciation (e.g., subspecies) or between different species, because there may be poor integration of the two genomes and hence physiological and developmental problems. This is illustrated by crosses between European newts of variable taxonomic affinity [summarized by Callan and Spurway (1951)]. *Triturus cristatus* and *T. vulgaris* are phylogenetically quite distinct (they are morphologically very different) and cannot be crossed except by artificial means. The hybrids are completely sterile. More closely related are *T. cristatus* and *T. marmoratus* (they are similar in size and morphology, the most striking difference being their coloring), in which natural hybridization does occur (Arnold and Burton, 1978). Hybrid females are fertile, but hybrid and backcrossed males show failure of spermatogenesis (White, 1946). Crosses between the three subspecies of *T. cristastus* (*cristatus, karelinii, carnifex*) also show breakdown in spermatogenesis, but to a much lesser degree than observed between the *Triturus* species. However, the F_2 and backcross hybrids suffer high larval and metamorphic mortality (Callan and Spurway, 1951). A reduction in fitness of hybrids between species is not universal (Arnold and Hodges, 1995), but greater effects might be expected than between different populations, as observed in *Triturus*.

Experimental evaluation of outbreeding depression should take the two aspects (physiological and ecological) into account; the two seem to be frequently "bundled together," leading to confusion. It has been suggested that the F_1 of crosses between different populations might exhibit heterosis, but the F_2 will suffer depression due to the breakdown of the gene complexes (Wallace, 1953). This hypothesis centers on physiological outbreeding depression. The hypothesis has been investigated in detail from the perspective of genetic architecture by Lynch (1991). Consider two populations that are brought together and mated at random. By the F_2 generation, the loci are in Hardy–Weinberg equilibrium; this generation is, therefore, the appropriate reference generation. Letting the trait value in this generation be m_2, we have for the F_1 and parental (P) generations (Lynch, 1991)

$$F_1 \quad m_2 + \delta_{1\times} + \delta_{2\times} \tag{8.37a}$$

$$P \quad m_2 - \delta_{1\times} + \delta_{2\times} + \alpha_{2\times} \tag{8.37b}$$

where $\delta_{1\times}$ is the mean dominance crossbreeding effect, $\delta_{2\times}$ is the mean crossbreeding effect for dominance $\times$ dominance epistasis, and $\alpha_{2\times}$ is the mean crossbreeding effect for additive $\times$ additive epistasis. The mean dominance contributes positively in the F_1 but negatively in the parental generation. Dominance $\times$ dominance epistasis has the same effect in both generations, whereas additive $\times$ additive epistasis acts only in the parental generation. Depending on the relative values of these parameters, any outcome is possible. Thus, *no general prediction can be made*.

Fitness, as measured by seed set, abortion, or germination, is sometimes greatest when plants from the most distant populations are crossed, sometimes declines with distance, and sometimes is greatest at an intermediate distance (Fig. 8.25). Most of these studies [number not specified by Waser (1993)] were undertaken under artificial (greenhouse) conditions and therefore represent the action of physiological outbreeding depression rather than ecological outbreeding depression. There is no obvious difference in the range of differences used in the three categories (Fig. 8.25) and, hence, the results are probably not a statistical artifact. In the light of Lynch's analysis, they are not unexpected.

Experiments with populations of nondomestic animal species have also demonstrated all possible outcomes: Outbreeding depression has been reported in *D. pseudoobscura* [wing length; Anderson (1968)], humans [fetal loss among Caucasians increased with the number of different countries in their immediate ancestry (Bresler, 1970); although I cannot find any major statistical error in this analysis, I find it unlikely], *Rana pipiens* [embryonic failure was reported for Vermont $\times$ Costa Rica crosses but not Vermont $\times$ Colorado cross; Moore

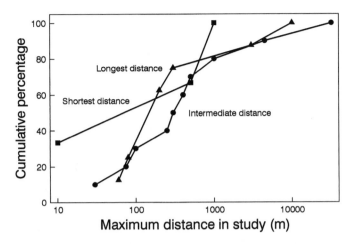

Figure 8.25 Distribution of distances over which studies of angiosperm species have found heterosis among crosses from the most distantly spaced individuals, at intermediate distances, and only at the shortest distances. There is no obvious tendency for the results to be due to differences in the distances studied. Sample sizes are 7 (longest distance), 3 (shortest distance), and 11 (intermediate distance; two species were represented by two studies each; for these, I used the study with the longer distance). Data from Table 9.4 of Waser (1993) and van Treuren et al. (1993, *Scabiosa columbaria*).

(1950)], and the primate *Callimico goeldii* [infant survival (Lacy et al., 1993); the presence of an effect depends upon the particular analysis and hence needs verification]. No outbreeding depression was found in *D. subobscura* crosses over a distance of 4000 km [body size, development time, larval survival; McFarquhar and Robertson (1963)], and heterosis was observed in crosses of *D. melanogaster* over a range of 300–3200 km [larval viability, fecundity, coldshock mortality; Ehiobu and Goddard (1990)]. Deleterious effects of crossing widely separated populations (100–1000s of kilometers) were not found in the waterstrider species *Limnoporus dissortis*, *L. notabilis* (Spence, 1990), *Aquarius remigis*, or *A. remigoides* (Gallant and Fairbairn, 1997).

The animal and plant data indicate that considerable variation is possible. This is not unexpected given the theoretical basis of outbreeding depression, but this basis cannot predict the relative frequency of the different possibilities. The issue of the importance of coadapted gene complexes thus becomes an issue that can only be solved empirically, and at the present, we have insufficient data to draw any general conclusions.

8.5 Summary

The effective population size of a population may differ from the census population size because of many factors, such as variation in sex ratio, temporal variation in census population size, nonrandom contribution of parents, selection, overlapping generations, or spatial subdivision of the population. In general, the effective population size is less than the census population size. Theoretical considerations suggest that under many, if not most, circumstances, the effective population size is unlikely to be less than one-quarter of the census population size. An important case in which there can be substantial difference between N and N_e is when there is very unequal contribution by parents to the next generation and marked fluctuations in population size. This is most likely to occur in plants and ectotherms, because in both cases, there can be orders of magnitude variation in fecundity.

Passage through a single generation bottleneck will generally lead to a change in allele frequencies and, hence, a change in components of genetic variance. Increases in heritability and decrease in mean trait value are most likely if there is dominance or epistasis but can still occur if there is only additive genetic variance. Experimental investigations of the effect of bottlenecks on the genetic constitution of a population must take the latter fact into account, but those few studies that have investigated bottlenecks have not adequately accounted for it in drawing conclusions. If a population is maintained over time at a relatively small size (say, <200), there will be a gradual erosion of additive genetic variance, although in some nonadditive genetic models, there might be an initial increase.

The presence of dominance will lead to a decline in trait value, a phenomenon known as inbreeding depression. There is abundant evidence of inbreeding depression in traits, such as viability, that are closely related to fitness. Two theories have been proposed to account for inbreeding depression, the partial dominance model, and the overdominance model. According to the partial dominance model, the dominance results from partially dominant deleterious alleles at low frequency, whereas the overdominance model posits that heterozygotes have the highest fitness. Although the experimental data are meager, they tend to support the partial dominance model. An important prediction of this model is that continued inbreeding will lead to a purging of the deleterious alleles and an increase in fitness. Thus, we might expect that organisms that undergo frequent inbreeding should show relatively little inbreeding depression. However, theoretical analyses have shown that the amount of inbreeding depression maintained in a population undergoing even self-fertilization can be quite substantial. The available data support the hypothesis that inbreeding depression decreases with the level of inbreeding in natural populations of plants. It has been suggested that selection will favor particular genetic combinations within a particular area and that outbreeding beyond this area can also lead to a depression of fitness as coadapted gene complexes are broken apart. The data indicate that outbreeding depression can occur in some cases but is far from universal, as suggested by more recent, rigorous analyses of outbreeding.

9

The Maintenance of Genetic Variation

For evolution to proceed according to the rules outlined in the previous chapters, there must be additive genetic variation. As shown in Chapter 4, directional selection tends to erode genetic variation, although it is perhaps unreasonable to expect selection to be continually acting in one direction. If selection is not acting continuously in one direction, is there a most fit genotype? If so, why is it that selection does not act to eliminate all deviant genotypes? This question is the focus of the present chapter: Specifically, what mechanisms have been proposed that would prevent the erosion of additive genetic variance, and what is the empirical evidence for their presence? The first section examines stabilizing selection, a type of selection that can clearly maintain a phenotype within narrow bounds and thought to frequently act in natural populations. Does stabilizing selection maintain genetic variation or lead to its erosion until only a single optimal genotype remains? Disruptive selection maintains phenotypic optima by an entirely opposite mode of action: Does this lead to the opposite result from stabilizing selection?

The ultimate source of genetic variation is mutation, and it has been proposed that much of the additive genetic variance is maintained through a balance between stabilizing selection and mutation; this possibility is explored in Section 9.3. An obvious candidate mechanism preserving genetic variation is heterozygote advantage. It is shown in Section 9.4 that moving from the single-locus, two-allele case to the multiple-locus, multiple-allele model greatly restricts the possibility for this mechanism to act. Antagonistic pleiotropy is another mechanism that was proposed many years ago but did not receive a detailed analytical treatment until the last decade. The viability of this mechanism is considered in Section 9.5. Section 9.6 examines the potential of frequency-dependent selection to preserve additive genetic variance, and, finally, the potential importance of environmental heterogeneity is considered in Section 9.7.

9.1   Stabilizing Selection

Stabilizing selection is defined as selection in which an intermediate optimum is favored (Fig. 9.1). This optimum can arise because selection acts directly on the focal trait, or because there is **balancing selection** in which selection on several traits or at different episodes oppose each other to generate apparent stabilizing selection on the focal trait. An example of this type of selection acting on body size in the water strider *Aquarius remigis* is shown in Fig. 9.2: Directional selection on daily fecundity favors an increased size, whereas directional selection on longevity favors a decreased size, the result being that an intermediate optimum body size maximizes lifetime fecundity.

Mathematically, stabilizing selection occurs when there is selection against extreme phenotypes. Waddington (1957) pointed out that this can occur through two processes: First, there may be the elimination of genotype combinations at the extremes and, second, there might be selection for genes that canalize development such that the phenotypic range is reduced (see Chapter 6). The first process he called **normalizing selection** and the second, **canalizing selection**. Most of the theoretical analysis of stabilizing selection has focused on the former process.

Stabilizing selection can be detected by plotting the force of selection, which is typically some component of fitness but not fitness itself, against the trait of interest; for details of the mechanics of this, see Lande and Arnold (1983) and Schluter (1988). A simpler technique is to compare the means and variances before and after selection: This method has to be applied carefully because directional selection will also lead to a reduction in variance (Endler, 1986). Stabilizing selection has been observed in a wide range of organisms under both domestic and field conditions (Table 9.1). Estimates of the percentage change in variance before and after selection show that in more than 20% of reported cases there is greater than a 35% reduction in variance, a change that indicates strong stabilizing selection (Fig. 9.3).

What are the consequences of stabilizing selection for the additive genetic variance? The predicted effects of this type of selection on genetic variation appear to depend rather critically on the assumptions of the genetic model used. To illustrate this, I present a selection of models that generate different predictions.

9.1.1   The Theory of Stabilizing Selection: The Infinitesimal Model

Starting with Haldane (1954), stabilizing selection has typically been modeled as a Gaussian-type curve

$$W(x) = \exp\left(-\frac{(x-\theta)^2}{2\gamma}\right) \qquad \textbf{(9.1)}$$

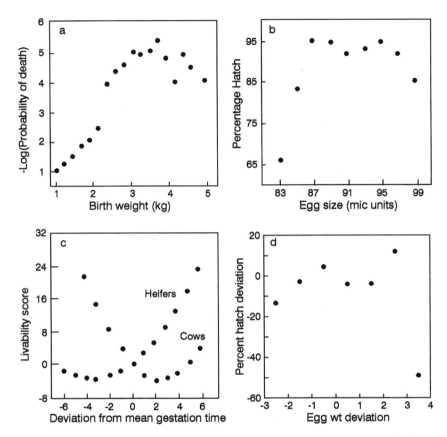

Figure 9.1 Some examples of stabilizing selection on offspring survival. (a) Survival in human infants (Van Valen and Mellin, 1967); (b) hatching success in *D. melanogaster* (Curtsinger, 1976a, 1976b); (c) calf "livability" (1 = calf lived, 2 = calf dead, fitted data points shown) for heifers and cows (Martinez et al., 1983); (d) percentage hatching rate in chickens (Lerner and Gunns, 1952; data given in standard deviation units).

where $W(x)$ is the relative fitness of trait x, θ is the optimal value, and γ is a parameter that measures the strength of the stabilizing selection (as γ increases, the strength of stabilizing selection decreases). This model is sometimes referred to as **noroptimal selection**.

Assuming that at generation t before selection, the trait is normally distributed with mean $\mu(t)$ and phenotypic variance $V_P(t)$ {i.e., the probability density function, $f(x)$, is $\exp(-[x - \mu(t)]^2/2V_P(t))([2\pi V_P(t)]^{-1/2})$}, then from standard probability theory, the probability density function of x after selection, $f'(x)$, is

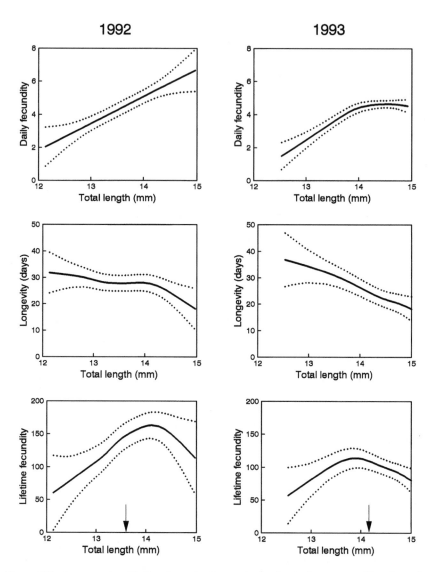

Figure 9.2 An example of balancing selection on body size in the water strider, *Aquarius remigis*. Daily fecundity increases with body size while longevity decreases, producing a concave function between lifetime fecundity and body size. Solid lines are cubic spline estimates and dashed lines are bootstrapped 95% confidence regions. Arrows in the bottom panels indicate the mean total length of all females present in the reproductive season of that year. [From Preziosi and Fairbairn (1997).]

Table 9.1 Some Examples of Stabilizing Selection

Species	Trait	Fitness Measure	Reference
Measurements in the Laboratory or "Nonwild" Conditions			
Drosophila melanogaster	Egg size	Survival	Curtsinger (1976a, 1976b), Roff (1976)
	Pupation height	Survival	Joshi and Mueller (1993)
			Rendel (1943)
Poecilia reticulata	Orange area	Mating success	Schluter (1988)
Chicken	Egg size	Survival	Lerner and Gunns (1952)
Cattle	Gestation	Calf survival	Philipsson (1976), Martinez et al. (1983)
Homo sapiens	Birth weight	Survival	Karn and Penrose (1951), Van Valen and Mellin (1967)
Measurement in Wild Populations			
Tetraopes tetraophthalmus	Body size	Mating success	Mason (1964), Scheiring (1977)
	Wing length	Mating success	Mason (1969)
Libellula luctuosa	Body size	Fertilization success, mating success	Moore (1990)
Eurosta solidaginis	Gall size	Survival	Weis et al. (1992)
Passer domesticus	Body size	Survival	Bumpus (1899), Lande and Arnold (1983)
Melospiza melodia	Morphology	Winter survival (δ, $\female$), reproduction ($\female$)	Schluter and Smith (1986)

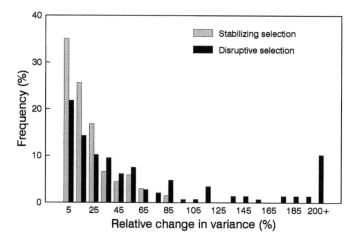

Figure 9.3 Frequency distribution of the relative change in variance due to stabilizing (stippled bars) or disruptive (solid bars) selection ($= 100 \times |V - V'|/V$). Values have been corrected for effects of directional selection. Sample sizes are 137 and 147 for stabilizing and disruptive selection, respectively. [Data from Table 7.1 of Endler (1986).]

$$
\begin{aligned}
f'(x) &= \frac{W(x)f(x)}{\displaystyle\int W(x)f(x)\,dx} \\
&= \exp\!\left(-\frac{(x - \theta)^2}{2\gamma}\right) \exp\!\left(-\frac{[x - \mu(t)]^2}{2V_P(t)}\right) \left[\int W(x)f(x)\,dx\right]^{-1}
\end{aligned}
\tag{9.2}
$$

From the symmetry of $f(x)$ and $W(x)$ it is obvious that the population mean will converge to θ. If the phenotypic distribution is not symmetric, then at equilibrium there will persist a difference between the optimum and θ (Zhivotovsky and Feldman, 1992). The additive genetic variance declines initially, following the recursive equation (Bulmer, 1985, p. 152):

$$
V_A(t + 1) = \tfrac{1}{2} V_A(t) \frac{\gamma + V_D + V_E}{V_A(t) + \gamma + V_D + V_E} + \tfrac{1}{2} V_A(0)
\tag{9.3}
$$

and eventually converges to

$$
V_A(\infty) = \frac{V_A(0) - C + [(C + V_A(0))^2 + 8V_A(0)C]^{1/2}}{4}
\tag{9.4}
$$

where $C = \gamma + V_D + V_E$, V_D is the dominance variance, V_E is the environmental variance, and V_A is the additive genetic variance. The change in additive genetic variance is brought about entirely by the linkage disequilibrium induced by selection; hence, if selection is relaxed, the variance returns to its initial value (Bulmer, 1971a; Tallis, 1987). In the presence of physical linkage between loci, the change in the variance will be larger, although no exact solutions can be obtained (Bulmer, 1985, pp. 158–160).

9.1.2 The Theory of Stabilizing Selection: A Single-Locus, Two-Allele Additive Model

According to the infinitesimal model, the additive genetic variance reaches a limiting value greater than zero. However, consideration of individual loci shows that in the finite case, the variance will actually decline to zero (Robertson, 1956; Latter, 1960). Suppose at some locus A there are two alleles contributing either 0 or $+\frac{1}{2}a$ to the phenotypic value; thus, $A_1A_1 = a$, $A_1A_2 = \frac{1}{2}a$, and $A_2A_2 = 0$. Letting the frequency of A_1 be p and A_2 be q $(= 1 - p)$, and the phenotypic variance about each genotype be V_P, the change in frequency of allele A_1 at generation t, $\Delta p(t)$, is given by

$$\Delta p(t) = \frac{a^2 p(t)q(t)[p(t) - q(t)]}{8[V_P(t) + \gamma]} \qquad (9.5)$$

When $q(t) = p(t) = \frac{1}{2}$, the change is zero and the system is at an unstable equilibrium, in that any deviation from $\frac{1}{2}$ will lead to a directional change in p away from $\frac{1}{2}$, the direction being dependent on the sign of the term $p(t) - q(t)$. Consequently, the allelic frequency will move to fixation at either 0 or 1: This qualitative result is intuitively obvious from considerations of symmetry. Thus, in contrast to the infinitesimal model, the finite model predicts that genetic variation will not be maintained by stabilizing selection.

Suggesting that heterozygotes can sometimes be less variable than homozygotes, Curnow (1964) analyzed the above model for the case in which the phenotypic variance of the heterozygote is equal to cV_P. For $c < 1$, it is possible for a stable equilibrium at $p = q = 0.5$, the parameter space over which this occurs being a function of the value of a and the intensity of selection. From this analysis and consideration of complete dominance, Curnow suggested the following rule (unproven) for the case of strong selection on loci having genetic effects on the mean value of the character that are small compared to the phenotypic variation:

> The selection of individuals in any region near the population mean will favor the least variable genotype irrespective of its position on the scale of mean values. Therefore, one or other allele will eventually become fixed unless there is overdominance on the scale of variability with the heterozygote less variable than either homozygote. (Curnow, 1964, p. 351)

9.1.3 The Theory of Stabilizing Selection: Multiple Loci, Two Alleles per Locus, No Dominance or Epistasis

Bulmer (1985, p. 167) analyzed a model in which there are n unlinked loci, each with two alleles. He assumed no dominance or epistasis. Although an equilibrium allele frequency can be specified, this equilibrium might not be stable in that, as previously, slight perturbations will not be followed by a return to the equilibrium state but a progression to fixation. Letting the equilibrium allele frequency be p^* and the perturbation at the ith locus at some arbitrary generation t be $e_i(t)$, we have $p_i(t) = p^* + e_i(t)$, and the deviation from the equilibrium in the next generation is then $e_i(t + 1) = e_i(t) + \Delta p_i(t)$. Two parameters measure the stability of the equilibrium:

$$\bar{e} = \frac{\Sigma e_j}{n} = \bar{p} - p^* \tag{9.6a}$$

the deviation of the average allele frequency over all loci ($=\bar{p}$) from the equilibrium value, and

$$d_i = e_i - \bar{e} = p_i - p^* \tag{9.6b}$$

the deviation of the ith allele frequency from the average allele frequency. The changes in these parameters in the following generation is

$$\bar{e}(t + 1) \approx \left(1 + \frac{\Delta V_P(t)}{V_P(t)^2} V_G^*\right) \bar{e}(t) \tag{9.7a}$$

$$d_i(t + 1) \approx \left(1 - \frac{V_A^*}{V_A(0)} \frac{\Delta V_P(t)}{V_P(t)^2} \frac{V_G^*}{(2n)}\right) d_i(t) \tag{9.7b}$$

where V_G^* and V_A^* are respectively the genetic and additive genetic variances at equilibrium, $V_A(0)$ is the initial additive genetic variance, V_P is the phenotypic variance, and ΔV_P is the change in the phenotypic variance. Stabilizing selection reduces the phenotypic variance (i.e., $\Delta V_P < 0$) and, hence, $\bar{e}$ is stable [$\bar{e}(t + 1)/\bar{e}(t) < 1$], but d_i is unstable [$d_i(t + 1)/d_i(t) > 1$]. Consequently, the mean allele frequency will return to its equilibrium value after the perturbation, but the individual allele frequencies will move away from p^*, some toward 0 and some toward 1, the result being a reduction in additive genetic variance.

The above result suggests that stabilizing selection will lead to the erosion of additive genetic variance. A critical assumption of this model appears to be that each locus contributes equally to the phenotype; violation of this assumption could lead to stable equilibria (Gale and Kearsey, 1968; Kearsey and Gale, 1968). Suppose, as shown in Table 9.2, there are two loci, the contributions at each locus being strictly additive both within loci and between loci but in which the contribution of one locus is greater than the other (in the present case, locus *B* contributes more than locus *A*). Further, assume that fitness is maximal at the double heterozygote and decreases by 1 for a unit change of value on either side of the optimum. In this model, stable equilibria are possible, contingent on the value of recombination; as the recombination rate approaches 0.5 (no linkage), the disparity between loci must increase for an equilibrium to be possible. Although it can be argued that this model is simplistic and unrealistic (an argument that might be applied to the majority of theoretical genetic models), it makes the point that the details of the model can be critical for its dynamics and that we must take great care in extrapolating from any particular model to the general case.

9.1.4 The Theory of Stabilizing Selection: Multiple Loci, Two Alleles per Locus, Dominance but No Epistasis

Bulmer (1971b) considered the case in which there is dominance, so that A_1A_1 = a, $A_1A_2 = da$, $A_2A_2 = -a$. Without loss of generality we can divide throughout by a, giving the same model as discussed in previous chapters. Assuming n loci, all with equal effect, the condition for stability at allele frequency $p*$ is that either $p* - q*$ and d have the same sign and $8p*q* < (d^2 - 1)/d^2$, or $p* - q*$ and d have opposite signs and $(d^2 - 1)/d^2 < 8p*q* < 2(d^2 - 1)/d^2$ (Fig. 9.4). As can be seen from the figure, these conditions require that there is overdominance. Although the region of stability increases with the value of d, no intermediate equilibrium is possible over most of the parameter space. This conclusion conflicts with that arrived at by Kojima (1959), who analyzed a two-locus model with a quadratic fitness function (i.e., fitness decreases as the square of the distance from the optimum phenotype): From his analysis, Kojima concluded that equilibria at intermediate allele frequencies are possible even with partial dominance. However, as noted by Lewontin (1964, p. 761), "to maintain successively larger numbers of loci in stable equilibrium, the optimum phenotype must be successively a larger and larger proportion of the extreme phenotype." Extending the analysis of Kojima, Lewontin (1964, p. 764) concluded, "that selection based upon squared deviations from an optimum cannot maintain much variance for a character although it may maintain large number of loci segregating. However, when large numbers of loci are segregating each is maintained so close to fixation that random events are sure to reduce the number of segregating loci to a very few where the net selection per locus becomes more substantial." This is illustrated by the simulation model of Mani et al. (1990) discussed in Section 9.6. Introduction of linkage into this model produces a quasistable equilibrium of allele frequencies which can last for a long period, during which linkage dis-

Table 9.2 *The Two-Locus Additive Genetic Model Analyzed by Gale and Kearsey (1968)*

	A_1A_1	A_1A_2	A_2A_2
Contribution	2	1	0
	B_1B_1	B_1B_2	B_2B_2
Contribution	$2+2c$	$1+c$	0
Phenotypic values			
	A_1A_1	A_1A_2	A_2A_2
B_1B_1	$4+2c$	$3+2c$	$2+2c$
B_1B_2	$3+c$	$2+c$	$1+c$
B_2B_2	2	1	0
Fitness Values			
B_1B_1	$1-c$	$2-c$	$3-c$
B_1B_2	2	3	2
B_2B_2	$3-c$	$2-c$	$1-c$

Note: The top rows show the contribution to the phenotype made by the individual loci ($0<c<1$). Below these are the total phenotypic values, obtained by summing the individual contributions. The last entries show the fitness value associated with each genotype (see text for further details).

equilibrium is generated (Lewontin, 1964) and, further, increases the region of stability (Singh and Lewontin, 1966).

9.1.5 The Theory of Stabilizing Selection: Multiple Loci, Two Alleles per Locus, Epistasis

Using a simulation model, Fraser (1960) studied an epistatic model consisting of interactions among 40 loci with variable degrees of dominance. Although Fraser did not find any equilibrium states, the rate of erosion of additive genetic variance was considerably slowed in the presence of epistatic interactions.

With the exception of the infinitesimal model, the models so far discussed appear to make it unlikely that additive genetic variance will be maintained under stabilizing selection; rather, there is an overall erosion of additive genetic variance. However, Gimelfarb (1986) has cast doubt on this view with a very simple model which includes epistasis. His model supposes that there are n loci, each with two alleles, with the genotypic value, X, of an individual being given by

$$X = (1 - z) \sum_{i=1}^{n} X_i + z \prod_{i=1}^{n} X_i \qquad (9.8)$$

where X_i is the effect of the genotypic value of the ith locus. Suppose the two alleles at a particular locus A are A_1 and A_2; the genotypic values are $A_1A_1 = 0$,

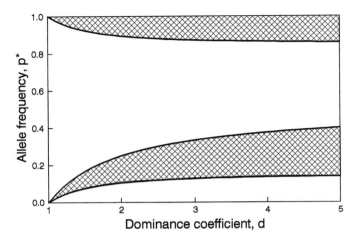

Figure 9.4 Combinations of equilibrium allele frequency (p^*) and dominance coefficient (d) for which stabilizing selection maintains additive genetic variance for the model considered in Section 9.1.4. Cross-hatched areas are those combinations producing equilibria: The upper region satisfies the condition that $p^* - q^*$ and d have the same sign, and $8p^*q^* < (d^2 - 1)/d^2$, whereas the lower region satisfies the condition that $p^* - q^*$ and d have opposite signs, and $(d^2 - 1)/d^2 < 8p^*q^* < 2(d^2 - 1)/d^2$.

$A_1A_2 = 1$, and $A_2A_2 = 2$. Table 9.3 shows the genotypic values obtained for a two-locus model. When $z = 0$ the model is strictly additive, and when $z = 1$ it is strictly multiplicative. A quadratic fitness function was used with the range being set such that fitness was never negative:

$$W(X) = 1 - S(X - \theta)^2 \frac{4}{(X_{max} - X_{min})^2} \qquad (9.9)$$

where $W(X)$ is the fitness associated with a trait value X, θ is the optimum value [always set equal to the middle of the genotypic range; i.e., $\theta = \frac{1}{2}(X_{max} - X_{min})$, where X_{max} and X_{min} are the maximum and minimum genotypic values], and S is a constant which measures the intensity of stabilizing selection (strength increases with S). Linkage, designated by the symbol C_L, was also incorporated into the analysis. Analysis of this model shows that over a wide range of parameter values stable equilibria are possible (Table 9.4). Furthermore, when stabilizing selection is relaxed, most of the genetic variation appears as additive genetic (this is also evident from the "heritability" estimated from offspring-on-parent regression in the population undergoing stabilizing selection, although, in this case, the value is inflated slightly because of nonlinearity introduced by the epistasis). The appearance of most of the genetic variation in this epistatic model is not in itself

Table 9.3   The Epistatic Model of Gimelfarb (1986) When the Trait Value Is Determined by Two Loci, a and B; the Genotypic Value Is Determined by the Formula $X = (1-z)(x_A + x_B) + zx_Ax_B$

Locus Genotype	Effect of Locus Genotype (x_A, x_B)
A_1A_1, B_1B_1	0
A_1A_2, B_1B_2	1
A_2A_2, B_2B_2	2

	Genotypic Value of Individual		
	A_1A_1	A_1A_2	A_2A_2
A_1A_1	0	$1-z$	$2-2z$
A_1A_2	$1-z$	$2-z$	$3-z$
A_2A_2	$2-2z$	$3-z$	4

surprising (see Chapter 2), but what is surprising is that equilibria are so readily obtained. Particularly disconcerting is that the usual genetic analyses would not pick up the epistatic effects and one would conclude that an additive model is appropriate, leading to the paradoxical situation perhaps of observing the maintenance of additive genetic variance in an additive model in the face of stabilizing selection; such an observation could lead one to postulate some other cause, such as mutation, being responsible for the variation.

This analysis indicates quite clearly the distinction that should be made between the statistical and functional description of additive and epistatic models. As Gimelfarb (1986, p. 218) notes,

> It is important, however, to recognize the difference between epistasis as a non-additive action of loci controlling a character and "epistasis" measured by the component of the total genotypic variance attributable to the nonadditive action of the loci. The "epistatic component of variance" depends on a particular distribution of genotypes, and, hence, "epistasis" measured by such a component does not represent a property of a quantitative character, but rather a characteristic of a particular population. Therefore, the fact that very little variation is attributable to non-additivity in the action of loci does not necessarily mean that the loci do indeed act additively.

9.1.6   Stabilizing Selection: Observations

The simple theory presented above (i.e., models with additive and dominance variance components) suggests that stabilizing selection will erode the phenotypic variance due to the erosion of the additive genetic variance. Waddington's model

Table 9.4   Heritability (%) of a Trait Determined by the Epistatic Model of Gimelfarb (1986) When Released from Stabilizing Selection, as Functions of the Strength of This Stabilizing Selection (S) and Linkage (C_L)

	Number of Loci = 2				Number of Loci = 4			
	$C_L = 0.5$		$C_L = 0.05$		$C_L = 0.5$		$C_L = 0.05$	
z	$S=0.1$	$S=0.9$	$S=0.1$	$S=0.9$	$S=0.1$	$S=0.9$	$S=0.1$	$S=0.9$
0.3				98				
0.4				96				88
0.5				94		85	86	86
0.6				93	84	85	85	86
0.7		92		92	84	85	85	85
0.8	91	91		91	84	85	85	85
0.9	90	91		91	85	85	85	86
1.0	90	90		90	85	86	86	86

Note: Empty cells indicate that no stable polymorphisms were found for the specified combination of parameters.

of canalization (see Chapter 6) predicts that stabilizing selection will lead to a reduction in the environmental component of the variance (Thoday, 1959; Prout, 1962; Kaufman et al., 1977).

Stabilizing selection experiments are almost universal in showing a reduction in phenotypic variance and a reduction in heritability (Table 9.5). There has been no satisfactory statistical demonstration of a decline in the environmental variance, but the genetic variance has been shown to decrease in three studies (Prout, 1962; Gibson and Bradley, 1974; Kaufman et al., 1977), and the general trend for h^2 to decrease indicates that additive genetic variance is eroded (Table 9.5). These results are somewhat confounded because there is also typically a component of directional selection, which itself reduces additive genetic variance (see Chapter 4).

9.2   Disruptive Selection

In **disruptive selection**, the two extreme tails of the phenotype are favored, thereby maintaining an intermediate optimum (Fig. 9.1). As with stabilizing selection, disruptive selection has been observed in a large number of studies of wild populations (Fig. 9.3). The magnitude of the effect of disruptive selection may be somewhat inflated by the correction for directional selection (Endler, 1986).

Table 9.5 Results of Stabilizing Selection Experiments

Species	Trait	Gen[a]	V_P	Ratio[a]	Reference
Mouse	6-wk body wt	?	0	1	Falconer and Roberts (1956)[b]
D. melanogaster	Abdominal bristles	14–19	—	0.57	Falconer (1957)[c]
	Sternopleural bristles	42	—	0.60	Thoday (1959)
	Sternopleural bristles	19	—	0.55	Gibson and Bradley (1974)
	Development time	20–36	—	0.42	Prout (1962)
	Wing vein	6, 36	—	0.86, 0.46	Scharloo (1964), Scharloo et al. (1967)
	"Escape" reaction	20	—, 0[d]	ng[e]	Grant and Mettler (1969)
	Wing length	10	—	0.63	Tantawy and Tayel (1970)
	Thorax length	19	0	1.11	Bos and Scharloo (1973a)
Tribolium casteneum	Pupal wt.	95	—	0.87	Kaufman et al. (1977)
	Development time	7	—	ng	Soliman (1982)

[a]Heritability estimated at (or over) generation(s) shown in column "Gen" / Heritability estimated at beginning of selection.

[b]Cited by Prout (1962).

[c]Based on reanalysis by Bulmer (1976).

[d]Phenotypic variance decreased in one line but showed no change in the other.

[e]ng = genetic parameters not given.

9.2.1 The Theory of Disruptive Selection: The Infinitesimal Model

According to the infinitesimal model, the change in the additive variance is given by (Bulmer, 1985, p. 155)

$$V_A(t + 1) = \frac{1}{2} V_A(t) [1 + (c - 1)h^2(t)] + \frac{1}{2} V_A(0) \tag{9.10}$$

where c is the fractional change in the phenotypic variance after selection. At equilibrium, the additive genetic variance, $V_A(\infty)$, is given by the positive solution to

$$
\begin{aligned}
&V_A(\infty) \\
&= \frac{V_A(0) - V_D - V_E \pm \{[V_A(0) - V_D - V_E]^2 + 4(2 - c)(V_D + V_E)\}^{1/2}}{2(2 - c)}
\end{aligned}
\tag{9.11}
$$

When $c < 2$, there is a unique positive solution; when $c > 2$, there is an equilibrium variance, which is greater than $V_A(0)$ if

$$h^2(0) < \frac{1}{2} \quad \text{and} \quad 4h^2(0)[1 - h^2(0)] < \frac{1}{c - 1} \tag{9.12}$$

otherwise, V_A will increase without limit. For disruptive selection, the normal curve is truncated at the values $+z$ and $-z$, only individuals lying to the extremes of these values contributing to the next generation. For simplicity, and without loss of generality, assume that the phenotypic distribution has a mean of 0 and variance of 1; the value of c is then $1 + ze^{-z^2/2}/p_z\sqrt{2\pi}$, where p_z is the proportion of the population lying to the extreme of $\pm z$. The value of c is less than the critical value of 2 when $p_z > 0.226$, meaning that when p_z is greater than 0.226, the variance will increase to a finite value; when p_z is less than 0.226, the variance increases without limit. Bulmer (1985, p. 160) considers linkage to be unimportant, although its theoretical effect is to increase the amount of disequilibrium generated at steady state (Sorensen and Hill, 1983).

9.2.2 The Theory of Disruptive Selection: Multiple Loci, Two Alleles per Locus, No Epistasis

The model to be considered is the same as outlined in Section 9.1.2. Robertson (1956) concluded from his analysis that disruptive selection could lead to stable equilibria in the absence of dominance. However, this conclusion is based on a very restrictive mathematical assumption; removal of this assumption leads to the prediction that no stable equilibria are generally possible when only additive effects occur (Bulmer, 1971b). When dominance is present, the condition for a stable equilibrium is that

$$2n < \frac{1 + 8d^2p^*q^* - d^2}{[1 + (p^* - q^*)d]^3} \qquad (9.13)$$

where n is the number of loci. The additive and dominance genetic variances are simply equal to n times the additive and dominance variances per locus; therefore, $V_A = n\{2p^*q^*[1 + (p^* - q^*)d]^2\}$, $V_D = n(2p^*q^*)^2$, and $h^2 = n\{2p^*q^*[1 + (p^* - q^*)]^2/V_P\}$. Rearranging, we get

$$2n = \frac{h^2V_P}{p^*q^*[1 + (p^* - q^*)d]^2} \qquad (9.14)$$

Assuming that $V_E = 0$ and equating Eqs. (9.13) and (9.14), we arrive at

$$h^2 < \frac{(1 + 8d^2p^*q^* - d^2)p^*q^*}{V_P[1 + (p^* - q^*)d]} \qquad (9.15)$$

Because $V_P = n(V_A + V_D)$, the maximum heritability decreases as the number of loci increases. Whereas it is possible to obtain heritabilities greater than 0.1 when the number of loci is small (say, 10), the parameter space over which this occurs diminishes dramatically as n is increased, and when the number of loci is 50, the region of parameter space is very small indeed (Fig. 9.5). The potential number of combinations is decreased if the loci have unequal effects (Bulmer, 1971b).

Both the infinitesimal and finite locus models predict that disruptive selection is unlikely to lead to the long-term maintenance of genetic variation. The foregoing analysis suggests that there will be an initial increase in the additive genetic variance as linkage disequilibrium is created by selection, followed by a decline as loci become fixed. Simulations have demonstrated the first event, but the number of generations and/or number of loci have apparently been too small (generations < 30, loci < 20) to show the second (Bulmer, 1976; Sorenson and Hill, 1983).

9.2.3 Disruptive Selection: Observations

An increase in heritability due to a buildup of linkage disequilibrium has been verified by Sorenson and Hill (1983) using abdominal bristle number in *D. melanogaster*. Disruptive selection was practiced for three generations, thereafter the two lines being allowed to mate at random. Heritabilities were estimated by offspring–parent regression at generations 1–3 (both lines), generation 10 (line 1),

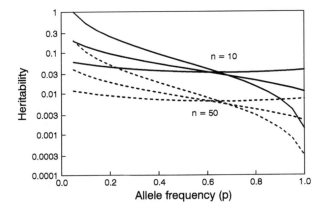

Figure 9.5 The maximum heritability versus allele frequency (p), number of loci (n), and dominance coefficient ($d = 0.9, 0.5, 0.1$) predicted by the model of disruptive selection described in Section 9.2.2.

and generation 8 (line 2). As predicted, there was an increase in heritability over the first three generations, followed by a reduction after the relaxation of selection (Fig. 9.6). Sorenson and Hill were able to accurately mimic the observed response using a simulation model comprising 10 identical, diallelic loci, on each of three chromosomes with recombination fraction 0.1 between adjacent loci and initial allele frequencies of 0.5 (Fig. 9.6).

A common feature of disruptive selection experiments is an increase in the phenotypic variance and an increase in heritability (Table 9.6). This increase can occur even after 40 generations of selection (e.g., sternopleural bristles). These experiments indicate that short-term increases in heritability will occur due to disruptive selection, although the effect in the long term (>100 generations) remains to be investigated.

9.3 Mutation–Selection Balance

The ultimate source of genetic variation is mutation; it is then natural that a question receiving considerable attention is, "Can the observed amount of additive genetic variation in a trait be explained as a consequence of a balance between mutation, which creates variation, and selection, which removes variation?" The type of selection generally studied is stabilizing selection. One of the greatest difficulties in resolving the question is that the answer depends very much on the type of mathematical approximation used (Lande, 1975; Turelli, 1984), and it is only relatively recently that simulation modeling has been extensively used to address the problem (e.g. Burger, et al., 1989; Houle, 1989a; Foley, 1992; Burger

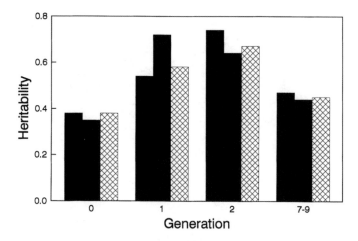

Figure 9.6 The observed (solid bars) and predicted (cross-hatched bars) changes in heritability during disruptive selection and after relaxation of selection on abdominal bristle number in *D. melanogaster*. [Data from Sorenson and Hill (1983).]

and Lande, 1994). The mathematical analyses are complex and I present here only a very brief overview; for a more mathematical but relatively intelligible review, see Bulmer (1989), from which I have drawn extensively.

The models can be classed into three categories, all assuming that there is neither dominance nor epistasis:

1. The diallelic model assumes n loci each with two alleles per locus, the values of the two homozygotes at the ith locus being designated a_i and $-a_i$ (the heterozygote thus has a value of 0), and an infinite population size. Assuming weak, noroptimal (stabilizing) selection, the change in the allelic frequency at the ith locus is given by

$$\Delta p_i(\text{selection}) = \frac{p_i q_i [a_i^2(p_i - 0.5) - a_i(M - \theta)]}{\gamma + V_E} \tag{9.16}$$

where M is the mean genotypic value [$= 2 \Sigma a_j(p_j - 0.5)$], and θ and γ are the parameters of the stabilizing selection function [Eq. (9.1)]. The change due to selection is opposed by the change due to mutation

$$\Delta p_i(\text{mutation}) = -2\mu_i(p_i - 0.5) \tag{9.17}$$

where μ_i is the mutation rate at the ith locus. At equilibrium, $\Delta p_i(\text{mutation}) + \Delta p_i(\text{selection}) = 0$, in which case either

Table 9.6 *Results of Disruptive Selection Experiments*

Species	Trait	Gen	V_P	Ratio[a]	Reference
Drosophila melanogaster	Development time	20–36	+	1.7	Prout (1962)
	Wing length	10	+	1.3	Tantawy and Tayel (1970) [see also Alicchio and Palenzona (1974)]
	Body size	19	+	1.4	Bos and Scharloo (1973a, 1973b, 1974)
	Abdominal bristles	3	+	1.9	Sorenson and Hill (1983)
	Sternopleural bristles	15	+	1.7	Barker and Cummins (1969) [see also Spiess and Wilke (1984)]
	Sternopleural bristles	42	+	1.8	Thoday (1959)
	Escape behavior	20	+	ng	Grant and Mettler (1969)
Mouse	6-wk body weight	ng[b]	0	1	Falconer and Robertson (1956)

[a]Heritability estimated at generation shown in column "Gen" / Heritability estimated at begining of selection.

[b]Cited by Prout (1962), generations not given.

$$p_i(1 - p_i)a_i^2 = 2(\gamma + V_E)\mu_i \quad \text{or} \quad p_i = \tfrac{1}{2} \qquad (9.18)$$

The left-hand case has two real roots:

$$p_i = \tfrac{1}{2} \pm \tfrac{1}{2}\left(1 - \frac{8\mu_i V_s}{a_i^2}\right)^{1/2} \qquad (9.19)$$

where $V_s = \gamma + V_E$. Equation (9.19) cannot be satisfied if the last term under the square root sign is greater than 1; Bulmer (1989) speculated that the allele frequency at these loci would move to $\tfrac{1}{2}$, whereas at the remainder of the loci, some would move to the smaller root and some to the larger such that $M = \theta$. The upper bound to the genetic variance is given by the assumption that allele frequencies at all loci move away from $\tfrac{1}{2}$, and so from the left-hand side of Eq. (9.18), we have

$$V_A(\text{max}) = 2 \sum 2(\gamma + V_E)\mu_i = 4(\gamma + V_E)\mu_g = 4V_s\mu_g \qquad (9.20)$$

The equilibrium additive genetic variance does not depend on the allelic contribution to the trait (a_i) but does depend on the strength of stabilizing selection (γ), the environmental variance (V_E), and the total mutation rate per gamete at all loci contributing to the trait (μ_g). If the mutation rate is the same at all n loci, then $V_A = 4n\mu V_s$.

A troubling finding is that in addition to the equilibria attained by the system moving to a state at which $M = \theta$, there are also other equilibria within this region but not exactly at the point $M = \theta$ (Barton, 1986). However, if drift is significant, it is likely that the equilibrium genetic variance for most populations will be at least approximately predicted (assuming the other assumptions are upheld) by Eq. (9.20) (Barton, 1986).

2. The continuum-of-alleles model assumes that there are n loci, at which mutation can produce an infinite series of alleles, the average effect being zero (i.e., there is symmetry of effects). This can be illustrated assuming a normal distribution of effects; the probability that an allele at locus i with effect x_i mutates to an allele with effect y_i is

$$\frac{\exp\left[-\frac{1}{2}\left(\frac{y_i - x_i}{\alpha}\right)^2\right]}{\alpha\sqrt{2\pi}} \qquad (9.21)$$

where α^2 is the variance of the mutational effect. There is no exact solution for the continuum-of-alleles model, but if it can be assumed that the variance due to a new mutation is much smaller than the existing genetic variance at the locus (i.e., $\alpha^2 \ll V_A$), then

$$V_A = [2nV_m (\gamma + V_E)]^{1/2} = (2nV_mV_s)^{1/2} \qquad (9.22)$$

where V_m is the average amount of new genetic variance introduced per zygote per generation by mutation ($=2 \sum \mu_i\alpha_i^2$). The above equation is based on the assumption that parameter values are the same at all loci; if this is not the case, then n must be replaced by n_e, the effective number of loci (see Bulmer, 1989, p. 764).

3. The house-of-cards approximation is not a specific model but a modification introduced by Turelli (1984) to the continuum-of-alleles model. Turelli argued that the above assumption regarding relative genetic variances is incorrect and that the variance associated with new mutations will generally overwhelm existing genetic variation at a locus. As a consequence, each locus will comprise one common "wild-type" allele in high frequency and rare mutant alleles with large effect (Lande, 1995). Correcting the continuum-of-alleles model for this assumption leads to the same prediction as the diallelic model.

Burger et al. (1989) extended the analysis to include the effect of finite population size, producing what they term the stochastic house-of-cards approximation, V_A (SHC),

$$V_A(\text{SHC}) = \frac{4n\mu V_s}{1 + V_s/N_e\alpha^2} \qquad (9.23)$$

Lynch and Hill (1986) showed that for a neutral trait, the additive genetic variance, $V_A(N)$, is $2N_eV_m$. The stochastic house-of-cards approximation can then be written as a function of the house-of-cards estimate, $V_A(\text{HC}) = 4n\mu V_s$, and the neutral prediction,

$$V_A(\text{SHC}) = \frac{V_A(\text{HC})V_A(N)}{V_A(\text{HC}) + V_A(N)} \qquad (9.24)$$

The above is one-half the harmonic mean of $V_A(N)$ and $V_A(\text{HC})$. Simulation modeling has demonstrated that this model is an excellent predictor of the genetic variance when the assumptions of the model are upheld (Burger et. al., 1989; Burger and Lande, 1994). There are, however, several assumptions that require further investigation. The first is that dominance and epistasis are assumed to be absent. Burger et al. (1989) argue that the difference will not be greater than a factor of 2–3. The work of Lynch and Hill (1986), earlier theoretical analyses of mutation–selection balance models, and the simulation modeling of Burger (1989, see this paper for references to previous analyses) is used by Burger et al. to argue that linkage will not dramatically alter their predictions. A far more difficult issue is that of pleiotropic effects. From his analysis of the effect of pleiotropy, Turelli (1985, pp. 188–189) concluded, "no simple message emerges from my analysis of mutation–selection balance with hidden pleiotropic effects . . . The difficulties imposed by pleiotropy may well preclude accurate predictions concerning mutation–selection balance for polygenic traits." A similar conclusion was reached using slightly different approaches by Wagner (1989), Barton (1990), Keightley and Hill (1990), and Slatkin and Frank (1990). Nevertheless, it would seem that, despite the analytical difficulties, the effect of pleiotropy will be to decrease the additive genetic variance, although by an amount that cannot be readily determined (Turelli, 1985).

For simplicity, it is convenient to scale the parameters to the environmental variance: The predicted heritability is then

$$h^2(\text{SHC}) = 4n\mu \left(4n\mu + \frac{V_E}{N_e\alpha^2} + \frac{V_E}{V_s} \right)^{-1} \tag{9.25}$$

Recalling that $V_m = 2n\mu\alpha^2$ and letting the gametic mutation rate, $n\mu$, be designated as μ_g, Eq. (9.25) can also be written as

$$h^2(\text{SHC}) = 4\mu_g \left(4\mu_g + \frac{2\mu_g}{N_e h_m^2} + \frac{V_E}{V_s} \right)^{-1} \tag{9.26}$$

where $h_m^2 = V_m/V_E$, termed the **mutational heritability**, because it is approximately the expected heritability of an initially homozygous population following one generation of mutation (Lynch, 1988b). From a survey of the meager literature available, Turelli (1984) suggested that a typical set of parameter values are $\mu_g = 0.01$, $V_s/V_E = 20$, and $h_m^2 = 0.001$, with possible lower values of $\mu_g = 0.002$ and $V_s/V_E = 5$. Further analysis by Lynch (1988b) and others has shown that the mutational heritability can vary enormously, the range of mean values alone being 0.00002–0.0161, with an observed upper value of 0.0716 (Table 9.7). Data compiled by Endler (1986) can be used to crudely estimate V_s/V_E. The change in phenotypic variance before, V_P, and after, V_P', selection, predicted by the infinitesimal model, is (Bulmer, 1985, p. 151)

$$V_P' = V_P - \frac{V_P^2}{V_P + \gamma} \tag{9.27}$$

Letting $V_P = cV_E$, [hence, $c = 1/(1 - h^2)$] and rearranging the above, we obtain

$$\frac{V_s}{V_E} = 1 + \frac{c}{R_P} = 1 + \frac{1}{R_P (1 - h^2)} \tag{9.28}$$

where $R_P = |V - V'|/V$ (i.e., the observed relative change in phenotypic variance). The value of V_s/V_E is strongly influence by the assumed heritability and R_P (Table

Table 9.7   Estimates of Mutational Heritability ($h_m{}^2 = V_m/V_E$) for Different Types of Traits in Different Taxa

Taxa	Traits	$h_m{}^2 \times 10^4$	n^a	Range[b]	Ref
D. melanogaster	Morphology	15.6	17	0–102	1, 2
	Physiology	7.5	2	6–9	1, 3
	Viability	0.2	4	0.1–0.4	1
T. casteneum	Pupal weight	36.0	3	0–134	1, 4
Mouse	Morphology	160.6	5	0–716	1
	6-wk body weight	160.0	1	130–240	5
Plants	Veg. and reprod.[c]	35.3	4	2–54	1

[a]Number of individual estimates per study.

[b]Range in values both within and between studies.

[c]Vegetative and reproductive structures not distinguished.

[d]References: (1) Lynch (1988b, review of studies prior to 1988); (2) Caballero et al. (1991), Mackay et al. (1992), Santiago et al. (1992), Lopez and Lopez-Fanjul (1993a); (3) Weber and Diggins (1990); (4) Enfield and Braskerud (1989); (5) Keightley and Hill [1992; different methods of estimation give values that are lower (14–100); the range shown is the estimated 95% confidence interval].

9.8), and the possible range of values is larger than given by Turelli (1984). The equilibrium heritability maintained at mutation–selection balance is decreased with a decrease in the value of V_s/V_E and in this regard it is notable that 14.6% of observed values of R_P suggest a value of V_s/V_E less than 5 (Table 9.8). Using the ranges for μ_g and V_s/V_E given by Turelli and three values of h_m^2 that encompass the observed range (0.000025, 0.0050, 0.0200), I estimated the equilibrium heritabilities for three population sizes ($N_e = 10, 100, 1000$; Table 9.9). For a low value of μ_g (0.002), very little additive genetic variance is maintained at mutation–selection balance, regardless of the value of the other parameters. The maximum value of $h^2 = 0.14$ is below that typically observed for morphological traits, which are assumed to be under relatively weak selection (but which frequently are not; e.g., the Darwin finch study examined in Chapter 4). Heritabilities in the range of 0.4, which is typical for morphological traits, require $\mu_g = 0.01$, $V_s/V_E = 20$, $h_m^2 > 0.005$, and $N_e > 100$. Whether such values are typical of organisms in natural populations can only be decided by the gathering of much more data than presently available. Endler's compilation of selection values for natural populations does suggest that V_s/V_E can frequently be considerably less than 20 (Table 9.8; Barton, 1990). The gametic mutation rate also appears high if one accepts the estimates of per locus mutation rate of $\mu < 10^{-4}$. Three resolutions to this problem have been proposed [reviewed by Turelli (1984)]: (1) the number of loci controlling a trait is in the range 10^3–10^4, rather than 10–10^2 as indicated by mapping techniques (see Chapter 1); (2) mutational effects are not uniform, there being many more that produce only minor changes. For evidence

Table 9.8 Estimates of V_s/V_E Obtained Using Equation 9.28 and Data from Table 7.1 of Endler (1986)

Midpoint of Relative Change[a]	Percentage	V_s/V_E When Heritability =	
		0.25	0.50
0.05	35.0	26.3	39.0
0.15	25.5	8.5	12.3
0.25	16.8	5.0	7.0
0.35	6.6	3.5	4.7
0.45	4.4	2.6	3.4
0.55	5.8	2.1	2.6
0.65	2.9	1.7	2.1
0.75	0.0	1.4	1.7
0.85	1.5	1.2	1.4

[a]Relative change $= R_p = \dfrac{|V - V'|}{V}$

Table 9.9 Predicted Equilibrium Heritabilities at Mutation-Selection Balance Using the Stochastic House-of-Cards Model

μ_g	V_s/V_E	$h_m^2 \times 10^4$	$h^2 (\times 100)$ for $N_e =$		
			10	100	1000
0.002	5	0.25	0.05	0.4	2.2
		50.0	2.8	3.7	3.8
		200.0	3.5	3.8	3.8
	20	0.25	0.05	0.5	3.7
		50.0	5.8	12.1	13.6
		200.0	10.3	13.3	13.8
0.01	5	0.25	0.05	0.5	3.8
		50.0	6.2	14.3	16.4
		200.0	11.8	16.0	16.6
	20	0.25	0.05	0.5	4.5
		50.0	8.2	30.8	42.6
		200.0	21.0	40.0	44.0

of this in *D. melanogaster*, see Lopez and Lopez-Fanjul (1993b) and Mackey et al. (1992), and for a general review, Crow (1993); (3) the observed mutations are the product of unusual genetic events and are not representative of most mutations; (4) the high estimates for the gametic mutation rates are artifacts of the experimental design and the actual rates are much lower. Whereas the question of the

amount of observed additive genetic variation that is being maintained by mutation–selection balance remains open, the evidence is more and more suggestive
that factors other than mutation must play a critical role.

9.4 Heterozygous Advantage

For a single locus with two alleles it is readily apparent that polymorphism can
be maintained if the heterozygote has a higher fitness than either homozygote.
This raises two questions: (1) Can heterozygous advantage explain the maintenance of additive genetic variance in quantitative traits? (2) Is there any evidence
for heterozygous advantage for metric traits? As in the previous models in this
chapter, the theoretical answer changes with the model used.

9.4.1 The Theory of Heterozygous Advantage for Quantitative Traits

Assume that there are two alleles at each of n loci, the genotypic values at each
locus being $A_1A_1 = -1, A_1A_2 = 0$, and $A_2A_2 = 1$. The additive genetic variance
is $V_A = 2 \Sigma p_i q_i$, where p_i is the frequency of allele A_1 at the ith locus. Selection
consists of two elements; stabilizing selection as specified by Eq. (9.1), plus a
small selective advantage s of the heterozygote over either homozygote. This
model was first analyzed by Robertson (1962) and later in more detail by Bulmer
(1973). Assuming an infinite population size and no mutation, the criterion for a
stable equilibrium is

$$s > \frac{1}{2(\gamma + V_P)} \tag{9.29}$$

With a finite population size, N, and a mutation rate of μ, the additive genetic
variance is approximately (Bulmer, 1973)

$$V_A = nH \approx n\left(\frac{2\sqrt{\alpha}}{\beta\sqrt{\pi}} \frac{e^{-\alpha}}{\text{erf}(\sqrt{\alpha})} + 2\right)^{-1} \tag{9.30a}$$

where H is the expected heterozygosity per locus and

$$\alpha = N\left(s - \frac{1}{2\gamma}\right), \quad \beta = 4N\mu, \quad \text{erf}(y) = \frac{2}{\sqrt{\pi}} \int_0^y e^{-x^2} \, dx \tag{9.30b}$$

a small heterozygote advantage (e.g., $s = 0.01$) can maintain quite a considerable
amount of additive genetic variance in a population (Fig. 9.7).

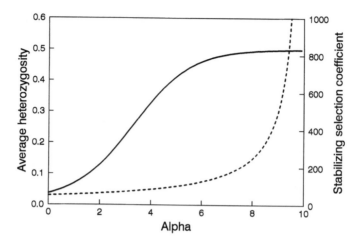

Figure 9.7 The predicted average heterozygosity as a function of α for the heterozygous advantage model described in Section 9.4.1. The dashed line shows the minimum value of γ required for stability when $N = 1000$ and $s = 0.01$.

Gillespie (1984) extended the analysis to the case of multiple alleles per locus, using the quadratic fitness function

$$W(x) = 1 - \frac{(\theta - x)^2}{2c_1} \tag{9.31}$$

where θ is the optimum value and c_1 measures the intensity of selection (it is thus conceptually the same as γ); without loss of generality, let $\theta = 0$. For further simplicity, I shall consider just a single locus, extension to multiple loci on the assumption of additivity being immediate. Homozygotes suffer an additional decrease in fitness of c_2 over heterozygotes; note that this decrease is irrespective of the value of the homozygote, which implies some sort of overall fitness attribute of heterozygosity per se. At equilibrium, the additive genetic variance is

$$V_A = \frac{S_2}{K} - \frac{1}{2c_1c_2}\left(S_4 - \frac{S_2^2}{K}\right) \tag{9.32a}$$

with

$$S_2 = \sum_{i=1}^{K} g_i^2, \quad S_4 = \sum_{k=1}^{K} g_k^4 \tag{9.32b}$$

where K is the number of alleles at the locus and g_i is the genetic contribution of the ith allele. The actual equilibrium number of alleles is determined from a further set of equations, which need not be discussed here. A decrease in the force of stabilizing selection (an increase in c_1) or an increase of the cost of homozygosity (increased c_2) will, for a given K, increase the additive genetic variance. How many alleles can be maintained at equilibrium? Gillespie provides the following illustrative example: Suppose the alleles take only integer values, the range of values being $-K, -(K-1), \ldots, -1, 0, 1, \ldots, (K-1), K$. At equilibrium,

$$c_1 c_2 > \frac{K^2 (K-1)(2K+1)}{3} \tag{9.33}$$

The values, to the nearest integer, on the right-hand side of the above equation for $K = 1, \ldots, 10$ are, respectively, 0, 7, 42, 144, 367, 780, 1470, 2539, 4104, and 6300. Because of the symmetry, there is guaranteed to be three alleles ($K = 1$, allelic values of $-1, 0, 1$). To increase K from 2 to 3, requires a six-fold increase in the product $c_1 c_2$, whereas an increase in K from 2 to 10 requires a 900-fold increase. For this example at least, the conditions for maintaining a large number of alleles at a locus by heterozygotic advantage appear to be quite restrictive.

The above conclusion was also arrived at by Lewontin et al. (1978) using a numerical approach [see also Gillespie (1977) for a similar analysis]. Lewontin drew at random from a uniform distribution fitness values of genotypes, subject to the restrictions that the allelic frequencies sum to 1 and that there was pairwise heterosis, meaning that the fitnesses rank as $W_{ii} < W_{ij} > W_{jj}$. Using the conditions derived by Kimura (1956), the stability of the random set was determined, and the whole procedure replicated 10,000 times. A second experiment was also run in which the condition was imposed that each heterozygote must have a greater fitness than all homozygotes. By definition, all combinations are stable when the number of alleles is two, but the proportion of feasible combinations drops to below 10% for either model when the number of alleles exceeds five (Table 9.10). This does not mean that the maintenance of polymorphic loci in nature by heterosis does not occur, but that if it does occur, it is because of considerable restriction in the parameter space. A region of parameter space in which a relatively large number of alleles (20) can be maintained is specified by the fitness relationships $W_{ij} = 1 + \varepsilon$ and $W_{ii} = 1 - s + \varepsilon$, where ε is a uniform distribution with mean 0 and standard deviation σ and $s/\sigma > 10$. However, within this region, the allele frequencies are nearly uniform (i.e., $p_i = 1/K$), which does not correspond to the distribution of electrophoretic variation (Lewontin et al., 1978) nor, probably, to the distribution of alleles at quantitative trait loci (Chapter 1).

Turelli and Ginzburg (1983) approached the problem from a somewhat different perspective: they hypothesized that for those classes of fitnesses that produce stable multilocus polymorphisms, the average fitness of a genotype increases with

Table 9.10   Proportion of Stable Equilibria for Different Numbers of Alleles (K)
Generated at Random According to Rules in Which Heterozygosity Has a Selective
Advantage

No. of alleles, K	Pairwise Heterosis	Total Heterosis
2	1.00000	1.00000
3	0.5224	0.7120
4	0.1259	0.3433
5	0.0116	0.1041
6	0.00003	0.0137
7	0	0.0011
8	0	0

Source: Data are from Lewontin (1978).

the level of heterozygosity. Although they were able to show that this hypothesis does not always hold, examination of randomly constructed fitness sets did suggest that it would be the prevalent pattern, at least for the case of diallelic loci.

9.4.2   The Maintenance of Genetic Variance by Heterozygous Advantage: Observations

There have been a large number of studies that have related heterozygosity measured at electrophoretically detectable loci with quantitative characters (Table 9.11). Three principle hypotheses have been advanced for a correlation between a metric trait and heterozygosity [for others, see Zouros and Mallet (1989)]: (1) **true overdominance hypothesis**—the allozyme loci have a direct effect on the quantitative trait [the term **functional overdominance** is sometimes used, but this term was originally coined by Frydenberg (1963) to apply to the statistical phenomenon and includes associative overdominance]; (2) **associative overdominance hypothesis**—allozyme loci are themselves neutral but are linked to loci which are themselves determinants of the quantitative trait. The linked loci under selection can show overdominance or be subject to an input of deleterious mutations [the partial dominance hypothesis for inbreeding depression; Ohta and Kimura (1970, 1971), Ohta (1971), and Charlesworth (1991)]; and (3) **inbreeding depression hypothesis**—heterozygosity at the allozyme loci are general indicators of the overall level of heterozygosity (Mitton and Pierce, 1980; Chakraborty and Ryman, 1983) and that individuals with high levels of homozygosity suffer inbreeding depression.

Heterozygosity at loci coding for enzymes might be favored because of the metabolic buffering it confers (Haldane, 1954; Lerner, 1954; Langridge, 1962; Fincham, 1972; Berger, 1976), but direct evidence for this appears to be lacking (Clarke, 1979; Hoffman and Parsons, 1991). Pogson and Zouros (1994) tested

Table 9.11 Survey of Studies Examining the Relationship Between Electrophoretically Detected Heterozygosity and Quantitative Characters

Species	Common Name	Trait	Slope	Ref.[a]
		Morphological Traits—Variability		
Apis mellifera	Honeybee	Fluctuating asymmetry	−, 0	1
Uta stansburiana	Side-blotched lizard	Fluctuating asymmetry	−	2
Trachydosaurus rugosus	Sleepy lizard	Fluctuating asymmetry	0	3
Salmo gairdneri	Rainbow trout	Fluctuating asymmetry	−, 0	4
Salmo clarki	Cutthroat trout	Fluctuating asymmetry	−	5
Salvelinus fontinalis	Brook trout	Fluctuating asymmetry	−	6
Poeciliopsis monacha	Fish	Fluctuating asymmetry	−	7
Elliptio complanata	Bivalve	Fluctuating asymmetry	−	8
Lampsilis radiata	Bivalve	Fluctuating asymmetry	−	9
Thomomys bottae	Pocket gopher	Fluctuating asymmetry	0	10
Mus musculus	House mouse	Asymmetry	0, 0	11
Odocoileus virginianus	White-tailed deer	Asymmetry	0	12
Papio hamadryas	Baboon	Variability	0	13
Drosophila melanogaster	Fruit fly	Asymmetry	0	14
Fundulus heteroclitus	Killifish	Variance	−	15
Poeciliopsis lucida	Fish	Variability	0	16
Oncorhynchus keta	Chum salmon	Variability	+	17
Danaus plexippus	Monarch butterfly	Variance	−	18
Zonotrichia capensis	Rufous-collared sparrow	Variability	0	19
Passer domesticus	House sparrow	Variance	−	20

(continued)

Table 9.11 Continued

Species	Common Name	Trait	Slope	Ref.[a]
		Morphological Traits—Others		
Cerion bendalli	West Indian land snail	PC score (11 traits)	0	21
Drosophila melanogaster	Fruit fly	Wing area	0	22
Hyla cinerea	Green treefrog	Body size	0	23
Pseudocris clarkii	Spotted chorus frog	Body weight (2)	0	24
Salmo gairdneri	Rainbow trout	Weight at hatching, egg size	2+	25
Salvelinus fontinalis	Brook trout	Weight at maturity, egg size	0, 0	26
Peromyscus polionotus	Oldfield mouse	Body weight	+	27
Plantago lanceolata	Plant	5 Traits	3+, 20	28
		Life History Traits		
Odocoileus virginianus	White-tailed deer	Fetal growth rate	+	29
Crassostrea virginica	American oyster	Growth rate	+	30
Ostrea edulis	European oyster	Growth rate, survival	+, +	31
Mytilus edulis	Blue mussel	Growth rate, fecundity of adults <49 mm, fecundity of adults >49 mm, survival, timing of spawning (2 traits)	+, 0, + +, 0, 0	32
Mercenaria mercenaria	Hard clam	Growth rate, survival	0, +	33
Mulinia lateralis	Coot clam	Growth rate	+	34
Placopecten magellanicus	Deep-sea scallop	Growth rate	+	35
Thais haemastoma	Gastropod	Growth rate	+	36
Littorea littorea	Periwinkle	Growth rate	0	37

(continued)

Table 9.11 Continued

Species	Common Name	Trait	Slope	Ref.[a]
Macoma balthica	Mollusc	Growth rate (2 tide levels)	+, 0	38
Drosophila melanogaster	Fruit fly	Growth, development rate	0, 0	39
Pinus rigida	Pitch pine	Growth rate	0	40
Pinus contorta	Lodgepole pine	Growth rate	0	41
Pinus ponderosa	Ponderosa pine	Growth rate, cone production, survival	0, +, +	42
Pinus attenuata	Knobcone pine	Growth rate, 2 reproductive traits	+, −, −	43
Pinus radiata	Radiata pine	Growth rate	+	44
Populus tremuloides	Quaking aspen	Growth rate	+	45
Avena barbata	Wildoat	Survival	+	50
Liatris cylindracea	Herb	4 Traits	4+	51
Gentiana pneumonanthe	Plant	PC1, PC2, PC3	+, 0, 0	52
Bufo boreas	Boreal toad	Survival	+	53
Rana sylvatica	Wood frog	Growth rate	0	54
Hyla cinerea	Green treefrog	Clutch size, # of hatched young, ♂ mating success	+, +, 0	55
Pseudocris clarkii	Spotted chorus frog	Early larval wt, larval growth rate	0, +	56
Salmo gairdneri	Rainbow trout	Growth rate, embryonic growth rate	0, −	57
Salvelinus fontinalis	Brook trout	4 Traits	0	58
Poecilia reticulata	Guppy	Survival	+	59
Poecilia latipinna	Guppy	Brood size	0	60
Gambusia holbrooki	Mosquito fish	Survival	+	61
Pleuronectes platessa	Plaice	Growth rate	0	62
Junco hyemalis	Dark-eyed junco	Survival	+	63

(continued)

Table 9.11 Continued

Species	Common Name	Trait	Slope	Ref.[a]
		Behavioral Traits		
Peromyscus polionotus	Oldfield mouse	Aggression, exploratory	+, +	64
Junco hyemalis	Dark-eyed junco	Dominance	+	65
		Physiological Traits		
Littorina littorea	Marine snail	Standard VO_2	0	66
Crassostrea virginica	American oyster	Standard VO_2	–	67
Mytilus edulis	Blue mussel	Glycogen content	0	68
Ambystoma tigrinum	Tiger salamander	Standard VO_2, active VO_2	–, +	69
Gambusia holbrooki	Mosquito fish	Mercury tolerance	+	70
Salmo gairdneri	Rainbow trout	Standard VO_2	–	71
Peromyscus polionotus	Oldfield mouse	Ability to maintain wt under stress	+	72

[a]References: (1) Bruckner (1976), Clarke et al. (1992); (2) Soule (1979); (3) Sarre and Dearn (1991); (4) Leary et al. (1983, 1984), Ferguson (1986); (5) Leary et al. (1984); (6) Leary et al. (1984); (7) Vrijenhoek and Lerman (1982); (8) Kat (1982); (9) Kat (1982); (10) Patterson (1990); (11) Bader (1965), Wooten and Smith (1986); (12) Smith et al. (1983); (13) Bamshad et al. (1994); (14) Houle (1989b); (15) Mitton (1978); (16) Angus and Schultz (1983); (17) Beacham and Withler (1985); (18) Eanes (1978); (19) Handford (1980); (20) Fleischer et al. (1983); (21) Booth et al. (1990); (22) Houle (1989b); (23) McAlpine (1993); (24) Whitehurst and Pierce (1991); (25) Danzmann et al. (1988); (26) Hutchings and Ferguson (1992); (27) Garten (1976); (28) Wolff and Haeck (1990); (29) Cothran et al. (1983); (30) Singh and Zouros (1978), Zouros et al. (1980); (31) Alvarez et al. (1989); (32) Koehn and Gaffney (1984), Diehl and Koehn (1985), Rodhouse et al. (1986), Zouros et al. (1988), Gentili and Beaumont (1988), Gaffney (1990); (33) Slattery et al. (1993); (34) Koehn et al. (1988); (35) Pogson and Zouros (1994); (36) Garton (1984); (37) Fevolden and Garner (1987); (38) Green et al. (1983); (39) Houle (1989b); (40) Ledig et al. (1983), Bush et al. (1987); (41) Mitton and Grant (1984); (42) Linhart et al. (1979), Farris and Mitton (1984); (43) Straus (1986); (44) Stauss and Libby (1987); (45) Mitton and Grant (1980); (50) Clegg and Allard (1973); (51) Schaal and Levin (1976); (52) Oostermeijer et al. (1994); (53) Samollow and Soule (1983); (54) Wright and Guttman (1995); (55) McAlpine (1993); (56) Whitehurst and Pierce (1991); (57) Danzmann et al. (1986, 1988); (58) Hutchings and Ferguson (1992), Ferguson et al. (1995); (59) Beardmore and Shami (1979); (60) Travis (1989); (61) Smith et al. (1989); (62) McAndrew et al. (1986); (63) Baker and Fox (1978); (64) Garten (1976, 1977); (65) Baker and Fox (1978); (66) Foltz et al. (1993); (67) Koehn and Shumway (1982); (68) Zouros (1990); (69) Mitton et al. (1986); (70) Smith et al. (1989); (71) Danzmann et al. (1988); (72) Teska et al. (1990).

for a causal relationship between heterozygosity and growth rate in the deep-sea scallop by examining the pattern between RFLP (restricted fragment length polymorphism) heterozygosity and growth rate. They argued that if the observed pattern with electrophoretic loci resulted from associative overdominance, then the same pattern should be observed with other assumed neutral markers. No relationship was observed between RFLP heterozygosity and growth rate, providing support for a causative relationship between the electrophoretic loci and growth rate. Only further studies can determine whether this is or is not a common phenomenon.

The true overdominance hypothesis could also be distinguished from the other two hypotheses by examining the relationship among heterozygosity, the quantitative trait, and population size: There should be no effect of population size if the true overdominance hypothesis is correct, but a loss of the correlation between heterozygosity and the quantitative trait in very large populations where linkage disequilibrium and inbreeding depression should be absent. Houle (1989b) tested this by examining the correlation between heterozygosity and three metric traits — wing area, growth, and development rate — in a large outbred population of *Drosophila melanogaster*. As there appears to be little or no inbreeding depression associated with morphological traits in *D. melanogaster* (see Chapter 8), the wing area data are not useful as tests of the hypotheses, but the life history traits can be because they do show inbreeding depression. Houle found no significant correlations and so rejected the true overdominance hypothesis. However, a more convincing demonstration would be to begin with a species which in the wild did show a relationship, demonstrate that the wild population was relatively small, and that an increase in population size (possibly lab reared) caused the pattern to disappear. I know of no such experiment.

If the variance in a trait is entirely due to the additive action of alleles, then the question of heterozygous advantage does not arise (fitness itself is excluded because pleiotropic effects can confer heterozygote advantage even when the genetic control is strictly additive). Because most genetic variance in morphological traits appears to be additive, and a considerable portion is due to dominance variance in life history traits (Chapter 2), it follows that heterozygous advantage, if important, will be found in fitness-related traits. The analysis of inbreeding depression supports this viewpoint (Chapter 8). Although there are presently insufficient data to distinguish between the overdominance and partial dominance hypotheses, what data do exist favors the partial dominance model (Section 8.4.4) and thus argues against heterozygote superiority.

From the above argument we would expect significant correlations between morphological traits and heterozygosity to be less frequent than those between fitness-related traits. To test this hypothesis, I compared the data on morphological traits (excluding variability measures) with the life history data presented in Table 9.11. To avoid bias introduced by different numbers of observations per species, for each species I scored a positive correlation as $+1$, no relationship as 0, and a negative correlation as -1 and then computed the mean score per species. As predicted, the mean score for the morphological traits is less than that for the life history traits (0.325 versus 0.575), but the difference is not significant ($t = 1.39$, $n_1 = 8$, $n_2 = 31$, $P = .08$, one-tailed test; $\chi^2 = 1.76$, $P = .09$, Mann–Whitney test, one-tailed).

Fluctuating asymmetry, and the more general phenomenon of developmental stability, presents an interesting case in itself. After reviewing the data on developmental stability both from crosses within and between species, Clarke (1993) concluded that the partial dominance model (hypothesis 2 or 3) is more likely to be correct. His reasons for this conclusion are as follows: (1) The number of nonsignificant results is as great as the number of significant associations. This is shown in Table 9.11 where there are nine significant negative correlations, eight nonsignificant correlations, one significant positive correlation, and two ambiguous results (different results from different studies). Whereas this is consistent with the partial dominance model, it is also consistent with a weak causal relationship between heterozygosity and developmental stability. (2) Founder events resulting in increased homozygosity and inbreeding depression have been demonstrated in at least three studies. (3) Some studies (e.g., that comparing cheetah populations) are statistically unsound. (4) Heterozygosity levels are typically very low (approximately 20%). (5) Developmental instability in hybrids between populations and species can readily be explained due to outbreeding depression. I do not find any or the total weight of these arguments compelling, and suggest that the question remains open.

9.5 Antagonistic Pleiotropy

The possibility that antagonistic pleiotropy might preserve genetic variation was introduced in Chapter 3. Although the suggestion that antagonistic pleiotropy can be important was made some time ago [Hazel, 1943; and by Falconer in the first edition (1960) of his book "Introduction to Quantitative Genetics"], it was only relatively recently that the mathematical justification for the contention has been explored (Rose, 1982, 1985; Curtsinger et al., 1994). We begin by considering a single diallelic locus which determines the value of two traits, X and Y (Table 9.12). Dominance relationships are determined by d_X and d_Y, and selection intensity by the coefficients S_X and S_Y. The particular set of parameter combinations guarantees that antagonistic pleiotopy must occur (i.e., the fitness of trait X is maximal when the genotype is A_1A_1, whereas the fitness of trait Y is maximal when the genotype is A_2A_2). Further, overdominance or underdominance is not possible for either component separately. Four categories, dependent on the values of d_X and d_Y, can be distinguished (Fig. 9.8): (1) *Additivity.* When $d_X = d_Y = 0.5$, the heterozygotes fall midway between the two homozygotes; (2) *Beneficial reversal.* If $d_X < 0.5$ and $d_Y < 0.5$, the dominance is in the direction of increased fitness; (3) *Deleterious reversal.* When $d_X > 0.5$ and $d_Y > 0.5$, the opposite

Table 9.12 Antagonistic Pleiotropy Model with Multiplicative Fitness

| | Genotype | | |
Fitness Components	A_1A_1	A_1A_2	A_2A_2
Trait X	1	$1 - d_X S_X$	$1 - S_X$
Trait Y	$1 - S_Y$	$1 - d_Y S_Y$	1
Combined	$1 - S_Y$	$(1 - d_X S_X)(1 - d_Y S_Y)$	$1 - S_X$

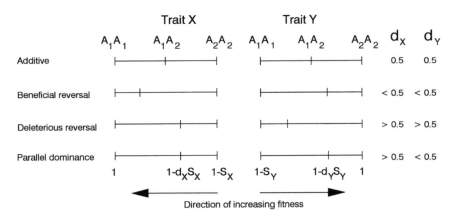

Figure 9.8 Pictorial representation of the antagonistic pleiotropy model.

condition holds, dominance being in the direction of decreased fitness. (4) *Parallel dominance.* If $d_X > 0.5$ and $d_Y < 0.5$, or $d_X < 0.5$ and $d_Y > 0.5$, one allele (A_1, A_2, respectively) is dominant to the other for both traits, leading to dominance toward increased fitness of one trait and dominance toward decreased fitness of the other. The overall fitness could be due to the multiplicative effect of the fitness components, as would occur if the traits were survival and fecundity, or the components could act additively, as would be appropriate if one trait were fecundity at one age and the other trait were fecundity at another age. As both the qualitative and quantitative results are very similar, I shall discuss only the multiplicative case. The requirement for a stable equilibrium is that the fitness of the heterozygote exceed that of either homozygote (Rose, 1982). For the present model, this requires

$$(1 - d_X S_X)(1 - d_Y S_Y) > 1 - S_X, \quad 1 - S_Y \qquad (9.34a)$$

Rearranging gives

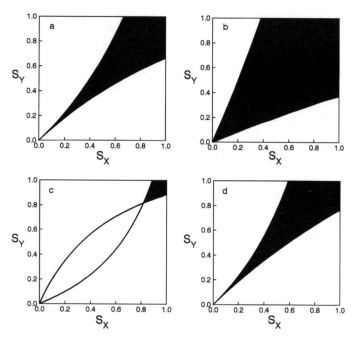

Figure 9.9 Combinations (solid area) of S_X and S_Y for which stable polymorphic equilibria are obtained for the multiplicative antagonistic pleiotropy model. (a) Additive case, $d_X = 0.5$, $d_Y = 0.5$; (b) beneficial reversal (deleterious alleles partially recessive), $d_X = 0.3$, $d_Y = 0.3$; (c) deleterious reversal (deleterious alleles partially dominant), $d_X = 0.7$, $d_Y = 0.7$; (d) parallel dominance (the same allele partially dominant in both traits), $d_X = 0.7$, $d_Y = 0.3$.

$$\frac{d_X S_X}{1 + d_X d_Y S_X - d_Y} < S_Y < \frac{d_X S_X - S_X}{d_X d_Y S_X - d_Y} \qquad \textbf{(9.34b)}$$

Combinations of d_X and d_Y for which stable equilibria are found are most frequent when deleterious alleles are partially recessive (beneficial reversal) and least likely when deleterious alleles are partially dominant (deleterious reversal, Fig. 9.9). Numerical evaluation of Eqs. (9.34a) and (9.34b) across the full range of parameter values (all lie between 0 and 1) reflects the picture portrayed in Fig. 9.9; 67% of combinations in which both deleterious alleles are partially recessive ($d_X < 0.5$, $d_Y < 0.5$) give stable combinations, whereas 25% of combinations showing parallel dominance ($d_X < 0.5$, $d_Y > 0.5$, and vice versa) are stable, and only 2% of combinations in which both deleterious alleles are dominant ($d_X > 0.5$, $d_Y > 0.5$) are stable (Curtsinger et al., 1994). Overall, 30% of combinations in the multiplicative fitness model produce stable equilibria; 25% of the combinations in the additive model do so.

Table 9.13 Average Ratio of Dominance to Additive Genetic Variance at Equilibrium for the Antagonistic Pleiotropy Model with Multiplicative Fitness

	Average V_D/V_A		
	Trait X	Trait Y	Larger Ratio
Full space	0.51	0.51	0.93
$S_X < 0.5, S_Y > 0$	0.56	0.56	1.02
$S_X < 0.1, S_Y > 0$	0.58	0.58	1.07
$S_X < 0.05, S_Y > 0$	0.58	0.58	1.08

Source: From Curtsinger et al. (1994).

That equilibria are most likely with dominance of beneficial alleles raises two questions: (1) What level of dominance variance is expected? (2) Do we observe this level? Curtsinger et al. (1994) addressed the first question by generating 456,000 random combinations of parameter values and computing the equilibrium genetic variances for those cases in which a stable polymorphism was obtained. On average, the dominance variance is approximately one-half the additive genetic variance and equal to the additive genetic variance if one considers only the larger of the two ratios for each stable equilibrium (Table 9.13). These data led Curtsinger et al. (1994, p. 221) to conclude, "if antagonism of fitness components often plays a role in maintaining polymorphisms, then the dominance variance for fitness components should, on average, be about half as large as the additive genetic variance for those same fitness components." They concluded, primarily from the data reported in Mousseau and Roff (1987), that dominance variance typically comprises a very small fraction of the total variance and, hence, that antagonistic pleiotropy is unlikely to play a major role in the maintenance of genetic variation. There are two reasons why this conclusion is not merited. First, inbreeding depression experiments suggest that dominance variance due to partially recessive alleles is very common for life history traits but possibly not for morphological traits. This is supported by the direct estimates reported in Chapter 2. The data summarized by Mousseau and Roff (1987) consisted almost entirely of morphological traits and, hence, cannot be used as an adequate test of the antagonistic pleiotropy hypothesis. The second reason for rejecting the conclusion of Rose (1985) and Curtsinger et al. (1994) is that the direct estimates of the ratio of dominance to additive genetic variances in life history traits are in accord with the prediction made by Curtsinger et al. (1994): Of the 20 estimates obtained from the literature, 70% (14) have a ratio of dominance to additive variance greater than 0.5, and in 65% (13) of cases V_D is greater larger than V_A (Table 9.14).

The above theoretical analysis is based on a single-locus model; undoubtedly, there are more loci generally involved and a thorough analysis of the multilocus case remains to be conducted. A qualitative examination of the two-locus case by

Table 9.14   Estimates of V_D/V_A *for Life History Traits*

Species	Trait(s)	V_D/V_A	Ref.[c]
D. melanogaster	Viability	0.03, 0.05	1
Muscidifurax raptor (wasp)	Reproductive traits (6),[a] development time	0.19, ∞[b], 11.5, 0.39, 5.25, ∞[b], 1.0	2
Muscidifurax raptorellus	Reproductive (2)	0.82, 1.56	3
Tribolium casteneum	Development rate	1.38	4
Tribolium confusum	Development rate	3.35	4
Hyla crucifer (frog)	Growth rate, larval period	4.26, 5.25	5
Gerbera hybrida (plant)	Flowering time	0.14	6
Papaver somniferum (poppy)	Flowering time, days to maturity	1.70, 2.33	7
Nicotiana rustica (plant)	Flowering time	0.14	8
Picea mariana (spruce)	Survival	∞[b]	9

Note: According to the analysis by Curtsinger et al. (1994), this ratio must exceed, on average, ½ for antagonistic pleiotropy to be a plausible mechanism for the maintenance of additive genetic variance (see Table 9.13).

[a]Traits, in order, are listed under the relevant reference.

[b]$V_A = 0$.

[c]References: (1) Suh and Mukai (1991); (2) Antolin (1992b)—sex ratio (6 days), sex ratio (10 days), lifetime fecundity, 6-day fecundity, 10 day fecundity, reproductive life span; (3) Legner (1991)—% gregarious oviposition, eggs per gregarious oviposition; (4) Dawson (1965); (5) Travis et al. (1987); (6) Harding et al. (1991); (7) Shukla and Khanna (1992); (8) Jinks et al. (1969); (9) Mullin et al. (1992).

Curtsinger et al. (1994) suggests that the conditions for stable polymorphism might be more restrictive. Given that it now seems that antagonistic pleiotropy is a plausible mechanism for the maintenance of genetic variation, more work on this problem is warranted.

9.6   Frequency-Dependent Selection

Frequency-dependent selection occurs when the fitness of a phenotype or genotype varies with the phenotypic or genotypic composition of the population (Ayala and Campbell, 1974; Gromko, 1977; DeBenedictis, 1978). Such selection can readily maintain genetic variation at a single locus (Wright, 1948; Haldane and Jayakar, 1963; Clarke and O'Donald, 1964; Clarke, 1964; Anderson, 1969). This may be illustrated using the model outlined in Table 9.15, which is a special

Table 9.15 *Simple Single-Locus Frequency-Dependent Model, Analyzed by Hedrick (1972, 1973)*

| Genotype | Freq. | Genotype Competed Against | | | Mean Fitness |
		A_1A_1	A_1A_2	A_2A_2	
A_1A_1	p^2	1	$1 + d_2s$	$1 + s$	$\overline{W}_{11}$
A_1A_2	$2pq$	$1 + d_1s$	1	$1 + (1-d_1)s$	$\overline{W}_{12}$
A_2A_2	q^2	$1 + s$	$1 + (1-d_2)s$	1	$\overline{W}_{22}$

Note:

$$\overline{W}_{11} = p^2 + 2pq(1+d_2s) + (1+s)q^2$$
$$\overline{W}_{12} = (1+d_1s)p^2 + 2pq + [(1+(1-d_1)]q^2$$
$$\overline{W}_{22} = (1+s)p^2 + [1+(1-d_2)s]2pq + q^2$$
$$\overline{W} = \overline{W}_{11}p^2 + \overline{W}_{12}2pq + \overline{W}_{22}q^2$$

case of the generalized frequency-dependent selection model for a single diallelic locus analyzed by Cockerham et al. (1972). The parameter d_1 measures the degree of dominance of the competing genotype, d_2 measures the degree of dominance of the genotype competed against, and s is the selection coefficient. The allele frequency, p', after selection is

$$p' = \frac{p(p\overline{W}_{11} + q\overline{W}_{12})}{\overline{W}} \qquad (9.35)$$

and hence the change in allele frequency is given by

$$\Delta p = p' - p = \frac{pq[p(\overline{W}_{11} - \overline{W}_{12}) + q(\overline{W}_{12} - \overline{W}_{22})]}{\overline{W}} \qquad (9.36)$$

Figure 9.10 shows the change in allele frequency for four combinations of d_1 and d_2. At equilibrium, $\Delta p = 0$, and all in the cases shown, the equilibria at intermediate frequencies of p are stable in that small deviations from the equilibrium are opposed by selection. The strength of the selection pushing the allele frequency back to the equilibrium value depends on the steepness of the curve; those which are shallow (e.g., $d_1 = d_2 = 0.5$, in Fig. 9.10) are relatively weak and, in this case, in small populations drift could readily cause fixation of alleles even in the presence of frequency-dependent selection. The time required for fixation to occur is least when d_1 and d_2 are approximately equal, values which Hedrick (1973) suggest are biologically most feasible.

If there is neither dominance nor epistasis, and n unlinked diallelic loci, the conditions for stability are (Bulmer, 1985, p. 169)

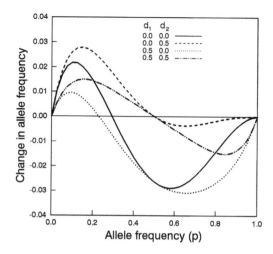

Figure 9.10 The change in allele frequency for the frequency-dependent model described in Table 9.15. In all cases, $s = 0.5$.

$$\bar{e}(t + 1) \approx \left[1 + V_G^* \left(\frac{\Delta V_P(t)}{V_P(t)} + C\right)\right] \bar{e}(t) \qquad (9.37a)$$

$$d_i(t + 1) \approx \left(1 - \frac{V_A^*}{V_A(0)} \frac{\Delta V_P(t)}{V_P(t)} \frac{V_G^*}{2n}\right) d_i(t) \qquad (9.37b)$$

where

$$C = \frac{\text{Cov}(x, \partial W/\partial m)}{V_P(t)\bar{W}} \qquad (9.37c)$$

The above equations are identical to Eqs. (9.7a) and (9.7b), with the additional component C, which is the covariance between the trait value x and the rate of change in its fitness with respect to the mean, m, divided by the phenotypic variance and mean fitness. The conditions for a stable equilibrium are (1) $\Delta V_P(t)/V_P(t)^2 > 0$, which is to say that the change in phenotypic variance due to selection is positive, (2) $\Delta V_P(t)/V_P(t)^2 + C < 0$, which makes the system stable with respect to the mean. The second condition requires that C is negative, meaning that the frequency-dependent selection for rare phenotypes must be sufficiently strong to counter the tendency of disruptive selection (indicated by the first condition) to destabilize the mean.

The above general theory can be illustrated by a model of competition between individuals (Bulmer, 1974, 1985). The fitness of an individual with a phenotypic value of x is given by

$$W(x) = [c_1 - c_2 F(x)] \exp\left(-\frac{(x - \theta)^2}{2\gamma}\right) \tag{9.38}$$

where c_1 and c_2 are positive constants and $F(x)$ is the function describing the effect of competition between individuals; as $F(x)$ increases, the fitness of an individual with phenotypic value x decreases. Bulmer assumed that the effect of competition between two individuals of phenotypes x and x' to be a Gaussian function and that phenotypic values were distributed normally with mean m and variance V_P, giving

$$F(x) = N\left(\frac{\alpha}{\alpha + V_P}\right)^{1/2} \exp\left(-\frac{(x - m)^2}{2(\alpha + V_P)}\right) \tag{9.39}$$

The term α determines the strength of the interaction, larger values leading to reduced competition, and N is population size. The second term in Eq. (9.39) represents stabilizing selection tending to push the mean value to θ. At equilibrium, $m = \theta$ and $W = 1$. In the above model, selection always forces the population to the mean value, but the genetic variance is not necessarily preserved. Specifically, if stabilizing selection (which reduces variance) is stronger than the effect of competition, a condition which occurs when $\alpha/\gamma > c_1 - 1$, then all of the variance will be eliminated. However, the condition $\alpha/\gamma < c_1 - 1$ does not guarantee the preservation of variance. The actual numerical conditions under which variation is preserved need not be specified here (see Bulmer, 1985, p. 172); we may simply note that they are not excessively restrictive. Slatkin (1979) extended Bulmer's analysis showing that the qualitative results do not depend on the underlying genetic model provided that the mean and variance are not constrained.

Using a simulation model, Mani et al. (1990) explored the combined effect of mutation, stabilizing, and frequency-dependent selection on a genetic system in which there are n (≤ 12) loci, each with up to 32 alleles which act additively, the ith allele contributing an amount i to the genotypic value. The environmental variance was assumed to be zero and linkage between loci allowed. The fitness of an individual with value x is equal to

$$W(x) = W_S(x)W_F(f_x) \tag{9.40}$$

where W_S is stabilizing selection, W_F is frequency-dependent selection, and f_x is the frequency of x. The stabilizing selection component is defined as

$$W_S(x) = \exp\left(-\frac{(x - \theta)^2}{2\gamma}\right) \qquad (9.41)$$

To provide a convenient scale for the effect of the strength of stabilizing selection (γ), Mani et al. (1990) defined the parameter α as $\exp(-m^2/2\gamma)$, which varies between 0 and 1 as γ varies between 0 and ∞. As $\alpha \to 1$, selection becomes increasingly weak. The frequency-dependent component is defined as

$$W_F(x) = a + (1 - a)\frac{(1 - f_x)^b}{c^b f_x^b + (1 - f_x)^b} \qquad (9.42)$$

where a, b, and c are constants. The shape of the function is determined by b and c; when $b = c = 1$, the function is linear, $b < 1$ gives a concave curve, and $b > 1$ gives a convex curve. Frequency dependence is suppressed when $a = 1$. The particular values of c and b used in the simulations are not stated in the article. The results did not differ significantly for the linked and unlinked cases and I shall present the results only for the latter (Fig. 9.11). With the exception of no frequency dependence ($a = 1$), the number of alleles maintained at equilibrium appears to be independent of the strength of the frequency-dependent selection and only weakly related to the strength of stabilizing selection. Despite beginning with 32 alleles per locus, at equilibrium the number of alleles per locus is only 1.5–3.0. Nevertheless, these simulations show very clearly that frequency-dependent selection can play a major role in the maintenance of genetic variation.

Frequency-dependent selection has been shown to operate on many discrete traits, such as color polymorphisms and predation risk (Ayala and Campbell, 1974; Clarke, 1969, 1979; Endler, 1988), but its importance with respect to traits showing continuous variation has not been investigated sufficiently to determine if it may be important for such traits.

9.7 Environmental Heterogeneity

9.7.1 Environmental Heterogeneity: Theory

There has been considerable investigation of the conditions necessary for temporal or spatial heterogeneity to generate stable single-locus, two-allele polymorphisms [see reviews by Felsenstein (1976), Hedrick et al. (1976), and Hedrick (1986)]. For an environment that is temporally variable, the necessary condition is that the heterozygote has a higher geometric mean fitness than either homozygote. As far as I am aware there has been no formal analysis of temporal variation and the maintenance of genetic variation in quantitative characters. However, given the restrictive conditions for a single locus, it would seem unlikely that temporal variation alone would preserve additive genetic variation. An intuitive argument also suggests this: In each generation, there is selection against extremes at one

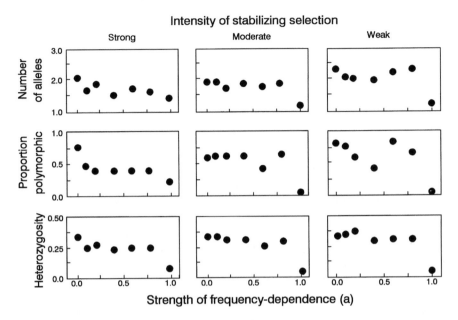

Figure 9.11 The effect of frequency-dependent selection on genetic equilibrium properties. Population size is 1000, the number of loci is 5, and the mutation rate is 10^{-5} per locus per generation. Each simulation was run for 30,000 generations, the statistics being calculated on the last 200 generations, with 10 replicates per parameter combination. Three values of the stabilizing selection parameter δ were used, 0.0 (strong), 0.6 (moderate) and 0.99 (weak). For details of the model, see text. [Modified from Mani et al (1990).]

tail of the distribution, which tail being selected against varying over time, and hence a steady erosion of genetic variation. Alternatively, we may think of temporal variation as stabilizing selection in which the optimal value fluctuates; because stabilizing selection erodes genetic variance, there is no reason to suppose that fluctuation in the optimal value will be itself sufficient to prevent this erosion.

The above conclusion assumes nonoverlapping generations: When generations overlap, genetic variation can be maintained (Ellner and Hairston, 1994; Ellner, 1996). With a Gaussian selection function, the requirement for genetic variance to be maintained is

$$\frac{\delta \, \text{Var}(\theta)}{\gamma} > 1 \qquad (9.43)$$

where $\text{Var}(\theta)$ is the variance in the optimal value, γ is the stabilizing selection coefficient [cf. Eq. (9.1)], and δ is a measure of generation overlap.

Table 9.16   Two-Patch Model Analyzed by Levene (1953)

		Fitness in patch #	
Genotypes	Frequencies	1	2
A_1A_1	p^2	W_1	W_2
A_1A_2	$2pq$	1	1
A_2A_2	q^2	V_1	V_2

Spatial variation can also maintain genetic variation. Again, an intuitive argument suggests this: Genetic variation lost in one patch can be restored by migration of individuals from another patch in which a different selection regime operates. This intuitive argument also applies to the above case of overlapping generations, "patches" being replaced by "generations." The first analysis of spatial variation was that of Levene (1953) who assumed the single-locus model shown in Table 9.16. Mating is at random and the offspring are distributed randomly among the habitat patches, at which time selection occurs. Selection is "soft" in that each patch contributes a fixed fraction of the next generation regardless of how many individuals get eliminated from the patch. A sufficient, but not necessary, condition for a stable equilibrium to occur is that the weighted harmonic means of the fitnesses are less than 1; that is,

$$\left(\sum \frac{c_i}{V_i}\right)^{-1} < 1, \qquad \left(\sum \frac{c_i}{W_i}\right)^{-1} < 1 \tag{9.44}$$

where c_i is the fraction of the population occurring in the ith patch. In Fig. 9.12 are plotted the change in allele frequency for two parameter combinations, one which satisfies the above relationships and one which does not, although still producing a stable equilibrium. In the former case, there is a single global equilibrium at an intermediate allele frequency, whereas in the latter, there are two equilibria, one which is locally stable and one which is unstable.

The stability of an additive genetic model in a two-patch universe was investigated by Bulmer (1971c, 1985). Within each patch there is stabilizing selection

$$W(x) = \exp\left(-\frac{(x - \theta_i)^2}{2\gamma}\right) \tag{9.45}$$

with the optimal value θ differing between patches ($\theta_1 > \theta_2$). The equilibrium heritability is

$$h^2 = \frac{(V_P + \gamma)}{V_P}\left[\left(\theta_1 - \theta_2\right)\left(\frac{m(1 - m)}{V_P + \gamma}\right)^{1/2} - 2m\right)(1 - 2m)^{-1}\right] \tag{9.46}$$

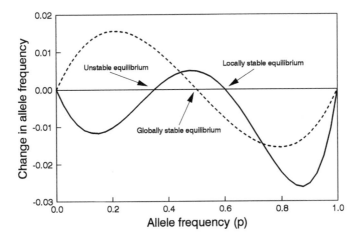

Figure 9.12 The change in allele frequency as a function of allele frequency for the two-patch model analyzed by Levene (1953). The solid line shows a combination of values for which the sufficient conditions given in Eq. 9.44 do not hold: $W_1 = 2$, $V_1 = 1.1$, $W_2 = 0.5$, $V_2 = 1.1$. The dashed line shows a combination in which the conditions are met: $W_1 = \frac{3}{2}$, $V_1 = \frac{2}{3}$, $W_2 = \frac{2}{3}$, $W_2 = \frac{3}{2}$. In both cases $c_1 = c_2 = 0.5$ (i.e., equal representation in patches).

where m is the migration rate between patches and V_P is the phenotypic variance. If the equilibrium value falls outside the permissible range 0–1, the equilibrium is unstable and the additive genetic variance decreases. The condition for a stable equilibrium is (Bulmer, 1985, p. 182)

$$(\theta_1 - \theta_2)^2 > \frac{4m(V_E + \gamma)}{1 - m} \tag{9.47}$$

Note the similarity between the above equation and Eq. (9.43): The term $(\theta_1 - \theta_2)^2$ is equivalent to $\text{Var}(\theta)$, $V_E + \gamma$ is equivalent to γ, and $(1 - m)/4m$ is equivalent to δ. Heritabilities maintained by migration between two patches are shown in Fig. 9.13. Clearly, significant genetic variation can be maintained, although the results suggest that the phenotypic variance must be significantly smaller than the difference between the optima and the stabilizing selection coefficient (γ). Clinal models also suggest that large amounts of additive genetic variance can be maintained by spatial heterogeneity (Felsenstein, 1977; Slatkin, 1978).

A critical assumption of the above analyses is that an organism cannot jointly satisfy both optima. However, under the infinitesimal model, unless the genetic correlation between environments is exactly ± 1, a reaction norm can evolve which will lead to the character value being θ_1 in environment 1 and θ_2 in envi-

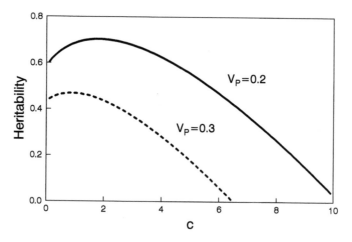

Figure 9.13 Heritability maintained in a two-patch model analyzed by Bulmer (1971c). For details see text. The parameter c is the ratio between the stabilizing selection coefficient and the phenotypic variance ($= \gamma/V_P$). Migration rate $= 10\%$ between patches per generation. Difference in optimal mean value in each patch, $\theta_1 - \theta_2 = 1.0$.

ronment 2 (see Chapter 6). In this case, there would be subsequent total erosion of additive genetic variance; the same argument applies to the case of overlapping generations. One way to circumvent this problem is to assume that there are no or imperfect cues which permit the evolution of a reaction norm. Gillespie and Turelli (1989) proposed a solution of this type, using a model in which there cannot be a single genotype that is most fit in all environments. The details of the model are as follows: The phenotype of an individual is given by

$$X = G + E + Z \qquad (9.48)$$

where G is the average phenotype produced by its genotype averaged over all environments, E is an environmental effect that is independent of the genotype (normally distributed with mean 0 and variance V_E), and Z is a genotype $\times$ environment effect. There are n loci with a finite but unspecified number of alleles. Assuming neither dominance nor epistasis, $G + Z$ is simply the sum of the individual contributions of the alleles. Labeling the two alleles at the ith locus by the subscripts j and k, and for mathematical convenience making some symmetry assumptions, the variance of $G + Z$ for a particular genotype is

$$\text{Var}(G + Z \mid \text{genotype}) = V_{G+Z} = \\ 2nV_Z[1 + r_w + p_H(1 - r_w) + 2(n - 1)r_b] \qquad (9.49)$$

where V_Z is the variance of the genotype–environment effect, r_w is the correlation between these effects within loci, r_b is the correlation between these effects between loci, and p_H is the proportion of homozygous loci. A consequence of this particular formulation is that the variance of the average phenotype produced by a given genotype across all environments is a decreasing function of the number of heterozygous loci. The next step is to assume that the fitness of a particular phenotype is determined by a single stabilizing selection function (i.e., there is a single phenotype that is optimal in all environments). With this assumption it can be shown that the mean fitness of a genotype is an increasing function of the number of heterozygous loci. Because of the overdominance averaged across all environments, selection will tend to preserve genetic variation. The important assumption of this model is that increasing heterozygosity "buffers" the organism against environmental perturbations. The data given in Section 9.4 are pertinent to this question but cannot resolve it. Experimental evaluation is obviously merited, but defining the appropriate range of environments and measuring the relevant parameters will be a monumental task.

9.7.2 *Environmental Heterogeneity and Genetic Variation: Observations*

There has, to my knowledge, been only one experimental investigation of the effect of environmental heterogeneity on quantitative genetic variation, that of Mackay (1980, 1981) using *D. melanogaster*. The experiment comprised four treatments (two replicates per treatment) to population cages: (1) control—weekly addition of two bottles of control medium (C); (2) spatial variation—weekly addition of one bottle of C medium and one bottle of medium plus 15% alcohol (medium A); (3) short-term temporal variation—alternately, two bottles of C and two bottles of A, (4) long-term temporal variation—alternation of A and C media every 4 weeks. The experiment was run for 1 year (approximately 20–25 generations). The results are rather difficult to decipher: All environmental treatments increased the heritabilities and additive genetic variances of sterno-pleural bristle number and body weight, relative to the control, but had no effect on the genetic variability of abdominal bristle number (Fig. 9.14). Because it is not possible to isolate the forces of selection, it is not possible to suggest with any confidence why such odd results were obtained. The fact that temporal variation appeared to maintain even higher levels of additive genetic variance than spatial variation suggests that heterozygote advantage might be important (Mackay, 1981). However, there is a possible trivial reason for the results obtained. From later analysis, Mackay concluded that the difference in genetic variances was a result of a decline in the genetic variance within the control lines rather than an increase in the treated lines; therefore, it is conceivable that there was simply less directional selection imposed by the A medium. Instead of two replicates of C medium for the control lines, a better control would have been four lines—two with C medium and two with A medium.

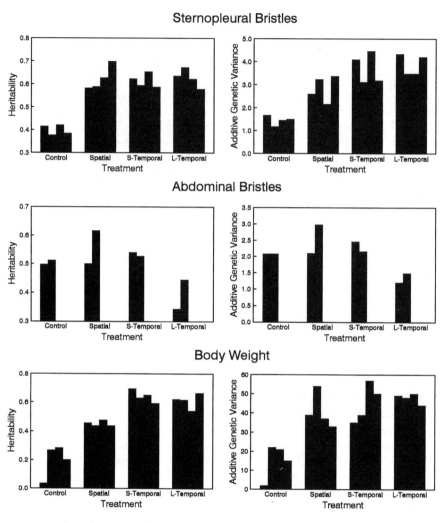

Figure 9.14 Heritability and additive genetic variance estimates for three morphological traits in *D. melanogaster* populations subject to different types of environmental variation (S-Temporal = short-term temporal, L-Temporal = long-term temporal). Standard errors are not shown but varied between 5% and 10% of the estimates. The first two bars in each cluster are the results for flies raised in the C medium; the second two bars are for flies raised in the A medium. (A and C not distinguished for the abdominal bristles in the original publication and only two values, not four, given.)

I concur with the observation by Mackay (1980, p. 1221) that, "The study of evolution in heterogeneous environments is rich in theoretical models and expectations, but relatively poor in well-designed experimental analyses by which empirical evaluation of the theory is possible. My results show the necessity of experimental approaches in this field." An important lesson from Mackay's study is that further experiments should be designed such that the selective factors are known.

9.8 Summary

Whereas it is clear from the fact that evolution has continued over many millions of years that genetic variation must be maintained in natural populations, the processes by which this occurs are still far from being understood. Directional selection, stabilizing selection, and disruptive selection all erode additive genetic variance, although the experimental data on disruptive selection are not conclusive. Significant additive variation can be maintained by a balance between mutation and stabilizing selection, but only under a restrictive set of conditions. There is also inconsistency between gametic and per locus mutation rates; this inconsistency needs resolution before further progress can be made on the plausibility of mutation–selection balance as being a significant factor preserving additive genetic variation. Heterozygous advantage does not look like a good prospect. The observation that the allozyme heterozygosity is sometimes correlated with phenotypic variation in traits is insufficient evidence for the importance of heterozygosity per se, because the relationship can also be accounted for by the associative overdominance or inbreeding depression hypotheses. Antagonistic pleiotropy has been discounted in the literature because a relatively high ratio of dominance to additive genetic variance is required: however, such a ratio is indeed found in life history traits and, hence, antagonistic pleiotropy remains an attractive model. Frequency-dependent selection and environmental heterogeneity are both mechanisms that can maintain a high heritability. There is a dearth of experimental evidence testing either of these hypotheses. The one experiment examining the role of environmental heterogeneity produced results that were difficult to reconcile with theoretical expectations.

10

A Summing Up

The foregoing chapters have reviewed the fundamental elements of evolutionary quantitative genetics. In this chapter I shall present an overview and outline where, in my opinion, the most serious lacunae in our knowledge lie. To this end the sections are divided into a series of questions.

10.1 Are the Basic Assumptions of Quantitative Genetics Reasonable?

There can be no doubting that genetic variation for quantitative traits exists and that it can be described by the application of Mendelian genetic principles to the offspring–parent regression. But to arrive at a mathematical formulation that is analytically, statistically, and experimentally tractable, it has been necessary to make a number of simplifying assumptions, two of the most important being that (1) there are a large number of loci and (2) epistatic effects, and in many cases dominance, can be ignored. How realistic or important are these assumptions?

10.1.1 How Many Loci?

To apply the machinery of biometrical analysis, we assume that the number of loci is infinite. Provided that the number is at least "large," this is reasonable. There are two lines of evidence bearing on the actual number: first, direct estimates and, second, observations on the response to selection. Direct estimates using statistical methods as exemplified by the Wright method have indicated anything from very few (2–5) to many (>100, Table 1.2). However, the statistical problems associated with this method cast great doubt on these numbers; certainly they are likely to be considerable underestimates. The use of quantitative trait locus (QTL) analysis promised to resolve this problem, but simulation models suggest that, unless sample sizes are considerably larger than presently used, the magnitude of QTL effects may be overestimated, thereby drastically decreasing the estimated number of loci. Equally disturbing is the finding that nonadditive effects can be detected spuriously. Given these caveats, the available data based on large samples suggests that the number of loci is at least in the tens if not hundreds (Table 1.3).

As noted by Barton and Turelli (1989, p. 341): "Breeders will be content to identify a few loci with large effects. In contrast the dynamics of additive variance under selection depends on a variance-effective number that accounts for variation within a population . . . Ideally we would like to know the joint distribution of frequencies and effects (on both morphology and fitness) of alleles segregating in natural populations." The present QTL analyses are designed primarily to identify alleles with major effects in highly domesticated animals and plants. Indeed, at present, the technique is largely restricted to such systems and because the organisms are highly inbred there are likely to be relatively few loci segregating. Consequently, even without the statistical problems, it is not clear what relevance such data have to natural populations.

Observations on response to selection both in domestic and nondomestic organisms provide evidence for a relatively large number of loci. It can readily be shown that with only a few loci determining a trait, say <20, selection will quickly erode additive genetic variance and response to selection will cease. Although a plateau in response is frequently seen after approximately 10 generations of selection (Table 4.3), response does not cease and pedigree analysis of such populations typically demonstrates the continued presence of additive genetic variance (Chapter 4). These data argue for a number of loci much greater than 10, although whether hundreds are required is not clear.

To summarize, although we do not yet have an adequate description of the number of loci and the distribution of allelic effects, the available data indicate that the number is probably large enough to satisfy, to a reasonable approximation, the statistical assumption of quantitative genetics that covariances among relatives can be described using the normal distribution.

10.1.2 How Important Are Nonadditive Effects?

The simplest quantitative genetic models assume that phenotypic variance is composed of environmental variance and additive genetic variance. The incorporation of dominance can be readily achieved in the statistical analysis of effects, although incorporation into theoretical models can present analytical difficulties. The recent review by Crnokrak and Roff (1995) shows that for morphological traits, the assumption of no dominance is warranted, but for life history traits, which are traits of considerable interest to evolutionary biologists, there is a significant contribution by dominance effects (Table 2.10). This conclusion is also supported by observations on inbreeding depression (Chapter 8). Thus, models of life history evolution that fail to consider dominance might be in error.

Nonadditive effects can be decomposed into both additive and nonadditive genetic variance, and much of the variance can actually appear as additive variance (Chapter 2). However, there is a considerable difference between two models that have the same additive genetic variance, one in which the variance is due entirely to additive loci and one in which the loci show both dominance and epistasis. This is particularly well illustrated by changes in genetic variance resulting from population bottlenecks; in the latter case, changes in allele frequencies can cause considerable "release" of additive genetic variation (Fig. 8.13). The detection of epistasis can be very difficult precisely because so much of the effect appears in the additive term. QTL analyses have in some instances indicated epistatic interactions, but until the possibility of statistical artifacts has been eliminated, such results must be viewed with caution. Recently, Cheverud and Routman (1995) have proposed a parameterization of physiological epistasis that allows measurement of the epistatic effect independent of its effect on the genetic variance components; further work in this area is required.

10.2   Is Heritability A Useful Parameter?

Heritability in the narrow sense has long been a central parameter in quantitative genetic analysis. As a convenient measure of the degree to which phenotypic variation can be decomposed into additive genetic variance and remainder variance (nonadditive + environmental) it has obvious utility. Houle (1992), among others, has suggested that too much attention might have been paid to heritability because, by itself, it cannot tell us how much a trait will respond to selection. Response depends on both h^2 and the selection differential; consequently, comparison between traits using solely heritability might be misleading. In its place, Houle suggested that we focus on the variances themselves and proposed a new statistic—evolvability. Unfortunately, this statistic is not scale independent and, hence, cannot be used to compare traits measured in different units (Section 4.2). At present, the best solution is to report in publications the heritability and sufficient information such that the variance components can also be calculated. Heritability is still a very concise indication of the likelihood that a trait will respond to selection and therefore remains useful.

10.3   How Should Heritability Be Estimated?

The particular experimental design will be a compromise between the constraints imposed by the natural history of the organism and the optimal experimental approach. The standard errors on h^2 are depressingly large (Fig. 2.7) and the raising of a sufficient number of families is a challenge in itself. Given this problem, one should choose a statistical design that is well characterized, the half-sib or mean offspring on mid-parent methods being the best in this regard. If neither of these methods is possible, a maximum likelihood approach should be taken, using whatever pedigree design is feasible. A potential shortcoming of the half-sib and offspring–parent regression is that both maternal and dominance effects cannot be separated. Various diallel designs can be used in this case and, depending on their simplicity, analyzed by ANOVA or maximum likelihood. The separation of maternal and dominance effects is not a trivial matter, but in the case of many life history traits, such analyses will be necessary.

10.4 Are Laboratory Estimates of Heritability Useful?

Conventional wisdom has it that heritability estimates measured in the laboratory are gross underestimates of field values. This argument rests on the assumption that V_A remains constant but that V_E is considerably larger in the field than the laboratory, thereby increasing V_P and decreasing h^2. Measurements of V_P in the laboratory and field do not bear this out, the increase being only about 20%. Further, direct comparison of laboratory and field values of h^2 show little difference (Fig. 2.15). Thus, laboratory estimates can be useful indicators of the heritabilities occurring in field populations. Nevertheless, it is certainly better to make the estimates in the most relevant environment and more studies on field populations are needed. Estimates from several different populations of the same species or from the same population at different times are particularly needed to empirically assess the stability of h^2. Because of the standard errors typically obtained, large sample sizes will be required to differentiate between estimates.

10.5 How Does Heritability Vary with Trait Type?

Life history traits have, on average, smaller heritabilities than morphological traits (Table 2.8). Two explanations have been proposed, one based on greater intensity of selection on life history traits, the other based on the assumption that life history traits are "downstream" from morphological traits (Fig. 2.13). These two hypotheses are not mutually exclusive and the present data cannot conclusively distinguish between them, although the relative ratios of dominance to additive genetic variance argue that the former is important. To resolve the issue, we need experimental dissection of the interaction between traits. By this I do not simply mean the statistical analysis of the variance–covariance matrices but rather an analysis of the functional relationship between the traits. We need to understand more about how morphology, physiology, and behavior interact to generate suites of traits. This is particularly important because experiments on selection on two traits simultaneously have not been noticeably predictable. Taking a strictly quantitative genetic approach can be misleading, as illustrated by the selection experiment of Reeve and Fairbairn (1996) on size dimorphism in *Drosophila melanogaster*: Selection on body size in one sex did not produce the predicted change in the other sex, but the observed change could be produced if it was assumed that selection actually operated on development time not body size per se.

10.6 The Genetic Correlation: From the Sublime to the Ridiculous?

Whereas estimation of heritability is experimentally difficult, the estimation of the genetic correlation is experimentally *extremely* difficult and published estimates generally have such large confidence limits that their worth is questionable. Fortunately, at least the optimal experimental design is the same for both h^2 and r_A. Given that we wish to understand how natural selection changes suites of traits in wild populations, we have no option but to attempt to estimate these two parameters (or the component genetic variances and covariances). For morphological traits, the phenotypic correlations appear to be reasonable estimates of the genetic correlation and might be substituted in large-scale comparative analyses, following the protocol suggested in Section 3.5. There is some doubt that correlations involving life history traits can be so estimated, and more empirical data are needed to more fully answer this question.

Attention has recently been given to the question of whether the genetic variance–covariance matrix remains stable. I have argued in Section 3.6 that the relevant question is not one of equality but rather of the difference between populations, species, and so forth. It is a problem of interval estimation rather than hypothesis testing. If one does wish to compare matrices, there are a number of proposed methods, but none are wholly satisfactory and there is considerable scope for more theoretical work in this area. A survey of the empirical comparisons undertaken to date reveals inconsistency in that different species might be less alike than different genera (Table 3.14).

10.7 Directional Selection on a Single Trait: Is It Predictable?

Quantitative genetic theory can be used to make long-term predictions, provided it is assumed that the genetic parameters do not change. But quantitative genetic theory itself predicts that these values will change as a result of changes in allele frequencies. Therefore, the finding from directional selection experiments that predictions are robust only for 10–20 generations is not an indictment against quantitative genetic theory itself but against our inability to add the necessary modifications. Measurements of selection intensities in field populations show that they can be as strong as those used in artificial selection experiments (Fig. 4.15) and, hence, it cannot be argued that the assumption of parameter stability is warranted on the grounds of weak selection in the wild.

There is considerable merit in testing the ability of quantitative genetic models to predict short-term response in field populations, although long-term predictions are probably not possible, except in a very qualitative sense. At present, there are too few studies to make an adequate assessment of the accuracy of short-term predictions in natural populations, but those which have been attempted have had notable success (Section 4.8). The requirements for this type of study are very daunting, requiring estimates of selection intensities and genetic parameters, both of which require very large sample sizes.

10.8   Can We Go from One to Several Traits?

From theoretical considerations we expect that although h^2 might remain more or less constant for 10–20 generations, the genetic correlation will change more quickly, making predictions reliable for an even shorter period. Selection experiments on a single trait have shown variable results with respect to correlated traits; in some cases, excellent fits between prediction and observation were obtained and in others very poor fits (Section 5.2.2). Experiments in which there was simultaneous selection on two traits have been generally qualitatively correct but quantitatively unreliable (Section 5.3.2). Prediction of morphological change in the medium ground finch, *Geospiza fortis*, over two short periods (1976–1977 and 1984–1986) was extraordinarily successful (Grant and Grant, 1995) and suggests that further field tests are warranted. Whether it will be possible to make useful predictions over longer periods remains to be demonstrated, although the artificial selection experiments suggests that this is unlikely, at least if selection is directional, always with same sign: It is possible that predictions would be better in the case of fluctuating selection, but this is only a conjecture.

10.9   Phenotypic Plasticity: An Experimental Nuisance?

From the breeder's perspective, phenotypic plasticity is a nuisance, because it does not readily permit extension of estimates from one environment to another. However, phenotypic plasticity occupies a central place in evolutionary theory. Given, as is readily apparent, that the natural world is spatially and temporally variable, we would expect organisms to show adaptive responses to immediate conditions, developmentally, physiologically, and behaviorally. Considerable advance has been made on this problem by the use of optimality models, but advancement within the quantitative genetic framework has been relatively slow. The recognition of the importance of phenotypic plasticity in adaptive evolution has considerably increased interest in this area, and it is clear that it is an expanding field of research.

Phenotypic plasticity can be analyzed using the character state approach, in which the same trait in two different environments is viewed as two traits genetically correlated, or the reaction norm method which sees a function relating trait and environmental values, with the coefficients of the function being inherited. The relationship between the parameters of the two approaches is complex and each gives somewhat different information (Section 6.1.2); hence, it is useful to use both approaches in an analysis. Statistical estimation of parameters when there are two environments is relatively straightforward, but extension to multiple, discrete environments requires further study.

From theoretical considerations, we expect that phenotypic plasticity should readily evolve, a prediction amply borne out by experimental studies. Not unexpectedly, most of the interest in phenotypic plasticity in natural populations has focused on the evolutionary response to spatial and temporal variation. An important aspect of this work is that of the cost of being plastic (i.e., given that the environment provides only probabilistic cues about its present and future state, what is the optimal response?). At one extreme, an organism can develop into a particular type with a certain probability irrespective of the environments, whereas at the other extreme, adoption of a particular morph can be a strict function of some environmental factor. The first case might be expected when the habitat is variable but unpredictable, whereas the second is expected when there is high predictability of the environment. Several theoretical analyses of this problem have been attempted (see Sections 6.2.3–6.2.7), but tests of the predictions have been generally unreliable, primarily because the data were not typically collected to test the specific hypothesis. This is an area where considerable empirical and theoretical work is still needed.

The obverse to the above focus on phenotypic plasticity is the focus on developmental stability or canalization. Despite a long history of the concept, its importance still needs to be demonstrated, particularly the extent to which lack of variation is due to genes that cause canalization of development versus simply fixation of alleles. Experiments on stabilizing selection suggest the latter is more important in the case of strong selection for an intermediate optimum. There is some evidence that traits closely connected to fitness are more buffered against random environmental noise than traits less related to fitness, but problems of finding an appropriate scale of comparison have hampered analysis (Section 6.3).

10.10 Parental Effects: Another Nuisance?

As with phenotypic plasticity, sex-related effects, specifically maternal effects, have been considered an unfortunate nuisance by breeders. In sexually reproducing organisms, there is probably not a single instance in which some type of maternal effect is not present. Despite this, there have been very few analyses of maternal effects in nondomestic organisms within a quantitative genetic framework. The original approach developed by Willham (Section 7.3.1) was based on ANOVA, but a more recent matrix formulation by Kirkpatrick and Lande (1989) is more flexible and more easily constructed when several interacting traits are involved (Fig. 7.2, Table 7.8). The principle result of parental effects is the introduction of a lag in the selection response (Fig. 7.3). There are a variety of methods that can be used to detect the presence of maternal/paternal effects, but not all provide sufficient data to estimate parameter values for either the Willham or Kirkpatrick–Lande models. One of the most detailed studies, that on a wild population of the collared flycatcher (Schluter and Gustafsson, 1993), used a variant of the Kirkpatrick–Lande approach and was able to demonstrate statistically, at least, the presence of a significant maternal effect on clutch size, operating through the mother's condition. As illustrated by this study, it is likely that maternal effect models will generally have to be "tailored" to the natural history of the organism. At present, we are in the situation of knowing that maternal effects are very common and that they can be very important in modulating the life history (e.g., diapause propensity) but have insufficient studies incorporating a genetic basis to the trait to assess how significant the genetic architecture is in either the short-term or long-term trait trajectory. It is an open question as to whether an optimality approach is the only feasible approach in most cases. Both theoretical and empirical studies are required to address this issue.

10.11 Should We Worry About Population Size?

The simple answer is YES. It can be readily shown that if populations sizes are relatively small (<100), then drift can be the most significant factor causing changes in allele frequencies. Of course, if natural populations were invariably very large (>1000), then we would not have to worry (and life would be much simpler). However, estimates show that effective population sizes are often (at least as indicated by the published data, which are a nonrandom sampling of organisms) less than 100 (Table 8.3, Fig. 8.2). Estimates larger than this might be overestimates, as such estimates generally do not take into account temporal fluctuations in population size.

Population bottlenecks are important both because they significantly reduce the long-term effective population size and because changes in allele frequencies due to sampling variation can generate considerable increases in additive genetic variance. This effect is potentially important in two respects: First, it can alter short-term responses to selection and, second it can cause the population to cross the fitness surface to climb another adaptive peak. Empirical studies of genetic changes due to bottlenecks have been unsatisfactory because they have not taken into account that changes in additive genetic variance can occur even if there are no nonadditive effects. To adequately address this problem requires many more bottlenecked lines than typically used (100 versus 2–10 actually used). Because of the very large labor requirements, prior to such an experiment an assessment should be made of the nonadditive component of the genetic variance (see below).

10.12 Inbreeding Effects: Partial Dominance or Overdominance?

Inbreeding arises because of nonrandom mating, which can occur due to small population size or the natural history of the organism (e.g., restricted migration, selfing, etc.). Decline of a trait will occur due to inbreeding, providing that there is dominance—the phenomenon being called inbreeding depression. Such effects have been observed in many life history traits and less frequently in morphological traits, as would be expected given the results reported in Chapter 2. The detection of inbreeding depression is relatively easy but does require larger sample sizes than typically used (see Section 8.4.2). Although there are many estimates for vertebrates and plants, there are surprisingly few for invertebrates, and more estimates are required for this group.

There are two theories for inbreeding depression: the overdominance model and the partial dominance model. The former held sway for many years, but presently, the consensus favors the latter. The two models can be distinguished by inbreeding lines: The partial dominance model predicts that in some lines the deleterious alleles will be purged, leaving lines which are fitter than the original population, whereas the overdominance model predicts a steady decline due to loss of heterozygosity. Although the few experiments following this protocol support the partial dominance model, the number of species examined and the number of lines used are too few to make any general statements (see Section 8.4.4).

10.13 What Maintains Genetic Variation in Populations?

This is clearly a central question of evolutionary quantitative genetics, and one for which we have no satisfactory answer, although a number of plausible models. Because traits appear to remain within fairly narrow bounds for long periods of time, it is generally supposed that stabilizing selection must be prevalent in natural populations. Unfortunately, most (but not all) quantitative genetic models predict that stabilizing selection will erode additive genetic variance. Stabilizing selection experiments on *Drosophila, Tribolium,* and *Mus* have generally supported this prediction (Table 9.5). Disruptive selection—the opposite of stabilizing selection—is also predicted to lead to the loss of genetic variation (Section 9.2). In this case, artificial disruptive selection experiments typically show an increase in additive genetic variance (Table 9.6), although this may be due to linkage disequilibrium. The discrepancy between theory and observation needs to be resolved and, as disruptive selection has been observed not infrequently in natural populations (Fig. 9.3), more longer-term selection experiments are warranted.

Mutation–selection balance models have been extremely popular and can, given the right parameter values, maintain high levels of additive genetic variance (Table 9.9). The estimates of mutation rates are exceedingly variable (Table 9.7), as are those of V_S/V_E (Table 9.8), both of which are critical parameters. Of particular concern is the discrepancy between per locus mutation rates and gametic mutation rates (Barton and Turelli, 1989). Although mutation–selection balance can account for some of the observed additive genetic variance, it appears unlikely that this can be an important mechanism for traits such as those related to fitness, which are under strong selection.

Heterozygosity would initially seem to be an excellent candidate mechanism for the preservation of additive genetic variation. Theory does predict that heterozygous advantage can maintain significant genetic variation across loci but not within a locus (Section 9.4). Experimental evaluation of this possibility has been made by correlating electrophoretic variation with quantitative traits (Table 9.11). This approach cannot effectively distinguish among true overdominance, associative overdominance, or inbreeding depression (Section 9.4.2). It seems likely that the importance of these three effects varies depending on the trait. Because of the difficulties of separating the three models, it seems unlikely that much progress will be made on this problem in the near future. What is not required are more studies that correlate electrophoretic variation with quantitative traits.

Antagonistic pleiotropy was initially advanced as a likely mechanism but discounted lately because of the requirement for a large contribution of dominance variance (Section 9.5). This rejection is premature, as the data demonstrate quite definitively that for a life history trait, dominance variance contributes significantly to the total genetic variance (Chapter 2). In fact, the measured amount of dominance variance is well within the range required to maintain substantial amounts of additive genetic variance (Table 9.14). Most of the analyses are predicated on the two-locus model; the requirements when large number of loci are involved requires further study. Given that a large negative genetic correlation is measured between two traits and that both are shown to have large dominance variance components, how do we prove that it is the antagonistic pleiotropy that is maintaining the genetic variation? At present, I know of no satisfactory experimental or statistical method to answer this question, although it importance cannot be overstressed.

Frequency-dependent selection can maintain additive genetic variance, although the simulations of Mani et al (1990) suggest that the number of alleles per locus will be small (Fig. 9.11). Field and laboratory studies have shown that frequency dependence is often found for polymorphic traits, but its importance for traits that show continuous variation remains to be demonstrated.

The final factor that can maintain additive genetic variance is environmental heterogeneity. From theoretical considerations, temporal variation alone should not preserve additive genetic variance (although it might slow the rate of erosion), but spatial and spatiotemporal variation are potentially important factors. The only experimental test of these predictions, that of Mackay (1981), produced results which are extremely difficult to interpret (Fig. 9.14), in part because the actual selective factors could not be isolated. More experiments are required, but ones in which the pattern of selection is clearly understood and which can therefore be simultaneously modeled.

10.14   Is Quantitative Genetic Analysis a Viable Approach to the Understanding of Evolution?

Obviously I believe it is a viable approach. But we should not ask too much of quantitative genetics. In order to proceed, it has been necessary to make simplifying assumptions. These have generated experimental studies that have been fruitful. With the advent of increased computing power it has been possible to relax these assumptions and examine alternate models, at least for a limited number of situations. These have stimulated further research—the question of bottlenecks and antagonistic pleiotropy being excellent examples. There is still considerable scope for improvement of the theoretical models and, more importantly, the testing of a quantitative genetic predictions in natural populations.

Glossary of Terms

Additive genetic variance: The genetic variance that can be statistically associated with the linear relationship between mean offspring and mid-parent values. If there are no nonadditive effects, the entire genetic variance is additive genetic variance. However, even if the actual action of alleles is entirely nonadditive, in a mechanistic sense, there can still be an additive component in a statistical sense.

Additivity: The assumption that the effect of loci and alleles can be represented by the sum of the individual effects. Additivity does not preclude nonadditive effects; it simply statistically divides effects into these two components.

Allele: The basic element of inheritance. Each character is composed of a series of loci, and at each locus there may be one to several alleles, each allele contributing a different amount to the genotype and, hence, the phenotype.

Allelic sensitivity: The expression of the genes is altered by changes in external conditions.

Antagonistic pleiotropy: Occurs when the genetic correlation between traits is such that selection on one trait is opposed by the consequent selection on the second trait (e.g., a negative correlation between longevity and age at reproduction).

Associative overdominance: Increased fitness not due to heterozygosity at the detected (electrophoretic) loci but to linked loci which are under selection.

Autoregulatory morphogenesis: Reaction norm for a trait that occurs as discrete morphs.

Average effect: Under random mating, the average effect of a particular allele is the mean deviation from the population mean caused by the presence of that allele in an individual, where the second allele is drawn at random from the population.

Balancing selection: Selection on several traits or at different episodes that results in stabilizing selection on the focal trait.

Beneficial reversal of dominance: Occurs when the direction of dominance in antagonistic pleiotropy is in the direction of increased fitness.

Bonferroni correction: Correction made when a series of statistical tests are performed. The correction adjusts the individual probability values such that the overall probability remains at 5%.

Bootstrap: A robust statistical method.

Breeding value: The value of an individual as measured by the mean value of its progeny.

Canalization: The production of the same phenotype regardless of the environment or production of the same phenotype by several different genotypes.

Canalizing selection: Selection for genes that canalize development such that the phenotypic range is reduced.

Close inbreeding: Matings between first-degree kin.

Coefficient of inbreeding: Probability that two alleles at a locus are identical by descent.

Common garden experiment: The elimination of environmental effects by the use of a single environment to raise individuals originating from different geographic populations, clones, and so forth.

Continuum-of-alleles model: A relatively small to moderate number of loci with high mutation rate giving a Gaussian distribution of small phenotypic effects at each locus.

Correlated response to selection: Change in trait values due to selection on another trait.

Cross-fostering: Switching offspring between parents to assess the importance of parental effects on the offspring phenotype.

Degree-days: Developmental processes in ectotherms depending on temperature. To take this into account, time is measured on a physiological scale, the basic unit being the degree-day (or day-degree). The number of degree-days corresponding to one day in "real" time is equal to temperature (in °C) minus the threshold temperature (= temperature at which the process ceases).

Deleterious reversal of dominance: Occurs when the direction of dominance in antagonistic pleiotropy is in the direction of decreased fitness.

Dependent development: See *Reaction norm.*

Developmental conversion: See *Autoregulatory morphogenesis.*

Diallel cross: The set of all possible matings between several genotypes.

Directional dominance: Dominance values across loci tend to have the same sign.

Directional selection: Occurs when the mean value of the parents which contribute to the next generation differs from that of the population from which the parents were drawn.

Dominance: The phenomenon in which the heterozygotic value does not fall midway between the two homozygotes.

Dominance deviation: The deviation from the additive model caused by the action of dominance. Note that the dominance deviation is a statistical phenomenon.

Dosage compensation: The production of the same product in both sexes despite different dosages due to heterogamy.

Drift: See *Genetic drift*.

Effective population size: A measure of the number of individuals making a genetic contribution to the next generation.

Environmental correlation: The correlation between two traits that cannot be ascribed to the genetic correlation.

Epistasis: The nonadditive interaction between loci.

Evolutionary momentum: Continued change in a trait value due to parental effects after selection on the trait has ceased.

Evolvability: The ability to respond to selection. For truncation selection, evolvability is the response to selection standardized by the intensity of selection and phenotypic mean.

Family selection: Selection of entire families based on the mean value of the family.

Fisher's Fundamental Theorem: The change in fitness caused by natural selection is equal to the additive genetic variance in fitness.

Frequency-dependent selection: The fitness of a phenotype of genotype varies with the phenotypic or genotypic constitution of the population.

Fluctuating asymmetry: Nondirectional deviations from bilateral symmetry.

Full-sibs: Individuals that have the same mother and father.

Functional overdominance: Fitness increase due to true overdominance or associative dominance.

Gametic phase disequilibrium: See *Linkage disequilibrium*.

Gaussian allelic model: See *Continuum-of-alleles model*.

Gaussian function: See *Normal distribution*. The term is sometimes used to designate a curve of the general form $\exp(-x^2)$.

Gene: A somewhat vague term, sometimes used when referring to a locus and sometimes when referring to an allele. I have not generally used the word in this book.

Genetic correlation: Association of different traits due to the correlation between the loci controlling the two traits. This can arise through linkage disequilibrium or pleiotropy.

Genetic drift: Variation in allelic frequencies caused by random sampling from a finite population.

Genomic imprinting: An epigenetic gamete-of-origin dependent modification of the genome.

Genotype: The genetic constitution of an organism.

Grandparent effect: Name given to the observation that genetically controlled maternal effects will be correlated with the grandfather rather than the father.

$G \times E$: Genotype by environment interaction.

Half-sibs: Individuals sharing a single common parent, generally the father.

Hardy–Weinberg equilibrium: The equilibrium attained at a single locus after one generation of random mating.

Heritability in the broad sense: Ratio of the total genetic variance to the phenotypic variance.

Heritability in the narrow sense: Also referred to as "heritability," designated by the symbol h^2 (sometimes subscripted);

$$h^2 = \frac{\text{Additive genetic variance}}{\text{Phenotypic variance}} = \frac{V_A}{V_P}$$

Heterosis: Originally this term referred to the phenomenon of hybrid vigor without respect to the mechanism generating this vigor. Today, it is generally used to denote that heterozygous genotypes have a higher fitness than homozygous genotypes.

Heterozygosity: The presence of alternate alleles at a given locus in a diploid organism (e.g., A_1A_2).

Heterozygous advantage: See *Heterosis*.

Homozygosity: In a diploid organism, the presence of the same allele at a given locus (e.g., A_1A_1).

House-of-cards model: In contrast to the continuum-of-alleles model, this model postulates a single "wild-type" allele in high frequency and mutations of large effect.

Identity by descent: Two genes are said to be identical by descent if they are descended from the same mutational event.

Inbreeding: The mating of related individuals.

Inbreeding depression: The decline in a trait value resulting from inbreeding.

Inbreeding effective population size: That population size which under the assumptions of random mating, Poisson distribution of family size, and constant population size will give the same probability of two genes coming from a common parent as actually observed.

Infinitesimal model: The statistical model underlying quantitative genetic analysis. The fundamental assumption is that the phenotype can be decomposed into additive and nonadditive genetic effects, and an environmental component based on the principle of multivariate normality.

Intensity of selection: The selection differential in phenotypic standard deviation units.

Jackknife: A robust statistical procedure for the estimation of means and variances.

Jinks–Connolly rule: A rule concerning the direction of response in two environments; when the direction of selection and effect of the environment of selection are opposed in their effects on the deviation of the resulting phenotype, the resulting selection lines are less sensitive to environmental variation than when the direction of selection and the effect of the environment are reinforcing. The rule does not follow from theory but has been found to hold in practice.

Liability: The underlying continuously trait in the threshold model.

Life history trait: A trait that is directly connected with fitness, such as development time, fecundity, and viability. Traits such as body size might be connected with fitness (e.g., success in obtaining mates), but a priori no such designation can be made.

Linkage: The presence of two loci on the same chromosome. Because of recombination, the two loci may still be in linkage equilibrium.

Linkage disequilibrium: The nonrandom association between two loci. This can be caused by physical linkage due to the two loci being on the same chromosome or to disassortative mating.

Locus: The unit of inheritance. Each character is made up of one to many loci, at which there are one to many alleles.

Macroptery: Presence of long (generally functional) wings.

Mantel test: A randomization test for the comparison of the correlation between two variables.

Mass selection: Selection of individuals from a single population. This is in contrast to family selection.

Maternal effect: The phenotype of the offspring is determined in part by the phenotype of the mother, in addition to her genotype.

Maximum likelihood: A statistical method based on the assumption that the "best" estimates are those which maximize the probability of obtaining the observed data set.

Microptery: Short, nonfunctional wings.

Mutation: In quantitative genetics, the change from one allelic type to another (e.g., $A_1 \rightarrow A_2$).

Mutation–selection balance hypothesis: Hypothesis that additive genetic variance is maintained by a balance between mutation which increases variance and stabilizing selection which decreases variances.

Norm of reaction: See *Reaction norm*.

Normal distribution: A variable x that is distributed normally has the probability density function (i.e., the probability of observing a value x) of

$$\frac{1}{\sigma\sqrt{2\pi}} \exp\left[-\frac{1}{2}\left(\frac{x-\mu}{\sigma}\right)^2 \right]$$

where μ is the mean and σ the standard deviation.

Normalizing selection: Elimination of genotype combinations that produce extreme phenotypes.

Noroptimal selection: Stabilizing selection in which the relative fitness is modeled using a Gaussian curve.

North Carolina design: Three breeding designs for the estimation of genetic parameters.

Outbreeding: The mating of unrelated individuals.

Outbreeding depression: The depression in fitness obtained by the crossing of unrelated individuals. Applies most specifically to individuals drawn from different habitats.

Overdominance: The value of the heterozygote exceeds that of either homozygote.

Overdominance hypothesis: Hypothesis that inbreeding depression is caused because heterozygotes have higher fitness. See *Partial dominance hypothesis*.

Panmictic index: 1 − rate of inbreeding.

Parallel dominance: Occurs when the direction of dominance in antagonistic pleiotropy is in the direction of increased fitness for one trait but decreased fitness for the other trait.

Partial dominance hypothesis: Hypothesis that inbreeding depression is caused by the increased frequency of deleterious partially dominant (recessive) alleles. See *Overdominance hypothesis*.

Phenotype: The phenotypic appearance of an organism, consisting of the interaction between the genotype and environment.

Phenotypic modulation: See *Reaction norm*.

Phenotypic plasticity: The change in the average phenotype expressed by a genotype in different environments.

Pleiotropy: The influence of a locus on more than one trait.

Power: The probability of rejecting the null hypothesis given that it is false.

Proband method: Statistical method of estimating the heritability of a threshold trait.

Quantitative trait locus: See *QTL*.

QTL: Quantitative trait locus; a region of a chromosome that has a significant effect on a quantitative trait.

Randomization test: A robust statistical test based on the randomization of the data set and recalculation of the test statistic. The randomization is done many times and from this, the probability of the observed value of the test statistic is obtained.

Reaction norm: The systematic change in phenotypic expression of a given genotype across a systematic change in the environment.

Realized heritability: Heritability estimated using response to selection.

Regulatory plasticity: The expression of genes is altered by other genes which detect the change in external conditions.

Repeatability: A measure of the stability of repeated measurements. Repeatability is the intraclass correlation coefficient. Repeatability is equal to $(V_G + V_{Eg})/V_P$, where V_{Eg} is the general environmental variance. It is the upper limit of heritability. However, one should distinguish a repeatability due to errors of measurement, as might occur when making multiple measurements on the same body part, from variation that is not due to the observer but results from variation in the trait itself, as might occur in repeated behaviors. The former is not as informative as the latter. Repeatabilities of clutch size show a remarkable similarity to the heritabilities [see Table 9.7 of Roff (1992)]. For further discussion on repeatability see Lessells and Boag (1987), Falconer (1989, pp. 139, 141), and Boake (1994).

Restrictive maximum likelihood: Use of the method of maximum likelihood with the additional assumption that certain values (e.g., negative variances) are not possible.

Selection differential: Difference between the mean value of the parents and the mean value of the population from which the parents were drawn.

Selection intensity: See *Intensity of selection.*

Sex-limited expression: A trait expressed in only one sex even though the genes for the trait can be found in both sexes (e.g., secondary sexual characteristics).

Sex linkage: Some or all of the loci controlling a trait are carried on the sex chromosome.

Stabilizing selection: Selection for an intermediate optimum.

Threshold trait: A trait which appears phenotypically as two or more discrete morphs but which can be modeled assuming a normally distributed underlying trait, the phenotypic manifestation being a consequence of a threshold effect. For a dimorphic trait, there is a single threshold—individuals above the threshold displaying one morph, individuals below the threshold displaying the alternate morph.

Triple test cross: A breeding design for the estimation of genetic parameters.

True overdominance: Fitness of genotype due to the fitness advantage conveyed by heterozygosity. See also *Functional overdominance* and *Associative overdominance.*

Truncation selection: Selection in which individuals exceeding a particular value are chosen. This value can be fixed or set such that a fixed percentage of the population is selected.

Variance effective population size: An alternative definition of the effective population size based on the variance in allele frequencies.

Within-family selection: Selection of the most deviant individuals within a series of families.

Glossary of Symbols

Table 1. List of Symbols Used in Chapter 1; Parameters May Be Subscripted

Symbol	Definition
a	Value of genotype A_1A_1 ($-a$ for A_2A_2)
b	Slope of regression
c	Constant
d	Value of genotype A_1A_2
h^2	Heritability in the narrow sense
m	Mean value
n	Number of loci
A_i	Additive contribution of locus i; A also used as symbol for allele A
D_i	Dominance contribution at locus i
G	Genotypic value
R	Divergence between two selected lines
V	Variance component
X	Trait value
δ	Parameter in Wright's method of estimating number of loci

Table 2. List of Symbols Used in Chapter 2; Parameters May Be Subscripted

Symbol	Definition
a	Value of genotype Aa ($-a$ for genotype aa)
b	Slope of regression
c	Constant
d	Value of genotype Aa
e	Error term
h^2	Heritability in the narrow sense
i	Parameter in Falconer's proband method
k	Probabilities of identity by descent [Eq. (2.15)]; coefficients in ANOVA
m	Mean value
n	Number of loci [Eq. (2.11)]; number in a family in ANOVA

p	Proportion of allele A; proportion of a particular morph (threshold trait)
q	$1 - p$
r	Correlation between parents; coefficient of relatedness [Eq. (2.54)]
r_A	Additive genetic correlation
t	Intraclass correlation (subscripts: S = Sire, D = Dam)
w	Weighting factor in offspring–parent regression
z	Ordinate on the standardized normal curve corresponding to probability p
A	Breeding value; also used as symbol for allele A
Cov	Covariance
D	Dominance deviation [Eqs. (2.4) and (2.11)]; number of dams per sire in ANOVA; $D_G = V_D/(V_D + V_A)$, $D_P = V_D/V_P$
E	Environmental value
G	Genotypic value
H^2	Heritability uncorrected for assortative mating
L	Likelihood
MS	Mean squares
N	Number of families
P	Probability
S	Number of sires in ANOVA
SE	Standard error
T	Total number of individuals in ANOVA
V	Variance component (subscripts: A = additive, D = dominance, E = environmental, G = genetic, P = phenotypic)
X, Y	Trait values
μ	Expected mean value; μ_2 = expected value of squared values
ρ	Correlation between full-sibs

Table 3. List of Symbols Used in Chapter 3; Parameters May Be Subscripted

Symbol	Definition
a	Value of Genotype Aa ($-a$ for genotype aa)
c	Constant; crossover rate [Eq. (3.1)]
h^2	Heritability in the narrow sense
k	Coefficient or number of traits
n	Number in a family; n_D = number of dams per sire [Eq. (3.26b)]; number of pseudovalues in jackknife [Eq. (3.32)]; sample size
p	Proportion of a particular allele at a given locus
q	$1 - p$
r	Correlation (subscripts, A = additive genetic, E = environmental, m = family mean, P = phenotypic)
z	Transformation for the correlation coefficient [Eq. (3.17)]
Cov	Covariance
D	Log-likelihood ratio statistic; D_T = Total number of dams
MC	Mean cross-products
N	Number of families

P	Probability
S	Number of sires
SE	Standard error
T	Total number of individuals
V	Variance component (subscripts: A = additive, D = dominance, E = environmental, G = genetic, P = phenotypic)
X, Y	Trait values (subscripts: f = female, m = male)

Table 4. List of Symbols Used in Chapter 4; Parameters May Be Subscripted

Symbol	Definition
b	Slope of regression
c	Constant
h^2	Heritability in the narrow sense
i	Selection intensity
m	Mean value; m_i = number of selected morph in a sample
n	Number in a family; number of loci; sample size
p	Proportion selected; $p(x, t)$ = frequency distribution of trait x at time t; proportion of dominant allele [Eq. (4.24)]
r	Coefficient of relatedness
t	Time or generation; intraclass correlation
w	Weighting factor
x	Trait value
z	Ordinate of the standard normal distribution
C	$1 - 1/(2N)$, where N = population size
L	Likelihood
M	Number of individuals measured
N	Number of families; population size
P	Probability
R	Response to selection
S	Selection differential
SE	Standard error
T	Threshold value
V	Variance component (subscripts: A = additive, P = phenotypic)
W	Fitness
X, Y	Trait values
μ	Mean; mutation rate [Eq. (4.27)]
σ	Standard deviation

Table 5.   List of Symbols Used in Chapter 5; Parameters May Be Subscripted

Symbol	Definition
c	Constant; parameters a, b, c, d used in model of Bohren et al. (see Table 5.1)
f	Fraction of resources shunted to trait X
h^2	Heritability in the narrow sense
i	Selection intensity
k	$i(i-z)$
r	Correlation (subscripts: A = additive genetic, P = phenotypic)
v	Proportion of nonpleiotropic mutations
x, y	Trait values
z	Ordinate of the standard normal distribution
E	Environmental component
CR	Correlated response
Cov	Covariance
R	Response to selection
S	Selection differential
T	Pool size in partition of resources model
V	Variance component (subscripts: A = additive, E = environmental, P = phenotypic)
X, Y	Trait values
β	Selection gradient
μ	Mean

Table 6.   List of Symbols Used in Chapter 6; Parameters May Be Subscripted

Symbol	Definition
b	Regression slope
c	Constant
d	Difference
e	Error term
f	Frequency of a particular type of habitat
g	Parameter in Tienderen model
h^2	Heritability in the narrow sense
i	Selection intensity
k	Number of dams per sire
m	Mean value
n	Family size
r	Correlation (subscripts: A = additive genetic, G = broad sense, P = phenotypic)
C	Coefficient in Gaussian fitness function
CR	Correlated response
Cov	Covariance
E	Environmental variable

F_i	Deviation from mean in family i of Weis and Gorman model
H^2	Broad sense heritability
MS	Mean square
R	Response to selection
S	Selection differential; S_{ij} is survival of family i in environment j
V	Variance component (subscripts: A = additive, E = environmental, P = phenotypic)
W	Fitness function
X, Y	Trait values
β	Slope in Weis and Gorman model
γ	Parameter measuring strength of stabilizing selection
ϵ	Error term
μ	Mean
θ	Optimum value; cost-free reaction in Tienderen model

Table 7. List of Symbols Used in Chapter 7; Parameters May Be Subscripted

Symbol	Definition
b	Slope in Schluter and Gustafsson model
c	Constant
e	Environmental deviation
h^2	Heritability in the narrow sense
i	Selection intensity
m	Maternal-effect coefficient
n	Observations per cross between lines
r	Correlation (subscripts: A = additive genetic, P = phenotypic); coefficient of relationship
x	Trait value
A	Additive genetic effect (subscripts: O = direct, M = maternal)
Cov	Covariance
E	Environmental variable
G	Effect attributable to line
M	Maternal-effect coefficient
Mat	Extranuclear maternal effect
N	Number of lines
Nuc	Nuclear contribution
Pat	Extranuclear paternal effect
R	Response to selection
S	Selection differential
V	Variance component (subscripts: A = additive, E = environmental, P = phenotypic)
X, Y	Trait values
β	Selection gradient
μ	Mean

Table 8. List of Symbols Used in Chapter 8; Parameters May Be Subscripted

Symbol	Definition
a, b	Values in epistatic model of Goodnight (Table 8.6)
c	Constant
d	Dominance value; migration rate (Table 8.1)
h^2	Heritability in the narrow sense
i	Selection intensity; value in epistatic model (Table 8.6)
k	Mean offspring per female
m	Mean
n	Number of a particular genotype (subscripted)
p	Proportion (subscripts: m = male, f = female)
s	Selection coefficient
t	Generation
w	Trait value (subscripts: I = inbred, O = outbred)
C	Coefficient
D	Dominance coefficient
F	Inbreeding coefficient
H	Heterozygosity
K	Number of first-degree kin
M	Mean trait value
N	Number of parents; census population size; N_e = effective population size
P	Probability of close inbreeding
T	Generation interval
U	Mutation rate per diploid genome per generation
V	Variance component (subscripts: A = additive, E = environmental, P = phenotypic)
X	Trait value
δ	Coefficient of inbreeding depression; cross breeding effects in outbreeding
μ	Statistical moments (subscripted)

Table 9. List of Symbols Used in Chapter 9; Parameters May Be Subscripted

Symbol	Definition
a	Genotypic value; constant in frequency-dependent model
b	Constant
c	Constant; fraction of population in a particular patch
d	Deviation of the average allele frequency; dominance coefficient
e	Deviation from equilibrium
g	Genetic contribution of a locus
h^2	Heritability in the narrow sense; h_m^2 = mutational heritability
m	Migration rate
n	Number of loci
p	Proportion of an allele; p_H = proportion of homozygous loci
q	$1 - q$

r	Correlation
s	Selection coefficient
t	Generation
x	Trait value
C	Constant; C_L = linkage
H	Expected heterozygosity per locus
K	Number of alleles per locus
M	Mean value
S	Summation [Eq. (9.32b)]; selection coefficient
V	Variance component (subscripts: A = additive, E = environmental, P = phenotypic); fitness in Levene's two-patch model
W	Fitness
X	Trait value
α	Standard deviation of the mutational effect; constant
β	Constant [Eq. (9.30a)]
γ	Strength of stabilizing selection
δ	Generation overlap
θ	Optimal value
μ	Mean or mutation rate

References

Aastveit, A. H. 1990. Use of bootstrapping for estimation of standard deviation and confidence n intervals of genetic variance- and covariance-components. *Biometrics Journal* 32: 515–527.

Agren, J., and D. W. Schemske. 1993. Outcrossing rate and inbreeding depression in two annual monoecious herbs, *Begonia hirsuta* and *B. semiovata*. *Evolution* 47: 125–135.

Ahlschwede, W. T., and O. W. Robison. 1971. Prenatal and postnatal influences on growth and backfat in swine. *Journal of Animal Science* 32: 10–16.

Ahn, S. N., C. N. Bollich, A. M. McClung, and S. D. Tanksley. 1993. RFLP analysis of genomic regions associated with cooked-kernel elongation in rice. *Theoretical and Applied Genetics* 87: 27–32.

Ajmone-Marsan, P., G. Monfredini, W. F. Ludwig, A. E. Melchinger, P. Franceschini, G. Pagnotto, and M. Motto. 1995. In an elite cross of maize a major quantitative trait locus controls one-fourth of the genetic variation for grain yield. *Theoretical and Applied Genetics* 90: 415–424.

Alatalo, R. V., and L. Gustaffsson. 1988. Genetic component of morphological differentiation in coal tits under competitive release. *Evolution* 42: 200–203.

Alatalo, R. V., and A. Lundberg. 1986. Heritability and selection on tarsus length in the pied flycatcher (*Ficedula hypoleuca*). *Evolution* 40: 574–583.

Alatalo, R. V., L. Gustafsson, and A. Lundberg. 1990. Phenotypic selection on heritable size traits: environmental variance and genetic response. *American Naturalist* 135: 464–471.

Alicchio, R., and L. D. Palenzona. 1974. Phenotypic variability and divergence in disruptive selection. *Theoretical and Applied Genetics* 45: 122–125.

Allard, R. W. 1965. Genetic systems associated with colonizing ability in predominately self-pollinated species. pp. 44–75. *In* H. Baker and G. Scribbins (editors). *The Genetics of Colonizing Species*. Academic Press, New York.

Altukhov, Y. P. 1981. The stock concept from the viewpoint of population genetics. Canadian Journal of Fisheries and Aquatic Sciences 38: 1523–1538.

Alvarez, G., C. Zapata, R. Amaro, and A. Guerra. 1989. Multilocus heterozygosity at protein loci and fitness in the European oyster, *Ostrea edulis L. Heredity* 63: 359–372.

Anderson, J. L. 1975. Phenotypic correlations among relatives and variability in reproductive performance in populations of the vole *Microtus townsendii*. Ph.D. Thesis, University of British Columbia, Vancouver, Canada.

Anderson, W. W. 1968. Further evidence for coadaptation in crosses between geographic populations of *Drosophila pseudoobscura*. *Genetic Research* 12: 317–330.

Anderson, W. W. 1969. Polymorphism resulting from the mating advantage of rare male genotypes. *Proceedings of the National Academy of Sciences* 64: 190–197.

Andersson, S., and B. Widen. 1993. Reaction norm variation in a rare plant, *Senecio integrifolius* (Asteraceae). *Heredity* 73: 598–607.

Angus, R. A., and R. J. Schultz. 1983. Meristic variation in homozygous and heterozygous fish. *Copeia* 1983: 287–299.

Antolin, M. F. 1992a. Sex ratio variation in a parasitic wasp I. Reaction norms. *Evolution* 46: 1496–1510.

Antolin, M. F. 1992b. Sex ratio variation in a parasitic wasp II. Diallel cross. *Evolution* 46: 1511–1524.

Antonovics, J. 1968. Evolution in closely adjacent plant populations. V. Evolution of self-fertility. *Heredity* 23: 219–238.

Antonovics, J., and J. Schmitt. 1986. Paternal and maternal effects on propagule size in *Anthoxanthum odoratum*. *Oecologia* 69: 277–282.

Armstrong, M. J., and P. A. Shelton. 1990. Clupeoid life-history styles in variable environments. *Environmental Biology of Fishes* 28: 77–85.

Arnold, E. N., and J. A. Burton. 1978. *A Field Guide to the Reptiles and Amphibians of Britain and Europe*. Collins, London.

Arnold, M. L., and S. A. Hodges. 1995. Are natural hybrids fit or unfit relative to their parents? *Trends in Ecology and Evolution* 10: 67–71.

Arnold, S. J. 1981. Behavioral variation in natural populations. I. Phenotypic, genetic and environmental correlations between chemoreceptive responses to prey in the garter snake, *Thamnophis elegans*. *Evolution* 35: 489–509.

Arnold, S. J. 1988. Quantitative genetics and selection in natural populations: microevolution of vertebral numbers in the garter snake *Thamnophis elegans*. pp. 619–636. *In* B. S. Weir, E. J. Eisen, M. M. Goodman, and G. Namkoong (editors), *Proceedings of the Second International Conference on Quantitative Genetics*. Sinauer Associates, Sunderland, MA.

Arnold, S. J. 1992. Constraints on phenotypic evolution. *American Naturalist* 140(Suppl.): s85–s107.

Arthur, P. F., M. F. Liu, and M. Makarechian. 1994. Estimates of direct and maternal heritabilities for growth and lifetime production traits in beef cows. p. 228. *In* C. Smith, J. S. Gavora, B. Benkel, J. Chesnais, W. Fairfull, J. P. Gibson, B. W. Kennedy, and E. B. Burnside (editors), *Proceedings of the 5th World Congress on Genetics Applied to Livestock Production* Dept. Animal and Poultry Science, University of Guelph, Guelph, Ontario, Canada.

Arvesen, J. N., and T. H. Schmitz. 1970. Robust procedures for variance component problems using the jackknife. *Biometrics* 26:677–686.

Aspi, J., and A. Hoikkala. 1993. Laboratory and natural heritabilities of male courtship song characters in *Drosophila montana* and *D. littoralis*. *Heredity* 70: 400–406.

Atchley, W. R., and J. J. Rutledge. 1980. Genetic components of size and shape. I. Dynamics of components of phenotypic variability and covariability during ontogeny in the laboratory rat. *Evolution* 34: 1161–1173.

Aulstad, D., and A. Kittelsen. 1971. Abnormal body curvatures of rainbow trout (*Salmo gairdneri*) inbred fry. *Journal of the Fisheries Research Board of Canada* 28: 1918–1920.

Avery, P. J., and W. G. Hill. 1977. Variability in genetic parameters among small populations. *Genetic Research* 29: 193–213.

Ayala, J., and C. A. Campbell. 1974. Frequency dependent selection. *Annual Review of Ecology and Systematics* 5: 115–138.

Backes, G., A. Graner, B. Foroughi-Wehr, G. Fischbeck, G. Wenzel, and A. Jahoor. 1995. Localization of quantitative trait loci (QTL) for agronomic important characters by the use of a RFLP map in barley (*Hordeum vulgare* L.). *Theoretical and Applied Genetics* 90: 294–302.

Bader, R. S. 1965. Fluctuating asymmetry in the dentition of the house mouse. *Growth* 29: 291–300.

Baker, M. C. 1981. Effective population size in a songbird: some possible implications. *Heredity* 46: 209–218.

Baker, M. C., and S. F. Fox. 1978. Dominance, survival, and enzyme polymorphism in dark-eyed juncos, *Junco hyemalis*. *Evolution* 32: 697–711.

Baker, R. L., E. H. Cox, and A. H. Carter. 1984. Direct and correlated responses to selection for weaning weight, post-weaning weight gain and six-week weight in mice. *Theoretical and Applied Genetics* 67: 113–122.

Baker, B. S., M. Gorman, and I. Marin. 1994. Dosage compensation in *Drosophila*. *Annual Review of Genetics* 28: 491–521.

Bakker, T. C. M., and A. Pomiankowski. 1995. The genetic basis of female mate preferences. *Journal of Evolutionary Biology* 8: 129–171.

Bamshad, M., M. H. Crawford, D. O'Rourke, and L. B. Jorde. 1994. Biochemical heterozygosity and morphological variation in a colony of *Papio hamadryas hamadryas* baboons. *Evolution* 48: 1211–1221.

Baptist, R., and A. Robertson. 1976. Asymmetrical responses to automatic selection for body size in *Drosophila melanogaster*. *Theoretical and Applied Genetics* 47: 209–213.

Barbato, G. F. 1991. Genetic architecture of growth curve parameters in chickens. Theoretical and Applied Genetics 83: 24–32.

Barbato, G. F., and R. Vasilatos-Younken. 1991. Sex-linked and maternal effects on growth in chickens. *Poultry Science* 70: 709–718.

Barker, J. S. F. 1979. Inter-locus interactions: a review of experimental evidence. *Theoretical Population Biology* 16: 323–346.

Barker, J. S. F., and L. J. Cummins. 1969. Disruptive selection for sternopleural bristle number in *Drosophila melanogaster*. *Genetics* 61: 697–712.

Barnes, P. T. 1984. A maternal effect influencing larval viability in *Drosophila melanogaster. Journal of Heredity* 75: 288–292.

Barrett, S. C. H., and D. Charlesworth. 1991. Effects of a change in the level of inbreeding on the genetic load. *Nature* 352: 522–524.

Barrett, S. C. H., and B. C. Husband. 1990. The genetics of plant migration and colonization, pp. 254–277. *In* A. H. D. Brown, M. T. Clegg, A. L. Kahler, and B. S. Weir (editors). *Plant Population Genetics, Breeding, and Genetic Resources.* Sinauer Associates Inc., Sunderland, MA.

Barrowclough, G. F. 1980. Gene flow, effective population sizes, and genetic variance components in birds. *Evolution* 34: 789–798.

Barrowclough, G. F. 1983. Biochemical studies of microevolutionary processes. pp. 223–262. *In* A. H. Bush and G. A. Clark (editors). *Perspective in Ornithology.* Cambridge University Press, New York.

Barrowclough, G. F., and S. L. Coates. 1985. The demography and population genetics of owls, with special reference to the conservation of the spotted owl (*Strix occidentalis*). pp. 74–85. *In* R. J. Guttierez and A. B. Carey (editors). *Ecology and Management of the Spotted Owl in the Pacific Northwest.* U.S. Forest Service, Portland, Oregon.

Barton, N. H. 1986. The maintenance of polygenic variation through a balance between mutation and stabilizing selection. *Genetic Research* 47: 209–216.

Barton, N. H. 1990. Pleiotropic models of quantitative variation. *Genetics* 124: 773–782.

Barton, N. H., and M. Turelli. 1989. Evolutionary quantitative genetics: how little do we know? *Annual Review of Genetics* 23: 337–370.

Barton, N. H., and M. Turelli. 1991. Natural and sexual selection on many loci. *Genetics* 127: 229–255.

Baskin, J. M., and C. C. Baskin. 1973. Plant population differences in dormancy and germination characteristics of seeds: heredity or environment? *American Midland Naturalist* 90: 493–498.

Beacham, T. D., and R. E. Withler. 1985. Heterozygosity and morphological variability of chum salmon (*Oncorhynchus keta*) in southern British Columbia. *Heredity* 54: 313–322.

Beardmore, J. A., and S. A. Shami. 1979. Heterozygosity and the optimum phenotype under stabilising selection. *Aquilo Ser Zoologica* 20: 100–110.

Beavis, W. D. 1994. The power and deceit of QTL experiments: lessons from comparative QTL studies. *Proceedings of the Forty-Ninth Annual Corn and Sorghum Industry Research Conference* 1994: 250–266.

Beavis, W. D., D. Grant, M. Albertsen, and R. Fincher. 1991. Quantitative trait loci for plant height in four maize populations and their associations with qualitative genetic loci. *Theoretical and Applied Genetics* 83: 141–145.

Becker, W. A. 1985. *Manual of Quantitative Genetics.* McNaughton and Gunn Inc., Ann Arbor, MI.

Becker, W. A., S. P. Sinha, and T. P. Bogyo. 1964. The quantitative genetic relationship of sexual dimorphism of birds (abstract). *Genetics* 50: 235.

Begon, M., C. B. Krimbas, and M. Loukas. 1980. The genetics of *Drosophila subobscura* populations XV. Effective size of a natural population estimated by three independent methods. *Heredity* 45: 335–350.

Bell, A. E., and M. J. Burris. 1973. Simultaneous selection for two correlated traits in *Tribolium. Genetic Research* 21: 24–46.

Berger, E. 1976. Heterosis and the maintenance of enzyme polymorphism. *American Naturalist* 110: 823–839.

Berthold, P. 1973. Relationships between migratory restlessness and migration distance in six *Sylvia* species. *Ibis* 115: 594–599.

Berthold, P. 1988. Evolutionary aspects of migratory behavior in European warblers. *Journal of Evolutionary Biology* 1: 195–209.

Berthold, P. 1995. Microevolution of migratory behaviour illustrated by the blackcap *Sylvia atricapilla*: 1993 Witherby Lecture. *Bird Study* 42: 89–100.

Berthold, P., and F. Pulido. 1994. Heritability of migratory activity in a natural bird population. *Proceedings of the Royal Society of London* B 257: 311–315.

Berthold, P., and U. Querner. 1981. Genetic basis of migratory behavior in European warblers. *Science* 212: 77–79.

Berthold, P., A. J. Helbig, G. Mohr, and U. Querner. 1992. Rapid microevolution of migratory behaviour in a wild bird species. *Nature* 360: 668–669.

Berven, K. A., and T. A. Grudzien. 1990. Dispersal in the wood frog (*Rana sylvatica*): implications for genetic population structure. *Evolution* 44: 2047–2056.

Bevington, J. 1986. Geographic differences in the seed germination of paper birch (*Betual papyrifera*). *American Journal of Botany* 73: 564–573.

Biebach, H. 1983. Genetic determination of partial migration in the European robin (*Erithacus rubecula*). *Auk* 100: 601–606.

Bijlsma, R., N. J. Ouberg, and R. van Treuren. 1994. On genetic erosion and population extinction in plants: a case study in *Scabiosa columbaria* and *Salvia pratensis*. pp. 255–271. *In* V. Loeschcke, J. Tomiuk, and S. K. Jain (editors). *Conservation Genetics*. Birkauser-Verlag, Basel.

Birch, L. C., Th. Dobzhansky, P. O. Elliott, and R. C. Lewontin. 1963. Relative fitness of geographic races of *Drosophila serrata*. *Evolution* 17: 72–83.

Birchler, J., and J. Hart. 1987. Interaction of endosperm size factors in maize. *Genetics* 117: 309–317.

Blakley, N., and S. R. Goodner. 1978. Size-dependent timing of metamorphosis in milkweed bugs (*Oncopeltus*) and its life history implications. *Biological Bulletin* 155: 499–510.

Blanckenhorn, W. U., and D. J. Fairbairn. 1995. Life history adaptation along a latitudinal cline in the water strider *Aquarius remigis* (Heteroptera: Gerridae). *Journal of Evolutionary Biology* 8: 21–41.

Boake, C. R. B. (editor). 1994. *Quantitative Genetic Studies of Behavioral Evolution*. University of Chicago Press, Chicago.

Bohidar, N. R. 1964. Derivation and estimation of variance and covariance components associated with covariance between relatives under sexlinked transmission. *Biometrics* 20: 505–521.

Bohren, B. B., and H. E. McKean. 1961. Relative efficiencies of heritability estimates based on regression of offspring on parent. *Biometrics* 17: 481–491.

Bohren, B. B., W. G. Hill, and A. Robertson. 1966. Some observations on asymmetrical correlated responses to selection. *Genetic Research* 7: 44–57.

Bondari, K. 1983. Response to bidirectional selection for body weight in channel catfish. *Aquaculture* 33: 73–81.

Bondari, K., and R. A. Dunham. 1987. Effects of inbreeding on economic traits of channel catfish. *Theoretical and Applied Genetics* 74: 1–9.

Bondari, K., R. Willham, and A. Freeman. 1978. Estimates of direct and maternal genetic correlations for pupa weight and family size in *Tribolium*. *Journal of Animal Science* 47: 358–365.

Boonstra, R., and P. T. Boag. 1987. A test of the Chitty hypothesis: inheritance of life history traits in meadow voles *Microtus pennsylvanicus*. *Evolution* 4: 929–947.

Booth, C. L., D. S. Woodruff, and S. J. Gould. 1990. Lack of significant associations between allozyme heterozygosity and phenotypic traits in the land snail *Cerion*. *Evolution* 44: 210–213.

Borlase, S. C., D. A. Loebel, R. Frankham, R. K. Nurthen, D. A. Briscoe, and G. E. Daggard. 1993. Modeling problems in conservation genetics using captive *Drosophila* populations: consequences of equalization of family sizes. *Conservation Biology* 7: 122–131.

Bos, M., and W. Scharloo. 1973a. The effects of disruptive and stabilizing selection on body size in *Drosophila melanogaster*. I. Mean values and variances. *Genetics* 75: 679–693.

Bos, M., and W. Scharloo. 1973b. The effects of disruptive and stabilizing selection on body size in *Drosophila melanogaster*. II. Analysis of responses in the thorax selection lines. *Genetics* 75: 695–708.

Bos, M., and W. Scharloo. 1974. The effects of disruptive and stabilizing selection on body size in *Drosophila melanogaster*. III. Genetic analysis of the two lines with different reactions to disruptive selection with mating of opposite extremes. *Genetics* 45: 71–90.

Bowman, J. C., and D. S. Falconer. 1960. Inbreeding depression and heterosis of litter size in mice. *Genetic Research* 1: 262–274.

Bradford, M. J., and D. A. Roff. 1995. Genetic and phenotypic sources of life history variation along a cline in voltinism in the cricket *Allonemobius socius*. *Oecologia* 103: 319–326.

Bradford, M. J., and D. A. Roff. 1997. Seasonality, environmental uncertainty and insect dormancy: an empirical model of diapause strategies in the cricket *Allonemobius socius*. *Ecology* (in press).

Bradshaw, A. D. 1965. Evolutionary significance of phenotypic plasticity in plants. *Advances in Genetics* 13: 115–155.

Bradshaw, H. D. Jr., and R. F. Stettler. 1995. Molecular genetics of growth and development in populus. IV. Mapping QTLs with large effects on growth, form and phenology traits in a forest tree. *Genetics* 139: 963–973.

Bradshaw, W. E., C. Slatore, and C. M. Holzapfel. 1997. Size and development time: their genetic correlation and the acquisition and allocation of resources contributing to them in the pitcher-plant mosquito, *Wyeomyia smithii. Genetics* (in press).

Brakefield, P. M., and I. J. Saccheri. 1994. Guidelines in conservation genetics and the use of the population cage experiments with butterflies to investigate the effects of genetic drift and inbreeding. pp 165–179. *In* V. Loeschcke, J. Tomink and S. K. Jain (editors). *Conservation Genetics.* Birkauser Verlag Basel, Switzerland.

Breden, F. 1987. The effect of post-metamorphic dispersal on the population genetic structure of Fowler's toad, *Bufo woodhousei fowleri. Copeia* 1987: 386–395.

Breden, F., and K. Hornaday. 1994. Test of indirect models of selection in the Trinidad guppy. *Heredity* 73: 291–297.

Bresler, J. B. 1970. Outcrossings in Caucasians and fetal loss. *Social Biology* 17: 17–25.

Brewer, B. A., R. C. Lacy, M. L. Foster, and G. Alaks. 1990. Inbreeding depression in insular and central populations of *Peromyscus* mice. *Journal of Heredity* 81: 257–266.

Briscoe, D. A., J. M. Malpica, A. Robertson, G. J. Smith, R. Frankham, R. G. Banks, and J. S. F. Barker. 1992. Rapid loss of genetic variation in large captive populations of *Drosophila* flies: implications for the genetic management of captive populations. *Conservation Biology* 6: 416–425.

Brodie, E. D. III. 1993. Homogeneity of the genetic variance–covariance matrix for antipredator traits in two natural populations of the garter snake *Thamnophis ordinoides. Evolution* 47: 844–854.

Brown, S. W., and U. Nur. 1964. Heterochromatic chromosomes in the coccids. *Science* 145: 130–136.

Bruckner, D. 1976. The influence of genetic variability on wing symmetry in honeybees (*Apis mellifera*). *Evolution* 30: 100–108.

Brumby, P. J. 1960. The influence of the maternal environment on growth in mice. *Heredity* 14: 1–18.

Brumpton, R. J., and J. L. Jinks. 1977. Joint selection for both extremes of mean performance and of sensitivity to a macroenvironmental variable. I. Family selection. *Heredity* 38: 219–226.

Bryant, E. H. 1977. Morphometric adaptation of the housefly, *Musca domestica* L., in the United States. *Evolution* 31: 580–596.

Bryant, E. H., and L. M. Meffert. 1991. The effects of bottlenecks on genetic variation, fitness, and quantitative traits in the housefly. pp. 591–601. *In* E. C. Dudley (ed.). *The Unity of Evolutionary Biology, Vol. II*, Dioscorides Press, Portland, OR.

Bryant, E. H., S. A. McCommas, and L. M. Combs. 1986. The effect of an experimental bottleneck upon quantitative genetic variation in the housefly. *Genetics* 114: 1191–1211.

Bull, J. J. 1983. *Evolution of Sex Determining Mechanisms.* Benjamin/Cummings Publ. Co. Inc., Menlo Park, CA.

Bull, J. J., R. C. Vogt, and M. G. Bulmer. 1982. Heritability of sex ratio in turtles with environmental determination. *Evolution* 36: 333–341.

Bulmer, M. G. 1971a. The effect of selection on genetic variability. *American Naturalist* 105: 201–211.

Bulmer, M. G. 1971b. The stability of equilibria under selection. *Heredity* 27: 157–162.

Bulmer, M. G. 1971c. Stable equilibria under the two island model. *Heredity* 27: 321–330.

Bulmer, M. G. 1973. The maintenance of the genetic variability of polygenic characters by heterozygous advantage. *Genetic Research* 22: 9–12.

Bulmer, M. G. 1974. Density-dependent selection and character displacement. *The American Naturalist* 108: 45–58.

Bulmer, M. G. 1976. The effect of selection on genetic variability: a simulation study. *Genetic Research* 28: 101–117.

Bulmer, M. G. 1985. *The Mathematical Theory of Quantitative Genetics*. Claredon Press, Oxford.

Bulmer, M. G. 1989. Maintenance of genetic variability by mutation–selection balance: a child's guide through the jungle. *Genome* 31: 761–767.

Bumpus, H. C. 1899. The elimination of the unfit as illustrated by the introduced sparrow, *Passer domesticus*. Biological Lectures, Marine Biology Laboratory, Woods Hole, MA, pp. 209–226.

Burfening, P. J., D. D. Kress, and R. L. Friedrich. 1981. Calving ease and growth rate of Simmental-sired calves. III. Direct and maternal effects. *Journal of Animal Science* 53: 1210–1216.

Burger, R., and R. Lande. 1994. On the distribution of the mean and variance of a quantitative trait under mutation–selection–drift balance. *Genetics* 138: 901–912.

Burger, R., G. P. Wagner, and F. Stettinger. 1989. How much heritable variation can be maintained in finite populations by mutation–selection balance? *Evolution* 43: 1748–1766.

Buri, P. 1956. Gene frequency in small populations of mutant *Drosophila*. *Evolution* 10: 367–402.

Busack, C. A. 1983. Four generations of selection for high 56-day weight in the mosquitofish (*Gambusia affinis*). *Aquaculture* 33: 83–87.

Busack, C., and G. Gall. 1983. An initial description of the quantitative genetics of growth and reproduction in the mosquito fish. *Aquaculture* 32: 123–140.

Bush, R. M., P. E. Smouse, and F. T. Ledig. 1987. The fitness consequences of multiple-locus heterozygosity: the relationship between heterozygosity and growth rate in pitch pine (*Pinus rigida* Mill.). *Evolution* 41: 787–798.

Caballero, A. 1994. Developments in the prediction of effective population size. *Heredity* 73: 657–679.

Caballero, A. 1995. On the effective size of populations with separate sexes, with particular reference to sex-linked genes. *Genetics* 139: 1007–1011.

Caballero, A., and W. G. Hill. 1992. A note on the inbreeding effective population size. *Evolution* 46: 1969–1972.

Caballero, A., M. A. Toro, and C. López-Fanjul. 1991. The response to artificial selection from new mutations in *Drosophila melanogaster*. *Genetics* 127: 89–102.

Cacoyianni, Z., I. V. Kovacs, and A. A. Hoffmann. 1995. Laboratory adaptation and inbreeding in *Helicoverpa punctigera* (Lepidoptera: Noctuidae). *Australian Journal of Zoology* 43: 83–90.

Cadieu, N. 1983. Maternal influence and the effect of heterosis on the viability of *Drosophila melanogaster* during different pre-imaginal stages. *Canadian Journal of Zoology* 61: 1152–1155.

Caligari, P. D. S., and K. Mather. 1980. Dominance allele frequency and selection in a population of *D. melanogaster*. *Proceedings of the Royal Society of London* B 208: 163–187.

Callan, H. G., and H. Spurway. 1951. A study of meiosis in interracial hybrids of the newt, *Triturus cristatus*. *Journal of Genetics* 50: 235–249.

Campton, D. E., and G. A. E. Gall. 1988. Responses to selection for body size and age at sexual maturity in the mosquitofish, *Gambusia affinis*. *Aquaculture* 68: 221–241.

Carbonell, E. A., T. M. Gerig, E. Balansard, and M. J. Asins. 1992. Interval mapping in the analysis of nonadditive quantitative trait loci. *Biometrics* 48: 305–315.

Carbonell, E. A., M. J. Asins, M. Baselga, E. Balansard, and T. M. Gerig. 1993. Power studies in the estimation of genetic parameters and the localization of quantitative trait loci for backcross and doubled haploid populations. *Theoretical and Applied Genetics* 86: 411–416.

Carr, D. E., and C. B. Fenster. 1994. Levels of genetic variation and covariation for *Mimulus* (Scophulariaceae) floral traits. *Heredity* 72: 606–618.

Carrière, Y. 1994. Evolution of phenotypic variance: non-Mendelian parental influences on phenotypic and genetic components of life-history traits in a generalist herbivore. *Heredity* 72: 420–430.

Carrière, Y., and D. A. Roff. 1995. Change in genetic architecture resulting from the evolution of insecticide resistance: a theoretical and empirical analysis. *Heredity* 75: 618–629.

Carrière, Y., and B. D. Roitberg. 1994. Trade-offs in responses to host plants within a population of a generalist herbivore, *Choristoneura rosaceana*. *Entomologica Experimentalis Applicata* 72: 173–180.

Carrière, Y., and B. D. Roitberg. 1995a. Evolution of host-selection behaviour in insect herbivores: genetic variation and covariation in host acceptance within and between populations of the obliquebanded leafroller, *Choristoneura rosaceana* (Family: Tortricidae). *Heredity* 74: 357–368.

Carrière, Y., J.-P. Deland, D. A. Roff, and C. Vincent. 1994. Life-history costs associated with the evolution of insecticide resistance. *Proceedings of the Royal Society of London* B 258: 35–40.

Carroll, S. P., and P. S. Corneli. 1995. Divergence in male mating tactics between two populations of the soapberry bug. II. Genetic change and the evolution of a plastic reaction norm in a variable social environment. *Behavioral Ecology* 6: 46–56.

Castle, W. E. 1921. An improved method of estimating the number of genetic factors concerned in cases of blending inheritance. *Science* 54: 223.

Ch'ang, T. S., and A. L. Rae. 1972. The genetic basis of growth, reproduction, and maternal environment in Romney ewes. II. Genetic covariation between hogget characters, fertility, and maternal environment of the ewe. *Australian Journal of Agricultural Research* 23: 149–165.

Chai, C. K. 1956. Analysis of quantitative inheritance of body size in mice. II. Gene action and segregation. *Genetics* 41: 165–178.

Chakraborty, R., and J. V. Neel. 1989. Description and validation of a method for simultaneous estimation of effective population size and mutation rate from human population data. *Proceedings of the National Academy of Sciences* 86: 9407–9411.

Chakraborty, R., and N. Ryman. 1983. Relationship of mean and variance of genotypic values with heterozygosity per individual in a natural population. *Genetics* 103: 149–152.

Chandra, H. S., and S. W. Brown. 1975. Chromosome imprinting and the mammalian X chromosome. *Nature* 253: 165–168.

Chandraratna, M. F., and K. Sakai. 1960. A biometrical analysis of matriclinous inheritance of grain weight in rice. *Heredity* 14: 365–373.

Charlesworth, B., D. Charlesworth, and M. T. Morgan. 1990a. Genetic loads and estimates of mutation rates in highly inbred plant populations. *Nature* 347: 380–382.

Charlesworth, B., M. T. Morgan, and D. Charlesworth. 1991. Multilocus models of inbreeding depression with synergistic selection and partial self-fertilization. *Genetic Research* 57: 177–194.

Charlesworth, D. 1991. The apparent selection on neutral marker loci in partially inbreeding populations. *Genetic Research* 57: 159–175.

Charlesworth, D., and B. Charlesworth. 1987. Inbreeding depression and its evolutionary consequences. *Annual Review of Ecology and Systematics* 18: 237–268.

Charlesworth, D., and B. Charlesworth. 1990. Inbreeding depression with heterozygote advantage and its effect on selection for modifiers changing the outcrossing rate. *Evolution* 44: 870–888.

Charlesworth, D., E. E. Lyons, and L. B. Litchfield. 1994. Inbreeding depression in two highly inbreeding populations of *Leavenworthia*. *Proceedings of the Royal Society of London* B 258: 209–214.

Charlesworth, D., M. T. Morgan, and B. Charlesworth. 1990b. Inbreeding depression, genetic load, and the evolution of outcrossing rates in a multilocus system with no linkage. *Evolution* 44: 1469–1489.

Chen, X. 1993. Comparison of inbreeding and outbreeding in hermaphroditic *Arianta arbustorum* (L.) (land snail). *Heredity* 71: 456–461.

Chepko-Sade, B. D., W. M. Shields, J. Berger, Z. T. Halpin, W. T. Jones, L. L. Rogers, J. P. Rood, and A. T. Smith. 1987. The effects of dispersal and social structure on effective population size. p. 287–321. *In* B. D. Chepko-Sade and Z. T. Halpin (editors). *Mammalian Dispersal Pattern.* The University of Chicago Press, Chicago.

Chesser, R. K. Jr., O. E. Rhodes, D. W. Sugg, and A. Schnabel. 1993. Effective sizes for subdivided populations. *Genetics* 135: 1221–1232.

Cheung, T. K., and R. J. Parker. 1974. Effect of selection on heritability and genetic correlation of two quantitative traits in mice. *Canadian Journal of Genetics and Cytology* 16: 599–609.

Chevalet, C. 1988. Control of genetic drift in selected populations. pp. 379–394. *In* E. J. Eisen, M. M. Goodman, G. Namkoong, and B. S. Weir (editors). *Proceedings of the Second International Conference on Quantitative Genetics.* Sinauer Associates, Sunderland, MA.

Cheverud, J. M. 1984. Evolution by kin selection: a quantitative genetic model illustrated by maternal performance in mice. *Evolution* 38: 766–777.

Cheverud, J. M. 1988. A comparison of genetic and phenotypic correlations. *Evolution* 42: 958–968.

Cheverud, J. M. 1995. Morphological integration in the saddle-back tamarin (*Saguinus fuscicollis*) cranium. *American Naturalist* 145: 63–89.

Cheverud, J. M. 1996. Quantitative genetic analysis of cranial morphology in the cotton-top (*Saguinus oedipus*) and saddle-back (*S. fuscicollis*) tamarins. *Journal of Evolutionary Biology* 9: 5–42.

Cheverud, J. M., and A. J. Moore. 1994. Quantitative genetics and the role of the environment provided by relatives in behavioral evolution. pp. 67–100. *In* C. R. B. Boake (editor). *Quantitative Genetic Studies of Behavioral Evolution.* University of Chicago Press, Chicago.

Cheverud, J. M., and E. J. Routman. 1995. Epistasis and its contribution to genetic variance components. *Genetics* 139: 1455–1461.

Cheverud, J. M., J. J. Rutledge, and W. R. Atchley. 1983. Quantitative genetics of development: genetic correlations among age-specific trait values and the evolution of ontogeny. *Evolution* 37: 895–905.

Cheverud, J. M., M. M. Dow, and W. Leutenegger. 1985. The quantitative assessment of phylogenetic constraints in comparative analysis: sexual dimorphism in body weight among primates. *Evolution* 39: 1335–1341.

Cheverud, J., E. Routman, C. Jaquish, S. Tardiff, G. Peterson, N. Belfiore, and L. Foreman. 1994. Quantitative and molecular genetic variation in captive cotton-top tamarins (*Saguinus oedipus*). *Conservation Biology* 8: 95–105.

Cheverud, J. M., E. J. Routman, F. A. M. Duarte, B. van Swinderen, K. Cothran, and C. Perel. 1996. Quantitative trait loci for murine growth. *Genetics* 142: 1305–1319.

Chia, A. B., and E. Pollack. 1974. The inbreeding effective number and the effective number of alleles in a population that varies in size. *Theoretical Population Biology* 6: 149–172.

Chitty, D. 1967. The natural selection of self-regulatory behaviour in animal populations. *Proceedings of the Ecological Society of Australia* 2: 51–78.

Chitty, D. 1996. Do lemmings commit suicide? Beautiful hypotheses and ugly facts. Oxford University Press, Oxford.

Choo, J. 1975. Genetic studies on walking behavior in *Drosophila melanogaster*. I. Selection and hybridization analysis. *Canadian Journal of Genetics and Cytology* 17: 535–542.

Clarke, B. 1964. Frequency-dependent selection for the dominance of rare polymorphic genes. *Evolution* 18: 364–369.

Clarke, B. 1969. The evidence for apostatic selection. *Heredity* 24: 347–352.

Clarke, B. 1974. Frequency-dependent selection. *Heredity* 19: 201–206.

Clarke, B., and P. O'Donald. 1964. Frequency-dependent selection. *Heredity* 19: 201–206.

Clarke, B. C. 1979. The evolution of genetic diversity. *Proceedings of the Royal Society of London* B 205: 453–474.

Clarke, G. M. 1993. The genetic basis of developmental stability. I. Relationships between stability heterozygosity and genomic coadaptation. *Genetica* 89: 15–23.

Clarke, G. M., B. P. Oldroyd, and P. Hunt. 1992. The genetic basis of developmental stability in *Apis mellifera*: heterozygosity versus genic balance. *Evolution* 46: 753–762.

Clarke, J. M., J. Maynard Smith, and K. C. Sondhi. 1961. Asymmetrical response to selection for rate of development in *Drosophila subobscura*. *Genetic Research* 2: 70–81.

Clauss, M. J., and L. W. Aarssen. 1994. Phenotypic plasticity of size-fecundity relationships in *Arabidopsi thaliana*. *Journal of Ecology* 82: 447–455.

Clayton, G., and A. Robertson. 1955. Mutation and quantitative variation. *American Naturalist* 89: 151–158.

Clayton, G. A., and A. Robertson. 1957. An experimental check on quantitative genetical theory: II. The long-term effects of selection. *Journal of Genetics* 55: 152–170.

Clayton, G. A., G. R. Knight, J. A. Morris, and A. Robertson. 1957a. An experimental check on quantitative genetical theory. I. Short-term response to selection. *Journal of Genetics* 55: 131–151.

Clayton, G. A., G. R. Knight, J. A. Morris, and A. Robertson. 1957b. An experimental check on quantitative genetical theory III. Correlated responses. *Journal of Genetics* 55: 171–180.

Clegg, M. T., and R. W. Allard. 1973. Viability versus fecundity selection in the slender wild oat, *Avena barbata* L. *Science* 181: 667–668.

Cockerham, C. C. 1963. Estimation of genetic variances. pp. 53–93. *In* W. D. Hanson and H. F. Robinson (editors). *Statistical Genetics and Plant Breeding*. National Academy of Science Research Council, Washington, DC.

Cockerham, C. C., and B. S. Weir. 1977. Quadratic analyses of reciprocal crosses. *Biometrics* 33: 187–203.

Cockerham, C. C., P. M. Burrows, S. S. Young, and T. Prout. 1972. Frequency-dependent selection in random-mating populations. *American Naturalist* 106: 493–515.

Comstock, R. E., and F. D. Enfield. 1981. Gene number estimation when multiplicative genetic effects are assumed—growth in flour beetles and mice. *Theoretical and Applied Genetics* 59: 373–79.

Comstock, R. E., and H. F. Robinson. 1952. Estimation of average dominance of genes. pp. 494–516. *In* J. W. Gowen (editor). *Heterosis.* Iowa State College Press, Ames.

Connolly, V., and J. L. Jinks. 1975. The genetical architecture of general and specific environmental sensitivity. *Heredity* 35: 249–259.

Connor, J. L., and M. J. Bellucci. 1979. Natural selection resisting inbreeding depression in captive wild housemice (*Mus musculus*). *Evolution* 33: 929–940.

Conover, D. O., D. A. Van Voorhees, and A. Ehtisham. 1992. Sex ratio selection and the evolution of environmental sex determination in laboratory populations of *Menidia menidia.* *Evolution* 46: 1722–1730.

Corey, L. A., D. F. Matzinger, and C. C. Cockerham. 1976. Maternal and reciprocal effects on seedling characters in *Arabidopsis thaliana* (L.) Heynh. *Genetics* 82: 677–683.

Cothran, E. G., R. K. Chesser, M. H. Smith, and P. E. Johns. 1983. Influences of genetic variability and maternal factors on fetal growth in white-tailed deer. *Evolution* 37: 282–291.

Cowley, D. E., and W. R. Atchley. 1988. Quantitative genetics of *Drosophila melanogaster.* II. Heritabilities and genetic correlations between sexes for head and thorax traits. *Genetics* 119: 421–433.

Cowley, D. E., and W. R. Atchley. 1990. Development and quantitative genetics of correlation structure among body parts of *Drosophila melanogaster. American Naturalist* 135: 242–268.

Cowley, D. E., and W. R. Atchley. 1992. Comparison of quantitative genetic parameters. *Evolution* 46: 1965–1967.

Cowley, D. E., W. R. Atchley, and J. J. Rutledge. 1986. Quantitative genetics of *Drosophila melanogaster.* I. Sexual dimorphism in genetic parameters for wing traits. *Genetics* 114: 549–566.

Coyne, J. A., and E. Beecham. 1987. Heritability of two morphological characters within and among natural populations of *Drosophila melanogaster. Genetics* 117: 727–737.

Coyne, J. A., and H. A. Orr. 1989. Two rules of speciation. pp. 180–207. *In* D. Otte and J. A. Endler (editors). *Speciation and its Consequences.* Sinauer Associates Inc., Sunderland, MA.

Crespi, B. J. 1986. Territoriality and fighting in a colonial thrips, *Hoplothrips pedicularius,* and sexual dimorphism in Thysanoptera. *Ecological Entomology* 11: 119–130.

Crnokrak, P., and D. A. Roff. 1995. Dominance variance: associations with selection and fitness. *Heredity* 75: 530–540.

Crouse, H. V. 1960. The controlling element in sex chromosome behavior in *Sciara. Genetics* 45: 1429–1443.

Crow, J. F. 1993. Mutation, mean fitness, and genetic load. pp. 3–42. *In* D. Futuyma, and J. Antonovics (editors). *Oxford Surveys in Evolutionary Biology.* Oxford University Press, Oxford.

Crow, J. F., and C. Denniston. 1988. Inbreeding and effective population numbers. *Evolution* 42: 482–495.

Crow, J. F., and M. Kimura. 1970. *An Introduction to Population Genetics Theory.* Harper and Row, New York.

Crow, J. F., and N. E. Morton. 1955. Measurement of gene frequency drift in small populations. *Evolution* 9: 202–214.

Crusio, W. E., J. M. L. Kerbusch, and J. H. F. van Abeelen. 1984. The replicated diallel cross: A generalized method. *Behavior Genetics* 14: 81–104.

Curnow, R. N. 1964. The effect of continued selection of phenotypic intermediates on gene frequency. *Genetic Research* 5: 341–353.

Curtsinger, J. W. 1976a. Stabilizing selection in *Drosophila melanogaster*. *Journal of Heredity* 67: 59–60.

Curtsinger, J. W. 1976b. Stabilizing or directional selection on egg lengths?: A rejoinder. *Journal of Heredity* 67: 246–247.

Curtsinger, J. W., P. M. Service, and T. Prout. 1994. Antagonistic pleiotropy reversal of dominance and genetic polymorphism. *American Naturalist* 144: 210–228.

Danzmann, R. G., F. W. Allendorf, M. M. Ferguson, and K. L. Knudsen. 1986. Heterozygosity and developmental rate in a strain of rainbow trout (*Salmo gairdneri*). *Evolution* 40: 86–93.

Danzmann, R. G., M. M. Ferguson, and F. W. Allendorf. 1988. Heterozygosity and components of fitness in a strain of rainbow trout. *Biological Journal of the Linnean Society* 33: 285–304.

Darwin, C. 1868. *The Variation of Animals and Plants Under Domestication*. Vol. 1. Appleton, New York.

Darwin, C. 1876. *The Effects of Crossing and Self Fertilization in the Vegetable Kingdom*. John Murray, London.

Davenport, C. B. 1908. Degeneration, albinism and inbreeding. *Science* 28: 454–455.

David, J. R., and C. Bocquet. 1975. Evolution in a cosmopolitan species: genetic latitudinal clines in *Drosophila melanogaster* wild populations. *Experientia* 31: 164–166.

Dawson, P. S. 1965. Genetic homeostasis and developmental rate in *Tribolium*. *Genetics* 51: 873–885.

DeBenedictis, P. 1978. Frequency-dependent selection: what is the problem? *Evolution* 32: 915–916.

Deese, R. E., and M. Koger. 1962. Maternal effects on preweaning growth rate in cattle. *Journal of Animal Science* 26: 250–253.

DeFries, J. C., and R. W. Touchberry. 1961. A "maternal effect" on body weight in drosophila. *Genetics* 46: 1261–1266.

Dempster, E. R., and I. M. Lerner. 1950. Heritability of threshold characters. *Genetics* 35: 212–236.

Deng, H.-W., and T. T. Kibota. 1995. The importance of the environmental variance–covariance structure in predicting evolutionary responses. *Evolution* 49: 572–574.

Denlinger, D. L. 1970. Embryonic determination of pupal diapause induction in the flesh fly *Sarcophaga crassipalpus* Macquart. *American Zoologist* 10: 320–321.

Denlinger, D. L. 1971. Embryonic determination of pupal diapause induction in the flesh fly *Sarcophaga crassipalpus*. *Journal of Insect Physiology* 17: 1815–1822.

Denlinger, D. L. 1972. Induction and termination of pupal diapause in *Sarcophaga* (Diptera: Sarcophagidae). *Biological Bulletin* 142: 11–24.

Denno, R. F., G. K. Roderick, K. L. Olmstead, and H. G. Dobel. 1991. Density-related migration in planthoppers (Homoptera: Delphacidae): the role of habitat persistence. *American Naturalist* 138: 1513–1541.

Derr, J. A. 1980. The nature of variation in life history characters of *Dysdercus bimaculatus* (Heteroptera: Pyrrhocoridae), a colonizing species. *Evolution* 34: 548–557.

DeVincente, M. C., and S. D. Tanksley. 1993. QTL analysis of transgressive segregation in an interspecific tomato cross. *Genetics* 134: 585–596.

Dhondt, A. A. 1982. Heritability of blue tit tarsus length from normal and cross-fostered broods. *Evolution* 36: 418–419.

Dickerson, G. E. 1947. Composition of hog cargasses as influenced by heritable differences in rate and economy of gain. *Iowa Agricultural Experimental Station, Research Bulletin* 354: 489–524.

Dickerson, G. E. 1955. Genetic slippage in response to selection for multiple objectives. *Cold Spring Harbour Symposium on Quantitative Biology* 20: 213–223.

Diehl, W. J., and R. K. Koehn. 1985. Multiple-locus heterozygosity, mortality, and growth in a cohort of *Mytilus edulis*. *Marine Biology* 88: 265–271.

Dietz, E. J. 1983. Permutation tests for association between two distance measures. *Systematic Zoologist* 32: 21–26.

Dingle, H., and K. E. Evans. 1987. Responses in flight to selection on wing length in nonmigratory milkweed bugs, *Oncopeltus fasciatus*. *Entomologica Experimentalis Et Applicata* 45: 289–296.

Dingle, H., K. E. Evans, and J. O. Palmer. 1988. Responses to selection among life-history traits in a nonmigratory population of milkweed bugs (*Oncopeltus fasciatus*). *Evolution* 42: 79–92.

Dobson, A. J. 1983. *An Introduction to Statistical Modelling*. Chapman & Hall, London.

Dobzhansky, T., and H. Levene. 1955. Genetics of natural populations XXIV. Developmental homeostasis in natural populations of *Drosophila pseudoobscura*. *Genetics* 40: 797–808.

Dobzhansky, Th., and S. Wright. 1941. Genetics of natural populations. V. Relations between mutation rate and accumulation of lethals in populations of *Drosophila pseudoobscura*. *Genetics* 26: 23–51.

Dobzhansky, Th., and S. Wright. 1943. Genetics of natural populations. X. Dispersion rates in *Drosophila pseudoobscura*. *Genetics* 28: 304–340.

Dobzhansky, Th., B. Spassky, and J. Sved. 1969. Effects of selection and migration on geotactic and phototactic behaviour of Drosophila. II. *Proceedings of the Royal Society, B* 173: 191–207.

Doebley, J., and A. Stec. 1993. Inheritance of the morphological differences between maize and teosinte: comparison of results for two F2 populations. *Genetics* 134: 559–570.

Doebley, J., A. Bacigalupo, and A. Stec. 1994. Inheritance of kernel weight in two maize–teosinte hybrid populations: implications for crop evolution. *Journal of Heredity* 85: 191–195.

Dole, J., and K. Ritland. 1993. Inbreeding depression in two *Mimulus* taxa measured by multigenerational changes in the inbreeding coefficient. *Evolution* 47: 361–373.

dos Santos, E., H. P. Andreassen, and R. A. Ims. 1995. Differential inbreeding tolerance in two geographically distinct strains of root vole *Microtus oeconomus*. *Ecography* 18: 238–247.

Doums, C., F. Viard, A. Pernot, B. Delay, and P. Jarne. 1997. Inbreeding depression, neutral polymorphism and copulatory behavior in freshwater snails: a self-fertilization syndrome. *Evolution* (in press).

Drickamer, L. C. 1981. Selection for age of sexual maturation in mice and the consequences for population regulation. *Behavioral and Neural Biology* 31: 82–89.

Druger, M. 1962. Selection and body size in *Drosophila pseudoobscura* at different temperatures. *Genetics* 17: 209–222.

Dudash, M. R. 1990. Relative fitness of selfed and outcrossed progeny in a self-compatible, protandrous species, *Sabatia angularis* L. (Gentianaceae): a comparison in three environments. *Evolution* 44: 1129–1139.

Dudley, J. W. 1977. 76 generations of selection for oil and protein percentage in maize. pp. 459–473. *In* O. Kempthorne, T. B. Bailey, and E. Pollack (editors). *Proceedings of the International Conference on Quantitative Genetics.* Iowa State University Press, Ames.

Dumeril, A. 1867. Metamorphoses des batraciens urodeles e branchies exterieures du Mexique dits axolotls, observees a la Menagerie de Reptiles du Museum d'Histoire Naturelle. Annales De Sciences, *Naturelles Zoologie Et Biologie Animale* 7: 229–254.

Dutilleul, P., and C. Potvin. 1995. Among-environment heteroscedasticity and genetic autocorrelation: Implications for the study of phenotypic plasticity. *Genetics* 139: 1815–1829.

Eanes, W. F. 1978. Morphological variance and enzyme heterozygosity in the monarch butterfly. *Nature* 276: 263–264.

East, E. M. 1908. Inbreeding in corn. *Report of the Connecticut Agriculture Experimental Station* 1907: 419–429.

Easteal, S. 1985. The ecological genetics of introduced populations of the giant toad *Bufo marinus*. II. Effective population size. *Genetics* 110: 107–122.

Easteal, S., and R. B. Floyd. 1986. The ecological genetics of introduced populations of the giant toad, *Bufo marinus* (Amphibia: Anura): dispersal and neighbourhood size. *Biological Journal of the Linnean Society* 27: 17–45.

Eberhard, W. G. 1980. Horned Beetles. *Scientific American* March: 166–182.

Ebert, P. D., and J. S. Hyde. 1976. Selection for agonistic behavior in wild female *Mus musculus*. *Behavior Genetics* 6: 291–304.

Ebert, D., L. Yampolsky, and S. C. Stearns. 1993a. Genetics of life history in *Daphnia magna*. I. Heritabilities at two food levels. *Heredity* 70: 335–343.

Ebert, D., L. Yampolsky, and A. J. van Noordwijk. 1993b. Genetics of life history in *Daphnia magna*. II. Phenotypic plasticity. *Heredity* 70: 344–352.

Eck, H. J. van, J. M. E. Jacobs, P. Stams, J. Ton, W. J. Stiekema, and E. Jacobsen. 1994.

Multiple alleles for tuber shape in diploid potato detected by qualitative and quantitative genetic analysis using RFLPs. *Genetics* 137: 303–309.

Ecker, R., and A. Barzilay. 1993. Quantitative genetic analysis of growth rate in *Lisianthus*. *Plant Breeding* 111: 253–256.

Eckert, C. G., and S. C. H. Barrett. 1994. Inbreeding depression in partially self-fertilizing *Decodon verticillatus* (Lythraceae): population—genetic and experimental analyses. *Evolution* 48: 952–964.

Edgington, E. S. 1987. *Randomization Tests*. Marcel Dekker, Inc., New York.

Edwards, K. J. R., and Y. A. Emara. 1970. Variation in plant development within a population of *Lolium multiflorum*. *Heredity* 25: 179–194.

Edwards, M. D., C. W. Stuber, and J. F. Wendel. 1987. Molecular-marker-facilitated investigations of quantitative-trait loci in maize. I. Numbers, genomic distribution and types of gene action. *Genetics* 116: 113–125.

Edwards, M. D., T. Helentjaris, S. Wright, and C. W. Stuber. 1992. Molecular-marker-facilitated investigations of quantitative trait loci in maize 4. Analysis based on genome saturation with isozyme and restriction fragment length polymorphism markers. *Theoretical and Applied Genetics* 83: 765–774.

Ehiobu, N. G., and M. E. Goddard. 1990. Heterosis in crosses between geographically separated populations of *Drosophila melanogaster*. *Theoretical and Applied Genetics* 80: 569–575.

Ehiobu, N. G., M. E. Goddard, and J. F. Taylor. 1989. Effect of rate of inbreeding on inbreeding depression in *Drosophila melanogaster*. *Theoretical and Applied Genetics* 77: 123–127.

Eisen, E. J. 1967. Mating designs for estimating direct and maternal genetic variances and direct-maternal genetic covariances. *Canadian Journal of Genetics and Cytology* 9: 13–22.

Eisen, E. J. 1972. Long-term selection response for 12-day litter weight in mice. *Genetics* 72: 129–142.

Eisen, E. J., and B. S. Durrant. 1980. Genetic and maternal environmental factors influencing litter size and reproductive efficiency in mice. *Journal of Animal Science* 50: 428–441.

Eisen, E. J., and J. P. Hanrahan. 1972. Selection for sexual dimorphism in body weight of mice. *Australian Journal of Biological Science* 25: 1015–1024.

Eisen, E. J., and J. E. Legates. 1966. Genotype–sex interaction and the genetic correlation between the sexes for body weight in *Mus musculus*. *Genetics* 54: 611–623.

Eisen, E. J., B. B. Bohren, and H. E. McKean. 1966. Sex-linked and maternal effects in the diallel cross. *Australian Journal of Biological Science* 19: 1061–1071.

Eisen, E. J., J. E. Legates, and O. W. Robison. 1970. Selection for 12-day litter weight in mice. *Genetics* 64: 511–532.

Eklund, J., and G. E. Bradford. 1977. Genetic analysis of a strain of mice plateaued for litter size. *Genetics* 85: 529–542.

Ellner, S. 1996. Environmental fluctuations and the maintenance of genetic diversity in age or stage-structured populations. *Bulletin of Mathematical Biology* 58: 103–127.

Ellner, S., and N. G. Hairston, Jr. 1994. Role of overlapping generations in maintaining genetic variation in a fluctuating environment. *American Naturalist* 143: 403–417.

Elston, R. C. 1977. Estimating "Heritability" of a dichotomous trait. Response. *Biometrics* 33: 232–233.

Endler, J. A. 1986. *Natural Selection in the Wild*. Princeton University Press, Princeton, NJ.

Endler, J. A. 1988. Frequency-dependent predation, crypsis and aposematic coloration. *Philosophical Transactions of the Royal Society of London* B 319: 505–523.

Enfield, F. D., 1980. Long term effects of selection: the limits to response. pp. 69–86. *In* A. Robertson (editor). *Selection Experiments in Laboratory and Domestic Animals*. Commonwealth Agricultural Bureau, Slough, UK.

Enfield, F. D., and O. Braskerud. 1989. Mutational variance for pupa weight in *Tribolium castaneum*. *Theoretical and Applied Genetics* 77: 416–420.

Enfield, F. D., R. E. Comstock, and O. Braskerud. 1966. Selection for pupa weight in *Tribolium casteneum*. I. Parameters in base populations. *Genetics* 54: 523–533.

Enfield, F. D., D. T. North, R. Erickson, and L. Rotering. 1983. A selection response plateau for radiation resistance in the cotton boll weevil. *Theoretical and Applied Genetics* 65: 277–281.

Englert, D. C., and A. E. Bell. 1969. Components of growth in genetically diverse populations of *Tribolium casteneum*. *Canadian Journal of Genetics and Cytology* 11: 896–907.

Englert, D. C., and A. E. Bell. 1970. Selection for time of pupation in *Tribolium casteneum*. *Genetics* 64: 541–552.

Etges, W. J. 1989. Evolution of developmental homeostasis in *Drosophila mojavensis*. *Evolutionary Ecology* 3: 189–201.

Ewens, W. J. 1989. An interpretation and proof of the fundamental theorem of natural selection. *Theoretical Population Biology* 36: 167–180.

Ewens, W. J. 1992. An optimizing principle of natural selection in evolutionary population genetics. *Theoretical Population Biology* 42: 333–346.

Fairbairn, D. J. 1984. Microgeographic variation in body size and development time in the waterstrider, *Limnoporus notabilis*. *Oecologia* 61: 126–133.

Fairbairn, D. J. 1990. Factors influencing sexual size dimorphism in temperate waterstriders. *American Naturalist* 136: 61–86.

Fairfull, R., L. Haley, and J. Castell. 1981. The early growth of artificially reared American lobsters. Part 1: Genetic parameters and environments. *Theoretical and Applied Genetics* 60: 269–273.

Falconer, D. S. 1952. The problem of environment and selection. *American Naturalist* 86: 293–298.

Falconer, D. S. 1953. Selection for large and small size in mice. *Journal of Genetics* 51: 470–501.

Falconer, D. S. 1957. Selection for phenotypic intermediates in *Drosophila*. *Journal of Genetics* 55: 551–561.

Falconer, D. S. 1960. Selection of mice for growth on high and low planes of nutrition. *Genetic Research* 1: 91–113.

Falconer, D. S. 1965a. The inheritance of liability to certain diseases, estimated from the incidence among relatives. *Annals of Human Genetics* 29: 51–76.

Falconer, D. S. 1965b. Maternal effects and selection response. pp. 763–774. *In* S. J. Geerts (editor). *Genetics Today. Proceedings of the XI International Congress of Genetics.* Volume 3, Pergamon Press, Oxford.

Falconer, D. S. 1971. Improvement of litter size in a strain of mice at a selection limit. *Genetic Research* 17: 215–235.

Falconer, D. S. 1989. *Introduction to Quantitative Genetics.* Longmans, New York.

Falconer, D. S. 1990. Selection in different environments: effects on environmental sensitivity (reaction norm) and on mean performance. *Genetic Research* 56: 57–90.

Falconer, D. S., and A. Robertson. 1956. Selection for environmental variability of body size in mice. *Zeitschrift fuer Induktive Abstammungs-und Vererbungslehre* 87: 385–391.

Farris, M. A., and J. B. Mitton. 1984. Population density, outcrossing rate, and heterozygote superiority in ponderosa pine. *Evolution* 38: 1151–1154.

Fehr, W. R., and C. R. Weber. 1968. Mass selection by seed size and specific gravity in soybean populations. *Crop Science* 8: 551–554.

Felsenstein, J. 1976. The theoretical population genetics of variable selection and migration. *Annual Review of Genetics* 10: 253–280.

Felsenstein, J. 1977. Multivariate normal genetic models with a finite number of loci. pp. 227–246. *In* O. Kempthorne, T. B. Bailey, and E. Pollack (editors). *Proceedings of the the International Conference on Quantitative Genetics.* Iowa State University Press, Ames.

Ferguson, M. M. 1986. Developmental stability of rainbow trout hybrids: genomic coadaption or heterozygosity? *Evolution* 40: 323–330.

Ferguson, M. M., A. P. Liskauskas, and R. G. Danzmann. 1995. Genetic and environmental correlates of variation in body weight of brook trout (*Salvelinus fontinalis*). *Canadian Journal of Fisheries and Aquatic Sciences* 52: 307–314.

Festing, M. F., and A. W. Nordskog. 1967. Response to selection for body weight and egg weight in chickens. *Genetics* 55: 219–231.

Fevolden, S. E., and S. P. Garner. 1987. Environmental stress and allozyme variation in *Littorina littorea* (Prosobranchia). *Marine Ecology—Progress Series* 39: 129–36.

Fincham, J. R. S. 1972. Heterozygous advantage as a likely general basis for enzyme polymorphisms. *Heredity* 28: 387–391.

Findley, C. S., and F. Cooke. 1987. Repeatability and heritability of clutch size in lesser snow geese. *Evolution* 41: 453.

Fisher, R. A. 1918. The correlation between relatives on the supposition of Mendelian inheritance. *Transactions of the Royal Society of Edinburgh* 52: 399–433.

Fisher, R. A. 1930. *The Genetical Theory of Natural Selection.* Clarendon Press, Oxford.

Flavell, R., and M. O'Dell. 1990. Variation and inheritance of cytosine methylation patterns in wheat at the high molecular weight glutenin and ribosomal RNA gene loci. *Development* 110 (Supplement): 15–20.

Fleischer, R. C., R. F. Johnston, and W. J. Klitz. 1983. Allozymic hetrozygosity and morphological variation in house sparrows. *Nature* 304: 628–630.

Fleming, A. A. 1975. Effects of male cytoplasm on inheritance in hybrid maize. *Crop Science* 15: 570–573.

Flury, B., and H. Riedwyl. 1988. *Multivariate Statistics: A Practical Approach.* Chapman & Hall, London.

Flux, J. E. C., and M. M. Flux. 1982. Artificial selection and gene flow in wild starlings, *Sturnus vulgaris. Naturwissensch* 69: 96–97.

Foley, P. 1992. Small population genetic variability at loci under stabilizing selection. *Evolution* 46: 763–774.

Foltz, D. W., S. E. Shumway, and D. Crisp. 1993. Genetic structure and heterozygosity-related fitness effects in the marine snail *Littorina littorea. American Malacological Society* 10: 55–60.

Fondevila, A. 1973. Genotype–temperature interaction in *Drosophila melanogaster.* II. Body weight. *Genetics* 73: 125–134.

Fong, D. W. 1989. Morphological evolution of the amphipod *Gammarus minus* in caves: quantitative genetic analysis. *American Midland Naturalist* 121: 361–378.

Fox, C. W. 1993. The influence of maternal age and mating frequency on egg size and offspring performance in *Callosobruchus maculatus* (Coleoptera: Bruchidae). *Oecologia* 96: 139–146.

Fox, G. A. 1990. Components of flowering time variation in a desert annual. *Evolution* 44: 1404–1423.

Frahm, R. R., and K. Kojima. 1966. Comparison of selection responses on body weight under divergent larval density conditions in *Drosophila pseudoobscura. Genetics* 54: 625–637.

Frank, S. A., and M. Slatkin. 1992. Fisher's fundamental theorem of natural selection. *Trends in Ecology and Evolution* 7: 92–95.

Frankham, R. 1968. Sex and selection for a quantitative character in *Drosophila.* I. Single-sex selection. *Australian Journal of Biological Science* 21: 1215–1223.

Frankham, R. 1977a. The nature of quantitative genetic variation in *Drosophila.* III. Mechanism of dosage compensation for sex-linked abdominal bristle polygenes. *Genetics* 85: 185–191.

Frankham, R. 1977b. Optimum selection intensities in artificial selection programmes: an experimental evaluation. *Genetic Research* 30: 115–119.

Frankham, R., 1980. The founder effect and response to artificial selection in *Drosophila.* *In* A. Robertson (editor). *Proceedings of the International Symposium on Selection Experiments on Laboratory and Domestic Animals,* Commonwealth Agricultural Bureau, London.

Frankham, R. 1990. Are responses to artificial selection for reproductive fitness characters consistently asymmetrical? *Genetic Research* 56: 35–42.

Frankham, R. 1995a. Effective population size/adult population size ratios in wildlife: a review. *Genetic Research* 66: 95–107.

Frankham, R. 1995b. Conservation genetics. *Annual Review of Genetics* 29: 305–327.

Frankham, R., and R. K. Nurthen. 1981. Forging links between population and quantitative genetics. *Theoretical and Applied Genetics* 59: 251–263.

Frankham, R., L. P. Jones, and J. S. F. Barker. 1968. The effects of population size and selection intensity in selection for a quantitative character in *Drosophila*. I. Short-term response to selection. *Genetic Research* 12: 237–248.

Frankham, R., G. J. Smith, and D. A. Briscoe. 1993. Effects on heterozygosity and reproductive fitness of inbreeding with and without selection on fitness in *Drosophila melanogaster*. *Theoretical and Applied Genetics* 86: 1023–1027.

Fraser, A. S. 1960. Simulation of genetic systems by automatic digital computers. VI. Epistasis. *Australian Journal of Biological Science* 13: 150–162.

Freyre, R., S. Warnke, B. Sosinski, and D. S. Douches. 1994. Quantitative trait locus analysis of tuber dormancy in diploid potato (*Solanum* spp.). *Theoretical and Applied Genetics* 89: 474–480.

Fry, J. D. 1992. The mixed-model analysis of variance applied to quantitative genetics: biological meaning of the parameters. *Evolution* 46: 540–550.

Frydenberg, O. 1963. Population studies of a lethal mutant in *Drosophila melanogaster*. I. Behaviour in populations with discrete generations. *Hereditas* 50: 89–116.

Futuyma, D. J., and T. E. Philippi. 1987. Genetic variation and covariation in responses to host plants by *Alsophila pometria* (Lepidopetra: Geometridae). *Evolution* 41: 269–279.

Gabriel, W., and M. Lynch. 1992. The selective advantage of reaction norms for environmental tolerance. *Journal of Evolutionary Biology* 5: 41–60.

Gaffney, P. M. 1990. Enzyme heterozygosity, growth rate, and viability in *Mytilus edulis*: another look. *Evolution* 44: 204–210.

Gale, J. S., and M. J. Kearsey. 1968. Stable equilibria under stabilising selection in the absence of dominance. *Heredity* 23: 553–561.

Gall, G. A. E. 1971. Replicated selection for 21-day pupa weight of *Tribolium castaneum*. *Theoretical and Applied Genetics* 41: 164–173.

Gallant, S. L., and D. J. Fairbairn. 1997. Patterns of postmating reproductive isolation in a newly-discovered species pair: *Aquarius remigis* and *A. remigoides* (Hemiptera; Gerridae). *Heredity* (in press).

Gallardo, M. H., and N. Kohler. 1994. Demographic changes and genetic losses in populations of a subterranean rodent (*Ctenomys maulinus Brunneus*) affected by a natural catastrophe. *Zeitschrift fuer Saugetierkunde* 59: 358–365.

Gallardo, M. H., N. Kohler, and C. Araneda. 1995. Bottleneck effects in local populations of fossorial *Ctenomys* (Rodentia, Ctenomyidae) affected by vulcanism. *Heredity* 74: 638–646.

Gallego, A., and C. López-Fanjul. 1983. The number of loci affecting a quantitative trait in *Drosophila melanogaster* revealed by artificial selection. *Genetic Research* 42: 137–149.

Galton, F. 1889. *Natural Inheritance*. MacMillan and Co., London.

Gama, L. T., K. G. Boldman, and R. K. Johnson. 1991. Estimates of genetic parameters for direct and maternal effects on embryonic survival in swine. *Journal of Animal Science* 69: 4801–4809.

Garbutt, K., and J. R. Witcombe. 1986. The inheritance of seed dormancy in *Sinapis arvensis* L. *Heredity* 56: 25–31.

Garcia, N., C. López-Fanjul, and A. Garcia-Dorado. 1994. The genetics of viability in *Drosophila melanogaster*: effects of inbreeding and artificial selection. *Evolution* 48: 1277–1285.

Garten, C. T., Jr. 1976. Relationships between aggressive behavior and genic heterozygosity in the oldfield mouse, *Peromyscus polionotus*. *Evolution* 30: 59–72.

Garten, C. T., Jr. 1977. Relationship between exploratory behaviour and genic heterozygosity in the oldfield mouse. *Animal Behaviour* 25: 328–332.

Garton, D. W. 1984. Relationship between multiple locus heterozygosity and physiological energetics of growth in the estuarine gastropod. *Thais haemastoma*. *Physiological Zoology* 57: 530–543.

Gavrilets, S., and S. M. Scheiner. 1993. The genetics of phenotypic plasticity. V. Evolution of reaction norm shape. *Journal of Evolutionary Biology* 6: 31–48.

Gebhardt, M. D. 1991. A note on the application of diallel crosses for the analysis of genetic variation in natural populations. *Theoretical and Applied Genetics* 82: 54–56.

Gebhardt, M. D., and S. C. Stearns. 1993a. Phenotypic plasticity for life history traits in *Drosophila melanogaster*. Journal of Evolutionary Biology 6: 1–16.

Gebhardt, M. D., and S. C. Stearns. 1993b. Phenotypic plasticity for life history traits in *Drosophila melanogaster*. I. Effect on phenotypic and environmental correlations. *Journal of Evolutionary Biology* 6: 1–16.

Gebhardt-Henrich, S. G., and A. J. van Noordwijk. 1991. Nestling growth in the great tit. I. Heritability estimates under different environmental condition. *Journal of Evolutionary Biology* 3: 341–362.

Geiger, H. H. 1988. Epistasis and heterosis. pp. 395–399. *In* E. J. Eisen, M. M. Goodman, G. Namkoong, and B. S. Weir (editors). *Proceedings of the 2nd International Conference on Quantitative Genetics*, Sinauer Associates, Inc., Sunderland, MA.

Gentili, M. R., and A. R. Beaumont. 1988. Environmental stress, heterozygosity, and growth rate in *Mytilus edulis* L. *Journal of Experimental Marine Biology and Ecology* 120: 145–153.

Gerlai, R., W. E. Crusio, and V. Csanyi. 1990. Inheritance of species-specific behaviors in the paradise fish (*Macropodus opercularis*): a diallel study. *Behavior Genetics* 20: 487–498.

Gibbs, H. L. 1988. Heritability and selection on clutch size in Darwin's medium ground finches (*Geospiza fortis*). *Evolution* 42: 750–762.

Gibson, J. B., and B. P. Bradley. 1974. Stabilizing selection in constant and fluctuating environments. *Heredity* 33: 293–302.

Gilbert, N. E. 1958. Diallel cross in plant breeding. *Heredity* 12: 477–492.

Gill, D. E. 1978. The metapopulation dynamics of the red-spotted newt, *Notophthalmus viridescens* (Rafinesque). *Ecolological Monographs* 48: 145–166.

Gillespie, J. H. 1977. A general model to account for enzyme variation in natural populations. III. Multiple alleles. *Evolution* 31: 85–90.

Gillespie, J. H. 1984 Pleiotropic overdominance and the maintenance of genetic variation in polygenic characters. *Genetics* 107: 321–330.

Gillespie, J. H., and M. Turelli. 1989. Genotype–environment interactions and the maintenance of polygenic variation. *Genetics* 121: 129–138.

Gimelfarb, A. 1986. Additive variation maintained under stabilizing selection: a two-locus model of pleiotropy for two quantitative characters. *Genetics* 112: 717–725.

Gjerde, B., and T. Gjedrem. 1984. Estimates of phenotypic and genetic parameters for carcass traits in Atlantic salmon and rainbow trout. *Aquaculture* 36: 97–110.

Gjerde, B., K. Gunnes, and T. Ojedrem. 1983. Effect of inbreeding on survival and growth in rainbow trout. *Aquaculture* 34: 327–332.

Gomez-Raya, L., and E. B. Burnside. 1990. The effect of repeated cycles of selection on genetic variance, heritability, and response. *Theoretical and Applied Genetics* 79: 568–574.

Gomulkiewicz, R., and M. Kirkpatrick. 1992. Quantitative genetics and the evolution of reaction norms. *Evolution* 46: 390–411.

Goodnight, C. 1987. On the effect of founder events on epistatic genetic variance. *Evolution* 41: 80–91.

Goodnight, C. J. 1988. Epistasis and the effect of founder events on the additive genetic variance. *Evolution* 42: 441–454.

Goodnight, C. J. 1995. Epistasis and the increase in additive genetic variance: implications for phase 1 of Wright's shifting-balance process. *Evolution* 49: 502–511.

Goodwill, R. E., and R. D. Walker. 1978a. Epistatic contributions to quantitative traits in *Tribolium castaneum* I. Traits not closely related to fitness. *Theoretical and Applied Genetics* 51: 193–198.

Goodwill, R. E., and R. D. Walker, 1978b. Epistatic contributions to quantitative traits in *Tribolium castaneum* II. Traits closely related to fitness. *Theoretical and Applied Genetics* 51: 305–309.

Gotthard, K., S. Nylin, and C. Wiklund. 1994. Adaptive variation in growth rate: life history costs and consequences in the speckled wood butterfly *Pararge aegeria*. *Oecologia* 99: 281–289.

Gould, F. 1988. Genetics of pairwise and multispecies plant–herbivore coevolution. pp. 13–55. *In* K. C. Spencer (editor). *Chemical Mediation of Coevolution*. Academic Press, San Diego, CA.

Grant, B. R., and P. R. Grant. 1989. *Evolutionary Dynamics of a Natural Population*. University of Chicago Press, Chicago.

Grant, B. R., and P. R. Grant. 1993. Evolution of Darwin's finches caused by a rare climatic event. *Proceedings of the Royal Society of London* B 251: 111–117.

Grant, B. R., and L. E. Mettler. 1969. Disruptive and stabilizing selection on the "escape" behavior of *Drosophila melanogaster*. *Genetics* 62: 625–637.

Grant, P. R., and B. R. Grant. 1992. Demography and the genetically effective sizes of two populations of Darwin's finches. *Ecology* 73: 766–784.

Grant, P. R., and B. R. Grant. 1995. Predicting microevolutionary responses to directional selection on heritable variation. *Evolution* 49: 241–251.

Grattapaglia, D., F. L. Bertolucci, and R. R. Sederoff. 1995. Genetic mapping of QTLs controlling vegetative propagation in *Eucalyptus grandis* and *E. urophylla* using a pseudo-testcross strategy and RAPD markers. *Theoretical and Applied Genetics* 90: 933–947.

Graves, J. A. M. 1987. The evolution of mammalian sex chromosomes and dosage compensation: clues from marsupials and monotremes. *Trends in Genetics* 3: 252–256.

Gray, R. H. 1984. Effective breeding size and adaptive significance of color polymorphism in the cricket frog (*Acris crepitans*) in Illinois, U.S.A. *Amphibia–Reptilia* 5: 101–107.

Green, R. H., S. M. Singh, B. Hicks, and J. M. McCuaig. 1983. An Arctic intertidal population of *Macoma balthica* (Mollusca, Pelecypoda): genotypic and phenotypic components of population structure. *Canadian Journal of Fisheries and Aquatic Science* 40: 1360–1371.

Greenwood, P. J., and P. H. Harvey. 1978. Inbreeding and dispersal in the great tit. *Nature* 271: 52–54.

Griffing, B. 1956a. Concept of general and specific combining ability in relation to diallel crossing systems. *Australian Journal of Biological Science* 9: 463–493.

Griffing, B. 1956b. A generalised treatment of the use of diallel crosses in quantitative inheritance. *Heredity* 10: 31–50.

Griffing, B. 1965. Influence of sex on selection. I. Contribution of sex-linked genes. *Australian Journal of Biological Sciences* 18: 1157–1170.

Groeter, F. R., and H. Dingle. 1988. Genetic and maternal influences on life history plasticity in milkweed bugs (*Oncopeltus*): response to temperature. *Journal of Evolutionary Biology* 1: 317–333.

Groeters, F. R., and H. Dingle. 1987. Genetic and maternal influences on life history plasticity in response to photoperiod by milkweed bugs (*Oncopeltus fasciatus*). *American Naturalist* 129: 332–341.

Gromko, M. H. 1977. What is frequency-dependent selection? *Evolution* 31: 438–442.

Gromko, M. H. 1987. Genetic constraint on the evolution of courtship behavior in *Drosophila melanogaster*. *Heredity* 58: 435–441.

Gromko, M. H., A. Briot, S. C. Jensen, and H. H. Fukui. 1991. Selection on copulation-duration in *Drosphila melanogaster*: predictability of direct response versus unpredictability of correlated response. *Evolution* 45: 69–81.

Groover, A., M. Devey, T. Fiddler, J. Lee, R. Megraw, T. Mitchel-Olds, B. Sherman, S. Vujcic, C. Williams, and D. Neale. 1994. Identification of quantitative trait loci influencing wood specific gravity in an outbred pedigree of loblolly pine. *Genetics* 138: 1293–1300.

Grossman, M., and H. W. Norton. 1974. Simplification of the sampling variance of the correlation coefficients. *Theoretical and Applied Genetics* 44: 332.

Gupta, A. P., and R. C. Lewontin. 1982. A study of reaction norms in natural populations of *Drosophila pseudobscura*. *Evolution* 36: 934–948.

Gustafsson, L. 1986. Lifetime reproductive success and heritabilities: empirical support for Fisher's fundamental theorem. *American Naturalist* 128: 761–764.

Gustafsson, L., and J. Merila. 1994. Foster parent experiment reveals no genotype–environment correlation in the external morphology of *Ficedula albicollis*, the collared flycatcher. *Heredity* 73: 124–129.

Hagen, D. W. 1973. Inheritance of numbers of lateral plates and gill rakers in *Gasterosteus aculeatus*. *Heredity* 30: 303–312.

Haigh, G. R. 1983. Effects of inbreeding and social factors on the reproduction of young female *Peromyscus maniculatus bairdii*. *Journal of Mammology* 64: 48–54.

Hairston, N. G., Jr., and W. R. Munns, Jr. 1984. The timing of copepod diapause as an evolutionarily stable strategy. *American Naturalist* 123: 733–751.

Haldane, J. B. S. 1954. *The Biochemistry of Genetics*. Allen and Unwin, London.

Haldane, J. B. S. 1954. The measurement of natural selection. *Proceedings of the 9th International Congress of Genetics* (Caryologia, Supplement to Volume 6) 1: 480–487.

Haldane, J. B. S., and S. D. Jayakar. 1963. Polymorphism due to selection depending on the composition of a population. *Journal of Genetics* 58: 318–323.

Hamaker, H. C. 1978. Approximating the cumulative normal distribution and its inverse. *Applied Statistics* 27: 76–77.

Hammond, K., and F. W. Nicholas. 1972. The sampling variance of the correlation coefficients estimated from two-fold nested and offspring–parent regression analysis. *Theoretical and Applied Genetics* 42: 97–100.

Handford, P. 1980. Heterozygosity at enzyme loci and morphological variation. *Nature* 286: 261–262.

Hanrahan, J. P. 1976. Maternal effects and selection response with an application to sheep data. *Animal Production* 22: 359–369.

Hanrahan, J. P., and E. J. Eisen. 1973. Sexual dimorphism and direct and maternal genetic effects on body weights in mice. *Theoretical and Applied Genetics* 43: 39–45.

Hanrahan, J. P., and E. J. Eisen. 1974. Genetic variation in litter size and 12-day weight in mice and their relationships with post-weaning growth. *Animal Production* 19: 13–23.

Hard, J. J., W. E. Bradshaw, and C. M. Holzapfel. 1993. Genetic coordination of demography and phenology in the pitcher-plant mosquito, *Wyeomyia smithii*. *Journal of Evolutionary Biology* 6: 707–723.

Harding, J., H. Huang, and T. Byrne. 1991. Maternal, paternal, additive, and dominance components of variance in *Gerbera*. *Theoretical and Applied Genetics* 82: 756–760.

Harris, R. B., and F. W. Allendorf. 1989. Genetically effective population size of large mammals: an assessment of estimators. *Conservation Biology* 3: 181–191.

Hauser, T. P., C. Damgaard, and V. Loeschske. 1994. Effects of inbreeding in small plant populations: Expectations and implications for conservation. pp. 115–129 *In* V. Loeschske, J. Tomiuk, and S. K. Jain (editors). *Conservation Genetics*. Birkhauser Verlag, Basel.

Hayes, P. M., B. H. Liu, S. J. Knapp, F. Chen, B. Jones, T. Blake, J. Franckowiak, D. Rasmusson, M. Sorrells, S. E. Ullrich, D. Wesenberg, and A. Kleinhofs. 1993. Quantitative trait locus effects and environmental interaction in a sample of North American barley germ plasm. *Theoretical and Applied Genetics* 87: 392–401.

Hayman, B. I. 1954a. The theory and analysis of diallel crosses. *Genetics* 39: 789–809.

Hayman, B. I. 1954b. The analysis of variance of diallel tables. *Biometrics* 10: 235–244.

Hayman, B. I. 1957. Interaction, heterosis and diallel crosses. II. *Genetics* 42: 336–355.

Hayman, B. I. 1960. The theory and analysis of diallel crosses. III. *Genetics* 45: 155–172.

Hayward, M. D., and G. F. Nsowah. 1969. The genetic organisation of natural populations of *Lolium perenne* IV. Variation within populations. *Heredity* 24: 521–528.

Hazel, L. N. 1943. The genetic basis for constructing selection indices. *Genetics* 28: 476–490.

Hazel, W. N., R. S. Mock, and M. D. Johnson. 1990. A polygenic model for the evolution and maintenance of conditional strategies. *Proceedings of the Royal Society of London* B 242: 181–187.

Hedrick, P. W. 1972. Maintenance of genetic variation with a frequency-dependent selection model as compared to the overdominant model. *Genetics* 72: 771–775.

Hedrick, P. W. 1973. Genetic variation and the generalized frequency-dependent model. *American Naturalist* 107: 800–802.

Hedrick, P. W. 1986. Genetic polymorphism in heterogeneous environments: a decade later. *Annual Review of Ecology and Systematics* 17: 535–566.

Hedrick, P. W. 1994. Purging inbreeding depression and the probability of extinction: Full-sib mating. *Heredity* 73: 363–372.

Hedrick, P. W., M. E. Genevan, and E. P. Ewing. 1976. Genetic polymorphism in heterogeneous environments. *Annual Review of Ecology and Systematics* 7: 1–32.

Hegmann, J. P., and H. Dingle. 1982. Phenotypic and genetic covariance structure in Milkweed Bug life history traits. pp. 177–186. *In* H. Dingle and J. P. Hegmann (editors). *Evolution and Genetics of Life Histories*. Springer-Verlag, New York.

Heisler, I. L. 1984. A quantitative genetic model for the origin of mating preferences. *Evolution* 38: 1283–1295.

Helbig, A. J. 1992. Population differentiation of migratory directions in birds: comparison between ringing results and orientation behaviour of hand-raised migrants. *Oecologia* 90: 483–488.

Henderson, N. D. 1981. Genetic influences on locomotor activity in 11-day-old housemice. *Behavior Genetics* 11: 209–225.

Hewitt, J. K., and D. W. Fulker. 1981. Using the triple test cross to investigate the genetics of behavior in wild populations I. Methodological considerations. *Behavior Genetics* 11: 23–35.

Hey, J., and M. K. Gargiulo. 1985. Sex-ratio changes in *Leptopilina heterotoma* in response to inbreeding. *Journal of Heredity* 76: 209–211.

Heywood, J. S. 1986. The effect of plant size variation on genetic drift in populations of annuals. *American Naturalist* 127: 851–861.

Hill, J. L. 1974. Peromyscus: effect of early pairing on reproduction. *Science* 186: 1042–1044.

Hill, W. G. 1972a. Estimation of realised heritabilities from selection experiments. I. Divergent selection. *Biometrics* 28: 747–765.

Hill, W. G. 1972b. Estimation of realized heritabilities from selection experiments. II. Selection in one direction. *Biometrics* 28: 767–780.

Hill, W. G. 1977. Order statistics of correlated variables and implications in genetic selection programmes. II. Responses to selection. *Biometrics* 33: 703–712.

Hill, W. G. 1979. A note on effective population size with overlapping generations. *Genetics* 92: 317–322.

Hill, W. G. 1981. Estimation of effective population size from data on linkage disequilibrium. *Genetic Research* 38: 209–216.

Hill, W. G. 1982a. Predictions of response to artificial selection from new mutations. *Genetic Research* 40: 255–278.

Hill, W. G. 1982b. Rates of change in quantitative traits from fixation of new mutations. *Proceedings of the National Academy of Sciences* 79: 142–145.

Hill, W. G. 1985. Effects of population size on response to short and long term selection. *Zeitschrift Fur Tierzuchtung und Zuchtgsbiologie* 102: 161–173.

Hill, W. G., and P. J. Avery. 1978. On estimating number of genes by genotype assay. *Heredity* 40: 397–403.

Hill, W. G., and P. D. Keightley. 1988. Interrelations of mutation, population size, artificial and natural selection. pp. 57–70. *In* E. J. Eisen, M. M. Goodman, G. Namkoong, and B. S. Weir (editors). *Proceedings of the Second International Conference on Quantitative Genetics*. Sinauer Associates, Inc., Sunderland, MA.

Hill, W. G., and F. W. Nicholas. 1974. Estimation of heritability by both regression of offspring on parent and intra-class correlation of sibs in one experiment. *Biometrics* 30: 447–468.

Hill, W. G., and A. Robertson. 1968. The effects of inbreeding at loci with heterozygote advantage. *Genetics* 60: 615–628.

Hillesheim, E., and S. C. Stearns. 1991. The responses of *Drosophila melanogaster* to artificial selection on body weight and its phenotypic plasticity in two larval food environments. *Evolution* 45: 1909–1923.

Hodgkin, J. 1987. Sex determination and dosage compensation in *Caenorhabditis elegans*. *Annual Review of Genetics* 21: 133–154.

Hoffman, A. A., and P. A. Parsons. 1991. *Evolutionary Genetics and Environmental Stress*. Oxford University Press, Oxford.

Hohenboken, W. D., and J. S. Brinks. 1971. Relationship between direct and maternal effects on growth in Herefords: II. Partitioning of covariance between relatives. *Journal of Animal Science* 32: 26–34.

Hollingsworth, M. J., and J. M. Smith. 1955. The effects of inbreeding on rate of development and on fertility in *Drosophila subobscura*. *Genetics* 53: 295–315.

Holloway, G. J., and P. M. Brakefield. 1994. Artificial selection of reaction norms of wing pattern elements in *Bicyclus anynana*. *Heredity* 74: 91–99.

Holloway, G. J., P. W. de Jong, and M. Ottenheim. 1993. The genetics and cost of chemical defense in the two-spot ladybird (*Adalia bipunctata* L.). *Evolution* 47: 1229–1239.

Holt, S. B. 1968. *The Genetics of Dermal Ridges*. C. C. Thomas, Springfield, IL.

Holtsford, T. P., and N. C. Ellstrand. 1990. Inbreeding effects in *Clarkia tembloriensis* (Onagraceae) populations with different natural outcrossing rates. *Evolution* 44: 2031–2046.

Hooper, K. R., R. T. Roush, and W. Powell. 1993. Management of genetics of biological-control introductions. *Annual Review of Entomology* 38: 27–51.

Hopper, J. L., and J. D. Mathews. 1982. Extensions to multivariate normal models for pedigree analysis. *Annals of Human Genetics* 46: 373–383.

Hospital, F., and C. Chevalet. 1993. Effects of population size and linkage on optimal selection intensity. *Theoretical and Applied Genetics* 86: 775–780.

Houde, A. E. 1992. Sex-linked heritability of a sexually selected character in a natural population of *Poecilia reticulata* (Pisces: Poeciliidae) (guppies). *Heredity* 69: 229–235.

Houle, D. 1989a. The maintenance of polygenic variation in finite populations. *Evolution* 43: 1767–1780.

Houle, D. 1989b. Allozyme-associated heterosis in *Drosophila melanogaster*. *Genetics* 123: 789–801.

Houle, D. 1991. Genetic covariance of fitness correlates: what genetic correlations are made of and why it matters. *Evolution* 45: 630–648.

Houle, D. 1992. Comparing evolvability and variability of quantitative traits. *Genetics* 130: 195–204.

Howard, D. J. 1993. Small populations, inbreeding, and speciation. pp. 118–142. *In* N. W. Thornhill (editor). *The Natural History of Inbreeding and Outbreeding*. University of Chicago Press, Chicago.

Hudak, M. J., and M. H. Gromko. 1989. Response to selection for early and late development of sexual maturity in *Drosophila melanogaster*. *Animal Behaviour* 38: 344–351.

Hunter, P. E. 1959. Selection of *Drosophila melanogaster* for length of larval period. *Zeitschrift Für Vererbungslehre* 90: 7–28.

Husband, B. C., and S. C. H. Barrett. 1992. Effective population size and genetic drift in tristylous *Eichhornia paniculata* (Pontederiaceae). *Evolution* 46: 1875–1890.

Husband, B. C., and S. C. H. Barrett. 1995. Estimating effective population size: a reply to Nunney. *Evolution* 49: 392–394.

Husband, B. C., and D. W. Schemske. 1996. Evolution of the magnitude and timing of inbreeding depression in plants. *Evolution* 50: 54–70.

Hutchings, J. A., and M. M. Ferguson. 1992. The independence of enzyme heterozygosity and life-history traits in natural populations of *Salvelinus fontinalis* (brook trout). *Heredity* 69: 496–502.

Hyde, J. S., and T. F. Sawyer. 1980. Selection for agonistic behavior in wild female mice. *Behavior Genetics* 10: 349–359.

Ikeda, H., and O. Maruo. 1982. Directional selection for pulse repetition rate of the courtship sound and correlated responses occurring in several characters in *Drosophila mercatorum*. *Japanese Journal of Genetics* 57: 241–258.

Iwasa, Y., H. Ezoe, and A. Yamauchi. 1994. Evolutionarily stable seasonal timing of univoltine and bivoltine insects. pp. 69–89. *In* H. V. Danks (editor). *Insect Life-cycle Polymorphism*. Kluwer Academic Publishers, Dorchecht, The Netherlands.

Jacob, H. J., K. Lindpaintner, S. E. Lincoln, K. Kusumi, R. K. Bunker, Y.-P. Mao, D. Ganten, V. J. Dzau, and E. S. Lander. 1991. Genetic mapping of a gene causing hypertension in the stroke-prone spontaneously hypertensive rat. *Cell* 67: 213–224.

Jaenike, J. 1990. Host specialization in phytophagous insects. *Annual Review of Ecology and Systematics* 21: 243–273.

James, J. W. 1973. Covariances between relatives due to sex-linked genes. *Biometrics* 29: 584–588.

James, J. W. 1974. Genetic covariances under the partition of resources model. Appendix 1 in Sheridan and Barker (1974). *Australian Journal of Biological Science* 27: 99–101.

Jansen, R. C. 1994. Controlling the type I and type II errors in mapping quantitative trait loci. *Genetics* 138: 871–881.

Janssen, G. M., G. de Jong, E. N. G. Joosse, and W. Scharloo. 1988. A negative maternal effect in springtails. *Evolution* 42: 828–834.

Janzen, F. J. 1992. Heritable variation for sex ratio under environmental sex determination in the common snapping turtle (*Chelydra serpentina*). *Genetic* 131: 155–161.

Jarne, P., and B. Delay. 1990. Inbreeding depression and self-fertilization in *Lymnaea peregra* (Gastropoda: Pulmonata). *Heredity* 64: 169–175.

Jarne, P., L. Finot, B. Delay, and L. Thaler. 1991. Self-fertilization versus cross-fertilization in the hermaphroditic freshwater snail *Bulinus globosus*. *Evolution* 45: 1136–1146.

Jasieniuk, M., A. L. Brule-Babel, and I. N. Morrison. 1996. The evolution and genetics of herbicide resistance in agricultural weeds. *Weed Science* 44: 176–193.

Jenkins, N. L., and A. A. Hoffman. 1994. Genetic and maternal variation for heat resistance in *Drosophila* from the field. *Genetics* 137: 783–789.

Jernigan, R. W., D. C. Culver, and D. W. Fong. 1994. The dual role of selection and evolutionary history as reflected in genetic correlations. *Evolution* 48: 587–596.

Jimenez, J. A., K. A. Hughes, G. Alaks, L. Graham, and R. C. Lacy. 1994. An experimental study of inbreeding depression in a natural habitat. *Science* 266: 271–273.

Jinks, J. L. 1954. The analysis of continuous variation in a diallel cross of *Nicotiana rustica*. *Genetics* 39: 767–788.

Jinks, J. L., and P. L. Broadhurst. 1963. Diallel analysis of litter size and body weight in rats. *Heredity* 18: 319–336.

Jinks, J. L., and V. Connolly. 1973. Selection for specific and general response to environmental differences. *Heredity* 30: 33–40.

Jinks, J. L., and V. Connolly. 1975. Determination of the environmental sensitivity of selection lines by the selection environment. *Heredity* 34: 401–406.

Jinks, J. L., and H. S. Pooni. 1988. The genetic basis of environmental sensitivity. pp. 505–522. *In* E. J. Eisen, M. M. Goodman, G. Namkoong, and B. S. Weir (editors). *Proceedings of the Second International Conference on Quantitative Genetics.* Sinauer Associates, Sunderland, MA.

Jinks, J. L., and P. Towey. 1976. Estimating the number of genes in a polygenic system by genotype assay. *Heredity* 37: 69–81.

Jinks, J. L., J. M. Perkins, and E. L. Breese. 1969. A general method for the detection of additive, dominance and epistatic components of variation. II. Application of inbred lines. *Heredity* 24: 45–57.

Jinks, J. L., J. M. Perkins, and H. S. Pooni. 1973. The incidence of epistasis in normal and extreme environments. *Heredity* 31: 263–269.

Jinks, J. L., N. E. M. Jayasekara, and H. Boughey. 1977. Joint selection for both extremes of mean performance and of sensitivity to a macroenvironmental variable. II. Single-seed descent. *Heredity* 39: 345–355.

Joakimsen, O., and R. L. Baker. 1977. Selection for litter size in mice. *Acta Agriculturae Scandinavica* 27: 301–318.

Johnson, M. S., and R. Black. 1995. Neighbourhood size and the importance of barriers to gene flow in an intertidal snail. *Heredity* 75: 142–154.

Johnston, J. S. 1982. Genetic variation for anemotaxis (wind-directed movement) in laboratory and wild-caught populations of *Drosophilia*. *Behavior Genetics* 12: 281–293.

Johnston, M. O. 1992. Effects of cross- and self-fertilization on progeny fitness in *Lobelia cardinalis* and *L. siphilitica*. *Evolution* 46: 688–702.

Johnston, M. O., and D. J. Schoen. 1995. Mutation rates and dominance levels of genes affecting total fitness in two angiosperm species. *Science* 267: 226–229.

Johnston, M. O., and D. Schoen. 1994. On the measurement of inbreeding depression. *Evolution* 48: 1735–1741.

Jones, D. F. 1917. Dominance of linked factors as a means of accounting for heterosis. *Genetics* 466: 465–479.

Jones, L. P., R. Frankham, and J. S. F. Barker. 1968. The effects of population size and selection intensity in selection for a quantitative character in *Drosophila*. II. Long-term response to selection. *Genetic Research* 12: 249–266.

Jones, P. J. 1973. Some aspects of the feeding ecology of the Great Tit *Parus major* L. Ph.D. Thesis, University of Oxford, Oxford, UK.

Jong, de G. 1995. Phenotypic plasticity as a product of selection in a variable environment. *American Naturalist* 145: 493–512.

Jong, de G. 1990a. Quantitative genetics of reaction norms. *Journal of Evolutionary Biology* 3: 447–468.

Jong, de G. 1990b. Genotype-by-environment interaction and the genetic covariance between environments: multilocus genetics. *Genetica* 81: 171–177.

Jong, de G., and S. C. Stearns. 1991. Phenotypic plasticity and the expression of genetic variation. pp. 707–718. *In* E. C. Dudley (editor). *The Unity of Evolutionary Biology*, Vol. II. Dioscorides Press, Portland, OR.

Jong, de G., and A. J. van Noordwijk. 1992. Acquisition and allocation of resources: genetic (co)variances, selection, and life histories. *American Naturalist* 139: 749–770.

Joshi, A., and L. D. Mueller. 1993. Directional and stabilizing density-dependent natural selection for pupation height in *Drosophila melanogaster*. *Evolution* 47: 176–184.

Joshi, A., and J. N. Thompson. 1995. Trade-offs and the evolution of host specialization. *Evolutionary Ecology* 9: 82–92.

Jung, M., T. Weldekidan, D. Schaff, A. Paterson, S. Tingey, and J. Hawk. 1994. Generation-means analysis and quantitative trait locus mapping of anthracnose stalk rot genes in maize. *Theoretical and Applied Genetics* 89: 413–418.

Kallman, K. D. 1989. Genetic control of size at maturity in *Xiphophorus*. pp. 163–184. *In* G. K. Meffe, Jr. and F. F. Snelson (editors), *Ecology and Evolution of Livebearing Fish (Poeciliidae)*. Prentice-Hall, Inc., Englewood Cliffs, NJ.

Kambysellis, M. P., and W. B. Heed. 1971. Studies of oogenesis in natural populations of Drosophilidae. I. Relation of ovarian development and ecological habitats of the Hawaiian species. *American Naturalist* 105: 31–49.

Karn, M. N., and L. S. Penrose. 1951. Birth weight and gestation time in relation to maternal age, parity and infant survival. *Annals of Eugenics* 16: 147–164.

Kasule, F. K. 1992. A quantitative genetic analysis of reproductive allocation in the cotton stainer bug, *Dysdercus fasciatus*. *Heredity* 69: 141–149.

Kat, P. 1982. The relationship between heterozygosity for enzyme loci and developmental homeostasis in peripheral populations of aquatic bivalves (Unionidae). *American Naturalist* 119: 824–832.

Kaufman, P. K., F. D. Enfield, and R. E. Comstock. 1977. Stabilizing selection for pupa weight in *Tribolium casteneum*. *Genetics* 87: 327–341.

Kawecki, T. J., and S. C. Stearns. 1993. The evolution of life histories in spatially heterogeneous environments: optimal reaction norms revisited. *Evolutionary Ecology* 7: 155–174.

Keane, B. 1990. The effect of relatedness on reproductive success and mate choice in the white-footed mouse, *Peromyscus leucopus*. *Animal Behaviour* 39: 264–273.

Kearsey, M. J. 1980. The efficiency of the North Carolina expt. III and the selfing, backcrossing series for estimating additive and dominance variance. *Heredity* 45: 73–82.

Kearsey, M. J., and J. S. Gale. 1968. Stabilising selection in the absence of dominance: an additional note. *Heredity* 23: 617–620.

Kearsey, M. J., and J. L. Jinks. 1968. A general method of detecting additive, dominance and epistatic variation for metrical traits. I. Theory. *Heredity* 23: 403–409.

Keightley, P. D., and W. G. Hill. 1990. Variation maintained in quantitative traits with mutation–selection balance: pleiotropic side-effects on fitness traits. *Proceedings of the Royal Society of London* B 242: 95–100.

Keightley, P. D., and W. G. Hill. 1992. Quantitative genetic variation in body size of mice from new mutations. *Genetics* 131: 693–700.

Keim, P., B. W. Diers, and R. C. Shoemaker. 1990. Genetic analysis of soybean hard seededness with molecular markers. *Theoretical and Applied Genetics* 79: 465–469.

Kekic, V., and D. Marinkovic. 1974. Multiple-choice selection for light preference in *Drosophila subobscura*. *Behavior Genetics* 4: 285–300.

Kelly, C. A. 1993. Quantitative genetics of size and phenology of life-history traits in *Chamaecrista fasciculata*. *Evolution* 47: 88–97.

Kempthorne, O. 1956. The theory of the diallel cross. *Genetics* 41: 451–459.

Kempthorne, O., and R. N. Curnow. 1961. The partial diallel cross. *Heredity* 17: 229–250.

Kempthorne, O., and O. B. Tandon. 1953. The estimation of heritability by regression of offspring on parent. *Biometry* 9: 90–100.

Kennedy, B. W. 1981. Variance component estimation and prediction of breeding values. *Canadian Journal of Genetics and Cytology* 23: 565–578.

Kermicle, J., and M. Alleman. 1990. Gametic imprinting in maize in relation to the angiosperm life cycle. *Development* 110 (Supplement): 9–14.

Kerster, H. W. 1964. Neighborhood size in the rusty lizard, *Sceloporus olivaceus*. *Evolution* 18:445–457.

Kerver, W. J. M., and G. Rotman. 1987. Development of ethanol tolerance in relation to the alcohol dehydrogenase locus in *Drosophila melanogaster*. II. The influence of phenotypic adaptation and maternal effect on survival on alcohol supplemented media. *Heredity* 58: 239–248.

Kessler, S. 1969. The genetics of *Drosophila* mating behavior. II. The genetic architecture of mating speed in *Drosophila pseudobscura*. *Genetics* 62: 421–433.

Kidwell, J. F., and M. M. Kidwell. 1966. The effects of inbreeding on body weight and abdominal chaeta number in *Drosophila melanogaster*. *Canadian Journal of Genetics and Cytology* 8: 207–215.

Kimura, M. 1956. Rules for testing stability of a selective polymorphism. *Proceedings of the National Academy of Sciences* 42: 336–340.

Kincaid, H. L. 1976a. Effects of inbreeding on rainbow trout populations. *Transactions of the American Fisheries Society* 2: 273–280.

Kincaid, H. L. 1976b. Inbreeding in rainbow trout (*Salmo gairdneri*). *Journal of the Fisheries Research Board of Canada* 33: 2420–2426.

Kincaid, H. L., W. R. Bridges, and B. Limbach. 1977. Three generations of selection for growth rate in Fall-spawning rainbow trout. *Transactions of the American Fisheries Society* 106: 621–628.

Kindred, B. 1965. Selection for temperature sensitivity in scute *Drosophila*. *Genetics* 52: 723–728.

King, B. H. 1987. Offspring sex ratios in parasitoid wasps. *Quarterly Review of Biology* 62: 367–396.

King, B. H. 1988. Sex-ratio manipulation in response to host size by the parasitioid wasp *Spalangia cameroni*: a laboratory study. *Evolution* 42: 1190–1198.

Kinney, T. B., Jr. 1969. A summary of reported estimates of heritabilities and of genetic and phenotypic correlations for traits of chickens. *US Department of Agriculture Agricultural Handbook* 363: 1–49.

Kirkpatrick, M. 1982. Sexual selection and the evolution of female choice. *Evolution* 36: 1–12.

Kirkpatrick, M., and N. Heckman. 1989. A quantitative genetic model for growth, shape, reaction norms, and other infinite-dimensional characters. *Journal of Mathematical Biology* 27: 429–450.

Kirkpatrick, M., and R. Lande. 1989. The evolution of maternal characters. *Evolution* 43: 485–503.

Kirkpatrick, M., D. Lofsvold, and M. Bulmer. 1990. Analysis of the inheritance, selection and evolution of growth trajectories. *Genetics* 124: 979–993.

Kirpichnikov, V. S. 1981. *Genetic Basis of Fish Selection.* Springer-Verlag, New York.

Klein, T. W. 1974. Heritability and genetic correlation: Statistical power, population comparisons, and sample size. *Behavior Genetics* 4: 171–189.

Knapp, S. J., Jr., W. C. Bridges, and M. Yang. 1989. Nonparametric confidence estimators for heritability and expected selection response. *Genetics* 121: 891–898.

Knott, S. A., R. M. Sibly, R. H. Smith, and H. Moller. 1995. Maximum likelihood estimation of genetic parameters in life-history using the "animal model." *Ecology* 9: 122–126.

Koch, R. M., and R. T. Clark. 1955. Genetic and environmental relationships among economic characters in beef cattle. III. Evaluating maternal environment. *Journal of Animal Science* 14: 979–996.

Koehn, R. K., and P. M. Gaffney. 1984. Genetic heterozygosity and growth rate in *Mytilus edulis. Marine Biology* 82: 1–7.

Koehn, R. K., and S. E. Shumway. 1982. A genetic/physiological explanation for differential growth rate among individuals of the American oyster, *Crassostrea virginica* (Gmelin). *Marine Biology Letters* 3: 35–42.

Koehn, R. K., W. J. Diehl, and T. M. Scott. 1988. The differential contribution by individual enzymes of glycolysis and protein catabolism to the relationship between heterozygosity and growth rate in the coot clam. *Mulinia lateralis. Genetics* 118: 121–130.

Koenig, W. D., and R. L. Mumme. 1987. *Population Ecology of the Cooperatively Breeding Acorn Woodpecker.* Princeton University Press, Princeton, NJ.

Kohn, L. A. P., and W. R. Atchley. 1988. How similar are genetic correlation structures? Data from mice and rats. *Evolution* 42: 467–482.

Kojima, K. 1959. Stable equilibria for the optimum model. *Proceedings of the National Academy of Sciences* 45: 989–993.

Kojima, K. I. 1961. Effects of dominance and size of population on response to mass selection. *Genetic Research* 2: 177–188.

Koots, K. R., and J. P. Gibson. 1994. How precise are genetic correlation estimates? *Proceedings of the 5th World Congress on Genetics Applied to Livestock Production* 18: 353–356.

Koots, K. R., and J. P. Gibson. 1996. Realized sampling variances of estimates of genetic parameters and the difference between genetic and phenotypic correlations. *Genetics* 143: 1409–1416.

Krebs, C. J., and J. H. Myers. 1974. Population cycles in small mammals. *Advances in Ecological Research* 8: 267–399.

Krimbas, C. B., and S. Tsakas. 1971. The genetics of *Dacus oleae*. V. Changes of esterase polymorphism in a natural population following insecticide control—selection or drift? *Evolution* 25: 454–460.

Kuhlers, D. L., A. B. Chapman, and N. L. First. 1977. Estimates of maternal and grand-maternal influences on weights and gains of pigs. *Journal of Animal Science* 44: 181–188.

Kuiper, D., and P. J. Kuiper. 1988. Phenotypic plasticity in a physiological perspective. *Acta Oecologica Oecologia Plantarum* 9: 43–59.

Lacy, R. C., A. Petric, and M. Warneke. 1993. Inbreeding and outbreeding in captive populations of wild animal species. pp. 352–74, *In* N. W. Thornhill (editor), *The Natural History of Inbreeding and Outbreeding* University of Chicago Press, Chicago.

Lambio, A. L. 1981. Response to divergent selection for 4-week body weight, egg production and total plasma phosphorus in Japanese quail. *Dissertation Abstracts International* B 42: 2694.

Lande, R. 1975. The maintenance of genetic variation by mutation in a polygenic character with linked loci. *Genetic Research* 26: 221–235.

Lande, R. 1976. Natural selection and random genetic drift in phenotypic evolution. *Evolution* 30: 314–334.

Lande, R. 1977. Statistical tests for natural selection on quantitative characters. *Evolution* 31: 442–444.

Lande, R. 1979a. Quantitative genetic analysis of multivariate evolution applied to brain: body size allometry. *Evolution* 33: 402–416.

Lande, R. 1979b. Effective deme sizes during long-term evolution estimated from rates of chromosomal rearrangement. *Evolution* 33: 234–251.

Lande, R. 1980a. The genetic covariance between characters maintained by pleiotropic mutations. *Genetics* 94: 203–215.

Lande, R. 1980b. Sexual dimorphism, sexual selection, and adaptation in polygenic characters. *Evolution* 34: 292–305.

Lande, R. 1981a. The minimum number of genes contributing to quantitative variation between and within populations. *Genetics* 99: 541–553.

Lande, R. 1981b. Models of speciation by sexual selection on polygenic traits. *Proceedings of the National Academy of Sciences* 78: 3721–3725.

Lande, R. 1984. The genetic correlation between characters maintained by selection, linkage and inbreeding. *Genetic Research* 44: 309–320.

Lande, R. 1995. Mutation and conservation. *Conservation Biology* 9: 782–91.

Lande, R., and S. J. Arnold. 1983. The measurement of selection on correlated characters. *Evolution* 37: 1210–1226.

Lande, R., and M. Kirkpatrick. 1990. Selection response in traits with maternal inheritance. *Genetic Research* 55: 189–197.

Lande, R., and T. Price. 1989. Genetic correlations and maternal effect coefficients obtained from offspring-parent regression. *Genetics* 122: 915–922.

Lande, R., and D. W. Schemske. 1985. The evolution of self-fertilization and inbreeding depression in plants. I. Genetic models. *Evolution* 39: 24–40.

Langridge, J. 1962. A genetic and molecular basis for heterosis in *Arabidopsis* and *Drosophila*. *American Naturalist* 96: 5–27.

Langslow, D. R. 1979. Movements of blackcaps ringed in Britain and Ireland. *Bird Study* 26: 239–252.

Lannon, J. E. 1972. Estimating heritability and predicting response to selection for the Pacific oyster, *Crassostrea gigas*. *Proceedings of the National Shell-Fish Association* 62: 62–66.

Larsson, K., and P. Forslund. 1992. Genetic and social inheritance of body and egg size in the barnacle goose (*Branta leucopsis*). *Evolution* 46: 235–244.

Latta, R., and K. Ritland. 1994. The relationship between inbreeding depression and prior inbreeding among populations of four *Mimulus* taxa. *Evolution* 48: 806–817.

Latter, B. D. H. 1960. Natural selection for an intermediate optimum. *Australian Journal of Biological Sciences* 13: 30–35.

Latter, B. D. H. 1964. Selection for a threshold character in *Drosophila*. I. An analysis of the phenotypic variance on the underlying scale. *Genetic Research* 5: 198–210.

Latter, B. D. H., and J. C. Mulley. 1995. Genetic adaptation to captivity and inbreeding depression in small laboratory populations of *Drosophila melanogaster*. *Genetics* 139: 255–266.

Latter, B. D. H., and A. Robertson. 1960. Experimental design in the estimation of heritability by regression methods. *Biometrics* 16: 348–353.

Latter, B. D. H., and A. Robertson. 1962. The effects of inbreeding and artificial selection on reproductive fitness. *Genetic Research* 3: 110–138.

Latter, B. D. H., J. C. Mulley, D. Reid, and L. Pascoe. 1995. Reduced genetic load revealed by slow inbreeding in *Drosophila melanogaster*. *Genetics* 139: 287–297.

Laurie-Ahlberg, C. C., and B. S. Weir. 1979. Allozymic variation and linkage disequilibrium in some laboratory populations of *Drosophila melanogaster*. *Genetics* 92: 1295–1314.

Lawrence, C. W. 1964. Genetic studies on wild populations of *Melandrium III*. *Heredity* 19: 1–19.

Leamy, L. 1986. Directional selection and developmental stability: evidence from fluctuating asymmetry of dental characters in mice. *Heredity* 57: 381–88.

Leamy, L., and J. Cheverud. 1984. Quantitative genetics and the evolution of ontogeny. II. Genetic and environmental correlations among age-specific characters in randombred mice. *Growth* 48: 339–353.

Leary, R. F., F. W. Allendorf, and K. L. Knudsen. 1983. Developmental stability and enzyme heterozygosity in rainbow trout. *Nature* 301: 71–72.

Leary, R. F., F. W. Allendorf, and K. L. Knudsen. 1984. Superior developmental stability of heterozygotes at enzyme loci in salmonid fishes. *American Naturalist* 124: 540–551.

Leary, R. F., F. W. Allendorf, and K. L. Knudsen. 1985. Inheritance of meristic variation and the evolution of developmental stability in rainbow trout. *Evolution* 39: 308–314.

Leberg, P. L. 1992. Effects of population bottlenecks on genetic diversity as measured by allozyme electrophoresis. *Evolution* 46: 477–494.

Ledig, F. T., R. P. Guries, and B. A. Bonefeld. 1983. The relation of growth to heterozygosity in pitch pine. *Evolution* 37: 1227–1238.

Legner, E. F. 1991. Estimation of the number of active loci, dominance and heritability in polygenic inheritance of gregarious behavior in *Muscidifurax raptorellus* (Hymenoptera: Pteromalidae). *Entomophaga* 36: 1–18.

Lerner, I. M. 1950. *Population Genetics and Animal Improvement*. Cambridge University Press, Cambridge.

Lerner, I. M. 1954. *Genetic Homeostasis*. John Wiley & Sons, New York.

Lerner, I. M., and E. R. Dempster. 1951. Attenuation of genetic progress under continued selection in poultry. *Heredity* 5: 75–94.

Lerner, I. M., and C. A. Gunns. 1952. Egg size and reproductive fitness. *Poultry Science* 31: 537–544.

Lessells, C. M., and P. T. Boag. 1987. Unrepeatable repeatabilities: A common mistake. *Auk* 104: 116–121.

Lessells, C. M., F. Cooke, and R. F. Rockwell. 1989. Is there a trade-off between egg weight and clutch size in wild Lesser Snow Geese (*Anser c. caerulescens*)? *Journal of Evolutionary Biology* 2: 457–472.

Leutenegger, W., and J. M. Cheverud. 1985. Sexual dimorphism in primates: the effects of size. pp. 33–50. *In* W. L. Jungers (editor). *Size and Scaling in Primate Biology*. Plenum Press, New York.

Levene, H. 1953. Genetic equilibrium when more than one ecological niche is available. *American Naturalist* 87: 331–333.

Levin, D. 1991. The effect of inbreeding on seed survivorship in *Phlox*. *Evolution* 45: 1047–1049.

Levin, L. A., J. Zhu, and E. Creed. 1991. The genetic basis of life history characters in a polychaete exhibiting planktotrophy and lecthotrophy. *Evolution* 45: 380–397.

Lewontin, R. C. 1964. The interaction of selection and linkage. II. Optimum models. *Genetics* 50: 757–782.

Lewontin, R. C., L. R. Ginsburg, and S. D. Taljaparkar. 1978. Heterosis as an explanation for large amounts of genic polymorphism. *Genetics* 88: 149–170.

Linhart, Y. B., J. B. Mitton, D. M. Bowman, K. B. Sturgeon, and J. L. Hamrick. 1979. Genetic aspects of fertility differentials in ponderosa pine. *Genetic Research* 33: 237–242.

Lints, F. A., and M. Bourgois. 1982. A test of the genetic revolution hypothesis of speciation. *In* S. Lakovaara (editor). *Advances in Genetics, Development and Evolution of Drosophila*. Plenum Press, New York.

Lively, C. M. 1986. Canalization versus developmental conversion in a spatially variable environment. *American Naturalist* 128: 561–572.

Lloyd, D. G. 1979. Some reproductive factors affecting the selection of self-fertilization in plants. *American Naturalist* 113: 67–97.

Lloyd, D. G. 1984. Variation strategies of plants in heterogeneous environments. *Biological Journal of the Linnean Society* 21: 357–385.

Loebel, D. A., R. K. Nurthen, R. Frankham, D. A. Briscoe, and D. Craven. 1992. Modeling problems in conservation genetics using captive *Drosophila* populations: consequences of equalizing founder representation. *Zoo Biology* 11: 319–332.

Lofsvold, D. 1986. Quantitative genetics of morphological differentiation in *Peromyscus*. I. Tests of the homogeneity of genetic covariance structure among species and subspecies. *Evolution* 40: 559–573.

Lopez, M. A., and C. López-Fanjul. 1993a. Spontaneous mutation for a quantitative trait in *Drosophila melanogaster*. I. Response to artificial selection. *Genetic Research* 61: 107–116.

Lopez, M. A., and C. López-Fanjul. 1993b. Spontaneous mutation for a quantitative trait in *Drosophila melanogaster*. II. Distribution of mutant effects on the trait and fitness. *Genetic Research* 61: 117–126.

López-Fanjul, C., and W. G. Hill. 1973. Genetic differences between populations of *Drosophila melanogaster* for a quantitative trait. *Genetic Research* 22: 69–78.

Lush, J. L. 1948. *The Genetics of Populations*. Iowa State University, Ames.

Lynch, C. B. 1977. Inbreeding effects upon animals derived from a wild population of *Mus musculus*. *Evolution* 31: 526–537.

Lynch, C. B. 1994. Evolutionary inferences from genetic analyses of cold adaptation in laboratory and wild populations of the house mouse. pp. 278–304. *In* C. R. B. Boake (editor). *Quantitative Genetic Studies of Behavioral Evolution*. University of Chicago Press, Chicago.

Lynch, M. 1988a. Design and analysis of experiments on random drift and inbreeding depression. *Genetics* 120: 791–807.

Lynch, M. 1988b. The rate of polygenic mutation. *Genetic Research* 51: 137–148.

Lynch, M. 1991. The genetic interpretation of inbreeding depression and outbreeding depression. *Evolution* 45: 622–629.

Lynch, M., and W. Gabriel. 1987. Environmental tolerance. *American Naturalist* 129: 283–303.

Lynch, M., and W. G. Hill. 1986. Phenotypic evolution by neutral mutation. *Evolution* 40: 915–935.

Mackay, T. F. C. 1980. Genetic variance, fitness, and homeostasis in varying environments: an experimental check of the theory. *Evolution* 34: 1219–1222.

Mackay, T. F. C. 1981. Genetic variation in varying environments. *Genetic Research* 37: 79–93.

Mackay, T. F. C. 1985. A quantitative genetic analysis of fitness and its components in *Drosophila melanogaster*. *Genetic Research* 47: 59–70.

Mackay, T. F. C., R. F. Lyman, M. S. Jackson, C. Terzian, and W. G. Hill. 1992. Polygenic mutation in *Drosophila melanogaster*: estimates from divergence among inbred strains. *Evolution* 46: 300–316.

Mallet, A. L., and L. E. Haley. 1983. Effects of inbreeding on larval and spat performance in the American oyster. *Aquaculture* 33: 229–235.

Mangan, R. L. 1991. Analysis of genetic control of mating behavior in screwworm (Diptera: Calliphoridae) males through diallel crosses and artificial selection. *Theoretical and Applied Genetics* 81: 429–436.

Mangus, W. L., and J. S. Brinks. 1970. Relationships between direct and maternal effects on growth in Herefords: I. Environmental factors during preweaning growth. *Journal of Animal Science* 32: 17–25.

Mani, G. S., B. C. Clarke, and P. R. Sheltom. 1990. A model of quantitative traits under frequency-dependent balancing selection. *Proceedings of the Royal Society of London* B 240: 15–28.

Manly, B. F. J. 1986. *Multivariate Statistics: A Primer*. Chapman & Hall, London.

Manly, B. F. J. 1991. *Randomization and Monte Carlo Methods in Biology*. Chapman & Hall, London.

Manning, A. 1961. The effects of artificial selection for mating speed in *Drosophila melanogaster*. *Animal Behaviour* 9: 82–92.

Manning, A. 1963. Selection for mating speed in *Drosophila melanogaster* based on the behaviours of one sex. *Animal Behaviour* 11: 116–120.

Marien, D. 1958. Selection for developmental rate in *Drosophila pseudoobscura*. *Genetics* 56: 3–15.

Marshall, D. R., and S. K. Jain. 1968. Genetic polymorphism in natural populations of *Avena fatua* and *A. barbata*. *Nature* 221: 276–278.

Martin, G. A., and A. E. Bell. 1960. An experimental check on the accuracy of prediction of response during selection. pp. 178–187. *In* O. Kempthorne (editor). *Biometrical Genetics*. Pergamon Press, London.

Martin, N. G., J. L. Jinks, H. S. Berry, and D. Z. Loesch. 1982. A genetical analysis of diversity and asymmetry in finger ridge counts. *Heredity* 48: 393–405.

Martinez, M. L., A. E. Freeman, and P. J. Berger. 1983. Age of dam and direct and maternal effects on calf livability. *Journal of Dairy Science* 66: 1714–1720.

Maruyama, T., and P. A. Fuerst. 1985. Population bottlenecks and nonequilibrium models in population genetics: II. Number of alleles in a small population that was formed by a recent bottleneck. *Genetics* 111: 675–689.

Masaki, S., and E. Seno. 1990. Effect of selection on wing dimorphism in the ground cricket *Dianemobius fascipes* (Walker). *Boletin De Sanidad Vegetal, Plagas* (Fuera De Serie) 20: 381–393.

Mason, L. G. 1964. Stabilizing selection for mating fitness in natural populations of *Tetropes*. *Evolution* 18: 492–497.

Mason, L. G. 1969. Mating selection in the California Oak moth (Lepidoptera, Dioptidae). *Evolution* 23: 55–58.

Mather, K., and J. L. Jinks. 1982. *Biometrical Genetics*. Chapman & Hall, London.

Matos, C. A. P., D. L. Thomas, L. D. Young, and D. Gianola. 1994. Analysis of lamb survival using linear and threshold models with maternal effects. pp. 426–429. *In* J. S. Gavora, B. Benkel, J. Chesnais, W. Fairfull, J. P. Gibson, B. W. Kennedy, E. B. Burnside, and C. Smith (editors). *Proceedings of the 5th World Congress on Genetics Applied to Livestock Production*, Volume 18. University of Guelph, Guelph, Ontario.

Matsumura, M. 1996. Genetic Analysis of a threshold trait: density-dependent wing dimorphism in *Sogatella furcifera* (Horvath) (Hemiptera: Delphacidae), the whitebacked planthopper. *Heredity* 76: 229–237.

Maynard Smith, J. 1956. Fertility, mating behaviour and sexual selection in *Drosophila subobscura*. Genetics 54: 261–279.

Maynard Smith, J. 1978. *The Evolution of Sex*. Cambridge University Press, Cambridge.

Maynard Smith, J., R. Burian, S. Kauffman, P. Alberch, J. Campbell, B. Goodwin, R. Lande, D. Raup, and L. Wolpert. 1985. Developmental constraints and evolution. *Quarterly Review of Biology* 60: 265–287.

Mazer, S. 1987a. Parental effects on seed development and seed yield in *Raphanus raphanistrum*: implications for natural and sexual selection. *Evolution* 41: 355–371.

Mazer, S. J. 1987b. The quantitative genetics of life-history and fitness components in *Raphanus raphanistrum* L. (Brassiceae): ecological and evolutionary consequences of seed-weight variation. *American Naturalist* 130: 891–914.

Mazer, S. J., and C. T. Schick. 1991. Constancy of population parameters for life history and floral traits in *Raphanus sativus* L. I. Norms of reaction and the nature of genotype by environment interactions. *Heredity* 67: 143–156.

McAlpine, S. 1993. Genetic heterozygosity and reproductive success in the green treefrog, *Hyla cinerea*. *Heredity* 70: 553–558.

McAndrew, B. J., R. D. Ward, and J. A. Beardmore. 1986. Growth rate and heterozygosity in the plaice, *Pleuronectes platessa*. *Heredity* 57: 171–180.

McCall, C., D. M. Waller, and T. Mitchell-Olds. 1994. Effects of serial inbreeding on fitness components in *Impatiens capensis*. *Evolution* 48: 818–827.

McCarthy, J. C., and D. P. Doolittle. 1977. Effects of selection for independent changes in two highly correlated body weight traits of mice. *Genetic Research* 29: 133–145.

McCracken, G. F., and P. F. Brussard. 1980. Self-fertilization in the white-lipped land snail *Triodopsis albolabris*. *Biological Journal of the Linnean Society* 14: 429–434.

McFarquhar, A. M., and F. W. Robertson. 1963. The lack of evidence for co-adaptation in crosses between geographical races of *Drosophila subobscura* Coll. *Genetic Research* 4: 104–131.

McGrath, J. W., J. M. Cheverud, and J. E. Buikstra. 1984. Genetic correlations between sides and heritability of asymmetry for nonmetric traits in rhesus macaques on Cayo Santiago. *American Journal of Physical Anthropology* 64: 401–411.

McInnis, D. O., L. E. Wendel, and C. J. Whitten. 1983. Directional selection and heritability for pupal weight in the screwworm, *Cochliomyia hominivorax* (Diptera: Calliphoridae). *Annals of the Entomological Society of America* 76: 30–36.

McKenzie, J. A., and P. Batterham. 1994. The genetic, molecular and phenotypic consequences of selection for insecticide resistance. *Trends in Ecology and Evolution* 9: 166–169.

McLaren, I. 1976. Inheritance of demographic and production parameters in the marine copepod, *Eurytemora herdmani*. *Biological Bulletin* 151: 200–213.

McLaren, I., and C. Corkett. 1978. Unusual genetic variation in body size, development times, oil storage, and survivorship in the marine copepod *Pseudocalanus*. *Biological Bulletin* 155: 347–359.

Mercer, J. T., and W. G. Hill. 1984. Estimation of genetic parameters for skeletal defects in broiler chickens. *Heredity* 53: 193–203.

Merrell, D. J. 1968. A comparison of the estimated size and the "effective size" of breeding populations of the leopard frog, *Rana pipiens*. *Evolution* 22: 274–283.

Meyer, A. 1990. Morphometrics and allometry in the trophically polymorphic cichlid fish, *Cichlasoma citrinellum*: Alternative adaptations and ontogenetic changes in shape. *Journal of Zoology* 221: 237–260.

Meyer, H. H., and F. D. Enfield. 1975. Experimental evidence on limitations of the heritability parameter. *Theoretical and Applied Genetics* 45: 268–273.

Miller, R. G. 1974. The jackknife—a review. *Biometrika* 61: 1–15.

Millet, E., and M. J. Pinthus. 1980. Genotypic effects of the maternal tissues of wheat on its grain weight. *Theoretical and Applied Genetics* 58: 247–252.

Minvielle, F. 1979. Comparing the means of inbred lines with the base population: a model with overdominant loci. *Genetic Research* 33: 89–92.

Misra, R. K., and E. C. R. Reeve. 1964. Clines in body dimensions in populations of *Drosophilia subobscura*. *Genetic Research* 5: 240–256.

Misztal, I. 1994. Comparison of software packages in animal breeding. pp. 3–10. In J. S. Gavora, B. Benkel, J. Chesnais, W. Fairfull, J. P. Gibson, B. W. Kennedy, E. B. Burnside, and C. Smith (editors). 5th World Congress on Genetics Applied to Livestock Production, Vol. 22. University of Guelph, Guelph, Ontario, Canada.

Mitchell-Olds, T. 1991. Quantitative genetic changes in small populations. pp. 634–638. *In* E. C. Dudley (editor). *The Unity of Evolutionary Biology,* Vol. II. Dioscorides Press, Portland, OR.

Mitchell-Olds, T., and J. J. Rutledge. 1986. Quantitative genetics in natural plant populations: a review of the theory. *American Naturalist* 127: 379–402.

Mitchell-Olds, T., and D. M. Waller. 1985. Relative performance of selfed and outcrossed progeny in *impatiens capensis*. *Evolution* 39: 533–544.

Mitton, J. B. 1978. Relationship between heterozygosity for enzyme loci and variation of morphological characters in natural populations. *Nature* 273: 661–662.

Mitton, J. B., and M. C. Grant. 1980. Observations on the ecology and evolution of quaking aspen, *Populus tremuloides*, in the Colorado front range. *American Journal of Botany* 67: 202–9.

Mitton, J. B., and M. C. Grant. 1984. Associations among protein heterozygosity, growth rate, and developmental homeostasis. *Annual Review of Ecology and Systematics* 15: 479–499.

Mitton, J. B., and B. A. Pierce. 1980. The distribution of individual heterozygosity in natural populations. *Genetics* 95: 1043–1054.

Mitton, J. B., C. Carey, and T. D. Kocher. 1986. The relation of enzyme heterozygosity to standard and active oxygen consumption and body size of tiger salamanders, *Ambystoma tigrinum. Physiological Zoology* 59: 574–582.

Moav, R., and G. Wohlfarth. 1976. Two-way selection for growth rate in the common carp (*Cyprinus carpio* L.). *Genetics* 82: 83–101.

Moller, A. P. 1994. Sexual selection in the barn swallow (*Hirundo rustica*). IV. Patterns of fluctuating asymmetry and selection against asymmetry. *Evolution* 48: 658–670.

Moller, A. P., and R. Thornhill. 1997. A meta-analysis of the heritability of developmental stability. *Journal of Evolutionary Biology* (in press).

Moller, H., R. H. Smith, and R. M. Sibly. 1989. Evolutionary demography of a bruchid beetle. I. Quantitative genetical analysis of the female life history. *Functional Ecology* 3: 673–681.

Montalvo, A. M. 1994. Inbreeding depression and maternal effects in *Aquilegia caerulea* a partially selfing plant. *Ecology* 75: 2395–2409.

Montalvo, A. M., and R. G. Shaw. 1994. Quantitative genetics of sequential life-history and juvenile traits in the partially selfing perennial *Aquilegia caerulea. Evolution* 48: 828–841.

Moody, D. E., D. Pomp, S. Newman, and M. D. MacNeil. 1994. Characterization of DNA polymorphisms and their associations with growth and maternal traits in line 1 Hereford cattle. pp. 221–224. *In* J. S. Gavora, B. Benkel, J. Chesnais, W. Fairfull, J. P. Gibson, B. W. Kennedy, E. B. Burnside, and C. Smith (editors). *5th World Congress on Genetics Applied to Livestock Production,* Vol. 21. University of Guelph, Guelph, Ontario, Canada.

Moore, A. J. 1990. The evolution of sexual dimorphism by sexual selection: the separate effects of intrasexual selection and intersexual selection. *Evolution* 44: 319–331.

Moore, J. A. 1950. Further studies on *Rana pipiens* racial hybrids. *American Naturalist* 84: 247–254.

Moore, R. W., E. J. Eisen, and L. C. Ulberg. 1970. Prenatal and postnatal maternal influences on growth in mice selected for body weight. *Genetics* 64: 59–68.

Mori, K., and F. Nakasuji. 1990. Genetic analysis of the wing-form determination of the small brown planthopper, *Laodelphax striatellus* (Hemiptera: Delphacidae). *Researchers in Population Ecology* 32: 279–287.

Moriwaki, D., and Y. Fuyama. 1963. Responses to selection for rate of development in *D. melanogaster. Drosophila Information Service* 38: 74.

Morris, R. F. 1971. Observed and simulated changes in genetic quality in natural populations of *Hyphantria cunea. Canadian Entomologist* 103: 893–906.

Morris, R. F., and W. Fulton. 1970. Heritability of diapause intensity in *Hyphantria cunea* and correlated fitness responses. *Canadian Entomologist* 102: 927–938.

Morrison, W. W., and R. Milkman. 1978. Modification of heat resistance in *Drosophila* by selection. *Nature* 273: 49–50.

Moss, R., and A. Watson. 1982. Heritability of egg size, hatch weight, body weight and viability in red grouse (*Lagopus lagopus scoticus*). *Auk* 99: 683–686.

Mousseau, T. A., and H. Dingle. 1991. Maternal effects in insect life histories. *Annual Review of Entomology* 36: 511–534.

Mousseau, T. A., and D. A. Roff. 1987. Natural selection and the heritability of fitness components. *Heredity* 59: 181–198.

Mousseau, T. A., and D. A. Roff. 1989a. Adaptation to seasonality in a cricket: patterns of phenotypic and genotypic variation in body size and diapause expression along a cline in season length. *Evolution* 43: 1483–1496.

Mousseau, T. A., and D. A. Roff. 1989b. Geographic variability in the incidence and heritability of wing dimorphism in the striped ground cricket, *Allonemobius fasciatus*. *Heredity* 62: 315–318.

Mousseau, T. A., and D. A. Roff. 1995. Genetic and environmental contributions to geographic variation in the ovipositor length of a cricket. *Ecology* 76: 1473–1482.

Mrakovcic, M., and L. E. Haley. 1979. Inbreeding depression in the Zebra fish *Brachydanio rerio* (Hamilton Buchanan). *Journal of Fish Biology* 15: 323–327.

Mueller, J. P., and J. W. James. 1985. Phenotypic response to selection for traits with direct and maternal components when generations overlap. *Theoretical and Applied Genetics* 70: 123–127.

Muir, W. M. 1986. Estimation of response to selection and utilization of control populations for additional information and accuracy. *Biometrics* 42: 381–391.

Muir, W., W. E. Nyquist, and S. Xu. 1992. Alternative partitioning of the genotype-by-environment interaction. *Theoretical and Applied Genetics* 84: 193–200.

Mullin, T. J., E. K. Morgenstern, Y. S. Park, and D. P. Fowler. 1992. Genetic parameters from a clonally replicated test of black spruce (*Picea mariana*). *Canadian Journal of Forest Research* 22: 24–36.

Murray, J., and B. Clarke. 1967. Inheritance of shell size in *Partula*. *Heredity* 23: 189–198.

Nasholm, A., and O. Danell. 1994. Maternal effects on lamb weights. pp. 163–166. *In* J. S. Gavora, B. Benkel, J. Chesnais, W. Fairfull, J. P. Gibson, B. W. Kennedy, E. B. Burnside, and C. Smith (editors). *Proceedings of the 5th World Congress on Genetics Applied to Livestock Production*. Dept. Animal and Poultry Science, University of Guelph, Guelph, Ontario, Canada.

Naylor, A. F. 1964. Natural selection through maternal influence. *Heredity* 19: 509–511.

Nei, M., and F. Tajima. 1981. Genetic drift and estimation of effective population size. *Genetics* 98: 625–640.

Nei, M., T. Maruyama, and R. Chakraborty. 1975. The bottleneck effect and genetic variability in populations. *Evolution* 29: 1–10.

Nei, M., and Y. Imaizumi. 1966. Genetic structure of human populations: II. differentiation of blood group gene frequencies among isolated populations. *Heredity* 21: 461–472.

Nelson, J. R., G. A. Harris, and C. J. Goebel. 1970. Genetic vs. environmentally induced variation in medusahead (*Taeniatherum asperum* [Simonkai] Nevski). *Ecology* 51: 526–529.

Nestor, K. E., K. I. Brown, and C. R. Weaver. 1972. Egg quality and poult production in turkeys. 2. Inheritance and relationship among traits. *Poultry Science* 51: 147–158.

Newman, R. A. 1988. Genetic variation for larval anuran (*Scaphiopus couchii*) development time in an uncertain environment. *Evolution* 42: 763–773.

Nicholas, F. W. 1980. Size of population required for artificial selection. *Genetic Research* 35: 85–105.

Nielsen, B. V. H., and S. Andersen. 1987. Selection for growth on normal and reduced protein diets. *Genetic Research* 50: 7–15.

Nijhout, H. F. 1975. A threshold size for metamorphosis in the tobacco hornworm, *Manduca sexta* (L.). *Biological Bulletin* 149: 214–225.

Nijhout, H. F. 1979. Stretch-induced moulting in *Oncopeltus fasciatus*. *Journal of Insect Physiology* 25: 277–281.

Nijhout, H. F., and C. M. Williams. 1974a. Control of moulting and metamorphosis in the tobacco hornworm, *Manduca sexta* (L): cessation of juvenile hormone secretion as a trigger for pupation. *Journal of Experimental Biology* 61: 493–501.

Nijhout, H. F., and C. M. Williams. 1974b. Control of moulting and metamorphosis in the tobacco hornworm, *Manduca sexta* (L.): growth of the last-instar larva and the decision to pupate. *Journal of Experimental Biology* 61: 481–491.

Nodari, R. O., S. M. Tsai, P. Guzman, R. L. Gilbertson, and P. Gepts. 1993. Toward an integrated linkage map of common bean. III. Mapping genetic factors controlling host-bacteria interactions. *Genetics* 134: 341–350.

Noordwijk, A. J. V., and W. Scharloo. 1981. Inbreeding in an island population of the great tit. *Evolution* 35: 674–688.

Noordwijk, A. J. van, J. H. van Balen, and W. Sharloo. 1980. Heritability of ecologically important traits in the great tit. *Ardea* 68: 193–203.

Noordwijk, A. J. van, J. H. van Balen, and W. Scharloo. 1981. Genetic and environmental variation in clutch size of the great tit (*Parus major*). *Netherlands Journal of Zoology* 31: 342–372.

Nordskog, A. W. 1977. Success and failure of quantitative genetic theory in poultry. pp. 569–586. *In* E. Pollack, O. Kempthorne and T. B. Bailey, Jr. (editors). *Proceedings of the International Conference on Quantitative Genetics.* Iowa State University Press, Ames.

Nunney, L. 1991. The influence of age structure and fecundity on effective population size. *Proceedings of the Royal Society of London* B 246: 71–76.

Nunney, L. 1993. The influence of mating system and overlapping generations on effective population size. *Evolution* 47: 1329–1341.

Nunney, L. 1995. Measuring the ratio of effective population size to adult numbers using genetic and ecological data. *Evolution* 49: 389–392.

Nunney, L., and D. R. Elam. 1994. Estimating the effective population size of conserved populations. *Conservation Biology* 8: 175–184.

Ohta, T. 1971. Associative overdominance caused by linked detrimental mutations. *Genetic Research* 18: 277–286.

Ohta, T., and M. Kimura. 1970. Development of associative overdominance through linkage disequilibrium in finite populations. *Genetic Research* 16: 165–177.

Ohta, T., and M. Kimura. 1971. Behavior of neutral mutants influenced by associated overdominant loci in finite populations. *Genetics* 69: 247–260.

Ojanen, M., M. Orell, and R. A. Vaisanen. 1979. Role of heredity in egg size variation in the Great Tit *Parus major* and the Pied Flycatcher *Ficedula hypoleuca*. *Ornis Scandinavica* 10: 22–28.

Olausson, A., and K. Ronningen. 1975. Estimation of genetic parameters for threshold characters. *Acta Agricultural Scandinavica* 25: 201–208.

Ooijen, J. W. van. 1992. Accuracy of mapping quantitative trait loci in autogamous species. *Theoretical and Applied Genetics* 84: 803–811.

Oortmerssen, G. A., and C. M. Bakker. 1981. Artificial selection for short and long attack latencies in wild *Mus musculus domesticus*. *Behavior Genetics* 11: 115–126.

Oostermeijer, J. G. B., M. W. van Eijck, and J. C. M. den Nijs. 1994. Offspring fitness in relation to population size and genetic variation in the rare perennial plant species *Gentiana pneumonanthe* (Gentianaceae). *Oecologia* 97: 289–296.

Orozco, F. 1976. A dynamic study of genotype environment interaction with egg laying of *Tribolium casteneum*. *Heredity* 37: 157–171.

Orzack, S. H. 1985. Population dynamics in variable environments. V. The genetics of homeostasis revisted. *American Naturalist* 125: 550–572.

Oyama, K. 1994a. Differentiation in phenotypic plasticity among populations of *Arabi serrata* Fr&Sav (Brassicaceae). *Biological Journal of the Linnaean Society* 51: 417–432.

Oyama, K. 1994b. Ecological amplitude and differentiation among populations of *Arabis serrata* (Brassicaceae). *International Journal of Plant Science* 155: 220–234.

Palmer, J. O., and H. Dingle. 1986. Direct and correlated responses to selection among life-history traits in mikweed bugs (*Oncopeltus fasciatus*). *Evolution* 40: 767–777.

Palmer, A. R., and C. Strobeck. 1986. Fluctuating asymmetry: measurement, analysis, patterns. *Annual Review of Ecology and Systematics* 17: 391–421.

Pang, H., M. F. Liu, M. Makarechian, and R. T. Berg. 1994. Estimation of variance components due to direct and maternal effects for growth traits of young beef bulls in four breed groups. pp. 229–232. *In* J. S. Gavora, B. Benkel, J. Chesnais, W. Fairfull, J. P. Gibson, B. W. Kennedy, E. B. Burnside, and C. Smith (editors). *Proceedings of the 5th World Congress on Genetics Applied to Livestock Production*. Dept. Animal and Poultry Science, University of Guelph, Guelph, Ontario, Canada.

Pape, G. le, and J. M. Lassalle. 1981. Maternal effects in mice: influence of parents' and offspring's genotypes. *Behavior Genetics* 11: 379–384.

Parker, R. J., L. D. McGilliard, and J. L. Gill. 1969. Genetic correlation and response to selection in simulated populations. I. Additive model. *Theoretical and Applied Genetics* 39: 365–370.

Parker, R. J., L. D. McGilliard, and J. L. Gill. 1970a. Genetic correlation and response to selection in simulated populations. II. Model of complete dominance. *Theoretical and Applied Genetics* 40: 106–110.

Parker, R. J., L. D. McGilliard, and J. L. Gill. 1970b. Genetic correlation and response to selection in simulated populations. III. Correlated response to selection. *Theoretical and Applied Genetics* 40: 157–162.

Parkhurst, S. M., and P. M. Meneely. 1994. Sex determination and dosage compensation: lessons from flies and worms. *Science* 264: 924–932.

Parsons, P. A. 1964. A diallel cross for mating speeds in *Drosophila melanogaster*. *Genetica* 35: 141–151.

Paterson, A. H., E. S. Lander, J. D. Hewitt, S. Peterson, S. E. Lincoln, and S. D. Tanksley. 1988. Resolution of quantitative traits into Mendelian factors by using a complete linkage map of restriction fragment length polymorphisms. *Nature* 335: 721–726.

Paterson, A. H., S. Damon, J. D. Hewitt, D. Zamir, H. D. Rabinowitch, S. E. Lincoln, E. Lander, and D. Tanksley. 1991. Mendelian factors underlying quantitative traits in tomato: comparison across species, generations, and environments. *Genetics* 127: 181–197.

Patterson, B. D. 1990. Fluctuating asymmetry and allozymic heterozygosity among natural populations of pocket gophers (*Thomomys bottae*). *Biological Journal of the Linnean Society* 40: 21–36.

Patterson, H. D., and R. Thompson. 1971. Recovery of inter-block information when block sizes are unequal. *Biometrika* 58: 545–554.

Paulsen, S. M. 1996. Quantitative genetics of the wing color pattern in the buckeye butterfly (*Precis coenia* and *Precis evarete*): Evidence against the constancy of G. *Evolution* 50: 1585–1597.

Payne, R. B., and L. L. Payne. 1989. Heritability estimates and behaviour observations: extra-pair matings in indigo buntings. *Animal Behaviour* 38: 457–467.

Perkins, J. M., and J. L. Jinks. 1973. The assessment and specificity of environmental and genotype–environmental components of variability. *Heredity* 30: 111–126.

Perrin, N., and J. F. Rubin. 1990. On dome-shaped norms of reaction for size-at-age at maturity in fishes. *Functional Ecology* 4: 53–57.

Perrins, C. M., and P. J. Jones. 1974. The inheritance of clutch size in the great tit (*Parus major* L.). *Condor* 76: 225–229.

Peters, J, and S. T. Ball. 1990. Parental influences on expression of glucose-7-phosphate dehydrogenase, *Gópd*, in the mouse; a case of imprinting. *Genetic Research* 56: 245–252.

Peterson, K., and C. Sapienza. 1993. Imprinting the genome: imprinted genes, imprinting genes, and a hypothesis for their interaction. *Annual Review of Genetics* 27: 7–31.

Petras, M. L. 1967a. Studies of natural populations of *Mus*. I. Biochemical polymorphisms and their bearing on breeding structure. *Evolution* 21: 259–274.

Petras, M. L. 1967b. Studies of natural populations of *Mus*. II. Polymorphism at the *T* loci. *Evolution* 21: 466–478.

Philipsson, J. 1976. Studies on calving difficulty, stillbirth and associated factors in Swedish cattle breeds. *Acta Agriculturae Scandinavica* 26: 211–220.

Platenkamp, G. A. J., and R. G. Shaw. 1992. Environmental and genetic constraints on adaptive population differentiation in *Anthoxanthum odoratum*. *Evolution* 46: 341–352.

Pogson, G. H., and E. Zouros. 1994. Allozyme and RFLP heterozygosities as correlates of growth rate in the scallop *Placopecten magellanicus*: A test of the associative over-dominance hypothesis. *Genetics* 137: 221–231.

Polivanov, S. 1975. Response of *Drosophila persimilis* to phototactic and geotactic selection. *Behavior Genetics* 5: 255–267.

Pollak, E. 1983. A new method for estimating the effective population size from allele frequency changes. *Genetics* 104: 531–548.

Pomiankowski, A. N. 1988. The evolution of female mate preferences for male genetic quality. pp. 136–184. *In* P. H. Harvey and L. Partridge (editors). *Oxford Surveys in Evolutionary Biology*. Oxford University Press, Oxford.

Pomiankowski, A., and Y. Iwasa. 1993. Evolution of multiple sexual preferences by Fisher's runaway process of sexual selection. *Proceedings of the Royal Society of London* B 253: 173–181.

Pomp, D., M. A. Cushman, S. C. Foster, D. K. Drudik, M. Fortman, and E. J. Eisen. 1994. Identification of quantitative trait loci for body weight and body fat in mice. pp. 209–212. *In* J. S. Gavora, B. Benkel, J. Chesnais, W. Fairfull, J. P. Gibson, B. W. Kennedy, E. B. Burnside, and C. Smith (editors). *Proceedings of the 5th World Congress on Genetics Applied to Livestock Production*, Vol. 21. University of Guelph, Guelph, Ontario, Canada.

Pooni, H. S., J. L. Jinks, and N. E. M. Jayasekara. 1978. An investigation of gene action and genotype × environment interactions in two crosses of *Nicotiana rustica* by triple test cross and inbred line analysis. *Heredity* 41: 83–92.

Pooni, H. S., J. L. Jinks, and G. S. Pooni. 1980. A general method for the detection and estimation of additive, dominance and epistatic variation for metrical traits. IV. Triple test cross analysis for normal families and their selfs. *Heredity* 44: 177–192.

Pooni, H. S., J. L. Jinks, and R. K. Singh. 1984. Methods of analysis and the estimation of the genetic parameters from a diallel set of crosses. *Heredity* 52: 243–253.

Pooni, H. S., D. T. Coombs, P. S. Virk, and J. L. Jinks. 1987. Detection of epistasis and linkage of interacting genes in the presence of reciprocal differences. *Heredity* 58: 257–266.

Potti, J. 1993. Environmental ontogenetic and genetic variation in egg size of pied fly-catchers. *Canadian Journal of Zoology* 71: 1534–1542.

Potti, J., and S. Merino. 1994. Heritability estimates and maternal effects on tarsus length in pied flycatchers *Ficedula hypoleuca*. *Oecologia* 100: 331–338.

Potvin, C., and D. A. Roff. 1993. Distribution-free and robust statistical methods: viable alternatives to parametric statistics? *Ecology* 74: 1617–1628.

Pray, L. A., and C. J. Goodnight. 1995. Genetic variation in inbreeding depression in the red flour beetle *Tribolium castaneum*. *Evolution* 49: 176–188.

Pray, L. A., J. M. Schwartz, C. J. Goodnight, and L. Stevens. 1994. Environmental dependency of inbreeding depression: implications for conservation biology. *Conservation Biology* 8: 562–568.

Prevosti, A. 1955. Geographical variability in quantitative traits in populations of *Drosophila subobscura*. *Quantitative Biology* 20: 294–299.

Prevosti, A. 1967. Inversion heterozygosity and selection for wing length in *Drosophila subobscura*. *Genetic Research* 10: 81–93.

Preziosi, R. F., and D. J. Fairbairn. 1992. Genetic population structure and levels of gene flow in the stream dwelling waterstrider, *Aquarius* (= *Gerris*) *remigis* (Hemiptera: Gerridae). *Evolution* 46: 430–444.

Preziosi, R. F., and D. J. Fairbairn. 1997. Sexual size dimorphism and selection in the wild in the waterstrider *Aquarius remigis*: body size, components of body size, fecundity and female survival. *Evolution* (in press).

Price, D. K., and N. T. Burley. 1993. Constraints on the evolution of attractive traits: genetic (co)variance of zebra finch bill colour. *Heredity* 71: 405–412.

Price, G. R. 1972. Fisher's 'fundamental theorem' made clear. *Annals of Human Genetics* 36: 129–140.

Price, T. D. 1984. The evolution of sexual size dimorphism in Darwin's finches. *American Naturalist* 123: 500–518.

Price, T., and L. Liou. 1989. Selection on clutch size in birds. *American Naturalist* 134: 950–959.

Price, T., and D. Schluter. 1991. On the low heritability of life-history traits. *Evolution* 45: 853–861.

Price, T., M. Kirkpatrick, and S. J. Arnold. 1988. Directional selection and the evolution of breeding date in birds. *Science* 240: 798–799.

Prince, H. H., P. B. Siegel, and G. W. Cornwell. 1970. Inheritance of egg production and juvenile growth in mallards. *Auk* 87: 342–352.

Prout, T. 1954. Genetic drift in irradiated experimental populations of *Drosophila melanogaster*. *Genetics* 39: 529–545.

Prout, T. 1962. The effects of stabilising selection on the time development in *D. melanogaster*. *Genetic Research* 3: 364–382.

Prout, T. 1980. Some relationships between density-dependent selection and density dependent population growth. *Evolutionary Biology* 13: 1–68.

Prout, T., and J. S. F. Barker. 1989. Ecological aspects of the heritability of body size in *Drosophila buzzatii*. *Genetics* 123: 803–813.

Provine, W. B. 1971. *Origins of Theoretical Population Genetics*. University of Chicago Press, Chicago.

Quinn, J. A., and J. C. Colosi. 1977. Separating genotype from environment in germination ecology studies. *American Midland Naturalist* 97: 484–489.

Rabinowitz, D., J. K. Rapp, S. Cairns, and M. Mayer. 1989. The persistence of rare prairie grasses in Missouri: environmental variation buffered by reproductive output of sparse species. *American Naturalist* 134: 525–544.

Rahnefeld, G. W., R. E. Comstock, M. Singh, and S. R. Napuket. 1966. Genetic correlation between growth rate and litter size in mice. *Genetics* 54: 1423–1429.

Ralls, K., and J. Ballou. 1982. Effects of inbreeding on infant mortality in captive primates. *International Journal of Primatology* 3: 491–505.

Ralls, K., J. D. Ballou, and A. Templeton. 1988. Estimates of lethal equivalents and the cost of inbreeding in mammals. *Conservation Biology* 2: 185–193.

Ralls, K., K. Brugger, and J. Ballou. 1979. Inbreeding and juvenile mortality in small populations of ungulates. *Science* 206: 1101–1103.

Ramsey, P. R., and J. M. Wakeman. 1987. Population structure of *Sciaenops ocellatus* and *Cynoscion nebulosus* (Pisces: Sciaenidae): biochemical variation, genetic subdivision and dispersal. *Copeia* 1987: 682–695.

Rathie, K. A., and J. S. F. Barker. 1968. Effectiveness of regular cycles of intermittent artificial selection for a quantitative character in *Drosophila melanogaster*. *Australian Journal of Biological Science* 21: 1187–1213.

Rausher, M. D. 1988. Is coevolution dead? *Ecology* 69: 898–901.

Reeve, E. C., and F. W. Robertson. 1953. Studies in quantitative inheritance. II. Analysis of a strain of *Drosophila melanogaster* selected for long wings. *Journal of Genetics* 51: 276–316.

Reeve, E. C. R. 1955. The variance of the genetic correlation coefficient. *Biometrics* 11: 357–374.

Reeve, J. P., and D. J. Fairbairn. 1996. An empirical test of models for the evolution of sexual size dimorphism as a correlated response to selection on body size. *Evolution* 50: 1927–1938.

Reich, T., J. W. James, and C. A. Morris. 1972. The use of multiple thresholds in determining the mode of transmission of semi-continuous traits. *Annals of Human Genetics* 36: 163–184.

Reiter, R. S., J. G. Coors, M. R. Sussman, and W. H. Gabelman. 1991. Genetic analysis of tolerance to low-phosphorus stress in maize using restriction fragment length polymorphisms. *Theoretical and Applied Genetics* 82: 561–568.

Rendel, J. M. 1943. Variations in the weights of hatched and unhatched duck's eggs. *Biometrika* 33: 48–58.

Rendel, J. M. 1963. Correlation between the number of scutellar and abdominal bristles in *Drosophila melanogaster*. *Genetics* 48: 391–408.

Rendel, J. M. 1967. *Canalization and Gene Control*. Logos/Academic Press, London.

Reznick, D. N. 1981. "Grandfather effects": the genetics of offspring size in mosquito fish *Gambusia affinis*. *Evolution* 35: 941–953.

Reznick, D. N. 1982. Genetic determination of offspring size in the guppy (*Poecilia reticulata*). *American Naturalist* 120: 181–188.

Rice, W. R. 1989. Analyzing tables of statistical tests. *Evolution* 43: 223–225.

Richardson, R. H., K. Kojima, and H. L. Lucas. 1968. An analysis of short-term selection experiments. *Heredity* 23: 493–506.

Ricker, W. E. 1973. Linear regressions in fishery research. *Journal of the Fisheries Research Board of Canada* 30: 409–434.

Rise, M. L., W. N. Frankel, J. M. Coffin, and T. N. Seyfried. 1991. Genes for epilepsy mapped in mouse. *Science* 253: 669–672.

Riska, B. 1986. Some models for development, growth and morphometric correlation. *Evolution* 40: 1303–1311.

Riska, B. 1989. Composite traits, selection response and evolution. *Evolution* 43: 1172–1191.

Riska, B., W. R. Atchley, and J. J. Rutledge. 1984. A genetic analysis of targeted growth in mice. *Genetics* 107: 79–101.

Riska, B., J. J. Rutledge, and W. R. Atchley. 1985. Covariance between direct and maternal genetic effects in mice, with a model of persistent environmental influences. *Genetic Research* 45: 287–297.

Riska, B., T. Prout, and M. Turelli. 1989. Laboratory estimates of heritabilities and genetic correlations in nature. *Genetics* 123: 865–871.

Roach, D. A., and R. D. Wulff. 1987. Maternal effects in plants. *Annual Review of Ecology and Systematics* 18: 209–235.

Roberts, R. C. 1960. The effects on litter size of crossing lines of mice inbred without selection. *Genetic Research* 1: 239–252.

Roberts, R. C. 1966a. The limits to artificial selection for body weight in the mouse: I. The limits attained in earlier experiments. *Genetic Research* 8: 347–360.

Roberts, R. C. 1966b. The limits to artificial selection for body weight in the mouse. II. The genetic nature of the limits. *Genetic Research* 8: 361–377.

Robertson, A. 1951. The analysis of heterogeneity in the binomial distribution. *Annals of Eugenics* 16: 1–15.

Robertson, A. 1952. The effect of inbreeding on the variation due to recessive genes. *Genetics* 37: 189–207.

Robertson, A. 1956c. The effect of selection against extreme deviants based on deviation or on homozygosis. *Journal of Genetics* 54: 236–248.

Robertson, A. 1959a. Experimental design in the evaluation of genetic parameters. *Biometrics* 15: 219–226.

Robertson, A. 1959b. The sampling variance of the genetic correlation coefficient. *Biometrics* 15: 469–485.

Robertson, A., 1960a. Experimental design on the measurement of heritabilities and genetic correlations. pp. 101–106. *In* O. Kempthorne (editor). *Biometrical Genetics*. Pergamon Press, London.

Robertson, A. 1960b. A theory of limits in artificial selection. *Proceedings of the Royal Society of London* B 153: 234–249.

Robertson, A. 1962. Selection for heterozygotes in small populations. *Genetics* 47: 1291–1300.

Robertson, A. 1970a. Some optimum problems in individual selection. *Theoretical Population Biology* 1: 120–127.

Robertson, A. 1970b. A theory of limits in artificial selection with many linked loci. pp. 246–288, and *In* K. Kojima (editor). *Mathematical Topics in Population Genetics*. Springer-Verlag, Berlin.

Robertson, A. (editor). 1980. Selection experiments in laboratory and domestic animals. Proceedings of a symposium held at Harrogate, U.K., 21–22 July 1979. Commonwealth Agricultural Bureau, Slough, U.K.

Robertson, A., and W. G. Hill. 1983. Population and quantitative genetics of many linked loci in finite populations. *Proceedings of the Royal Society of London* B 219: 253–264.

Robertson, A. W., A. Diaz, and M. R. MacNair. 1994. The quantitative genetics of floral characters in *Mimulus guttatus*. *Heredity* 72: 300–311.

Robertson, F. W. 1955. Selection response and the properties of genetic variation. *Cold Spring Harbor Symposium in Quantitative Genetics* 20: 166–177.

Robertson, F. W. 1960a. The ecological genetics of growth in *Drosophila* 1. Body size and developmental time on different diets. *Genetic Research* 1: 288–304.

Robertson, F. W. 1960b. The ecological genetics of growth in *Drosophila* 3. Growth and competitive ability of strains selected on different diets. *Genetic Research* 1: 333–350.

Robertson, F. W. 1963. The ecological genetics of growth in *Drosophila* 6. The genetic correlation between the duration of the larval period and body size in relation to larval diet. *Genetic Research* 4: 74–92.

Robertson, F. W., and E. Reeve. 1952. Studies in quantitative inheritance. I. The effects of selection of wing and thorax length in *Drosophila melanogster*. *Journal of Genetics* 50: 414–448.

Robinson, D. L. 1994. Models which might explain negative correlations between direct and maternal genetic effects. pp. 378–381. *In* J. S. Gavora, B. Benkel, J. Chesnais, W. Fairfull, J. P. Gibson, B. W. Kennedy, E. B. Burnside, and C. Smith (editors). *Proceedings of the 5th World Congress on Genetics Applied to Livestock Production*. Dept. Animal and Poultry Science, University of Guelph, Guelph, Ontario, Canada.

Rockwell, R. F., and G. F. Barrowclough. 1995. Effective population size and lifetime reproductive success. *Conservation Biology* 9: 1225–1233.

Rodhouse, P. G., J. H. McDonald, R. I. E. Newell, and R. K. Koehn. 1986. Gamete production, somatic growth and multiple-locus enzyme heterozygosity. *Marine Biology* 90: 209–214.

Roff, D. A. 1976. Stabilizing selection in *Drosophila melanogaster*: a comment. *Journal of Heredity* 67: 245–246.

Roff, D. A. 1980. Optimizing development time in a seasonal environment: the "ups and downs" of clinal variation. *Oecologia* 45: 202–208.

Roff, D. A. 1983. Phenological adaptation in a seasonal environment: a theoretical perspective. pp. 253–270. *In* V. K. Brown and I. Hodek (editors). *Diapause and Life Cycle Strategies in Insects*. Dr. W. Junk, The Hague.

Roff, D. A. 1986a. The evolution of wing dimorphism in insects. *Evolution* 40: 1009–1020.

Roff, D. A. 1986b. The genetic basis of wing dimorphism in the sand cricket, *Gryllus firmus* and its relevance to the evolution of wing dimorphisms in insects. *Heredity* 57: 221–231.

Roff, D. A. 1990. Selection for changes in the incidence of wing dimorphism in *Gryllus firmus*. *Heredity* 65: 163–168.

Roff, D. A., 1992. *The Evolution of Life Histories: Theory and Analysis.* Chapman & Hall, New York.

Roff, D. A. 1994a. The evolution of dimorphic traits: effect of directional selection on heritability. *Heredity* 72: 36–41.

Roff, D. A. 1994b. Evidence that the magnitude of the trade-off in a dichotomous trait is frequency dependent. *Evolution* 48: 1650–1656.

Roff, D. A. 1994c. The evolution of dimorphic traits: predicting the genetic correlation between environments. *Genetics* 136: 395–401.

Roff, D. A. 1995a. Antagonistic and reinforcing pleiotropy: a study of differences in development time in wing dimorphic insects. *Journal of Evolutionary Biology* 8: 405–419.

Roff, D. A. 1995b. The estimation of genetic correlations from phenotypic correlations: a test of Cheverud's conjecture. *Heredity* 74: 481–490.

Roff, D. A. 1996a. The evolution of threshold traits in animals. *Quarterly Review of Biology* 71: 3–35.

Roff, D. A. 1996b. The evolution of genetic correlations: an analysis of patterns. *Evolution* (in press).

Roff, D. A., and M. J. Bradford. 1996. Quantitative genetics of the trade-off between fecundity and wing dimorphism in the cricket *Allonemobius socius. Heredity* 76: 178–185.

Roff, D. A., and D. J. Fairbairn. 1993. The evolution of alternative morphologies: fitness and wing morphology in male sand crickets. *Evolution* 47: 1572–1584.

Roff, D. A., and T. A. Mousseau. 1987. Quantitative genetics and fitness: lessons from *Drosophila. Heredity* 58: 103–118.

Roff, D. A., and R. Preziosi. 1994. The estimation of the genetic correlation: the use of the jackknife. *Heredity* 73: 544–548.

Ronningen, K. 1974. Monte carlo simulation of statistical-biological models which are of interest in animal breeding. *Acta Agriculturae Scandinavica* 24: 135–142.

Rose, M. R. 1982. Antagonistic pleitropy, dominance, and genetic variation. *Heredity* 48: 63–78.

Rose, M. R. 1985. Life history evolution with antagonistic pleiotropy and overlapping generations. *Theoretical Population Biology* 28: 342–358.

Ross, K. G., E. L. Vargo, L. Keller, and J. C. Trager. 1993. Effect of a founder event on variation in the genetic sex-determining system of the fire ant *Solenopsis invicta. Genetics* 135: 843–854.

Roush, R. T. 1986. Inbreeding depression and laboratory adaptation in *Heliothis virescens* (Lepidoptera: Noctuidae). *Annals of the Entomological Society of America* 79: 583–587.

Rowley, I., E. Russell, and M. Brooker. 1993. Inbreeding in birds. pp. 304–328. *In* N. W. Thornhill (editor). *The Natural History of Inbreeding and Outbreeding.* University of Chicago Press, Chicago.

Ruano, R. G., F. Orozco, and C. López-Fanjul. 1975. The effect of different selection intensities on selection response in egg-laying of *Tribolium castaneum*. *Genetic Research* 25: 17–27.

Ruiz, A., M. Santos, A. Barbadilla, J. E. Quezada-Diaz, E. Hasson, and A. Fontdevila. 1991. Genetic variance for body size in a natural population of *Drosophila buzzatii*. *Genetics* 128: 739–749.

Rutledge, J. J., E. J. Eisen, and J. E. Legates. 1973. An experimental evaluation of genetic correlation. *Genetics* 75: 709–726.

Rutledge, J. J., O. W. Robison, E. J. Eisen, and J. E. Legates. 1972. Dynamics of genetic and maternal effects in mice. *Journal of Animal Science* 35: 911–918.

Ryman, N. 1973. Two-way selection for body weight in the guppy-fish, *Lebistes reticularus*. *Hereditas* 74: 239–246.

Sager, R., and Z. Ramanis. 1973. The mechanism of maternal inheritance in *Chlamydomonas*: biochemical and genetic studies. *Theoretical and Applied Genetics* 43: 101–108.

Samollow, P. B., and M. Soule. 1983. A case of stress related heterozygote superiority in nature. *Evolution* 37: 646–649.

Sandstrom J. 1994. High variation in host adaptation among clones of the pea aphid *Acyrthosiphon pisum* on peas *Pisum sativum*. *Entomologica Experimentalis Applicata* 71: 245–256.

Sang, J. H. 1962. Selection for rate of larval development using *Drosophila melanogaster* cultured axenically on deficient diets. *Genetic Research* 3: 90–109.

Sang, J. H., and G. A. Clayton. 1957. Selection for larval development time in *Drosophila*. *Journal of Heredity* 48: 265–270.

Santiago, E., and A. Caballero. 1995. Effective size of populations under selection. *Genetics* 139: 1013–1030.

Santiago, E., J. Albornoz, A. Dominguez, M. A. Toro, and C. López-Fanjul. 1992. The distribution of spontaneous mutations on quantitative traits and fitness in *Drosophila melanogaster*. *Genetics* 132: 771–781.

Sarre, S., and J. M. Dearn. 1991. Morphological variation and fluctuating asymmetry among insular populations of the sleepy lizard, *Trachydosaurus rugosus* Gray (Squamata: Scincidae). *Australian Journal of Zoology* 39: 91–104.

Sax, K. 1923. The association of size differences with seed-coat pattern and pigmentation in *Phaseolus vulgaris*. *Genetics* 8: 552–560.

Schaal, B. 1980. Reproductive capacity and seed size in *Lupinus texensis*. *American Journal of Botany* 67: 703–709.

Schaal, B. A., and D. A. Levin. 1976. The demographic genetics of *Liatris cylindracea* Michx. (Compositae). *American Naturalist* 110: 191–206.

Scharloo, W. 1964. The effect of disruptive and stabilising selection on the expression of cubitus interruptus mutant in *Drosphila*. *Genetics* 50: 553–562.

Scharloo, W. 1984. Genetics of adaptive reactions. pp. 5–15. *In* K. Wohrmann and V. Loeschcke (editors). *Population Biology and Evolution*. Springer-Verlag, Berlin.

Scharloo, W., M. S. Hoogmoed, and A. Ter Kuile. 1967. Stabilizing and disruptive selection on a mutant character in *Drosophila*. I. The phenotypic variance and its components. *Genetics* 56: 709–726.

Scharloo, W., A. Zweep, K. A. Schuitema, and J. G. Wijnstra. 1972. Stabilizing and disruptive selection on a mutant character in *Drosophila*. IV. Selection on sensitivity to temperature. *Genetics* 71: 551–566.

Scheiner, S. M. 1993a. Genetics and evolution of phenotypic plasticity. *Annual Review of Ecology and Systematics* 24: 35–68.

Scheiner, S. M. 1993b. Plasticity as a selectable trait: reply to Via. *American Naturalist* 142: 371–373.

Scheiner, S. M., and C. J. Goodnight. 1984. The comparison of phenotypic plasticity and genetic variation in populations of the grass *Danthonia spicata*. *Evolution* 38: 845–855.

Scheiner, S. M., and R. F. Lyman. 1989. The genetics of phenotypic plasticity. I. Heritability. *Journal of Evolutionary Biology* 2: 95–107.

Scheiner, S. M., and R. F. Lyman. 1991. The genetics of phenotypic plasticity. II. Response to selection. *Journal of Evolutionary Biology* 4: 23–50.

Scheiner, S. M., R. L. Caplan, and R. F. Lyman. 1991. The genetics of phenotypic plasticity. III. Genetic correlations and fluctuating asymmetries. *Journal of Evolutionary Biology* 4: 51–68.

Scheiner, S. M., J. Gurevitch, and J. A. Teeri. 1984. A genetic analysis of the photosynthetic properties of populations of *Danthonia spicata* that have different growth responses to light level. *Oecologia* 64: 74–77.

Scheiring, J. E. 1977. Stabilizing selection for size as related to mating fitness in *Tetraopes*. *Evolution* 31: 447–449.

Schemske, D. W. 1983. Breeding system and habitat effects on fitness components in three neotropical *Costus* (Zingiberaceae). *Evolution* 37: 523–539.

Schemske, D. W., and R. Lande. 1985. The evolution of self-fertilization and inbreeding depression in plants. II. Empirical observations. *Evolution* 39: 41–52.

Schlichting, C. D. 1986. The evolution of phenotypic plasticity in plants. *Annual Review of Ecology and Systematics* 17: 667–693.

Schlichting, C. D., and D. A. Levin. 1990. Phenotypic plasticity in *Phlox*. III. Variation among natural populations of *P. drummondii*. *Journal of Evolutionary Biology* 3: 411–428.

Schlichting, C. D., and M. Pigliucci. 1993. Control of phenotypic plasticity via regulatory genes. *American Naturalist* 142: 366–370.

Schlichting, C. D., and M. Pigliucci. 1995. Gene regulation, quantitative genetics, and the evolution of reaction norms. *Evolutionary Ecology* 9: 154–168.

Schluter, D. 1988. Estimating the form of natural selection on a quantitative trait. *Evolution* 42: 849–861.

Schluter, D., and L. Gustafsson. 1993. Maternal inheritance of condition and clutch size in the collared flycatcher. *Evolution* 47: 658–667.

Schluter, D., and J. N. M. Smith. 1986. Natural selection on beak and body size in the song sparrow. *Evolution* 40: 221–231.

Schmalhausen, I. I. 1949. *Factors of Evolution: The Theory of Stabilizing Selection.* Blakiston, Philadelphia.

Schmitt, J., and D. W. Ehrhardt. 1987. A test of the sib-competition hypothesis for outcrossing advantage in *Impatiens capensis. Evolution* 41: 579–590.

Schmitt, J., and D. W. Ehrhardt. 1990. Enhancement of inbreeding depression by dominance and suppression in *Impatiens capensis. Evolution* 44: 269–278.

Schneider, J. C., and R. T. Roush. 1986. Genetic differences in oviposition preference between two populations of *Heliothis virescens.* pp. 163–172. *In* M. D. Huettel (editor). *Evolutionary Genetics of Invertebrate Behavior: Progress and Prospects.* Plenum, New York.

Schoen, D. J., and A. H. D. Brown. 1991. Intraspecific variatin in population gene diversity and effective population size correlates with the mating system in plants. *Proceedings of the National Academy of Sciences* 88: 4494–4497.

Schoen, D. J., G. Bell, and M. Lechowicz. 1994. The ecology and genetics of fitness in forest plants. IV. Quantitative genetics of fitness components in *Impatiens pallida* (Balsaminaceae). *American Journal of Botany* 81: 232–239.

Schuler, L. 1985. Selection for fertility in mice—the selection plateau and how to overcome it. *Theoretical and Applied Genetics* 70: 72–79.

Schwaegerle, K. E., and D. A. Levin. 1990. Quantitative genetics of seed size variation in *Phlox. Evolutionary Ecology* 4: 143–148.

Searle, A., and C. V. Beechey. 1990. Genome imprinting phenomena on mouse chromosome 7. *Genetic Research* 56: 237–244.

Sedcole, J. R. 1981. A review of the theories of heterosis. *Egyptian Journal of Genetics and Cytology* 10: 117–146.

Selander, R. K., D. W. Kaufman, and R. S. Ralin. 1974. Self-fertilization in the terrestrial snail *Rumina decollata. The Veliger* 16: 265–270.

Semlitsch, R. D., R. N. Harris and H. M. Wilbur 1990. Paedomorphosis in *Ambystoma talpoideum*: maintenance of population variation and alternative life-history pathways. *Evolution* 44: 1604–1613.

Semlitsch, R. D. 1993. Adaptive genetic variation in growth and development of tadpoles of the hybridogenetic *Rana esculenta* complex. *Evolution* 47: 1805–1818.

Sen, B. K., and A. Robertson. 1964. An experimental examination of methods for the simultaneous selection of two characters, using *Drosophila melanogaster. Genetics* 50: 199–209.

Service, P. M., and M. R. Rose. 1985. Genetic covariation among life-history components: the effects of novel environments. *Evolution* 39: 943–945.

Severson, D. W., V. Thathy, A. Mori, Y. Zhang, and B. M. Christensen. 1995. Restriction fragment length polymorphism mapping of quantitative trait loci for malaria parasite susceptibility in the mosquito *Aedes aegypti. Genetics* 139: 1711–1717.

Sewell, D., B. Burnet, and K. Connolly. 1975. Genetic analysis of larval feeding behaviour in *Drosophila melanogaster. Genetic Research* 24: 163–173.

Sharp, P. M. 1984. The effect of inbreeding on competitive male-mating ability in *Drosophila melanogaster. Genetics* 106: 601–612.

Shaw, D. V., 1989. Variation among heritability estimates for strawberries obtained by offspring-parent regression with relatives raised in separate environments. *Euphytica* 44: 157–162.

Shaw, R. G. 1987. Maximum-likelihood approaches applied to quantitative genetics of natural populations. *Evolution* 41: 812–826.

Shaw, R. G. 1991. The comparison of quantitative genetic parameters between populations. *Evolution* 45: 143–151.

Shaw, R. G. 1992. Comparison of quantitative genetic parameters: reply to Cowley and Atchley. *Evolution* 46: 1967–1969.

Shaw, R. G., and H. L. Billington. 1991. Comparison of variance components between two populations of *Holcus lanatus*: a reanalysis. *Evolution* 45: 1287–1289.

Shaw, R. G., and G. A. J. Platenkamp. 1993. Quantitative genetics of response to competitors in *Nemophila menziesii*. *Evolution* 47: 801–812.

Sheridan, A. K. 1988. Agreement between estimated and realised genetic parameters. *Animal Breeding Abstracts* 56: 877–889.

Sheridan, A. K., and J. S. F. Barker. 1974. Two-trait selection and the genetic correlation. II. Changes in the genetic correlation during two-trait selection. *Australian Journal of Biological Sciences* 27: 89–101.

Sheridan, A. K., R. Frankham, L. P. Jones, K. A. Rathie, and J. S. F. Barker. 1968. Partitioning of variance and estimation of genetic parameters for various bristle number characters of *Drosphila melanogaster*. *Theoretical and Applied Genetics* 38: 179–187.

Shields, W. M. 1993. The natural and unnatural history of inbreeding and outbreeding. pp. 143–169. *In* N. W. Thornhill (editor). *The Natural History of Inbreeding and Outbreeding*. University of Chicago Press, Chicago.

Shrimpton, A. E., and A. Robertson. 1988. The isolation of polygenic factors controlling bristle score in *Drosophila melanogaster*. II. Distribution of third chromosome bristle effects within chromosome sections. *Genetics* 118: 445–459.

Shukla, S., and K. R. Khanna. 1992. Genetical study for earliness in *Papaver somniferum L*. *Indian Journal of Genetics* 52: 33–38.

Siegal, P. B. 1980. Behavior genetic analysis in chickens and quail. pp. 1–13. *Proceedings of the 29th National Breeders' Roundtable*.

Siegel, P. B. 1965. Genetics of behavior: selection for mating ability in chickens. *Genetics* 52: 1269–1277.

Simmons, L. W., and P. I. Ward. 1991. The heritability of sexually dimorphic traits in the yellow dung fly *Scathophaga stercoraria* (L.). *Journal of Evolutionary Biology* 4: 593–601.

Simmons, M. J., C. R. Preston, and W. R. Engels. 1980. Pleiotropic effects on fitness of mutations affecting viability in *Drosophila melanogaster*. *Genetics* 94: 467–475.

Simons, A., and D. A. Roff. 1994. The effect of environmental variability on the heritabilities of traits of a field cricket. *Evolution* 48: 1637–1649.

Simons, A. M., and D. A. Roff. 1996. The effect of a variable environment on the genetic correlation structure in a field cricket. *Evolution* 50: 267–275.

Singh, M., and R. C. Lewontin. 1966. Stable equilibria under optimizing selection. *Proceedings of the National Academy of Sciences* 56: 1345–1348.

Singh, S. 1970. Inheritance of asymmetry in finger ridge counts. *Human Heredity* 20: 403–408.

Singh, S. M., and E. Zouros. 1978. Genetic variation associated with growth rate in the American oyster (*Crassostrea virginica*). *Evolution* 32: 342–353.

Sittmann, K., H. Abplanalp, and R. A. Fraser. 1966. Inbreeding depression in Japanese quail. *Genetics* 54: 371–379.

Sjogren, P. 1991. Genetic variation in relation to demography of peripheral pool frog populations (*Rana lessonae*). *Evolutionary Ecology* 5: 248–271.

Slatkin, M. 1978. Spatial patterns in the distribution of polygenic characters. *Journal of Theoretical Biology* 70: 213–228.

Slatkin, M. 1979. The evolutionary response to frequency- and density-dependent interactions. *American Naturalist* 114: 384–398.

Slatkin, M., and S. A. Frank. 1990. The quantitative genetic consequences of pleiotropy under stabilizing and directional selection. *Genetics* 125: 207–213.

Slattery, J. P., R. A. Lutz, and R. C. Vrijenhoek. 1993. Repeatability of correlations between heterozygosity, growth and survival in a natural population of the hard clam *Mercenaria mercenaria* L. *Journal of Experimental Marine Biology and Ecology* 165: 209–224.

Smith, A. T. 1993. The natural history of inbreeding and outbreeding in small mammals. pp. 329–351. *In* N. W. Thornhill (editor). *The Natural History of Inbreeding and Outbreeding*. University of Chicago Press, Chicago.

Smith, H. G. 1993. Heritability of tarsus length in cross-fostered broods of the European starling (*Sturnus vulgaris*). *Heredity* 71: 318–322.

Smith, H. G., and K. -J. Wettermark. 1995. Heritability of nestling growth in cross-fostered European starlings *Sturnus vulgaris*. *Genetics* 141: 657–665.

Smith, J. N. M., and A. A. Dhondt. 1980. Experimental confirmation of heritable morphological variation in a natural population of song sparrows. *Evolution* 34: 1155–1158.

Smith, K. P., and B. B. Bohren. 1974. Direct and correlated responses to selection for hatching time in the fowl. *British Poultry Science* 15: 597–604.

Smith, M. H., R. K. Chesser, E. G. Cothran, and P. E. Johns. 1983. Genetic variability and antler growth in a natural population of white-tailed deer. pp. 365–387. *In* R. D. Brown (editor). *Antler Development in Cervidae*. Caesar Kleberg Wildlife Research Institute, Texas A&M University, Kingsville, TX.

Smith, M. H., K. T. Scribner, J. D. Hernandez, and M. C. Wooten. 1989. Demographic, spatial, and temporal genetic variation in *Gambusia*. pp. 235–257. *In* G. K. Meffe, Jr. and F. F. Snelson (editors). *Ecology and Evolution of Livebearing Fishes (Poeciliidae)*. Prentice-Hall, Englewood Cliffs, NJ.

Smith, W. E., and J. E. Fitzsimmons. 1965. Maternal inheritance of seed weight in flax. *Canadian Journal of Genetics and Cytology* 7: 658–662.

Smith-Gill, S. J. 1983. Developmental plasticity: developmental conversion versus phenotypic modulation. *American Zoologist* 23: 47–55.

Snyder, R. J. 1991. Quantitative genetic analysis of life histories in two freshwater populations of the threespine stickleback. *Copeia* 1991: 526–529.

Sokal, R. R., and F. J. Rohlf. 1995. *Biometry*. W. H. Freeman and Co., San Francisco.

Sokoloff, A. 1965. Geographic variation of quantitative characters in populations of *Drosophila pseudoobscura*. *Evolution* 19: 300–310.

Sokoloff, A. 1966. Morphological variation in natural and experimental populations of *Drosophila pseudoobscura* and *Drosophila persimilis*. *Evolution* 20: 49–71.

Soliman, M. H. 1982. Directional and stabilizing selection for developmental time and correlated response in reproductive fitness in *Tribolium casteneum*. *Theoretical and Applied Genetics* 63: 111–116.

Soller, M., and A. Genizi. 1967. Optimal experimental designs for realised heritability estimates. *Biometrics* 23: 361–365.

Sorenson, D. A., and W. G. Hill. 1982. Effect of short term directional selection on genetic variability: experiments with *Drosophila melanogaster*. *Heredity* 48: 27–33.

Sorenson, D. A., and W. G. Hill. 1983. Effects of disruptive selection on genetic variance. *Theoretical and Applied Genetics* 65: 173–180.

Soule, M. E. 1979. Heterozygosity and developmental stability: another look. *Evolution* 33: 396–401.

Spence, J. R. 1990. Introgressive hybridization in Heteroptera: the example of *Limnoporus Stal* (Gerridae) species in western Canada. *Canadian Journal of Zoology* 68: 1770–1782.

Sperling, F. A. H. 1994. Sex-linked genes and species differences in Lepidoptera. *Canadian Entomologist* 126: 807–818.

Spickett, S. G., and J. M. Thoday. 1966. Regular responses to selection. 3. Interaction between located polygenes. *Genetic Research* 7: 96–121.

Spiess, E. B., and C. M. Wilke. 1984. Still another attempt to achieve assortative mating by disruptive selection in *Drosophila*. *Evolution* 38: 505–515.

Spilke, J., and E. Groeneveld. 1994. Comparison of four multivariate REML (co)variance component estimation packages. pp. 11–14. *In* J. S. Gavora, B. Benkel, J. Chesnais, W. Fairfull, J. P. Gibson, B. W. Kennedy, E. B. Burnside, and C. Smith (editors). *Proceedings of the 5th World Congress on Genetics Applied to Livestock Production*, Vol. 22. University of Guelph, Guelph, Ontario, Canada.

Spofford, J. 1976. Position-effect variegation in *Drosophila*. pp. 955–1018. *In* M. Ashburner and E. Novitski (editors). *The Genetics and Biology of Drosophila*. Academic Press, New York.

Spuhler, K. P., D. W. Crumpacker, J. S. Williams, and B. P. Bradley. 1978. Response to selection for mating speed and changes in gene arrangement frequencies in descendants from a single population of *Drosophila pseudoobscura*. *Genetics* 89: 729–749.

Stafford, J. 1956. The wintering of Blackcaps in the British Isles. *Bird Study* 3: 251–257.

Stalker, H. D., and H. L. Carson. 1947. Morphological variation in natural populations of *Drosophila robusta* Sturtevant. *Evolution* 1: 237–248.

Stanton, M. L. 1984. Developmental and genetic sources of seed weight variation in *Raphanus raphanistrum* L. (Brassicaceae). *American Journal of Botany* 71: 1090–1098.

Stearns, S. C. 1989. The evolutionary significance of phenotypic plasticity. *BioScience* 39: 436–445.

Stearns, S. C. 1992. *The Evolution of Life Histories*. Oxford University Press, New York.

Stearns, S. C., and T. J. Kawecki. 1994. Fitness sensitivity and the canalization of life-history traits. *Evolution* 48: 1438–1450.

Stearns, S. C., and J. C. Koella. 1986. The evolution of phenotypic plasticity in life-history traits: predictions of reaction norms for age and size at maturity. *Evolution* 40: 893–913.

Stearns, S. C., M. Kaiser, and T. J. Kawecki. 1995. The differential genetic and environmental canalization of fitness components in *Drosophila*. *Journal of Evolutionary Biology* 8: 539–558.

Strauss, S. H. 1986. Heterosis at allozyme loci under inbreeding and crossbreeding in *Pinus attenuata*. *Genetics* 113: 115–134.

Strauss, S. H., and W. J. Libby. 1987. Allozyme heterosis in radiata pine is poorly explained by overdominance. *American Naturalist* 130: 879–890.

Strong, L. C. 1978. Inbred mice in science. pp. 45–66. *In* H. C. Morse III (editor). *Origins of Inbred Mice*. Academic Press, New York.

Stuber, C. W., S. E. Lincoln, D. W. Wolff, T. Helentjaris, and E. S. Lander. 1992. Identification of genetic factors contributing to heterosis in a hybrid from two elite maize inbred lines using molecular markers. *Genetics* 132: 823–839.

Suh, D. S., and T. Mukai. 1991. The genetic structure of natural populations of *Drosophila melanogaster*. XXIV. Effects of hybrid dysgenesis on the components of genetic variance of viability. *Genetics* 127: 545–552.

Sultan, S. E. 1987. Evolutionary implications of phenotypic plasticity in plants. *Evolutionary Biology* 21: 127–178.

Sultan, S. E., and F. A. Bazzaz. 1993. Phenotypic plasticity in *Polygonum persicaria*. I. Diversity and uniformity in genotypic norms of reaction to light. *Evolution* 47: 1009–1031.

Sutherland, W. J. 1988. The heritability of migration. *Nature* 334: 471–472.

Swallow, W. H., and J. F. Monahan. 1984. Monte Carlo comparison of ANOVA, MIVQUE, REML, and ML estimators of variance components. *Technometrics* 26: 47–57.

Swan, A. A., and J. D. Hickson. 1994. Maternal effects in Australian merinos. pp. 143–146. *In* J. S. Gavora, B. Benkel, J. Chesnais, W. Fairfull, J. P. Gibson, B. W. Kennedy, E. B. Burnside, and C. Smith (editors). *Proceedings of the 5th World Congress on Genetics Applied to Livestock Production*. Dept. Animal and Poultry Science, University of Guelph, Guelph, Ontario, Canada.

Swiger, L. A., W. R. Harvey, D. O. Everson, and K. E. Gregory. 1964. The variance of intraclass correlation involving groups with one observation. *Biometrics* 20: 818–826.

Tachida, H., and C. C. Cockerham. 1989. Effects of identity disequilibrium and linkage on quantitative variation in finite populations. *Genetic Research* 53: 63–70.

Tallis, G. M. 1987. Ancestral covariance and the Bulmer effect. *Theoretical and Applied Genetics* 73: 815–820.

Tanaka, S., M. Matsuka, and T. Sakai. 1976. Effect of change in photoperiod on wing form in *Pteronemobius taprobanensis* Walker (Orthoptera: Gryllidae). *Applied Enterology and Zoology* 11: 27–32.

Tanksley, S. D. 1993. Mapping polygenes. *Annual Review of Genetics* 27: 205–233.

Tantawy, A. O. 1956a. Selection for long and short wing lengths in *Drosophila melanogaster* with different systems of mating. *Genetica* 28: 231–262.

Tantawy, A. O. 1956b. Response to selection and changes of genetic variability for wing length in *Drosophila melanogaster* with brother–sister matings. *Genetica* 28: 177–200.

Tantawy, A. O. 1957. Genetic variance of random-inbred lines of *Drosophila melanogaster* in relation to coefficients of inbreeding. *Genetics* 42: 121–136.

Tantawy, A. O. 1964. Studies on natural populations of *Drosophila*. III. Morphological and genetic differences in wing length in *Drosophila melanogaster* and *D. simulans* in relation to season. *Evolution* 18: 560–570.

Tantawy, A. O., and M. R. El-Helw. 1966. Studies on natural populations of *Drosophila*. V. Correlated response to selection in *Drosophila melanogaster*. *Genetics* 53: 97–110.

Tantawy, A. O., and M. R. El-Helw. 1970. Studies on natural populations of *Drosophila*. IX. Some fitness components and their heritabilities in natural and mutant populations of *Drosophila melanogaster*. *Genetics* 64: 79–91.

Tantawy, A. O., and G. S. Mallah. 1961. Studies on natural populations of *Drosophila*. I. Heat resistance and geographical variation in *Drosophila melanogaster* and *D. simulans*. *Evolution* 15: 1–14.

Tantawy, A. O., and F. A. Rakha. 1964. Studies on natural populations of *Drosophila*. IV. Genetic variances of and correlations between four characters in *D. melanogaster* and *D. simulans*. *Genetics* 50: 1349–1355.

Tantawy, A. O., and E. C. R. Reeve. 1956. Studies in quantitative inheritance. Zeitschrift fur Indukt 87: 648–667.

Tantawy, A. O., and A. A. Tayel. 1970. Studies on natural populations of *Drosophila*. X. Effects of disruptive and stabilizing selection on wing length and the correlated response in *Drosophila melanogaster*. *Genetics* 65: 121–132.

Taylor, C. E., and C. Condra. 1978. Genetic and environmental interaction in *Drosophila pseudoobscura*. *Journal of Heredity* 69: 63–64.

Taylor, C. E., and G. C. Gorman. 1975. Population genetics of a "colonizing" lizard: natural selection for allozyme morphs in *Anolis grahami*. *Heredity* 35: 241–247.

Templeton, A. R., and B. Read. 1983. The elimination of inbreeding depression in a captive herd of Speke's gazelle. pp. 241–261. *In* C. M. Schonewald-Cox, S. M. Chambers, B. MacBryde, and W. L. Thomas (editors). *Genetics and Conservation*. The Benjamin/Cummings Publishing Co., London.

Templeton, A. R., and B. Read. 1984. Factors eliminating inbreeding depression in a captive herd of Speke's gazelle (*Gazella spekei*). *Zoo Biology* 3: 177–199.

Teska, W. R., M. H. Smith, and J. M. Novak. 1990. Food quality, heterozygosity, and fitness correlates in *Peromyscus polionotus*. *Evolution* 44: 1318–1325.

Thessing, A., and J. Ekman. 1994. Selection on the genetical and environmental components of tarsal growth in juvenile willow tits (*Parus montanus*). *Journal of Evolutionary Biology* 7: 713–726.

Thoday, J. M. 1959. Effects of disruptive selection. I. Genetic flexibility. *Heredity* 13: 187–203.

Thoday, J. M. 1961. Location of polygenes. Nature 191: 368–370.

Thoday, J. M., and J. N. Thompson, Jr. 1976. The number of segregating genes implied by continuous variation. *Genetica* 46: 335–344.

Thomas, B. J., and R. Rothstein. 1991. Sex, maps, and imprinting. *Cell* 64: 1–3.

Thomas, R. H. 993. Ecology of body size in *Drosophila buzzatii*: Untangling the effects of temperature and nutrition. *Ecological Entomology* 18: 84–90.

Thomas, R. L. 1967. Inter-population variation in perennial ryegrass. *Heredity* 22: 481–498.

Thomas, R. L. 1969a. Inter-population variation in perennial ryegrass 2. Population variance coefficients. *Heredity* 29: 85–90.

Thomas, R. L. 1969b. Inter-population variation in perennial ryegrass 3. Interaction of heritable and environmental variation. *Heredity* 29: 91–100.

Thompson, E. A., and R. G. Shaw. 1990. Pedigree analysis for quantitative traits: variance components without matrix inversion. *Biometrics* 46: 399–412.

Thompson, R. 1976. The estimation of maternal genetic variances. *Biometrics* 32: 903–917.

Thompson, R. 1977a. The estimation of heritability with unbalanced data. I. Observations available on parents and offspring. *Biometrics* 33: 485–495.

Thompson, R. 1977b. The estimation of heritability with unbalanced data. II. Data available on more than two generations. *Biometrics* 33: 497–504.

Thompson, S. R., and S. K. Rook. 1988. Lack of a correlated response to canalizing selection in *Drosophila melanogaster*. *Journal of Heredity* 79: 385–386.

Thornhill, R., and P. Sauer. 1992. Genetic sire effects on the fighting ability of sons and daughters and mating success of sons in a scorpionfly. *Animal Behavior* 43: 255–264.

Tienderen, P. H. van. 1991. Evolution of generalists and specialists in spatially heterogeneous environments. *Evolution* 45: 1317–1331.

Tienderen, P. H. van, and G. de Jong. 1994. A general model of the relation between phenotypic selection and genetic response. *Journal of Evolutionary Biology* 7: 1–12.

Tienderen, P. H. van, and H. P. Koelewijn. 1994. Selection on reaction norms genetic correlations and constraints. *Genetic Research* 64: 115–125.

Tinkle, D. W. 1965. Population structure and effective size of a lizard population. *Evolution* 19: 569–573.

Travis, J., 1989. Ecological genetics of life history traits in poeciliid fishes. pp. 185–200. *In* G. K. Meffe, Jr. and F. F. Snelson (editors). *Ecology and Evolution of Livebearing Fishes (Poeciliidae)*. Prentice-Hall, Englewood Cliffs, NJ.

Travis, J., 1994a. Ecological genetics of life-history traits: variation and its evolutionary significance. pp. 171–204. *In* L. A. Real (editor). *Ecological Genetics*. Princeton University Press, Princeton, NJ.

Travis, J., 1994b. Evolution in the sailfin molly: the interplay of life-history variation and sexual selection. pp. 205–232. *In* L. A. Real (editor). *Ecological Genetics*. Princeton University Press, Princeton, NJ.

Travis, J., S. B. Emerson, and M. Blouin. 1987. A quantitative genetic analysis of larval life-history traits in *Hyla crucifer*. *Evolution* 41: 145–156.

Turelli, M. 1984. Heritable genetic variation via mutation-selection balance: Lerch's zeta meets the abdominal bristle. *Theoretical Population Biology* 25: 138–193.

Turelli, M. 1985. Effects of pleiotropy on predictions concerning mutation–selection balance for polygenic traits. *Genetics* 111: 165–195.

Turelli, M. 1988. Phenotypic evolution, constant covariance, and the maintenance of additive variance. *Evolution* 42: 1342–1347.

Turelli, M., and L. R. Ginzburg. 1983. Should individual fitness increase with heterozygosity? *Genetics* 104: 191–209.

Underhill, D. K. 1968. Heritability of dorsal spot number and snout-vent length in *Rana pipiens*. *Journal of Heredity* 59: 235–240.

Van Dijken, F. R., and W. Scharloo. 1979. Divergent selection on locomotor activity in *Drosophila melanogaster*. I. Selection response. *Behaviour Genetics* 9: 543–553.

Van Treuren, R., R. Bijlsma, N. J. Ouborg, and W. van Delden. 1993. The significance of genetic erosion in the process of extinction. IV. Inbreeding depression and heterosis effects caused by selfing and outcrossing in *Scabiosa columbaria*. *Evolution* 47: 1669–1680.

Van Valen, L., and G. W. Mellin. 1967. Selection in natural populations. 7. New York babies (Fetal life study). *Annals of Human Genetics* 31: 109–127.

Van Vleck, L. D. 1968. Selection bias in estimation of the genetic correlation. *Biometrics* 24: 951–962.

Van Vleck, L. D., and C. R. Henderson. 1961. Empirical sampling estimates of genetic correlations. *Biometrics* 17: 359–371.

Van Vleck, L. D., D. St. Louis, and J. I. Miller. 1977. Expected phenotypic response in weaning weight of beef calves from selection for direct and maternal genetic effects. *Journal of Animal Science* 44: 360–367.

Varmuza, S., and M. Mann. 1994. Genomic imprinting—defusing the ovarian time bomb. *Trends in Genetics* 10: 118–123.

Veldboom, L. R., and M. Lee. 1994. Molecular-marker-facilitated studies of morphological traits in maize. II: Determination of QTLs for grain yield and yield components. *Theoretical and Applied Genetics* 89: 451–458.

Verrier, E., J. J. Colleau, and J. L. Foulley. 1990. Predicting cumulated responses to directional selection in finite panmictic populations. *Theoretical and Applied Genetics* 79: 833–840.

Via, S. 1984. The quantitative genetics of polyphagy in an insect herbivore. II. Genetic correlations in larval performance within and among host plants. *Evolution* 38: 896–905.

Via, S. 1987. Genetic constraints on the evolution of phenotypic plasticity. pp. 47–71. *In*

V. Loeschcke (editor). *Genetic Constraints on Adaptive Evolution*. Springer-Verlag, Berlin.

Via, S. 1991. The genetic structure of host plant adaptation in a spatial patchwork: demographic variability among reciprocally transplanted pea aphid clones. *Evolution* 45: 827–852.

Via, S. 1993a. Adaptive phenotypic plasticity: target or by-product of selection in a variable environment? *American Naturalist* 142: 352–365.

Via, S. 1993b. Regulatory genes and reaction norms. *American Naturalist* 142: 374–378.

Via, S., and R. Lande. 1985. Genotype–environment interaction and the evolution of phenotypic plasticity. *Evolution* 39: 505–522.

Via, S., R. Gomulkiewicz, G. de Jong, S. M. Scheiner, C. D. Schlichting, and P. H. van Tienderen. 1995. Adaptive phenotypic plasticity: consensus and controversy. *Trends in Ecology and Evolution* 5: 212–217.

Villanueva, B., and B. W. Kennedy. 1990. Effect of selection on genetic parameters of correlated traits. *Theoretical and Applied Genetics* 80: 746–752.

Villanueva, B., and B. W. Kennedy. 1992. Asymmetrical correlated responses to selection under an infinitesimal genetic model. *Theoretical and Applied Genetics* 84: 323–329.

Voigt, R. L., C. O. Gardner, and O. J. Webster. 1966. Inheritance of seed size in Sorghum, *Sorghum vulgare* Pers. *Crop Science* 6: 582–586.

Vrijenhoek, R. C., and S. Lerman. 1982. Heterozygosity and developmental stability under sexual and asexual breeding systems. *Evolution* 36: 768–776.

Waddington, C. H. 1940. The genetic control of wing development in *Drosophila*. *Journal of Genetics* 41: 75–139.

Waddington, C. H. 1942. Canalization of development and the inheritance of acquired characters. *Nature* 150: 563–565.

Waddington, C. H. 1957. *The Strategy of the Genes*. George Allen & Unwin Ltd, London.

Waddington, C. H. 1960. Experiments on canalizing selection. *Genetic Research* 1: 140–150.

Waddington, C. H., and E. Robertson. 1966. Selection for developmental canalization. *Genetic Research* 7: 303–312.

Wagner, G. P. 1989. Multivariate mutation-selection balance with constrained pleiotropic effects. *Genetics* 122: 223–234.

Waldman, B., and J. S. McKinnon. 1993. Inbreeding and outbreeding in fishes, amphibians, and reptiles. pp. 250–282. *In* N. W. Thornhill (editor). *The Natural History of Inbreeding and Outbreeding*. University of Chicago Press, Chicago.

Waldvogel, M., and F. Gould. 1990. Variation in oviposition preference of *Heliothis virescens* in relation to macroevolutionary patterns of Heliothine range. *Evolution* 44: 1326–1337.

Walker, T. J. 1987. Wing dimorphism in *Gryllus rubens* (Orthoptera: Gryllidae). *Annals of the Entomological Society of America* 80: 547–560.

Wallace, B. 1953. On coadaptation in *Drosophila*. *American Naturalist* 87: 343–358.

Waples, R. S. 1989. A generalized approach for estimating effective population size from temporal changes in allele frequency. *Genetics* 121: 379–391.

Waser, N. M. 1993. Population structure, optimal outbreeding, and assortative mating in angiosperms. pp. 173–199. *In* N. W. Thornhill (editor). *The Natural History of Inbreeding and Outbreeding*. University of Chicago Press, Chicago.

Waser, N. M., and M. V. Price. 1989. Optimal outcrossing in *Ipomopsis aggregata*: seed set and offspring fitness. *Evolution* 43: 1097–1109.

Watanabe, T. K., and W. W. Anderson. 1976. Selection for geotaxis in *Drosophila melanogaster*: heritability, degree of dominance, and correlated responses to selection. *Behavior Genetics* 6: 71–86.

Wearden, S. 1969. Alternative analyses of the diallel cross. *Heredity* 19: 669–680.

Webb, K. L., and D. A. Roff. 1992. The quantitative genetics of sound production in *Gryllus firmus*. *Animal Behaviour* 44: 823–832.

Weber, K. E. 1990. Increased selection response in larger populations. I. Selection for wing-tip height in *Drosophila melanogaster* at three population sizes. *Genetics* 125: 579–584.

Weber, K. E., and L. T. Diggins. 1990. Increased selection response in larger populations. II. Selection for ethanol vapor resistance in *Drosophila melanogaster* at two population sizes. *Genetics* 125: 585–597.

Weigensberg, I., and D. A. Roff. 1997. Natural heritabilities: can they be reliably estimated in the laboratory. *Evolution* (in press).

Weigensberg, I., Y. Carrière, and D. A. Roff. 1997. Effects of male genetic contribution and paternal investment on egg and hatchling size in the cricket *Gryllus firmus*. *Journal of Evolutionary Biology* (in press).

Weis, A. E., W. G. Abrahamson, and M. C. Andersen. 1992. Variable selection on *Eurosta*'s gall size, I: the extent and nature of variation in phenotypic selection. *Evolution* 46: 1674–1697.

Weis, A. E., and W. L. Gorman. 1990. Measuring selection on reaction norms: an exploration of the *Eurosta–Solidago* system. *Evolution* 44: 820–831.

Weller, J. I., and M. Ron. 1994. Detection and mapping of quantitative loci in segregating populations: theory and experimental results. pp. 213–220. *In* J. S. Gavora, B. Benkel, J. Chesnais, W. Fairfull, J. P. Gibson, B. W. Kennedy, E. B. Burnside, and C. Smith (editors). *Proceedings of the 5th World Congress on Genetics Applied to Livestock Production*, Vol. 21. University of Guelph, Guelph, Ontario, Canada.

Werren, J. 1991. The paternal-sex-ratio chromosome of *Nasonia*. *American Naturalist* 137: 392–402.

Wesselingh, R. A., and T. J. de Jong. 1995. Bidirectional selection on threshold size for flowering in *Cynoglossum officinale* (hound's-tongue). *Heredity* 74: 415–424.

Wesselingh, R. A., T. J. de Jong, P. G. L. Klinkhamer, M. J. Van Dijk, and E. G. M. Schlatmann. 1993. Geographical variation in threshold size for flowering in *Cynoglossum officinale*. *Acta Botanica Neerlandica* 42: 81–91.

Westerman, C. H. 1970. Genotype–environment interaction and developmental regulation in *Arabidopsis thaliana*. II. Inbred lines; analysis. *Heredity* 26: 93–106.

Wethington, A. R., and R. T. Dillon, Jr. 1993. Reproductive development in the hermaphroditic freshwater snail *Physa* monitored with complementing albino lines. *Proceedings of the Royal Society of London* B 252: 109–114.

White, M. J. D. 1946. The spermatogenesis of hybrids between *Triturus cristatus* and *T. marmoratus* (Urodela). *Journal of Experimental Zoology* 102: 179–204.

Whitehurst, P. H., and B. A. Pierce. 1991. The relationship between allozyme variation and life-history traits of the spotted chorus frog, *Pseudacris clarkii. Copeia* 1991: 1032–1039.

Wiggins, D. A. 1989. Heritability of body size in cross-fostered tree swallow broods. *Evolution* 43: 1808–1811.

Wigglesworth, V. B. 1934. The physiology of ecdysis in *Rhodnius prolixus* (Hemiptera). II. Factors controlling moulting and metamorphosis. *Quarterly Journal of Microscopical Science* 77: 191–222.

Wilkinson, G. S., K. Fowler, and L. Partridge. 1990. Resistance of genetic correlation structure to directional selection in *Drosophila melanogaster. Evolution* 44: 1990–2003.

Willham, R. L. 1963. The covariance between relatives for characters composed of components contributed by related individuals. *Biometrics* 19: 18–27.

Willham, R. L. 1972. The role of maternal effects in animal breeding: III. Biometrical aspects of maternal effects in animals. *Journal of Animal Science* 35: 1288–1293.

Willis, J. H., and H. A. Orr. 1993. Increased heritable variation following population bottlenecks: the role of dominance. *Evolution* 47: 949–957.

Wilson, S. P., H. D. Goodale, W. H. Kyle, and E. F. Godfrey. 1971. Long term selection for body weight in mice. *Journal of Heredity* 62: 228–234.

Windig, J. J. 1994a. Reaction norms and the genetic basis of phenotypic plasticity in the wing pattern of the butterfly *Bicyclus anynana. Journal of Evolutionary Biology* 7: 665–695.

Windig, J. J. 1994b. Genetic correlations and reaction norms in wing pattern of the tropical butterfly *Bicyclus anynana. Heredity* 73: 459–470.

Wolff, K., and J. Haeck. 1990. Genetic analysis of ecologically relevant morphological variability in *Plantago lanceolata* L. VI. The relation between allozyme heterozygosity and some fitness components. *Journal of Evolutionary Biology* 3: 243–255.

Woltereck, R. 1909. Weitere experimentelle Untersuchungen uber Artveranderung, speziell uber das Wesen quantitativer Artunterschiede bei Daphniden. Verhandlungen der Deutschensch. *Zoologischen Gessellschaft* 1909: 110–172.

Wood, J. W. 1987. The genetic demography of the Gainj of Papua New Guinea. 2. Determinants of effective population size. *American Naturalist* 129: 165–187.

Woodring, J. P. 1983. Control of moulting in the house cricket, *Acheta domesticus. Journal of Insect Physiology* 29: 461–464.

Woolfenden, G. E., and J. W. Fitzpatrick. 1978. The inheritance of territory in group breeding birds. *BioScience* 28: 104–108.

Wooten, M. C., and M. H. Smith. 1986. Fluctuating asymmetry and genetic variability in a natural population of *Mus musculus. Journal of Mammology* 67: 725–732.

Wright, A. J. 1985. Diallel designs, analyses, and reference populations. *Heredity* 54: 307–311.

Wright, M. F., and S. I. Guttman. 1995. Lack of an association between heterozygosity and growth rate in the wood frog, *Rana sylvatica*. *Canadian Journal of Zoology* 73: 569–575.

Wright, S. 1931. Evolution in Mendelian populations. *Genetics* 16: 97–159.

Wright, S. 1938. Size of population and breeding structure in relation to evolution. *Science* 87: 430–431.

Wright, S. 1948. On the roles of directed and random changes in gene frequency in the genetics of natural populations. *Evolution* 2: 279–294.

Wright, S. 1968. *Evolution and Genetics of Populations, Vol. 1. Genetic and Biometric Foundations*. University of Chicago Press, Chicago.

Wyngaard, G. A. 1986. Genetic differentiation of life history traits in populations of *Mesocyclops edax* (Crustacea: Copepoda). *Biological Bulletin* 170: 279–295.

Wyngaard, G. A. 1988. Heritable life history variation in widely separated populations of *Mesocyclops edax* (Crustacea: Copepoda). *Biological Bulletin* 170: 296–304.

Yamada, Y. 1962. Genotype by environment interaction and genetic correlation of the same trait under different environments. *Japanese Journal of Genetics* 37: 498–509.

Yamada, Y., and A. E. Bell. 1969. Selection for larval growth in *Tribolium* under two levels of nutrition. *Genetic Research* 13: 175–195.

Yates, F. 1947. Analysis of data from all possible reciprocal crosses between a set of parental lines. *Heredity* 1: 287–301.

Yoo, B. H. 1980a. Long-term selection for a quantitative character in large replicate populations of *Drosophila melanogaster*. Part 1. Response to selection. *Genetic Research* 35: 1–17.

Yoo, B. H. 1980b. Long-term selection for a quantitative character in large replicate populations of *Drosophila melanogaster*. Part 3: The nature of residual genetic variability. *Theoretical and Applied Genetics* 57: 25–32.

Yoo, B. H., F. W. Nicholas, and K. A. Rathie. 1980. Long-term selection for a quantitative character in large replicate populations of *Drosophila melanogaster*. Part 4: Relaxed and reversed selection. *Theoretical and Applied Genetics* 57: 113–117.

Young, C. W., and J. E. Legates. 1965. Genetic, phenotypic, and maternal interrelationships of growth in mice. *Genetics* 52: 563–576.

Young, S. S. Y., and H. Weiler. 1960. Selection for two correlated traits by independent culling levels. *Journal of Genetics* 57: 329–338.

Yu, Y., J. Harding, T. Byrne, and T. Famula. 1993. Estimation of components of genetic variance and heritability for flowering time and yield in gerbera using Derivative-Free Restricted Maximum Likelihood (DFRML). *Theoretical and Applied Genetics* 86: 234–236.

Yule, G. U. 1902. Mendel's laws and their probable relations to intra-racial heredity. *New Phytologist* 1: 193–207, 222–238.

Yule, G. U. 1906. On the theory of inheritance of quantitative compound characters on the basis of Mendel's laws—a preliminary note. pp. 140–142. Report of the 3rd International Conference of Genetics.

Zangerl, A. R., and F. A. Bazzaz. 1983. Plasticity and genotype variation in photosynthetic behavior of an early and a late successional species of *Polygonum*. *Oecologia* 57: 270–273.

Zar, J. H. 1984. *Biostatistical Analysis*. Prentice-Hall, Englewood Cliffs, NJ.

Zeng, Z.-B., D. Houle, and C. C. Cockerham. 1990. How informative is Wright's estimator of the number of genes affecting a quantitative character? *Genetics* 126: 235–247.

Ziehe, M., and J. H. Roberds. 1989. Inbreeding depression due to overdominance in partially self-fertilizing plant populations. *Genetics* 121: 861–868.

Zhivotovsky, L. A., and M. W. Feldman. 1992. On the difference between mean and optimum of quantitative characters under selection. *Evolution* 46: 1574–1578.

Zouros, E. 1990. Heterozygosity and growth in marine bivalves: response to Koehn's remarks. *Evolution* 44: 216–218.

Zouros, E., and A. L. Mallet. 1989. Genetic explanations of the growth/heterozygosity correlation in marine mollusks. pp. 314–324. *In* E. Zouros and A. L. Mallet (editors). *Genetic Explanations of the Growth/Heterozygosity Correlation in Marine Mollusks*. University of Wales, Swansea.

Zouros, E., M. Romero-Dorey, and A. L. Mallet. 1988. Heterozygosity and growth in marine bivalves: further data and possible explanations. *Evolution* 42: 1332–1341.

Zouros, E. S., M. Singh, and H. E. Miles. 1980. Growth rate in oysters: an overdominant phenotype and its possible explanations. *Evolution* 34: 856–867.

Subject Index

Taxonomic Index